Operations Research

T Veerarajan

Dean (Retired)
Department of Mathematics
Velammal College of Engineering and Technology
Madurai

Universities Press

Operations Research

UNIVERSITIES PRESS (INDIA) PRIVATE LIMITED

Registered office
3-6-747/1/A & 3-6-754/1, Himayatnagar, Hyderabad 500 029, Telangana, India
info@universitiespress.com; www.universitiespress.com

Distributed by
Orient Blackswan Private Limited

Registered office
3-6-752 Himayatnagar, Hyderabad 500 029, Telangana, India

Other offices
Bengaluru, Bhopal, Chennai, Guwahati, Hyderabad, Jaipur, Kolkata, Lucknow, Mumbai, New Delhi, Noida, Patna, Vijayawada

ISBN: 978-93-86235-16-9

Typeset in Times New Roman PS 10/12 by
MacroTex Solutions, Chennai 600 091, Tamil Nadu, India

Printed in India at Rajiv Book Binding House, Delhi

Published by
Universities Press (India) Private Limited
3-6-747/1/A & 3-6-754/1, Himayatnagar, Hyderabad 500 029, Telangana, India
email: info@universitiespress.com

Dedicated to
my beloved wife and affectionate children

Preface

Operations Research has been written to provide a lucid introduction to the subject, its theoretical background and numerical applications. The book caters to the requirements of both undergraduate and postgraduate students who study operations research as part of their curriculum in engineering, management, arts and science courses.

The text emphasises the applications of Operations Research starting from the first principles and builds on the concepts step by step to kindle interest in beginners who have chosen this subject as part of their study. Basic concepts and problem-solving techniques are explained as clearly as possible without compromising on the rigour.

I do not claim originality in respect of the treatment of the subject. However, I have presented the topics in my own unique way that was evolved through years of teaching the subject at introductory and advanced levels. I have provided theoretical explanations of each concept and backed them with a sizeable number of examples besides including numerous problems for practice. Answers for these practice problems have also been given for the students' benefit.

Although utmost care has been taken to make the book error-free, there may be a few omissions that have escaped my attention. I shall be highly obliged to those who could evaluate the book critically, point out such discrepancies, if any, and make suggestions for the improvement of the book.

I am sure that this book will be useful for students and teachers alike.

I am grateful to Universities Press (India) Private Limited, Chennai, for their painstaking effort to bring out this book in record time.

It would not have been possible for me to take up this venture and accomplish it successfully if it had not been for the encouragement and co-operation of my wife and children. I gratefully acknowledge their love and support.

T Veerarajan

About the Author

T Veerarajan retired as Dean of Mathematics Department, Velammal College of Engineering and Technology, Madurai, Tamil Nadu. A teacher par excellence for over 50 years, he has taught mathematics and allied subjects to undergraduate and postgraduate students in various engineering colleges affiliated to Anna University, Chennai.

A gold medalist from Madras University, Professor Veerarajan has had an illustrious academic career and assisted a few students in their Ph.D. theses in mathematics. He has authored several textbooks in mathematics and allied subjects and published them with several renowned publishers. His book on Discrete Mathematics for B.E. and B.Tech. students was translated into the Spanish language and prescribed as a textbook by a few universities in the State of Mexico.

Contents

Chapter 0 OPERATIONS RESEARCH—INTRODUCTION AND PRE-REQUISITES

0.0 INTRODUCTION

During world war II Britian had very limited military resources that were to be allocated to various military operations in an effective manner. Hence the military management called upon scientists from various disciplines and urged them to discuss evolve and suggest ways and means to improve the execution of various military operations. By their joint efforts, experience and deliberations, they suggested certain approaches that showed remarkable progress. This new approach to systematic and scientific study of the military operations was called **Operations Research (OR)**.

The apparent success of OR in the military context attracted the attention of industrial managers who were seeking solution to their complex executive type problems. Consequently OR achieved recognition as a scientific subject worthly of academic study in the universities. Since then, the subject has gained more and more importance for students of economics, management, public administration, behavioural science, social work, mathematics, commerce and engineering.

0.1 MEANING AND SCOPE OF OR

Operations research can be defined as a scientific approach to problem solving for executive management that involves constructing mathematical, economic and statistical models of decision and control to treat situations of complexity and uncertainty. It analyses the relationships that determine the probable future consequence of decision choices ad devise appropriate measures of effectiveness in order to evaluate the relative merits of alternative actions.

In short, the objective of OR is to provide a scientific basis to the decision makers for solving the problems involving the operations of a system to give a solution in the best interest of the organisation or an optimum solution to the problem, OR is used in almost every walk of life wherever a decision is sought. For example in industrial situations, OR is useful to the directing anthority, production specialist, marketing execute, personal administrator ad financial controller.

0.2 PHASES OF OR

The basic pattern of the application of OR to a problem is divided into the following six major phases:

1. **Problem formulating phase:** This phase involves the formulation of the problem in an appropriate form involving the statement of the problem elements that include decision (controllable) variables and uncontrollable parameters and objective for defining a good or improved solution.

2. **Model constructing phase:** This phase involves identifying the decision variables and parameters, finding constraints or restrictions satisfied by them and stating clearly the objective function which defines the measure of effectiveness of the system as a mathematical function of its decision variables.
3. **Solution deriving phase:** Here mathematical calculations are made for obtaining the solution of the model .
4. **Model updating phase:** In this phase, the validity of the model used is checked.
5. **Solution controlling phase:** This phase provides a control over the solution by proper feed of the information on variables which deviated significantly.
6. **Result implementing phase:** This is the last phase and it deals with the implimentation of the tested results of the model.

0.3 MATHEMATICAL PRE-REQUISITES FOR THE PURSUIT OF STUDY OF OR

The following mathematical prerequisites are expected to be known to a reader who wants to pursue the study of OR. We will not elaborately discuss them in this book, but simply mention them. If a reader feels that he/she is not familiar with a particular idea, he/she has to consult standard text looks in mathematics.

0.3.1 Matrices

1. Definition
2. Some special types of matrices
3. Matrix addition, multiplication by a scalar
4. Matrix multiplication and its properties
5. Transpose of matrix and its properties
6. Partitioned matrices, addition, scalar multiple, product and transpose of partitioned matrices.
7. Determinant of a square matrix and properties of determinant functions.
8. Inverse of a square matrix and properties of the inverse of a matrix
9. Elementary transformations of a matrix
10. Inverse of a square matrix by elementary transformation
11. Inverse of a partitioned matrix
12. Rank of a matrix and its properties
13. Applications to the solution of a system of linear equations
14. Applications to a system of general linear equations
15. Quadratic forms

0.3.2 Vectors

1. Vector inequalities
2. Linear combination of vectors
3. Hyperplanes and hyperspheres

0.3.2.1 Line Segment in R^n

A line segment joining two points $X_1 = \{x_1^{(1)}, x_2^{(1)}, \ldots x_n^{(1)}\}$ and $X_2 = \{x_1^{(2)}, x_2^{(2)}, \ldots x_n^{(2)}\}$ is a collection of all point $X = \{x_1, x_2, \ldots x_n\}$, where $X = lX_1 + (1 - \lambda)X_2$, for all values λ such that $0 \le \lambda \le 1$.

0.3.2.2 Hyperplane in R^n

The set of points $\{x_1, x_2, \ldots x_n\}$ in R^n satisfying $c_1x_1 + c_2x_2 + \ldots + c_nx_n = z$ (where not all c_i's = 0) defines a *hyperplane* for given values of c_i's and z. If $z = 0$, the hyperplane is said to pass through the origin. In this case, the equation of the hyperplane becomes $CX = 0$, viz., $C_1x_1 + C_2x_2 + \ldots + C_nx_n = 0$

This implies that the vector C is orthogonal to every vector X the hyperplane.

If $z \neq 0$, and $P(X_1)$ and $Q(X_2)$ are any two points on the hyperplane, then $CX_1 =$ z and CX_2 or $C(X_1 - X_2) = 0$ means that C is normal to the vector $\overrightarrow{PQ} = X_1 - X_2$ and hence lies in the direction of the normal to the hyperplanes. Usually C is called the normal to the hyperplanes and $\pm\frac{C}{|C|}$ are called *unit normals.*

Two hyperplanes are said to be *parallel*, if they have the same unit normals.

The hyperplanes $C_1X = Z_1$ and $C_2X = Z_2$ are parallel if $C_1 = \lambda C_2$ for some $\lambda \neq 0$.

A set XCR^n is said to be an $\in$*-neighbourhood* of a point $X_0 \in R^n$, if there existance $\in \not< 0$ such that

$$\left[X \| X - X_0\right| < \in \left] Cx\right.$$

A point $X_0 \in R^n$, of a set XCR^n is said to be a *boundary point* of X, if every neighbourhood of X_0 contains a point that is not in X, viz., if X is not a neighbourhood of X_0. A set XCR^n is said to be *closed*, if every boundary point of X is a point of it. Using this definition, we can prove that a hyperplane is a closed set.

In R^n, every hyperplane $CX = Z$ determine two *open half spaces* defined by $X_1 = \{X|Cx > z\}$ and $X_2 = \{X|Cx > z\}$ and two *closed half spaces* defined by $X_3 = \{X|Cx \geq z\}$ and $X_4 = \{X|Cx \leq z\}$

The intersection of a finite number of closed half spaces is called a *polytope*. The hyperplanes producing the half spaces are called the *generating hyperplanes* of the polytope.

0.3.2.3 Hypersphere in R^n

A hypersphere in R^n with centre at $\bar{a}$ and radius $\in > 0$ is defined to be the set of points

$$S = \{X \| X - \bar{a}| = \in\}$$

viz., the equation of the hyperplane in R^n is

$$\sum_{i=1}^{n} (x_i - a_i)^2 = \in^2.$$

The interior of a hypersphere with centre $\bar{a}$ and radius $\in$ is defined as the set of points

$$S_i = \{X \| X - \bar{a}| = \in\}$$

0.3.3 Convex Sets and Their Properties

0.3.3.1 Definition

A subset XCR^n is called a *convex set* , if for every two points X_1 and X_2 in X, the line segment joining X_1 and X_2 is contained in the set viz., a convex subset has no 'holes' i it and its boundaries are bent away from the set, as illustrated in the following diagram:

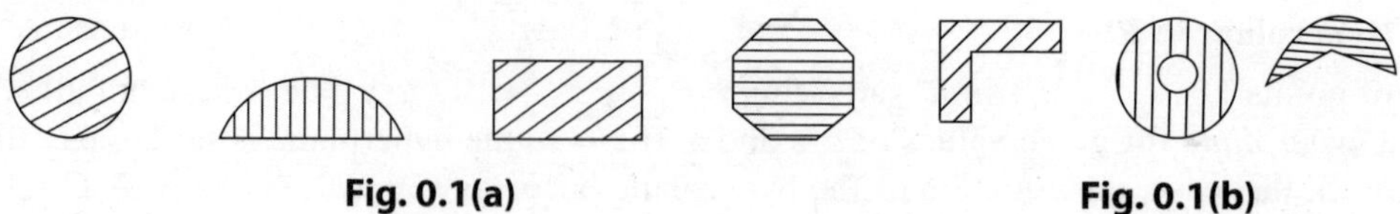

Fig. 0.1(a) Fig. 0.1(b)

Since the line segment joining the points X_1 and X_2 of a subset XCR^n was defined as the set

$$\{X|X = \lambda X_1 + (1-\lambda)X_2, 0 \le \lambda \le 1\}.$$

we note that X is a convex set if the point $\{\lambda X_1 + (1-\lambda)X_2\} \in X$, For $\forall X_1, X_2$. This diagram given in fig 1.1(b) do not possess this property.

0.3.3.2 Extreme Point of a Convex Set

An *extreme point* or *vertex* of a convex set is a point of the set which does not lie on any segment joining two other points of the set. viz., a point X of a convex set is an extreme of the set if there does not occur any pair of points X_1 and X_2 the set such that $X = \lambda X + (1-\lambda)X_2; 0 < \lambda < 1$

Note: An extreme point is a boundary point of the set, not all boundary points of a convex set need be extreme points.

0.3.3.3 Convex Combination

If $X_1, X_2, \ldots, X_K$ be a set of vectors, then a linear combination $X = \lambda_1 X_1 + \lambda_2 X_2 + \cdots \lambda_K X_K$ is called a *convex combination* of given vector, provided $\lambda_1, \lambda_2, \cdots \lambda_k \ge 0$ and $\sum_{i=1}^{n} i = 1$

Now it can be proved that the set of all *convex combinations* a finite number of linearly independent vectors $X_1, X_2, \ldots X_m$ is a convex set can also be proved that (i) a hyperplane is a convex set, (ii) closed half spaces are convex sets, (iii) the open half space convex sets and (iv) the intersection of two convex sets is a convex set. From the last result (iv), it follows the X_1 and X_2 are closed convex sets, then $X_1 \cap X_2$ is also a closed convex set.

0.3.3.4 Convex Polyhedron

The set of all convex combinations of a finite number of linearly independent vectors is called a *convex polyhedron*. The convex polyhedron generated by the finite set of linearly independent vectors $X_1, X_2,$ ¼..., X_m is the set

$$\left\{X \middle| X = \sum_{i=1}^{m} \lambda_i X_i; \lambda_i \ge 0; \sum_{i=1}^{m} \lambda_i = 1\right\}$$

A non-empty subset CR^n is called a *cone*, if for each X in C and $\lambda > 0$, the vector λX is also in C.

A cone is called a *convex cone*, if it is a convex set. Given a subset $Y \subset R^n$, the smallest convex set containing Y is called the *convex hull of Y* and is denoted by $< Y >$. The convex hull of a set Y is the intersection of all convex sets containing Y.

If A is any finite subset of vectors in R^n, then it can be proved that the convex hull of A is the set of all convex combinations of vectors in A. Using the above result, the convex polyhedron may also be defined as the convex hull set of a set of points S, consisting of a finite number of points in R^n.

A *simplex* in n-dimensions is a convex polyhedron having exactly $(n + 1)$ vertices. A simplex in 0, 1, 2 and 3 dimensions is respectively a point, a line segment, a triangle and a tetrahedron and so on.

0.3.3.5 Convex Function

If S is a non-empty convex subset of R^n, a function $f(X)$ on S is said to be *convex*, if for any two vectors X_1 and X_2 in S, $f[\lambda X_1 + (1-\lambda)X_2] \leq \lambda f(X_1) + (1-\lambda) f(X_2)$, for $0 \leq \lambda \leq 1$.

If $f[\lambda X_1 + (1-\lambda)X_2] \leq \lambda f(X_1) + (1-\lambda) f(X_2)$, for $0 < \lambda < 1$, then the function $f(X)$ on S is called a *strictly convex function*. Thus every strictly convex function is also a convex function. Similarly, concave and strictly concave functions are defined

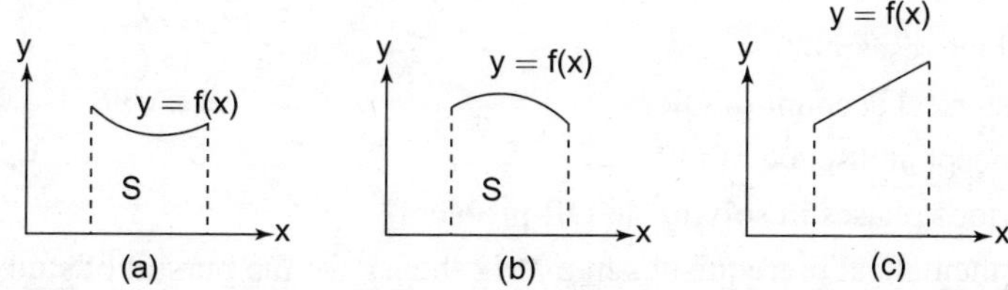

Fig. 0.2 (a) Strictly convex funciton; (b) Strictly concave function; (c) Both concave and convex function

0.3.3.6 Global and Local Minima (Maxima)

A *global minimum* of the function $f(X)$ is said to be attained at X_0, if $f(X_0) \leq f(X)$ for all X in the feasible region. [The set of vectors X satisfying the constraints $g_i(X) \leq 0$ ($i = 1, 2, \ldots, m$) is said to be the feasible region].

A *local minimum* of the function $f(X)$ is said to be attained at X_0 if there exists a positive $\in$ such that $f(X_0) \leq f(X)$ for all X in the feasible region which also satisfy the condition $|X_0 - X| \leq \in$.

If $f(X)$ has a local minimum on S, then this local minimum is also a global minimum on S.

Note: The word *extremum* (plural of extrema) is used to indicate either maximum or minimum. If the global minimum/maximum is attainable at two different points of S, then it is attainable at any infinite number of points of S.

0.3.4 Uncertainty and Probability

1. Sample space of a random experiment
2. Event
3. Apriori, Aposteriori and Axiomatic definitions of probability
4. Event – equality of, union of and intersection of events
5. Impossible and complementary events
6. Mutually exclusive events and addition theorem
7. Laws of probability and conditional probability
8. Independent events – totally and mutually, and product theorem
9. Random variables – discrete and continuous

10. P.M.F, P.D.F and distribution function
11. Expectation of a R.V
12. Mean, median and mode
13. Measures of variation, symmetry and hurtosis
14. Some standard probability distributions
    14.1 Binomial, Poisson and Geometric distributions
    14.2 Uniform, exponential, Erlang and Weibull distributions
15. Normal distribution

EXERCISE 0

Part A (Short answer questions)

1. What is operations research?
2. Give a brief historical account of OR?
3. Explain the concept and scope of OR?
4. Discuss the various phases in solving an OR problem?
5. List out the mathematical prerequisites in matrix theory for the pursuit of study of OR.
6. List out the prerequisites in theory of vector algebra for the pursuit of study of OR.
7 Explain the concept of (i) line segment (ii) the hyperplane and (iii) the hypersphere in R^n.
8. Explain briefly the concepts of convex set, extreme point of a convex set and convex polyhedron.
9. Explain the concept of convex function.
10. Explain the meaning of global and local extrema.
11. Explain the concepts of probability theory used in the study of OR.
12. Explain the uses of various probability distributions in the study of OR topics.

ANSWERS

Exercise 0

Part A

Since all the questions in Chapter 0 are more subjective than objective, they involve descriptive types of answers. Hence, the answers are not provided here.

However in all the remaining chapters (Chapters 1 to 16), answers are given to the exercise problems.

Chapter 1

LINEAR PROGRAMMING

1.0 INTRODUCTION

All industries and organisations are interested in maximising profit or minimising loss, viz., in utilising production capacity to the maximum and also reducing the cost of production to the minimum. With this objective in mind, it would be helpful to an industry if it had a scientific method by which it could decide upon the allocation of available resources in an efficient manner so as to manufacture the production satisfying not only the production requirements but also maximising its profit or minimising its production costs. Linear programming is just such a mathematical technique that helps technocrats or decision-making executives to arrive at a scientific decision regarding allocation of limited resources in an optimal manner under many constraints.

1.1 LINEAR PROGRAMMING PROBLEM (LPP)

A general LPP consists of a linear function which represents the objective of the problem such as maximising the profit or minimising the cost and which is to be optimised, and a set of simultaneous linear equations or inequalities that represent the constraints of the problem.

Note: The term 'linear' means that all the expressions in the objective function and in the constraints are linear and the term 'programming' refers to the process of determining a particular course or plan of action.

An LPP can be stated mathematically as follows.

Find the values of $x_1, x_2, x_3, \ldots, x_n$ which optimises the objective function

$$z = c_1x_1 + c_2x_2 + \ldots + c_nx_n \tag{1}$$

satisfying the structural constraints

$$a_{11}x_1 + a_{12}x_2 + \ldots + a_{1n}x_n \leq b_1$$

$$a_{21}x_1 + a_{22}x_2 + \ldots + a_{2n}x_n \leq b_2$$

$$\vdots$$

$$a_{m1}x_1 + a_{m2}x_2 + \ldots + a_{mn}x_n \leq b_m \tag{2}$$

and the non-negativity constraints

$$x_j \geq 0, j = 1, 2, \ldots, n \tag{3}$$

where all *a*s, *b*s and *c*s are known constants and all the *x*'s are unknown variables called decision variables

Note: [In the place of $\leq$ in the set (2), we may have $\geq$ or $=$ and all the bs assumed positive. If any b is negative, it is made positive by multiplying the corresponding constraint by -1 and reversing the corresponding inequality.]

The aforementioned LPP can be put in the matrix form as follows:

To maximise (or minimise) $Z = CX$

Subject to $$AX \begin{matrix} \geq \\ = \\ \leq \end{matrix} b$$

and $$X \geq 0,$$

where C is a row vector $(c_1\ c_2 \dots c_n)$,

X is a column vector $(x_1 x_2 \dots x_n)^T$

b is a column vector $(b_1\ b_2 \dots b_m)^T$ and

$$A = [a_{ij}] = \begin{bmatrix} a_{11} & a_{12} & a_{13} & \dots & a_{1n} \\ a_{21} & a_{22} & a_{23} & \dots & a_{2n} \\ \dots & \dots & \dots & \dots & \dots \\ a_{m1} & a_{m2} & a_{m3} & \dots & a_{mn} \end{bmatrix}$$

1.2 FORMULATION OF A LINEAR PROGRAMMING PROBLEM

As there are a variety of optimisation problems, viz., problems that deal with maximisation or minimisation of a function of several variables subject to certain constraints, it is necessary to recognise a problem which can be handled by linear programming techniques. In order to do this, we have to first formulate the mathematical model of the given problem.

The procedure for mathematical formulation of an LPP consists of the following steps:

1. The decision variables must be first identified.
2. The objective function to be optimised must be identified and expressed as a linear expression of the decision variables.
3. The constraints of the problem relating to availability of raw materials, machine hours, labour etc., market demand, storage capacity etc., must be expressed as linear inequalities or equations involving the decision variables.

As an example, let us consider the following problem and formulate the LPP model for it.

An industry desires to utilise the capacities of its lathe, shaping machine and milling machine to the maximum to make three products A, B and C. The available time per month in these machines are given here:

Machine:	**Lathe**	**Shaping**	**Milling**
Available time in hours per month:	200	100	180

The time taken in hours to process each unit of the product A, B and C on these machines is given in the following table.

Product	**Lathe**	**Shaping**	**Milling**
A	6	2	4
B	2	2	–
C	3	–	3

The profit per unit of the products A, B and C would be Rs. 50, Rs. 40 and Rs. 30 respectively. The objective is to maximise the total profit by a suitable product mix.

Clearly we have to decide as to how many units each of products A, B, C are to be produced so as to maximise the total profit.

We shall assume that x_1 units of A, x_2 units of B and x_3 units of C are to be produced. Then, the total profit would be $50x_1 + 40x_2 + 30x_3$. Hence, our objective would be to maximise

$$z = 50x_1 + 40x_2 + 30x_3$$

Now the total time required for processing the assumed units of the three products on lathe would be $(6x_1 + 2x_2 + 3x_3)$ hours which cannot exceed the total time available on lathe, namely, 200 hours. Thus, we have

$$6x_1 + 2x_2 + 3x_3 \leq 200.$$

Similarly, we can get

$$2x_1 + 2x_2 \leq 100$$

and

$$4x_1 + 3x_3 \leq 180.$$

Also x_1, x_2 and x_3 should be non-negative.

Thus, the LPP model for this problem is as follows.

To maximise

$$z = 50x_1 + 40x_2 + 30x_3,$$

Subject to

$$6x_1 + 2x_2 + 3x_3 \leq 200$$

$$2x_1 + 2x_2 \leq 100$$

$$4x_1 + 3x_3 \leq 180$$

and

$$x_1, x_2, x_3 \geq 0$$

1.3 SOLUTION OF A LINEAR PROGRAMMING PROBLEM

The following two methods are generally used to find the solution (viz., to find the optimal values of the decision variables) of an LPP:

1. Graphical Method Linear programming problems involving two variables can be solved graphically. However, the graphical method is impractical in LPPs with three or more variables.

2. Analytical Method This algebraic method is the most powerful tool for solving an LPP with any number of variables. It proceeds progressively and approaches the optimal solution systematically. This analytical method, also known as the simplex method, will be discussed in the next section.

We will now learn how to use the graphical method to solve an LPP.

Step by step procedure of the graphical method

1. The x_1 and x_2 axes (coordinate system) are fixed in the plane of the paper. The region in which the non-negative constraints $x_1 \geq 0$ and $x_2 \geq 0$ hold good in the plane area is the first quadrant.
2. The inequality symbols in the structural constraints are tentatively regarded as equality symbols and the straight line corresponding to them are drawn.
3. If the straight line drawn has the equation $ax_1 + bx_2 = c$, the inequality $ax_1 + bx_2 \leq c$, represents a region in the x_1 x_2-plane on one side of the line, either above the line or below the line. To identify

this region, we choose a test point in the plane whose x_1 and x_2 coordinates are obviously known (for example, the origin) and substitute its coordinates in the inequality $ax_1 + bx_2 \leq c$. If the inequality is satisfied, the required region is that in which the test point lies. Otherwise, the required region is that in which the test point does not lie. In a similar way, the region corresponding to each constraint is identified. Then the common region that satisfies all the constraints simultaneously is identified. This region is called the 'feasible region'. It contains all the solutions (viz., values of x_1 and x_2) which satisfy all the constraints of the problem. For clarity, this feasible region may be shaded or hatched.

4. An arbitrary value k is assigned to the objective function $z = c_1x_1 + c_2x_2$.
   If it is a maximisation problem, k may be taken as zero; a considerably big value may be assigned to k in the case of a minimisation problem. The straight line (preferably a dotted line) representing the objective function corresponding to the arbitrary value k is drawn.
5. The objective function line is then moved parallel to itself going from the origin in the case of maximisation and coming towards the origin in the case of minimisation, until it reaches the farthest corner (extreme point lying on the boundary lines) from the origin in the case of maximisation and the nearest corner close to the origin in the case of minimisation. If the objective function line is moved beyond these extreme corners (optimal points), it will leave the region of feasible solutions.
6. The coordinates of the optimal point are noted graphically. They give the optimal solution to the problem. Using these optimal values of x_1 and x_2 in the objective function, the optimal value of z is obtained.

Important special cases

1. If the feasible region cannot be uniquely determined due to the inconsistency of the structural constraints, it is an indication that the LPP does not have a solution.
2. If the objective function line can be moved parallel to itself indefinitely without making it leave the feasible region, it indicates that the LPP possesses an unbounded solution, viz., the optimal value of z occurs at the point at infinity.
3. In the extreme position, if the objective function line coincides with one of the boundary lines of the feasible region, it indicates that the LPP possesses many alternative optimal solutions. Not only the extreme points lying on this boundary lines but also any point lying between the extreme points on the boundary line will provide optimal solutions, giving the same optimal value of z.

All these special cases will be considered in the following worked out examples.

WORKED EXAMPLES 1(A)

Example 1

A company produces three products A, B and C with a profit contribution of Rs. 20, Rs. 25 and Rs. 15 per unit respectively. The resource requirements per unit of each of the products and total availability of the resources are given here:

Resource requirement	Product			Total availability
	A	B	C	
Man hours per unit	6	3	12	200 man hours
Machine hours per unit	2	5	4	350 machine hours
Material per unit in kg	1	2	1	100 kg

Formulate the problem as an LPP model.

SOLUTION

Let x_1 units of product A, x_2 units of product B and x_3 units of product C be produced so as to maximise the total profit.

The total profit due to the assumed units of the three products is $20x_1 + 25x_2 + 15x_3$. Hence, the objective is to

$$\text{Maximise } z = 20x_1 + 25x_2 + 15x_3,$$

Constraints:

The total number of man hours required to produce the assumed units of A, B and C is $6x_1 + 3x_2 + 12x_3$. Since only 200 man hours are available, we have the constraint.

$$6x_1 + 3x_2 + 12x_3 \leq 200$$

Similarly, the restrictions on the availability of machine hours and materials give the constraints

$$2x_1 + 5x_2 + 4x_3 \leq 350$$

and $$x_1 + 2x_2 + x_3 \leq 100$$

Thus, the LP model for the given problem is:

$$\text{Maximise } z = 20x_1 + 25x_2 + 15x_3,$$

Subject to

$$6x_1 + 3x_2 + 12x_3 \leq 200$$

$$2x_1 + 5x_2 + 4x_3 \leq 350$$

$$x_1 + 2x_2 + x_3 \leq 100$$

and $$x_1, x_2, x_3 \geq 0$$

Example 2

A feed mixing company purchases and mixes one or more of three types of grains G_1, G_2 and G_3, each containing varying amounts of three nutritional elements A, B and C as given in the following table.

Nutritional element	One unit weight of			Minimum total requirement for a week
	G_1	G_2	G_3	
A	2	4	6	125
B	0	2	5	24
C	3	1	3	80
Cost per unit weight in Rs	25	15	18	-

If the feed mix must meet the minimum nutritional requirements at the least cost, formulate the linear programming problem corresponding to this problem.

Solution

Let us assume that x_1, x_2 and x_3 units (of weight) of the grains G_1, G_2 and G_3 are purchased and mixed.

The total cost of the assumed units of weight of the grains is $25x_1 + 15x_2 + 18x_3$. As the total cost is to be the least, the objective is to

$$\text{Minimise } z = 25x_1 + 15x_2 + 18x_3$$

Constraints

The total amount of nutritional element A for the assumed weights of G_1, G_2, G_3 is $2x_1 + 4x_2 + 6x_3$.

Since the minimum requirement of A for a week is 125, we have the constraint

$$2x_1 + 4x_2 + 6x_3 \geq 125$$

Similarly, the minimum requirement of B and C give the constraints

$$2x_2 + 5x_3 \geq 24$$

and $$3x_1 + x_2 + 3x_3 \geq 80 \text{ respectively}$$

Thus, the LP model for the given problem is

$$\text{Minimise } z = 25x_1 + 15x_2 + 18x_3,$$

Subject to

$$2x_1 + 4x_2 + 6x_3 \geq 125$$

$$2x_2 + 5x_3 \geq 24$$

$$3x_1 + x_2 + 3x_3 \geq 80$$

and $$x_1, x_2, x_3 \geq 0$$

Example 3

A manufacturer produces three models M_1, M_2 and M_3 of a certain product. He uses two types of raw material A and B of which 2000 and 3000 units are available respectively. The raw material requirements per unit of the three models are given in the following table.

Raw material	Requirement per unit of		
	M_1	M_2	M_3
A	2	3	5
B	4	2	7

The labour time required for each unit of M_1 is twice that of M_2 and three times of M_3. The entire labour force of the factory can produce the equivalent of 700 units of M_1. The minimum market demand of the three models are 200, 200 and 150 units respectively. Assuming that the profits per unit of the three models are Rs. 30, Rs. 20 and Rs. 50 respectively, formulate the problem as an LP model in order to determine the number of units of each product that will maximise the total profit.

Solution

Let us assume that the number of units of models M_1, M_2 and M_3 to be produced are x_1, x_2 and x_3 respectively. As the total profit is maximum, the objective is to maximise.

$$z = 30x_1 + 20x_2 + 50x_3$$

Constraints

The total units of raw material A required for the production of the assumed units is $2x_1 + 3x_2 + 5x_3$.

Since only 2000 units of A are available, we have the constraint $2x_1 + 3x_2 + 5x_3 \leq 2000$. Similarly, availability constraint of B gives

$$4x_1 + 2x_2 + 7x_3 \leq 3000$$

If t is the labour time required for making 1 unit of M_1, the time required for making 1 unit of M_2 and 1 unit of M_3 are $\frac{t}{2}$ and $\frac{t}{3}$ respectively.

Hence, the total time required for making the assumed units of M_1, M_2 and M_3 is $tx_1 + \frac{t}{2}x_2 + \frac{t}{3}x_3$.

Since in the total time available, the entire labour force can produce 700 units of M_1, viz., is equal to $700t$, we have the constraint $tx_1 + \frac{t}{2}x_2 + \frac{t}{3}x_3 \leq 700t$

$$6x_1 + 3x_2 + 2x_3 \leq 4200$$

The marketing conditions give the constraint

$$x_1 \geq 200, x_2 \geq 200, x_3 \geq 150$$

Thus, the L.P model for the given problem is:

$$\text{Maximise } z = 30x_1 + 20x_2 + 50x_3$$

Subject to

$$2x_1 + 3x_2 + 5x_3 \leq 2000$$

$$4x_1 + 2x_2 + 7x_3 \leq 3000$$

$$6x_1 + 3x_2 + 2x_3 \leq 4200$$

and

$$x_1 \geq 200, x_2 \geq 200, x_3 \geq 150$$

Example 4

A soft drink plant has two bottling machines A and B. The following data regarding the bottling capabilities of the machines are available.

Machine	200 ml. bottles	300 ml. bottles
A	12/minute	5/minute
B	4/minute	10/minute

The machine can be run 8 hours per day, 5 days per week. Profit on a 200 ml bottle is Rs. 2.50 and on a 300 ml bottle is Rs. 4.50. Weekly production cannot exceed 5,000 litres and the market can absorb 10,000 two hundred ml bottles and 5,000 three hundred ml bottles per week.

The owner of the plant wishes to maximise the total profit, subject to all the production and marketing restrictions. Formulate this as an LP problem.

Solution

Let x_1 be the number of 200 ml bottles and x_2 be the number of 300 ml bottles needed to be produced in a week.

The total profit for the assumed numbers is

$$2.50x_1 + 3.50x_2.$$

Then the objective is to

$$\text{Maximise } z = 2.50x_1 + 3.50x_2$$

Constraints

The time required for processing the assumed numbers of bottles by machines A is $\frac{x_1}{12} + \frac{x_2}{5}$ minutes per week.

Since machine A can be run only for $5 \times 8 \times 60 = 2400$ minutes in a week, we have the constraint

$$\frac{x_1}{12} + \frac{x_2}{5} \leq 2400$$

or $$5x_1 + 12x_2 \leq 1{,}44{,}000$$

Similarly, we get the constraint with respect to machine B as

$$\frac{x_1}{4} + \frac{x_2}{10} \leq 2400$$

or $$5x_1 + 2x_2 \leq 48{,}000$$

Since weekly production cannot exceed 5000 litres or 50,00,000 ml, we have the constraint

$$200x_1 + 300x_2 \leq 50{,}00{,}000$$

or $$2x_1 + 3x_2 \leq 50{,}000$$

Also the marketing restrictions give the constraints

$$0 \leq x_1 \leq 10{,}000$$

and $$0 \leq x_2 \leq 5000$$

Thus, the LP model for the given problem is

$$\text{Maximise } z = 2.50x_1 + 3.50x_2,$$

Subject to $$5x_1 + 12x_2 \leq 1{,}44{,}000$$

$$5x_1 + 2x_2 \leq 48{,}000$$

$$2x_1 + 3x_2 \leq 50{,}000$$

$$0 \leq x_1 \leq 10{,}000$$

and $$0 \leq x_2 \leq 5000$$

Example 5

A lorry company has 5 booking offices at the localities A, B, C, D and E and 10 lorries for use. On a particular day, 4 lorries were parked in the shed at A and 6 in the shed at B. On the same day, there were demands for 2, 3 and 5 lorries in the localities C, D, E respectively. The distances in kilometers among the 5 localities are given in the following table.

From/To	C	D	E
A	2	3	8
B	3	7	5

How should the lorries be sent from A and B to C, D and E so that the total distance covered by the lorries may be minimum?

Formulate this problem as an L.P model.

SOLUTION

We observe that the total number of lorries available at A and B is equal to the total number of lorries required at C, D and E.

Let us assume that x lorries are sent from A to C and y lorries from A to D. Then, the numbers of lorries sent from A and B to C, D and E may be tabulated as follows:

From/To	C	D	E
A	x	y	$4-(x+y)$
B	$2-x$	$3-y$	$6-[(2-x)+(3-y)]$

Since the number of lorries sent are to be non-negative, we have the constraints

$$4-(x+y) \geq 0$$

$$2-x \geq 0$$

$$3-y \geq 0$$

$$x+y+1 \geq 0$$

$$x, y \geq 0$$

For the assumed numbers, the total distance covered by all the lorries is

$$2x+3y+8(4-x-y)+3(2-x)+7(3-y)+5(x+y+1) = 64-4x-7y$$

Thus, the LP model for the given problem is:

$$\text{Minimise } z = 64-4x-7y,$$

Subject to

$$x+y \leq 4$$

$$0 \leq x \leq 2$$

and

$$0 \leq y \leq 3$$

Note: This problem is a special type of LP problem, known as the transportation problem, which will be discussed later.

Example 6

A farmer can sell all the tomatoes, lettuce or radishes that he raises in his 100-acre farm. The price he can obtain is Rs. 4/- per kilogram for tomatoes, Rs. 3/- per head for lettuce and Rs. 8/- per kilogram for radishes. The average yield per acre is 2000 kg of tomatoes, 3000 heads of lettuce and 1000 kg of radishes. Fertiliser is available at Rs. 2/- per kg and the quantity required per acre is 100 kg each for tomatoes and lettuce and 50 kg for radishes. Labour required for cultivation per acre is 5 man-days for tomatoes and radishes and 6 man-days for lettuce. A total of 400 man-days of labour are available at Rs. 50/- per man-day. Formulate this problem as an LP model to maximise the farmer's total profit.

Solution

Let us assume that the farmer produces x_1 kg of tomatoes, x_2 units of lettuce and x_3 kg of radishes in his farm so as to maximise his total profit.

The total farm area required to produce the assumed quantities is

$$\frac{x_1}{2000}+\frac{x_2}{3000}+\frac{x_3}{1000}$$

Since the total farm area owned by the farmer is 100 acres, we have the constraint

$$\frac{x_1}{2000}+\frac{x_2}{3000}+\frac{x_3}{1000}\le 100 \tag{1}$$

Now the amount of fertiliser required to produce the assumed quantities of vegetables is

$$\left(\frac{x_1}{2000}\times 100+\frac{x_2}{3000}\times 100+\frac{x_3}{1000}\times 50\right)\text{kg}.$$

Since the cost of fertiliser is Rs. 2/- per kg; the total amount spent for fertilizer is

$$\text{Rs}\left(\frac{x_1}{10}+\frac{x_2}{15}+\frac{x_3}{10}\right) \tag{2}$$

Labour required in man-days for producing the assumed quantities of vegetables is

$$\left(\frac{x_1}{2000}\times 5+\frac{x_2}{3000}\times 6+\frac{x_3}{1000}\times 5\right).$$

Since the labour available is 400 man-days, we have the constraint

$$\frac{x_1}{400}+\frac{x_2}{500}+\frac{x_3}{400}\le 400 \tag{3}$$

Since the labour cost is Rs. 50/- per man-day, the total amount spent for labour is

$$\left(\frac{x_1}{400}+\frac{x_2}{500}+\frac{x_3}{400}\right)\times 50\,\text{Rs.}=\text{Rs}\left(\frac{x_1}{8}+\frac{x_2}{10}+\frac{x_3}{8}\right) \tag{4}$$

Since all the vegetables produced can be sold out, the total sales proceeds of the vegetables is

$$\text{Rs}\,(4x_1+3x_2+8x_3) \tag{5}$$

Hence, the total profit on the yield is given by

$$z = \text{Total sales proceeds} - \text{Total fertiliser cost} - \text{Total labour cost}$$

$$=(4x_1+3x_2+8x_3)-\left\{\left(\frac{x_1}{10}+\frac{x_2}{15}+\frac{x_3}{10}\right)+\left(\frac{x_1}{8}+\frac{x_2}{10}+\frac{x_3}{8}\right)\right\}$$

$$= \frac{1}{120}(453x_1 + 340x_2 + 933x_3)$$

Thus, the LP model for the given problem is:
To maximise

$$z = \frac{1}{120}(453x_1 + 340x_2 + 933x_3),$$

Subject to $$\frac{x_1}{2000} + \frac{x_2}{3000} + \frac{x_3}{1000} \leq 100$$

$$\frac{x_1}{400} + \frac{x_2}{500} + \frac{x_3}{400} \leq 400; \ x_1, x_2, x_3 \geq 0$$

Example 7

A company produces two types of leather belts, say A and B. Belt A is of a superior quality, while B is of an ordinary quality. Profit on the two types of belts are Rs. 12/- and Rs. 9/- per belt respectively. Each belt of type A requires twice as much time as required by a belt of type B. If all belts were of type B, the company could produce 500 belts per day, but the supply of leather is sufficient only for 400 belts per day. Only 200 fancy buckles required for belt A and only 300 ordinary buckles required for belt B are available per day.

Find graphically the numbers of belt A and belt B that the company should produce daily so as to maximise the total profit.

SOLUTION

Let the company produce x_1 belts of type A and x_1 belts of type B daily to maximise the profit that is equal to

$$z = 12x_1 + 9x_2$$

Let the time required to make one belt of type B be t. Then, the time required for making one belt of type A is $2t$.

Hence, the total time required to produce the assumed numbers of belts is $2t.x_1 + t.x_2$

Since the available time per day is $500t$,
we have

$$2t.x_1 + t.x_2 \leq 500t$$

i.e., $$2x_1 + x_2 \leq 500$$

Since the leather is sufficient to make only 400 belts per day, we have

$$x_1 + x_2 \leq 400$$

Also we have $x_1 \leq 200$ and $x_2 \leq 350$ due to the availability of buckles.

Thus, the LP model for the problem is:

To maximise $$z = 12x_1 + 9x_2$$

Subject to $$2x_1 + x_2 \leq 500$$

$$x_1 + x_2 \leq 400$$

$$0 \leq x_1 \leq 200$$

and $$0 \leq x_2 \leq 350$$

Graphical Solution

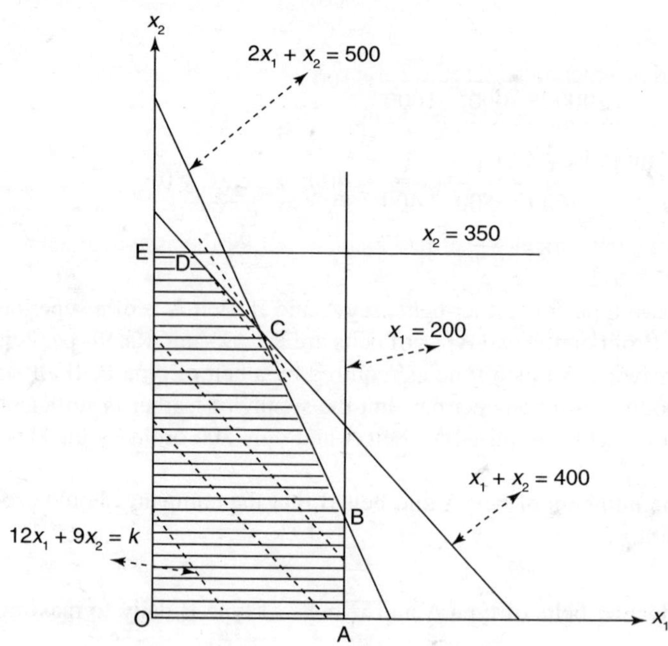

We fix the coordinate axis as Ox_1 and Ox_2. The region satisfying the non-negativity constraints $x_1 \geq 0$ and $x_2 \geq 0$ is the area in the first quadrant.

Then, we draw the lines

$$2x_1 + x_2 = 500, x_1 + x_2 = 400, x_1 = 200$$

and $$x_2 = 350.$$

The region satisfying all the constraints is identified and shaded. This shaded region OABCDE is the region of feasible solutions.

We then plot a dotted line representing a typical position of the objective function line by taking a convenient value for z. For example, by taking $z = 1080$, the line becomes

$$\frac{x_1}{90} + \frac{x_2}{120} = 1$$

Since the problem is one of maximisation, this dotted line is moved parallel to itself going away from the origin, until we reach the farthest position from the region. The farthest point is C. If we move the line beyond C, we will leave the region of feasible solution.

Thus, C is the point corresponding to the maximum value of z.

The coordinates of C are found out graphically as $x_1 = 100$ and $x_2 = 300$, which give the optimal solution of the problem. The maximum value of z is then obtained as $12 \times 100 + 9 \times 300 = 3900$.

Example 8

A person requires 10, 12 and 12 units of chemicals A, B, C respectively for his garden. A liquid product contains 5, 2 and 1 units of A, B, C respectively per jar. A dry product contains 1, 2 and 4 units of A, B and C per carton. If the cost of the liquid product is Rs. 30 per jar and that of the dry product is Rs. 20 per carton, find graphically how many of each should he purchase to maximise the cost and meet the requirements.

SOLUTION

Let x_1 jars of the liquid product and x_2 cartons of the dry product be purchased.

$\therefore$ the total cost of the items $= 30x_1 + 20x_2$.

The objective is to minimise

$$z = 30x_1 + 20x_2$$

The number of units of chemical A that will be obtained for the assumed quantities of the products

$$= 5x_1 + x_2$$

Since this must meet the requirement,

we have $5x_1 + x_2 \geq 10$

Similarly, $2x_1 + 2x_2 \geq 12$

and $x_1 + 4x_2 \geq 12,$

Also $x_1, x_2 \geq 0.$

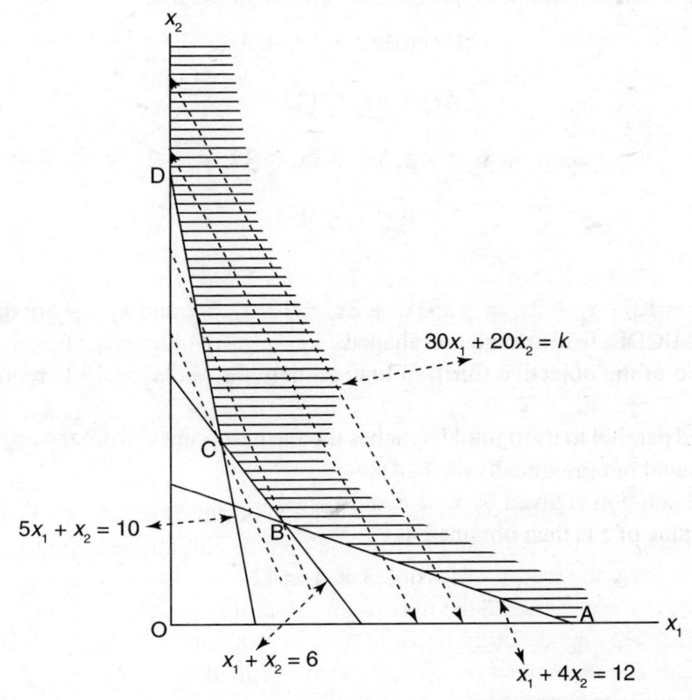

Thus, the LP model for the given problem is:

To minimise $z = 30x_1 + 20x_2$,

Subject to
$$5x_1 + x_2 \geq 10$$
$$x_1 + x_2 \geq 6$$
$$x_1 + 4x_2 \geq 12$$

and $x_1, x_2 \geq 0$

As in the previous example, we draw the lines $5x_1 + x_2 = 10$, $x_1 + x_2 = 6$ and $x_1 + 4x_2 = 12$ and identify the region of feasible solution as x_1ABCDx_2. It can be shaded as given in the figure.

Then, we draw a dotted line (sufficiently away from the origin) representing a typical position of the objective function line
$$30x_1 + 20x_2 = 260.$$

Since the LP problem is one of minimisation, this dotted line is moved parallel to itself going towards the origin until the line passes through the extreme point (nearest point to the origin) beyond which the dotted line will leave the region of feasible solution. We observe that the extreme point is C which is the minimum point and whose coordinates from the graph are (1, 5).

Thus, the optimal solution is $x_1 = 1$ and $x_2 = 5$ and the minimum value of
$$z = 30 \times 1 + 20 \times 5 = 130.$$

Example 9

Solve the following linear programming problem by graphical method:

$$\text{Maximise } z = 5x_1 + 3x_2,$$

Subject to
$$4x_1 - 3x_2 \leq 12$$
$$-x_1 + x_2 \geq -2,\ 3x_1 + 2x_2 \geq 12,\ x_1 \geq 2$$

and
$$0 \leq x_2 \leq 4$$

SOLUTION

The lines $4x_1 - 3x_2 = 12$, $-x_1 + 2x_2 = -2$, $3x_1 + 2x_2 = 12$, $x_1 = 2$ and $x_2 = 4$ are drawn. The region of feasible solutions, ABCDE, is identified and shaded.

A typical position of the objective function line, namely, $5x_1 + 3x_2 = 15$ is represented by a dotted line.

This line is moved parallel to itself until it reaches the farthest point C from the origin. The coordinates of the point C are found out graphically as (6, 4).

Thus the optimal solution is given by $x_1 = 6$ and $x_2 = 4$

The maximum value of z is then obtained as
$$5 \times 6 + 3 \times 4 = 42.$$

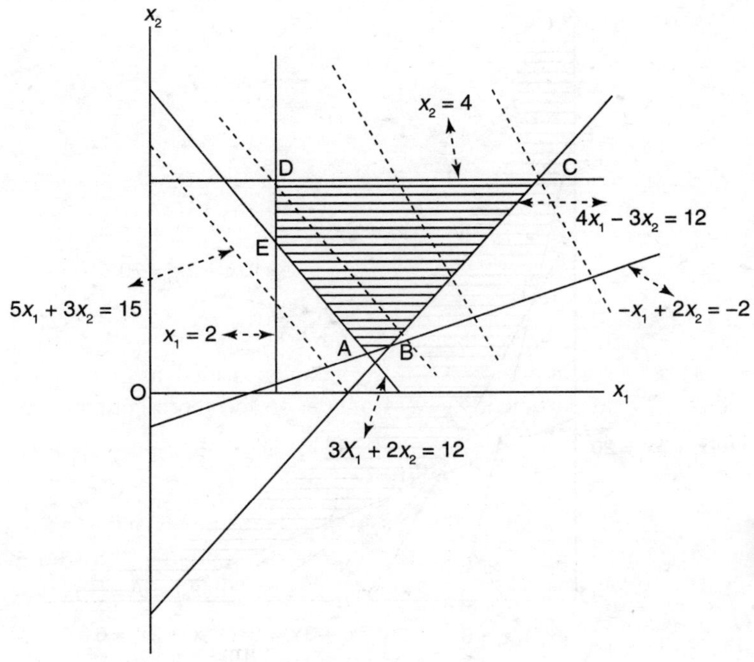

Example 10

Solve the following linear programming problem graphically:

$$\text{Minimise } z = 10x_1 + 7x_2$$

Subject to

$$4x_1 + x_2 \geq 8$$

$$7x_1 + 3x_2 \geq 21$$

$$x_1 + 2x_2 \geq 6$$

$$4x_1 + 5x_2 \geq 20$$

and

$$x_1, x_2 \geq 0$$

SOLUTION

The lines $4x_1 + x_2 = 8$, $7x_1 + 3x_2 = 21$, $x_1 + 2x_2 = 6$ and $4x_1 + 5x_2 = 20$ are drawn. The region of feasible solutions x_1ABCDEx_2 is identified and shaded.

A typical position of the objective function line, namely, $10x_1 + 7x_2 = 70$ is represented by a dotted line. This line is moved parallel to itself going towards the origin, until it passes through the extreme point C.

The coordinates of C are found out graphically as (1.9, 2.4) approximately.

Thus, the optimal solution is $x_1 = 1.9$ and $x_2 = 2.4$ approximately and the minimum value of z is 35.8.

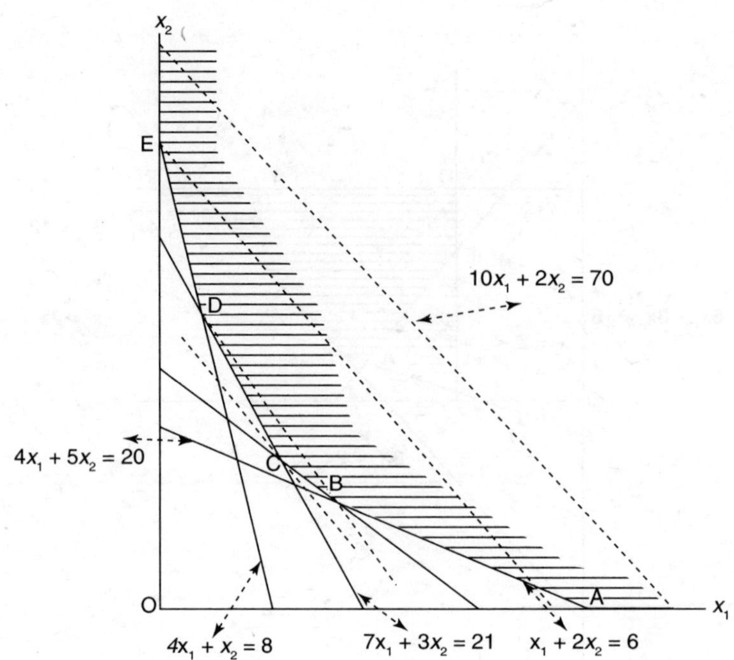

Example 11

Solve the following linear programming problem graphically:

$$\text{Maximise } z = x + 2y,$$

Subject to

$$-x + 2y \leq 8$$

$$x + 2y \leq 12$$

$$x - y \leq 3$$

and

$$x, y \geq 0$$

Solution

The lines $-x + 2y = 8$, $x + 2y = 12$, $x - y = 3$ are drawn. The region of feasible solutions OABCD is identified and shaded.

A typical position of the objective function line, namely, $x + 2y = 6$ is represented by a dotted line. This line is moved parallel to itself, going away from the origin.

In the extreme position, it coincides with the line BC, one of the boundaries of the region of feasible solutions. This indicates that the problem has more than one optimal solution (alternative optimal solutions).

From the graph, the coordinates of the points B and C are found to be (6, 3) and (2, 5) respectively

Thus, two optimal solutions for the problem are $x = 6, y = 3$ and $x = 2, y = 5$. The coordinates of any other point lying between B and C on the line segment BC (for example, $x = 4, y = 4$) will also provide another optimal solution.

The maximum value of $z = 12$.

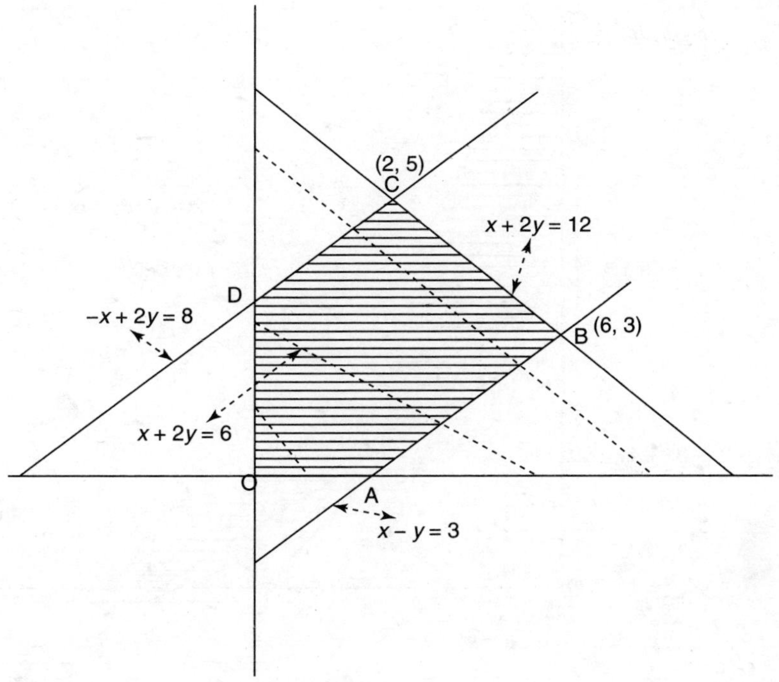

Example 12

Solve the following linear programming problem graphically:

$$\text{Minimise } z = 20x + 10y,$$

Subject to

$$3x + y \geq 30$$

$$2x + y \geq 24$$

$$3x + 4y \geq 60$$

and

$$x, y \geq 0$$

SOLUTION

The lines $3x + y = 30$, $2x + y = 24$, $3x + 4y = 60$ are drawn. The region of feasible solutions XABCDY is identified and shaded.

A typical position of the objective function line, namely, $20x + 10y = 60$ is represented by a dotted line.

This line is moved parallel to itself, going toward the origin.

In the extreme position, it coincides with the line BC, one of the boundaries of the region of feasible solutions.

Hence, the problem has more than one optimal solution.

From the graph, the coordinates of B and C are found to be (7.2, 9.6) and (6, 12) respectively. Thus, two optimal solutions for the problem are $x = 7.2$, $y = 9.6$ and $x = 6$, $y = 12$.

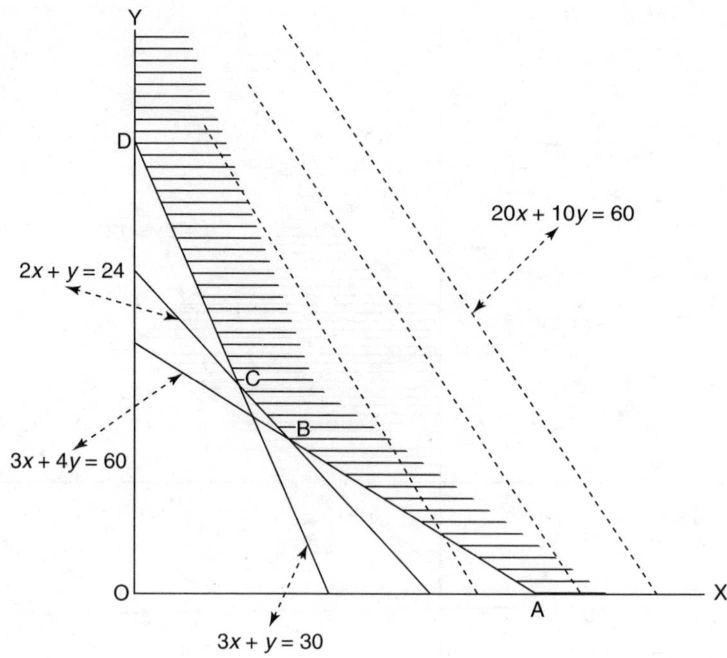

The coordinates of any other point lying between B and C on the line segment BC (for example, $x = 6.6$, $y = 10.8$) will also provide another optimal solution.

The maximum value of $z = 240$.

Example 13

Solve the following linear programming problem graphically:

$$\text{Maximise } z = 2x + 3y,$$

Subject to
$$-x + y \leq 15$$
$$-x + 2y \leq 40$$

and
$$x, y \geq 0$$

Solution

The lines $-x + y = 15$, and $-x + 2y = 40$ are drawn. The region of feasible solutions XOABC is identified and shaded.

A typical position of the objective function line, namely, $2x + 3y = 3$ is represented by a dotted line. This line is moved parallel to itself, going away from the origin.

We observe that the line can be moved indefinitely without making it leave the region of feasible solutions.

This is an indication that the LP problem has an unbounded solution which means that the optimal solution occurs at a point at infinity.

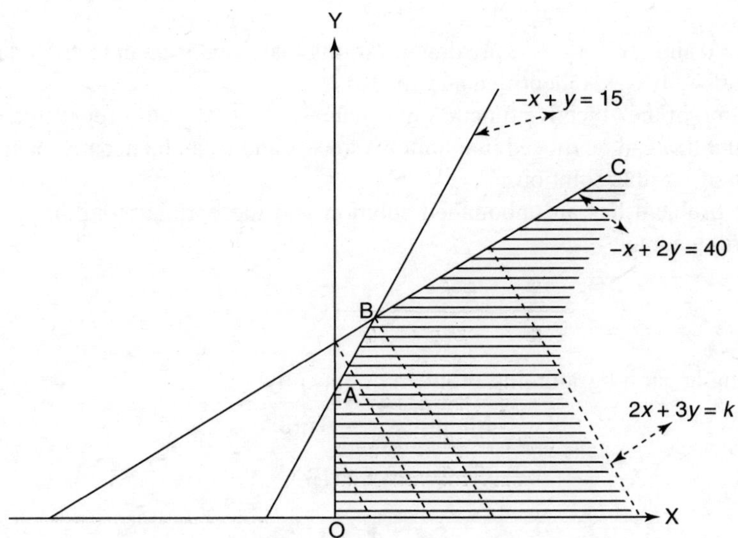

Example 14

Solve the following linear programming problem graphically:

$$\text{Minimise } z = 2x + 3y,$$

Subject to $$x - 2y \geq 0$$

$$-x + y \geq -3$$

and x and y are unrestricted in sign

Note: When x and y can take positive, zero or negative values, they are said to be unrestricted in sign.

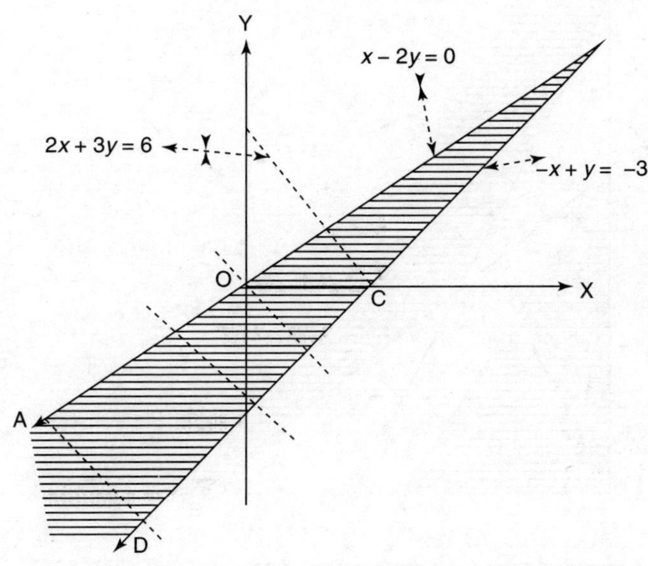

SOLUTION

The lines $x - 2y = 0$ and $-x + y = -3$ are drawn. Noting that x and y are unrestricted in sign, the region of feasible solutions AOBCD is identified and shaded.

A typical position of the objective function line, namely, $2x + 3y = 6$ is represented by a dotted line. We observe that the line can be moved indefinitely (since x and y can be negative also) without making it leave the region of feasible solutions.

Thus, the L.P. problem has an unbounded solution and the optimal solution occurs at a point at infinity in the third quadrant.

Example 15

Solve the following linear programming problem graphically:

$$\text{Maximise } z = 4x + 5y,$$

Subject to

$$2x + 3y \geq 12$$

$$x + y \leq 4$$

$$4x + 3y \geq 18$$

and

$$x, y \geq 0$$

The lines $2x + 3y = 12$, $x + y = 3$, $4x + 3y = 18$ are drawn. The region corresponding to $x + y \leq 4$; $x, y \geq 0$ (OAB) is identified and shaded.

The region corresponding to $2x + 3y \geq 12$, $4x + 3y \geq 18$, $x, y \geq 0$ (XEDCY) is identified and shaded.

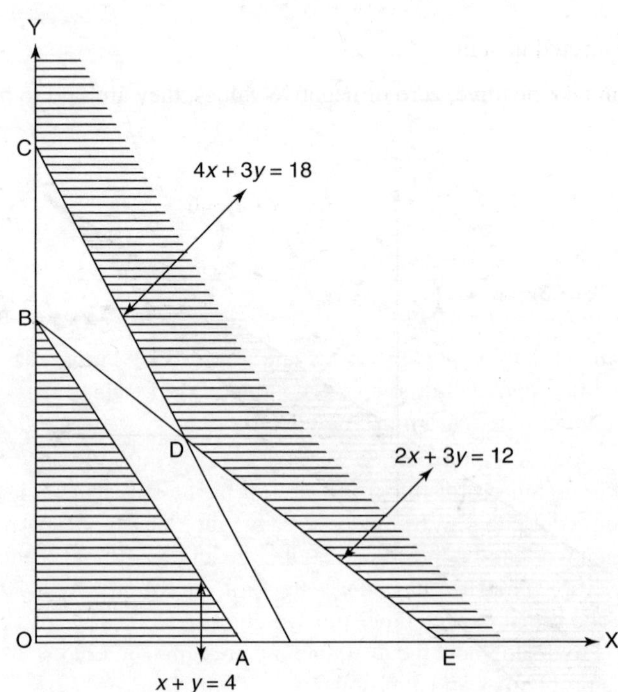

We observe that there is no common region satisfying all the constraints

$$2x + 3y \geq 12, x + y \leq 4, 4x + 3y \geq 18;$$

$$x, y \geq 0.$$

This means that there is no feasible solution for the given LP problem irrespective of the nature of the objective function.

EXERCISE 1(A)

Part A (Short answer questions)

1. What is a linear programming problem?
2. What do the words 'linear' and 'programming' mean in an LP problem?
3. State a linear programming problem in the mathematical format.
4. Give the step by step procedure for the graphical solution of an LP problem.
5. While solving an LP problem graphically, how will you identify that the problem has no feasible solution?
6. While solving an LP problem graphically, how will you find out that the problem has an unbounded solution?
7. While solving an LP problem graphically, how will you identify that the problem has alternative optimal solutions?

Part B

8. A company manufactures three products A, B and C. The profits for each unit of A, B, C are Rs. 30, Rs. 20 and Rs. 40 respectively. The company has two machines and the processing time required in minutes on each machine for each product is given in the following table:

Machine	Products		
	A	B	C
I	4	3	5
II	2	2	4

Machine I and II can be operated for 2000 and 2500 minutes respectively.
The company must manufacture 100 As, 200 Bs and 50 Cs, but no more than 150 As. It is required to find the number of units of each product the company has to manufacture in order to maximise the profit. Formulate the problem as an LP model.

9. A firm produces three types of cloth, say, A, B and C. Three kinds of wool, namely red wool, green wool and blue wool are required for the production of the clothes. One unit length of type A cloth requires 5 m of green wool and 4 m of blue wool. One unit length of type B cloth requires 3 m of red wool, 2 m of green wool and 2 m of blue wool. One unit length of type C cloth requires 2 m of red and 3 m of blue wool. The firm has only a stock of 8 m of red wool, 10 m of green wool and 15 m of blue wool. The profit expected per unit length of A is Rs.15/-, of B is Rs.25/- and of C is Rs.20/-. The firm has to decide how the available material may be used so as to maximise the total profit. Formulate the problem as an LP problem

10. A wholesale dealer deals in two kinds, A and B, of mixture of nuts. Each kg of mixture A contains 60 gms of almonds, 30 gms of cashew nuts and 30 gms of hazel nuts. Each kg mixture B contains 30 gms of almonds, 60 gms of cashew nuts and 180 gms of hazel nuts. The remainder of both mixtures is peanuts. The dealer wishes to use mixtures A and B to make a bag which will contain at least 240 gms of almonds, 300 gms of cashew nuts and 540 gms of hazel nuts. Mixture A costs Rs.20/- per kg and mixture B costs Rs.30/- per kg. Assuming that mixtures A and B are uniform, how many kgs of each should he use to minimise the cost of the bag? Formulate the problem as an LP model.
11. A person should consume at least 5 units of protein, 2 units of fat and 3 units of carbohydrate daily to maintain good health. These nutrients can be obtained by eating bread, butter and milk. It is known that one unit of bread contains 4 units of protein, 1 unit of fat and 2 units of carbohydrate. The corresponding figures for butter and milk are 3, 2, 1 and 3, 2, 1 respectively. The costs per unit of bread, butter and milk are respectively Rs.5, Rs.20 and Rs.6. Formulate an LP problem to determine the diet that satisfies the daily requirements at a minimum cost.
12. A soft drinks firm has two bottling plants P and Q. Each plant can produce three different soft drinks A, B and C. The capacities of the two plants in number of bottles per day are given in the following table:

		Plant	
		P	Q
	A	3000	1000
Drink	B	1000	1000
	C	2000	6000

A market survey indicates that during the month of May, there will be a demand for 24,000 bottles of A, 16,000 bottles of B and 48,000 bottles of C. The operating costs per day, of running plants P and Q are respectively Rs.3000/- and Rs.2000/-. How many days should the firm run each plant in May so that the production cost is minimised while still meeting the market demand? Formulate the problem as an LPP.
13. Four products are processed successively on two machines. The processing times in hours per unit of each product are tabulated for the two machines as follows.

Processing time per unit in hours

Machine	Product A	Product B	Product C	Product D
1	2	3	4	2
2	3	2	1	2

The total production cost per unit of each product is based directly on the processing time. The costs of operation per hour for machine 1 and 2 are Rs.100/- and Rs.150/- respectively. The total hours for which machines 1 and 2 can be operated for all the products are 500 and 380 respectively. If the sale price per unit for products A, B, C, D are Rs.650/-, Rs.700/-, Rs.550/- and Rs.450/- respectively, how should the products be produced so as to maximise the total net profit?
14. An oil refinery has to decide on the optimal mix of two blending processes 1 and 2 of which the inputs and outputs per production run are as follows:

Process	Input (units)		Output (units)	
	Crude A	Crude B	Gasoline X	Gasoline Y
1	5	3	5	8
2	4	5	4	4

The maximum amounts available of crudes A and B are 200 units and 150 units respectively. Market requires at least 100 units of gasoline X and 80 units of gasoline Y. The profits per production run from process 1 and process 2 are Rs.300/- and Rs.400/- respectively. Formulate the problem as an LP model so as to maximise the total profit of the oil refinery.

15. A manufacturer makes four products A, B, C and D, each of which must be processed in the same order in all the three departments P, Q and R. Figures of profit and rates of processing in the departments are given in the following table:

Product	Profit/unit (Rs.)	Processing time/units in weeks		
		P	Q	R
A	12	5	6	7
B	15	5	6	5
C	16	6	5	6
D	11	7	7	6

The company works for 50 weeks per year but the down time in the departments vary and the expected number of weeks of production for the three departments are 45, 42 and 48 respectively. Formulate the corresponding LP model.

16. A rubber company is engaged in producing three kinds of tyres, A, B and C. These tyres are produced at the company's two different plants with different production capacities. In a normal 8 hours working day, plant 1 produces 50, 100 and 100 tyres of type A, B and C respectively. Plant 2 produces 60, 60 and 200 tyres of type A, B and C respectively. The monthly demands for types A, B and C are 2400, 3000 and 7000 units respectively. The daily costs of operation of plants 1 and 2 are Rs.2400/- and Rs.3600/- respectively. Find the number of days of operation per month at the two plants to minimise the total cost while meeting the demands.

17. A manufacturer produces three models M_1, M_2 and M_3 of a certain product. He uses two types of raw materials A and B of which 2000 and 3000 units respectively are available. The raw material requirements per unit of the three models are given in the following table.

Raw material	Requirement per unit of model		
	M_1	M_2	M_3
A	2	3	5
B	4	4	7

The labour time for each unit of M_1 is twice that of M_2 and three times that of M_3. The labour force of the factory can produce the equivalent of 700 units of M_1. The minimum market demands of the three models M_1, M_2 and M_3 are 200, 300 and 250 units respectively. However, the ratio of the number of units produced must be 3:2:5. If the profits per unit of M_1, M_2 and M_3 are Rs.30/- Rs.40/- and Rs.50/- respectively, formulate the problem as an LP model so as to maximise the profit.

18. A transport company has offices in five localities A, B, C, D and E. On a particular day, the offices located at A and B has 6 and 8 spare trucks whereas offices at C, D and E require 4, 7 and 3 trucks respectively. The distances in km between the five localities are given as follows:

From/To	C	D	E
A	3	7	5
B	2	3	9

How should the trucks from A and B be sent to C, D and E so that the total distance covered by the trucks is minimum? Formulate the problem as an LP model.

19. A chemical industry produces two products A and B. Each unit of product A requires 3 hours on operation I and 3 hours on operation II, while each unit of product B requires 4 hours on operation I and 5 hours on operation II. Total times available for operations I and II are respectively 18 and 21 hours. The production of each unit of the product B also results in 3 units of a by-product C at no extra cost. Product A sells at a profit of Rs.3/- per unit and product B at a profit of Rs.8/- per unit. By-product C brings a unit profit of Rs.2/- and a destruction cost of Rs.1 per unit in case it cannot be sold. Forecasts indicate that all the units of A and B can be sold whereas not more than 5 units of C can be sold. How many units of A and B should be produced by the company, keeping C in mind, to maximise the profit? Formulate the problem as an LP model.

20. A company produces a product, each complete unit of which consists of 4 units of component A and 3 units of component B. The two components of A and B are manufactured from two different raw materials of which 100 and 200 units respectively are available. Production of components A and B is done in three departments. The following table gives the raw material requirements and the component yields per production run.

Department	Input units per run		Output units per run	
	Raw mat I	Raw mat II	Component A	Component B
1	7	5	6	4
2	4	8	5	8
3	2	7	7	3

Find the number of production runs for each department which will maximise the total number of complete units of the final product. Formulate this as an LP problem.

Solve the following LP problems graphically. Specify if the problem has many optimal solutions, an unbounded solution or no solution.

21. Old hens can be brought at Rs.40/- each and young ones at Rs.60/- each. The old hens lay 5 eggs per week and the young ones lay 1 egg daily. Each egg can be sold at Rs.1.50. Re.1 is required to feed a hen daily. If a person can spend Rs.1200/- for purchasing hens and spare space for

accommodating not more than 25 hens, how many hens of each kind should he purchase to maximise his profit?

22. A manufacturer of patent medicines is preparing a production plan on medicines A and B. There are sufficient ingredients available to make 20,000 bottles of A and 40,000 bottles of B, but there are only 45,000 bottles into which either of the medicines can be put. Also it takes 3 hours for preliminary work to fill 1000 bottles of A and 1 hour for preliminary work to fill 1000 bottles of B. There are 66 hours available for this operation. The profit is Rs.8/- per bottle of A and Rs.7/- per bottle of B. How should the manufacturer schedule production in order to maximise his profit?
23. A mining company is taking a certain kind of ore from two mines A and B. The ore is divided into 3 quality groups a, b and c. Every day, the company has to deliver 240 tonnes of a, 160 tonnes of b and 440 tonnes of c. The cost per day for running mine A is Rs.3000/- and for running mine B is Rs.2000/-. Each day, A will produce 60 tonnes of a, 20 tonnes of b and 40 tonnes of c. The corresponding figures for B are 20, 20 and 80. Find the most economical production plan.
24. Problem (9) in the exercise.
25. Maximise $z = 2x_1 + x_2$, subject to the constraints $x_1 + 2x_2 \leq 10$, $x_1 + x_2 \leq 6$, $x_1 - x_2 \leq 2$, $x_1 - 2x_2 \leq 1$ and $x_1, x_2 \geq 0$.
26. Maximise $z = 2x_1 + 3x_2$, subject to the constraints $x_1 + x_2 \leq 30$, $x_1 - x_2 \geq 0$, $0 \leq x_1 \leq 20$, $3 \leq x_2 \leq 12$ and $x_1, x_2 \geq 0$.
27. Maximise $z = 3x_1 + 4x_2$, subject to the constraints $5x_1 + 4x_2 \leq 200$, $3x_1 + 5x_2 \leq 150$, $5x_1 + 4x_2 \geq 100$, $8x_1 + 4x_2 \geq 80$ and $x_1, x_2 \geq 0$.
28. Minimise $z = 20x_1 + 10x_2$, subject to the constraints $x_1 + 2x_2 \leq 40$, $3x_1 + x_2 \geq 30$, $4x_1 + 3x_2 \geq 60$ and $x_1, x_2 \geq 0$.
29. Minimise $z = 4x + 3y$, subject to the constraints $2x + y \geq 10$, $-3x + 2y \leq 6$, $x + y \geq 6$ and $x, y \geq 0$.
30. Minimise $z = 12x_1 - 3x_2$, subject to the constraints $-x_1 + 5x_2 \leq 20$, $-2x_1 + x_2 \leq -2$, $2x_1 + x_2 \geq 6$ and $x_1, x_2 \geq 0$.
31. Maximise $z = x_1 + x_2$, subject to the constraints $-x_1 + 3x_2 \leq 10$, $x_1 + x_2 \leq 6$, $x_1 - x_2 \leq 2$ and $x_1, x_2 \geq 0$.
32. Minimise $z = 6x_1 + 2x_2$, subject to the constraints $x_1 + 2x_2 \leq 40$, $3x_1 + x_2 \geq 30$, $4x_1 + 3x_2 \geq 60$, and $x_1, x_2 \geq 0$.
33. Maximise $z = 3x_1 + 2x_2$, subject to the constraints $x_1 - x_2 \leq 1$, $x_1 + x_2 \geq 3$, and $x_1, x_2 \geq 0$.
34. Maximise $z = 3x_1 + 4x_2$, subject to the constraints $x_1 - x_2 \geq 0$, $-x_1 + 2x_2 \leq 2$, and $x_1, x_2 \geq 0$.
35. Maximise $z = 2x_1 + x_2$, subject to the constraints $3x_1 + 2x_2 \leq 12$, $5x_1 \leq 10$, $x_1 + x_2 \geq 8$, $-x_1 + x_2 \geq 4$ and $x_1, x_2 \geq 0$.

1.4 SIMPLEX METHOD

1.4.1 Introduction

In the previous section, we found graphically the optimal solution of an LP problem involving two variables only. If the LP problem involves more than two variables, the graphical method is impractical. In such cases, an iterative analytical procedure is used to solve the LP problem in a finite number of steps. This powerful and widely used method, called the Simplex method, was developed by George Dantzig in 1947.

The Simplex method provides an algorithm by which we move from one vertex of the region of feasible solutions to another in a systematic manner such that the value of the objective function at the succeeding vertex is more (for maximisation problems) or less (for minimisation problems) than at the preceding vertex. The process of moving from vertex to vertex is repeated until we reach the optimal vertex, which is accomplished in a finite number of steps.

1.4.2 Some Definitions and Notations

The standard form of an LP problem is defined as follows:

Maximise (or minimise)

$$z = c_1x_1 + c_2x_2 + c_3x_3 + \cdots + c_nx_n \tag{1}$$

Subject to the constraints

$$a_{11}x_1 + a_{12}x_2 + \cdots + a_{1n}x_n = b_1$$

$$a_{21}x_1 + a_{22}x_2 + \cdots + a_{2n}x_n = b_2$$

$$\cdots\cdots\cdots\cdots\cdots\cdots$$

$$a_{m1}x_1 + a_{m2}x_2 + \cdots + a_{mn}x_n = b_m \tag{2}$$

and

$$x_j \geq 0, j = 1, 2, \ldots, n \tag{3}$$

where a_{ij}s, b_is and c_js ($i = 1, 2, \ldots, m; j = 1, 2, \ldots, n$) are all constants.

The c_js in the objective function are called cost coefficients: the b_js occurring in the R.H.S of equations (2) are called stipulations; the a_{ij}s are called the structural coefficients.

Note: In the standard form of an LP problem, the constants b_i ($i = 1, 2, \ldots, m$) are taken as positive, which is a pre-requisite for the application of the Simplex method.

The Simplex method has been developed on the assumption that the constraints in equations (2) are equalities. However in an LP problem formulated from practical situations, the constraints will be mostly inequalities (either of the $\leq$ type or of the $\geq$ type). Hence, before the Simplex method is applied to solve an LP problem, the constraint inequalities must be converted into equalities by introducing some additional variables called slack and surplus variables.

Slack variable

The non-negative variable added to the left-hand side of a 'less than or equal to' inequality constraint in order to convert it into an equality is called a slack variable.

For example, if the constraint of the LPP is

$$a_{i1}x_1 + a_{i2}x_2 + \cdots + a_{in}x_n \leq b_i : i = 1, 2, \ldots, m.$$

and

$$\text{if } a_{i1}x_1 + a_{i2}x_2 + \cdots + a_{in}x_n + x_{n+i} = b_i : i = 1, 2, \ldots, m,$$

then the non-negative variables x_{n+i} are called slack variables.

The cost coefficient c_{n+i} corresponding to x_{n+i} will be assumed to be zero.

Surplus variable

The non-negative variable subtracted from the left-hand side of a 'greater than or equal to' inequality constraint in order to convert it into an equality is called a surplus variable.

For example, if the constraint of the LPP is

$$a_{i1}x_1 + a_{i2}x_2 + \cdots + a_{in}x_n \geq b_i : i = 1,2,\ldots,m.$$

and if $a_{i1}x_1 + a_{i2}x_2 + \cdots + a_{in}x_n - x_{n+i} = b_i : \mathrm{i} = 1,2,\ldots,\mathrm{m},$

then the non-negative variables x_{n+i} are called surplus variables.

The cost coefficient c_{n+i} corresponding to x_{n+i} will be assumed to be zero.

Now we shall introduce some more definitions and notations which will help us to understand the Simplex method. The standard form of an LPP can be put in matrix notation as:

Maximise (or minimise) $Z = CX$, subject to the constraints $AX = b$ and $X \geq 0$, where $C = (c_1, c_2, c_3,\ldots, c_n)$, $X = (x_1, x_2, x_3, \ldots, x_n)^T$, $A = (a_{ij})_{m\times n}$ and $b = (b_1, b_2, b_3,\ldots, b_m)^T$

The system of equations $AX = b$ is assumed to be linearly independent, i.e., the rank of A is assumed to be m.

If B is any $m \times m$ sub-matrix consisting of m linearly independent columns of A, then a solution got by putting $n - m$ variables (which do not correspond to the columns of B) to zero and solving the resulting system, is called a *basic solution* of the system $AX = b$. The $m \times m$ non-singular sub-matrix B is called a *basis matrix* or simply *basis*. The columns of B are called *basis vectors* and m variables corresponding to the columns of B are called *basic variables.*

The system of equations in the m basic variables can be put as $BX_B = b$, where X_B is the column of basic variables, viz., $(x_{B1}, x_{B2},\ldots, x_{Bm})^T$

Then, $X_B = B^{-1}b$

If the basic variables are non-negative, then the solution is called a *basic feasible solution* of the system $AX = b$.

A basic feasible solution (BFS) in which all the m basic variables are strictly positive is called a *non-degenerate BFS.* A basic feasible solution in which less than m basic variables are positive and others are zero each, is called a *degenerate BFS.*

1.4.3 Theorems

The Simplex procedure is based on the following theorems, which are just stated without proof.

THEOREM 1: If an LPP 'maximise (or minimise) $Z = CX$, subject to $AX = b, X \geq 0$' has at least one feasible solution, then it has at least one BFS also. (How to find the initial BFS for an LPP, is discussed later)

THEOREM 2: If an LPP has an optimal solution, then at least one BFS must be optimal.

THEOREM 3: *(Optimality criterion)* If $X_B = B^{-1}b$ is a BFS to the LPP 'maximise $Z = CX$, subject to $AX = b, X \geq 0$' $Z^* = C_BX_B$, where $C_B = (c_{B1}, c_{B2}, \ldots, c_{Bm})$, viz, the cost coefficients corresponding to the basic variables and $z_j - c_j \geq 0$ for every column A_j of A, but not in B, where $z_j = C_B A_j = \sum_{i=1}^{m} C_{Bi} a_{ij}$, then z^* is the optimum (maximum) value of z and this BFS, X_B is an optimal solution.

Note: If the LPP is 'minimise $Z = CX$ subject to $AX = b, X \geq 0$', then the optimality criterion is $z_j - c_j \leq 0$ and z^* is the minimum value of z.

THEOREM 4: If $X_B = B^{-1}b$ is a BFS of an LPP with $z = C_BX_B$ as the value of the objective function, if for any column A_j of A, but not in B, the condition $z_j - c_j < 0$ (for maximisation problem) or $z_j - c_j > 0$ (for minimisation problem) holds and if at least one $a_{ij} > 0$, $i = 1, 2, \ldots, m$, then it is possible to obtain a new BFS by replacing one of the columns in B by A_j and if the new value of the objective function is z', then $z' \geq z$. If the given BFS is non-degenerate, then $z' > z$.

THEOREM 5: (*Condition for unbound solution*) If the LPP is one of maximisation and if for any of its BFS there is a column A_j in A but not in B for which $z_j - c_j < 0$ and $a_{ij} \leq 0$ (i = 1, 2,…,m), then the problem has an unbounded solution.

THEOREM 6: (*Simplex criterion for selection of outgoing vector from the basis*) A new BFS can be obtained from the current BFS by removing the rth column vector of the basis matrix B by A_j, if $a_{ij} > 0$ and r is selected such that

$$\frac{x_{Br}}{a_{rj}} = \min_i \left(\frac{x_{Bi}}{a_{ij}}, a_{ij} > 0 \right)$$

Note: The ratios $\frac{x_{Bi}}{a_{ij}}$, $a_{ij} > 0$ are called replacement ratios)

THEOREM 7: (*Condition for alternative solution*) If, for some A_j of A, but not in B, $z_j - c_j = 0$ and $a_{ij} > 0$ for at least one i, then an alternative optimal solution will exist.

THEOREM 8: (*Condition for infeasible solution*) If the value of at least one artificial variable corresponding to an artificial vector in B is non-zero and the optimality criterion is satisfied at any stage of iteration, then there exists no feasible solution for the problem.

Note: The definition of artificial variable is given later

Initial Basic Feasible Solution

Case (1): When all the structural constraints of the LPP have a $\leq$ sign, we introduce m slack variables $x_{n+1}, x_{n+2}, \ldots, x_{n+m}$ and convert the inequalities into equalities. After conversion, the constraint equations are

$$\begin{aligned} a_{11}x_1 + a_{12}x_2 + \cdots + a_{1n}x_n + 1.x_{n+1} &= b_1 \\ a_{21}x_1 + a_{22}x_2 + \cdots + a_{2n}x_n + 1.x_{n+2} &= b_2 \\ &\ldots \\ a_{m1}x_1 + a_{m2}x_2 + \cdots + a_{mn}x_n + 1.x_{n+m} &= b_m \end{aligned}$$

In the matrix form, these equations can be put as

$$A = \begin{bmatrix} a_{11} & a_{12} & \ldots & a_{1n} & 1 & 0 & \ldots & 0 \\ a_{21} & a_{22} & \ldots & a_{2n} & 0 & 1 & \ldots & 0 \\ . & & & . & & & & \\ . & & & . & & & & \\ . & & & . & & & & \\ a_{m1} & a_{m2} & \ldots & a_{mn} & 0 & 0 & \ldots & 1 \end{bmatrix} \begin{bmatrix} x_1 \\ x_2 \\ . \\ . \\ . \\ x_{n+m} \end{bmatrix} = \begin{bmatrix} b_1 \\ b_2 \\ . \\ . \\ . \\ b_m \end{bmatrix}$$

In this case, the initial basis matrix B is chosen as I_m, the unit matrix of order m formed by the last m columns of the converted coefficient matrix.

Then the initial BFS is given by

$$X_B = B^{-1} b = I_m b = b \geq 0.$$

Thus, $\quad x_{B1}(= x_{n+1}) = b_1; x_{B2}(= x_{n+2}) = b_2: \ldots$

and $\quad x_{Bm}(= x_{n+m}) = b_m.$

Case (2): When all the structural constraints have $\geq$ sign, we introduce m surplus variables $x_{n+1}, x_{n+2}, \ldots, x_{n+m}$ and convert the inequalities which are given in the following matrix form:

$$\begin{bmatrix} a_{11} & a_{12} & \ldots & a_{1n} & -1 & 0 & \ldots & 0 \\ a_{21} & a_{22} & \ldots & a_{2n} & 0 & -1 & \ldots & 0 \\ . & & & & . & & & \\ . & & & & . & & & \\ . & & & & . & & & \\ a_{m1} & a_{m2} & \ldots & a_{mn} & 0 & 0 & 0 & \ldots & -1 \end{bmatrix} \begin{bmatrix} x_1 \\ x_2 \\ . \\ . \\ . \\ x_{n+m} \end{bmatrix} = \begin{bmatrix} b_1 \\ b_2 \\ . \\ . \\ . \\ b_m \end{bmatrix}$$

If we take the initial basis matrix B as $-I_m$, the initial BFS would become

$$X_B = B^{-1}b = -I_m b = -b \geq 0,$$

which is meaningless. To avoid this difficulty and to get the initial basis matrix B as I_m, we introduce one more non-negative variable to each constraint. These non-negative variables are called *artificial variables*.

After introducing surplus and artificial variables, the converted constraint equations in the matrix form would be

$$\begin{bmatrix} a_{11} & a_{12} & \ldots & a_{1n} & -1 & 0 & \ldots & 0 & 1 & 0.. \\ a_{21} & a_{22} & \ldots & a_{2n} & 0 & -1 & \ldots & 0 & 0 & 1.. \\ . & & & & . & & & & & \\ . & & & & . & & & & & \\ . & & & & . & & & & & \\ a_{m1} & a_{m2} & \ldots & a_{mn} & 0 & 0 & 0 & \ldots & -1 & 0 & 0.. \end{bmatrix} \begin{bmatrix} x_1 \\ x_2 \\ . \\ . \\ x_{n+m} \\ x_{n+m+1} \\ . \\ . \\ . \\ x_{n+m+m} \end{bmatrix} = \begin{bmatrix} b_1 \\ b_2 \\ . \\ . \\ . \\ b_m \end{bmatrix}$$

The artificial variables introduced are $x_{n+m+1}, x_{n+m+2}, \ldots, x_{n+m+m}$. In this case, the initial basis matrix B is chosen as I_m formed by the last m columns of the converted coefficient matrix corresponding to the artificial variables.

Then the initial BFS is given by

$$X_B = B^{-1} b = I_m b$$
$$= b \geq 0$$

Thus,

$$x_{B1}(= x_{n+m+1}) = b_1, x_{B2}(= x_{n+m+2}) = b_2, \ldots$$

and $$x_{Bm}(= x_{n+2m}) = b_m.$$

Case (3): When the structural constraints have $\leq$, $\geq$ and $=$ signs, we introduce slack, surplus and artificial variables and convert the constraints into equalities. In this case, the initial basis matrix B is chosen as I_m formed by the slack and artificial vectors.

Note:

1. In all the three cases, the initial BFS is given by $x_{B1} = b_1, x_{B2} = b_2, \ldots, x_{Bm} = b_m$
2. Sometimes, a column of I_m may be available in A even before introducing an artificial variable. In such situations, we need not introduce an artificial variable.
3. When we solve an LP problem by the 'big M-method', the cost coefficient corresponding to an artificial variable is taken as $-M$ or M according to whether the problem is one of maximisation or minimisation, when M is a large number.
4. When we solve an LP problem by the 'two phase simplex method', the cost coefficient corresponding to an artificial variable is taken as -1 or 1 according to whether the problem is one of maximisation or minimisation

Computational procedure of the Simplex method

Note: Most authors give the step by step procedure for solving only a maximisation problem. If the problem is one of minimisation, it is converted to a maximisation problem and then the maximisation algorithm is applied. However, we shall use a common procedure for both types of problems with a few changes here and there.]

Step 1: All the b_is (the constants in the RHS of the constraints) are made non-negative. If any of the b_is is negative, the corresponding constraint is multiplied by -1 so as to make the RHS constant positive.

Step 2: All the constraints are converted into equations by introducing slack, surplus and artificial variables as needed.

Step 3: The objective function is rewritten in terms of the decision variables and newly introduced slack, surplus and artificial variables with relevant cost coefficients.

Step 4: The initial Simplex table is constructed as follows:

(i) In the first row, the cost coefficients in the modified objective function are entered with the row heading c_j.

(ii) In the second row, the names of the column vectors A, viz, $A_1, A_2, A_3\ldots$ are entered.

(iii) In the subsequent rows, elements a_{ij} of the coefficient matrix A are entered so that $a_{1j}, a_{2j}, \ldots, a_{mj}$ are placed one below the other in the column headed by A_j.

(iv) Three columns are now introduced preceding the column headed by A_1, with column headings B, C_B and X_B. The names of initial basis vectors which form the initial basis (unit matrix) are entered one below the other in order in the column headed by B. The cost coefficients corresponding to the initial basic variables are entered in order in the column headed by C_B. The elements of the initial BFS given by $X_B = B^{-1}b$ are entered in order in the column headed by X_B (Note: For the initial BFS, $X_B = b$).

(v) The net evaluations $z_j - c_j (j = 1, 2, 3,\ldots)$ are computed, using the relation $z_j = \sum_{i=1}^{m} C_{Bi}a_{ij}$ and placed in the order in the next row of the table, called the index row.

Step 5: The signs of the index row entries are examined. If all $(z_j - c_j) \geq 0$ (for the maximisation problem and ≤ 0 for the minimisation problem), then the initial BFS X_B itself is the optimal solution. If not, viz., if at least one $(z_j - c_j) \leq 0$ (for the maximisation problem and ≥ 0 for the minimisation problem), we proceed to the next step.

Step 6: If there are more than one negative value of $z_j - c_j$, the most negative of them is chosen (for the maximisation problem). Similarly, if there are more than one positive value of $z_j - c_j$, the most positive of them is chosen (for the minimisation problem). Let it be $z_r - c_r$ for some $j = r$. If there is a tie, r is chosen arbitrarily.
Now if all a_{ir} (entries in the column headed by A_r) ≤ 0 $(i = 1,2,\ldots, m)$, there is an unbounded solution to the LP problem. On the other hand, if at least one $a_{ir} > 0$, then the vector A_r enters the basis B in the place of A_k (chosen in Step 7) which leaves the basis.

Step 7: The corresponding ratios $\frac{x_{Bi}}{a_{ir}}$, $a_{ir} > 0$ for $i = 1, 2,\ldots, m$ are computed and the minimum of these ratios is chosen. Let it be $\frac{x_{Bk}}{a_{kr}}$. Then the vector A_k has to leave the basis. If there is a tie for k, the leaving vector is selected in a different way (which will be considered later).

Step 8: The element a_{kr} which is at the intersection of the kth row (called key row) and the rth column (called key column) of the simplex table is called the pivot element. If the pivot element is not equal to 1, it is made 1 by dividing all the elements of the key row by a_{kr}.

Step 9: All the other elements of the key column are made zero each by suitable elementary matrix row operations.

Step 10: We go to steps 4(v) and repeat the procedure, until an optimal solution is obtained or we get the indication of an unbound solution.

WORKED EXAMPLES 1(B)

Example 1

Use Simplex method to solve the following LP problem.

$$\text{Maximise } z = 2x_1 + 5x_2,$$

Subject to

$$x_1 + 4x_2 \leq 24$$

$$3x_1 + x_2 \leq 21$$

$$x_1 + x_2 \leq 9$$

and

$$x_1, x_2 \geq 0$$

SOLUTION

Introducing the non-negative slack variable x_3, x_4 and x_5, the standard form of the given LP problem is got as

$$\text{Maximise } z = 2x_1 + 5x_2 + 0x_3 + 0x_4 + 0x_5,$$

Subject to

$$x_1 + 4x_2 + x_3 = 24$$

$$3x_1 + x_2 + x_4 = 21$$

$$x_1 + x_2 + x_5 = 9$$

and $$x_1, x_2, x_3, x_4, x_5 \geq 0$$

Since the column vectors $A_3 = (1, 0, 0)^T$, $A_4 = (0, 1, 0)^T$ and $A_5 = (0, 0, 1)^T$, corresponding to x_3, x_4, x_5, are linearly independent, the non-singular matrix $B = (A_3, A_4, A_5)$ is taken as the initial basis. The initial BFS is hence given by $X_B = B^{-1}b$, viz., $x_3 = 24$, $x_4 = 21$ and $x_5 = 9$. The initial simplex table is as follows.

Initial simplex table

		c_j	2	5	0	0	0	
B	C_B	X_B	A_1	A_2	A_3	A_4	A_5	RR
A_3	0	24	1	4	1	0	0	6 →
A_4	0	21	3	1	0	1	0	21
A_5	0	9	1	1	0	0	1	9
	$z = 0$	$z_j - c_j$	−2	−5	0	0	0	
				↑				

Note: The index row elements have been computed using $z = C_B X_B$ and $z_j - c_j = C_B A_j - c_j$. Since the index row elements $(z_j - c_j)$ are all non-positive, optimality has not been reached, viz., there is still scope for increasing the value of z, by changing the basis. Since there are two negative entries in the index row, namely, -2 and -5, let us consider the most negative of them, which is -5 that corresponds to A_2 (or the variable x_2). Since all the entries in A_2 column are positive, the solution can be revised by bringing in A_2 into the basis. Now the replacement ratios (RRs) have been computed by dividing the elements of X_B by the corresponding elements of A_2. Since the minimum replacement ratio corresponds to A_3 (or the variable x_3, A_3 is removed from the basis and replaced by A_2.

The revised Simplex table is got as follows: The elements of A_3 row are divided by the pivot element 4, so as to make it 1. Then the A_4 row and A_5 row are modified by using elementary row operations and the modified key row (A_3 row), so as to make the elements of A_2 column below 1 as zero, is as given here:

Existing A_4 row:	**21**	**3**	**1**	**0**	**1**	**0**
(−) Modified A_3 row:	6	1/4	1	1/4	0	0
Modified A_4 row:	15	11/4	0	−1/4	1	0
Existing A_5 row:	9	1	1	0	0	1
(−) Modified A_3 row:	6	1/4	1	1/4	0	0
Modified A_5 row:	3	3/4	0	−1/4	0	1

Revised Simplex table after first iteration

		c_j	2	5	0	0	0	
B	C_B	X_B	A_1	A_2	A_3	A_4	A_5	RR
A_2	5	6	1/4	1	1/4	0	0	24
A_4	0	15	11/4	0	−1/4	1	0	60/11
A_5	0	3	3/4	0	−1/4	0	1	4→
	$z = 30$	$z_j - c_j$	−3/4	0	5/4	0	0	
			↑					

Since one entry in the index row is negative, the optimality has not yet been reached. We revise the table by removing A_5 from the basis and bringing in A_1 into the basis.

Revised Simplex table after second iteration

		c_j	2	5	0	0	0
B	C_B	X_B	A_1	A_2	A_3	A_4	A_5
A_2	5	5	0	1	1/3	0	−1/3
A_4	0	4	0	0	2/3	1	−11/4
A_1	2	4	1	0	−1/3	0	4/3
	$z = 33$	$z_j - c_j$	0	0	1	0	1

Since all the $(z_j - c_j)$ values are ≥ 0, optimality has been obtained.
Maximum $z = 33$, corresponding to the optimal solution $x_1 = 4$ and $x_2 = 5$.

Example 2

Use Simplex method to solve the following LP problem

$$\text{Maximise } z = x_1 + x_2 + 4x_3 + 2x_4,$$

Subject to

$$x_1 + 3x_3 + x_4 \leq 4$$

$$x_3 + 2x_4 \leq 3$$

$$x_1 + 4x_2 + x_3 \leq 3$$

and

$$x_1, x_2, x_3, x_4 \geq 0$$

SOLUTION

Introducing the non-negative slack variables x_5, x_6 and x_7, the standard form of the given LP problem is got as

$$\text{Maximise } z = x_1 + x_2 + 4x_3 + 2x_4 + 0x_5 + 0x_6 + 0x_7$$

Subject to

$$x_1 + 3x_3 + x_4 + x_5 = 4$$

$$x_3 + 2x_4 + x_6 = 3$$

$$x_1 + 4x_2 + x_3 + x_7 = 3$$

and $$x_i \geq 0 \ (i = 1, 2, \ldots, 7)$$

Initial Simplex table

		C_j	1	1	4	2	0	0	0	
B	C_B	X_B	A_1	A_2	A_3	A_4	A_5	A_6	A_7	RR
A_5	0	4	1	0	3	1	1	0	0	4/3 →
A_6	0	3	0	0	1	2	0	1	0	3
A_7	0	3	1	4	1	0	0	0	1	3
	$Z = 0$	$Z_j - C_j$	−1	−1	−4	−2	0	0	0	
					↑					

Iteration I: A_3 enters and A_5 leaves the basis

		C_j	1	1	4	2	0	0	0	
B	C_B	X_B	A_1	A_2	A_3	A_4	A_5	A_6	A_7	R.R
A_3	0	4/3	1/3	0	1	1/3	1/3	0	0	-
A_6	0	5/3	−1/3	0	0	5/3	−1/3	1	0	-
A_7	0	5/3	2/3	4	0	−1/3	−1/3	0	1	5/12 →
	$z = 16/3$	$z_j - c_j$	1/3	−1	0	−2/3	4/3	0	0	
				↑						

Iteration II: A_2 enters and A_7 leaves the basis

		C_j	1	1	4	2	0	0	0	
B	C_B	X_B	A_1	A_2	A_3	A_4	A_5	A_6	A_7	RR
A_3	4	4/3	1/3	0	1	1/3	1/3	0	0	4
A_6	0	5/3	−1/3	0	0	5/3	−1/3	1	0	1 →
A_2	0	5/12	1/6	1	0	−1/12	−1/12	0	1/4	-
	$z = 23/4$	$z_j - c_j$	1/2	0	0	−3/4	5/4	0	1/4	-
						↑				

Note: The first two entries under A_2 column are already zero each, no row transformations are performed and the first two rows of the table corresponding to A_3 and A_6 are first carried over.

Iteration III: A_4 enters and A_6 leaves the basis

		c_j	1	1	4	2	0	0	0	
B	C_B	X_B	A_1	A_2	A_3	A_4	A_5	A_6	A_7	R.R
A_3	4	1	2/5	0	1	0	2/5	−1/5	0	-
A_4	2	1	−1/5	0	0	1	−1/5	3/5	0	-
A_2	1	1/2	3/20	1	0	0	−1/10	1/20	1/4	-
	$z = 13/2$	$z_j - c_j$	7/20	0	0	0	11/10	9/20	1/4	-

Since all the index row entries are ≥ 0, optimality has been attained.

Optimal solution is given by $x_1 = 0$, $x_2 = \frac{1}{2}$, $x_3 = 1$ and $x_4 = 1$ and maximum $z = \frac{13}{2}$

Note: Since x_1 is a non-basic variable, $x_1 = 0$

Example 3

Use Simplex method to solve the following LP problem.

$$\text{Maximise } z = 2x_1 - x_2,$$

Subject to

$$4x_1 - 3x_2 \leq 12$$

$$4x_1 + x_2 \leq 8$$

$$4x_1 - x_2 \leq 8$$

and

$$x_1 \geq 0 \text{ and } x_2 \text{ is unrestricted in sign.}$$

Since x_2 is an unrestricted variable, it may be positive, zero or negative. In such cases, we put $x_2 = x_2' - x_2''$, where x_2' and x_2'' are non-negative variables, so that the Simplex method may be applied to solve the problem (as all the decision variables x_1, x_2' and x_2'' are non-negative.)

Introducing the slack variables x_3, x_4 and x_5, the standard form of the given LP problem is got as

$$\text{Maximise } z = 2x_1 - x_2' + x_2' + 0x_3 + 0x_4 + 0x_5,$$

Subject to

$$4x_1 - 3x_2' + 3x_2'' + x_3 = 12$$

$$4x_1 + x_2' - x_2'' + x_4 = 8$$

$$4x_1 - x_2' + x_2'' + x_5 = 8$$

and

$$x_1, x_2', x_2'', x_3, x_4, x_5 \geq 0$$

Now the solution is provided as follows, where the initial and successive simplex tables have been given continuously:

		C_j	**2**	**−1**	**1**	**0**	**0**	**0**	
B	C_B	X_B	A_1	A_2'	A_2''	A_3	A_4	A_5	RR
A_3	0	12	4	−3	3	1	0	0	3
A_4	0	8	4	1	−1	0	1	0	2→
A_5	0	8	4	−1	1	0	0	1	2
	$z=0$	z_j-c_j	−2 ↑	1	−1	0	0	0	

		C_j	**2**	**−1**	**1**	**0**	**0**	**0**	
B	C_B	X_B	A_1	A_2'	A_2''	A_3	A_4	A_5	RR
A_3	0	4	0	−4	4	1	−1	0	1
A_1	2	2	1	1/4	−1/4	0	−1/4	0	-
A_5	0	0	0	−2	2	0	−1	1	0 →
	$z=4$	z_j-c_j	0	3/2	−3/2 ↑	0	1/2	0	
A_3	0	4	0	0	0	1	1	−2	4 →
A_1	2	2	1	0	0	0	1/8	1/8	16
A_2''	1	0	0	−1	1	0	−1/2	1/2	-
	$z=4$	z_j-c_j	0	0	0	0	−1/4↑	3/4	
A_4	0	4	0	0	0	1	1	−2	-
A_1	2	3/2	1	0	0	−1/8	0	3/8	-
A_2''	1	2	0	−1	1	1/2	0	−1/2	-
	$z=5$	z_j-c_j	0	0	0	1/4	0	1/4	

Since all $(z_j - c_j)$ values are ≥ 0, optimality has been attained.

Optimal solution is $x_1 = 3/2, x_2'' = 2$ and $x_2' = 0$ and maximum $z = 5$.

Hence, the optimal solution of the original problem is $x_1 = 3/2, x_2 = x_2' - x_2'' = -2$ and maximum $z = 5$.

Note on degeneracy:

Degeneracy is the phenomenon in which a basic feasible solution, either initially or at any stage of iteration is degenerate, viz., less than *m* basic variables are positive and others are zero each in the BFS.

In Example 3,, the BFSs after the first and second iterations are degenerate. After the first iteration, the BFS is given by $x_1 = 2, x_3 = 4$ and $x_5 = 0$, which is degenerate. After the second iteration, the BFS is given by $x_1 = 2, x_3 = 4$ and $x_2'' = 0$, which is also degenerate.

The fact that the BFS after the first iteration would be degenerate can be sensed before-hand, viz., while computing the replacement ratios in the initial Simplex table to decide which basic vector leaves the basis. If two or more minimum replacement ratios vectors [in Example 3, there is a tie between A_4 and A_5] are a tie, it is an indication that the next BFS would be degenerate.

In the present example, A_4 has been removed arbitrarily from the basis, which results in the revised BFS with $x_5 = 0$.

Sometimes, arbitrary removal of one of the tied vectors from the basis (due to two or more equal minimum replacement ratios) may lead to a problem known as *cycling*. Cycling is a phenomenon in the Simplex method of solving an LP problem in which we will start from some BFS and after a few iterations, we return to the same starting BFS without attaining the optimal solution. Fortunately, cycling will never occur in practical problems involving a few variables. However, arbitrary removal of a leaving vector may result in more number of iterations before the optimal solution is reached. This difficulty can be overcome by choosing the leaving vector using *Charne's perturbation technique* which is described here:

(i) When there is a tie for the leaving vector, we divide the elements of the first unit vector (first column of the basis) lying on the tied rows by the corresponding elements of the entering vector and get a new set of replacement ratios.

(ii) If the tie is broken, we choose the vector corresponding to the minimum replacement ratio as the leaving vector.

(iii) If the tie is not broken, the same procedure is repeated with respect to the successive unit vectors of the basis.

(iv) If the tie is not broken even after exhausting all unit vectors of the basis, the same procedure is repeated with respect to the other column vectors (other than the entering vector) of the Simplex table from left to right until the tie is broken.

Note: While applying Charne's method, artificial vectors should not be removed from the basis at any stage.

In Example 3, there was a tie for the leaving vector between A_4 and A_5 and so we removed the vector A_4 arbitrarily from the basis. Had we applied Charne's perturbation technique, we would have removed A_5 from the basis and got the optimal solution in two more iterations instead of three iterations as detailed below:

According to Crane's method, we divide the elements of A_3 (i.e., the first unit vector of the initial basis) lying on the tied rows A_4 and A_5 namely 0 and 0 by the corresponding elements of the entering vector A_1, namely 4 and 4. The new pair of replacement ratios are 0 and 0—still the tie is not broken. Hence, the elements of A_4 (i.e., the second unit vector in the initial basis) lying on A_4 and A_5 rows, namely 1 and 0 are divided by 4 and 4. The next pair of replacement ratios is 1/4 and 0. Since the minimum ratio is 0 corresponding to the A_5 row, we shall remove A_5 from the basis and proceed further.

Simplex table after I iteration

		c_j	2	−1	1	0	0	0	
B	C_B	X_B	A_1	A'_2	A''_2	A_3	A_4	A_5	R.R
A_3	0	4	0	−2	2	1	0	−1	2→
A_4	0	0	0	2	−2	0	1	−1	-
A_1	2	2	1	1/4	1/4	0	0	1/4	8
	$z = 4$	$z_j - c_j$	0	1/2	−1/2↑	0	0	1/2	

Simplex table after II iteration

		c_j	2	−1	1	0	0	0
B	C_B	X_B	A_1	A_2'	A_2''	A_3	A_4	A_5
A_2''	1	2	0	−1	1	1/2	0	−1/2
A_4	0	4	0	0	0	1	1	−2
A_1	2	3/2	1	0	0	−1/8	0	3/8
	$z=5$	z_j-c_j	0	0	0	1/4	0	1/4

Example 4

Solve the following LP problem by the Simplex method. Resolve the degeneracy by Charne's technique.

$$\text{Maximise } z = 3x_1 + x_2,$$

Subject to

$$6x_1 + x_2 \leq 8$$
$$2x_1 + x_2 \leq 4$$
$$6x_1 - x_2 \leq 8$$

and

$$x_1, x_2 \geq 0$$

Introducing the non-negative slack variables x_3, x_4 and x_5, the standard form of the given LP problem is obtained as given here.

$$\text{Maximise } z = 3x_1 + x_2 + 0x_3 + 0x_4 + 0x_5,$$

Subject to

$$6x_1 + x_2 + x_3 = 8$$
$$2x_1 + x_2 + x_4 = 4$$
$$6x_1 - x_2 + x_5 = 8$$

and

$$x_1, x_2, x_3, x_4, x_5 \geq 0$$

Solution

The initial and successive Simplex tables for the solution of the problem are given as follows. The degeneracy in the initial table has been resolved by using Charne's technique.

		c_j	3	1	0	0	0		
B	C_B	X_B	A_1	A_2	A_3	A_4	A_5	R.R1	R.R2
A_3	0	8	6	1	1	0	0	4/3	1/6
A_4	0	4	2	1	0	1	0	2	-
A_5	0	8	6	−1	0	0	1	4/3	0→
	$z=0$	z_j-c_j	−3	−1	0	0	0	-	
			↑						

		c_j	3	1	0	0	0	
B	C_B	X_B	A_1	A_2	A_3	A_4	A_5	R.R1
A_3	0	0	0	2	1	0	−1	0 →
A_4	0	4/3	0	4/3	0	1	−1/3	1
A_1	3	4/3	1	−1/6	0	0	1/6	-
	$z=4$	z_j-c_j	0	−3/2 ↑	0	0	1/2	
A_2	1	0	0	1	1/2	0	−1/2	-
A_4	0	4/3	0	0	−2/3	1	1/3	4 →
A_1	3	4/3	1	0	1/12	0	1/12	16
	$z=4$	z_j-c_j	0	0	5/4	0	−1/4 ↑	
A_2	1	2	0	1	−1/2	3/2	0	-
A_5	0	4	0	0	−2	3	1	-
A_1	3	1	1	0	1/4	−1/4	0	-
	$z=5$	z_j-c_j	0	0	1/4	3/4	0	

Hence, the optimal solution is $x_1 = 1$ and $x_2 = 2$ and the maximum $z = 5$.

Example 5

Solve the following LP problem by Simplex method.

$$\text{Maximise } z = 2x_1 + x_2,$$

Subject to

$$2x_1 - x_2 \leq 8$$

$$2x_1 + x_2 \leq 12$$

$$-x_1 + x_2 \leq 3$$

and

$$x_1, x_2 \geq 0$$

SOLUTION

Introducing the non-negative slack variables x_3, x_4 and x_5, the standard form of the given LP problem is obtained as given here

$$\text{Maximise } z = 2x_1 + x_2 + 0x_3 + 0x_4 + 0x_5,$$

Subject to

$$2x_1 - x_2 + x_3 = 8$$

$$2x_1 + x_2 + x_4 = 12$$

$$-x_1 + x_2 + x_5 = 3$$

and $$x_1, x_2, x_3, x_4, x_5 \geq 0$$

The initial and successive simplex tables for the solution of the problem are as follows.

		c_j	2	1	0	0	0	
B	C_B	X_B	A_1	A_2	A_3	A_4	A_5	RR
A_3	0	8	2	−1	1	0	0	4 →
A_4	0	12	2	1	0	1	0	6
A_5	0	3	−1	1	0	0	1	-
	$z = 0$	$z_j - c_j$	−2	−1	0	0	0	
			↑					

		c_j	2	1	0	0	0	
B	C_B	X_B	A_1	A_2	A_3	A_4	A_5	RR
A_1	2	4	1	−1/2	1/2	0	0	-
A_4	0	4	0	2	−1	1	0	2 →
A_5	0	7	0	1/2	1/2	70	1	14
	$z = 8$	$z_j - c_j$	0	−2 ↑	1	0	0	
A_1	2	5	1	0	1/4	1/4	0	20
A_2	1	2	0	1	−1/2	1/2	70	-
A_5	0	6	0	0	3/4	−1/4	1	8
	$z = 12$	$z_j - c_j$	0	0	0	1	0	

Since all the entries in the index row are non-negative, optimality has been reached. The optimal solution is given by $x_1 = 5$ and $x_2 = 2$ and the maximum $z = 12$.

An important note:

In the previous examples, we have observed that in the optimal table, the entries of the index row under the basis vectors alone were zero each. In the optimal table of the present example, the entries of the index row under the basis vectors, namely A_1, A_2 and A_5 are zero each as usual. However, there is an extra zero in the index row under the vector A_3.

This is an indication that the given LP problem has an alternative optimal solution.

To find the alternative optimal solution, we revise the final table by making the vector A_3 under which the extra zero occurs enter the basis by the usual Simplex procedure. The revised table is given here.

B	C_B	X_B	A_1	A_2	A_3	A_4	A_5
A_1	2	3	1	0	0	1/3	−1/3
A_2	1	6	0	1	0	1/3	2/3
A_3	0	8	0	0	1	−1/3	4/3
	$z = 12$	$z_j - c_j$	0	0	0	1	0

Thus, two optimal solutions of the given LP problem are $x_1 = 5$, $x_2 = 2$ (and $x_5 = 6$) and $x_1 = 3$, $x_2 = 6$ (and $x_3 = 8$) and the maximum value of z in both cases = 12.

Note: The presence of the extra zero under A_5 in the revised table indicates that there is an alternative optimal solution. If we revise this table so as to bring in A_5 into the basis, we will get the previous optimal table. The student is advised to verify this. Thus, the two optimal solutions are alternative to each other. Also as observed in the graphical method, any convex combination of the two solutions will also be an optimal solution of the given problem.

viz., $$x_1 = 5\lambda + 3(1 - \lambda)$$

and $$x_2 = 2\lambda + 6(1 - \lambda),$$

where $0 < \lambda < 1$, will also be an optimal solution of the problem.

All these alternative optimal solutions give the same maximum value of z, namely 12.

Example 6

Solve the following LP problem by the Simplex method.

$$\text{Maximise } z = 3x_1 + x_2,$$

Subject to $$x_1 - x_2 \le 1$$

$$3x_1 - 2x_2 \le 6$$

and $$x_1, x_2 \ge 0$$

Solution

Introducing the non-negative slack variables x_3, x_4, the standard form of the given LP problem is obtained as given here.

$$\text{Maximise } z = 3x_1 + x_2 + 0x_3 + 0x_4,$$

Subject to $$x_1 - x_2 + x_3 = 1$$

$$3x_1 - 2x_2 + x_4 = 6$$

and $$x_1, x_2, x_3, x_4 \ge 0$$

The initial and successive Simplex tables for the solution of the problem are given here.

		c_j	3	1	0	0	
B	C_B	X_B	A_1	A_2	A_3	A_4	RR
A_3	0	1	1	−1	1	0	1 →
A_4	0	6	3	−2	0	1	2
	$z = 0$	$z_j - c_j$	−3 ↑	−1	0	0	
A_1	3	1	1	−1	1	0	-
A_4	0	3	0	1	−3	1	3 →
	$z = 3$	$z_j - c_j$	0	−4 ↑	3	0	
A_1	3	4	1	0	−2	1	-
A_2	1	3	0	1	−3	1	-
	$z = 15$	$z_j - c_j$	0	0	−9	4	

Since there is a negative entry in the index row, there is a scope for further increasing the value of z by making the vector A_3 enter the basis.

However, both the entries of A_3 column are negative and so the replacement ratios could not be found out. This means that there is no vector to leave the basis. When this is the situation, it is an indication that the LP problem has an unbounded solution.

Example 7

Use the 'big M' method or penalty method to solve the following LP problem.

$$\text{Maximise } z = 4x_1 + 5x_2,$$

Subject to the constraints

$$x_1 + x_2 \le 4$$

$$2x_1 + 3x_2 \ge 12$$

$$4x_1 + 3x_2 \ge 18$$

and

$$x_1, x_2 \ge 0$$

Solution

Introducing the slack variables x_3, surplus variables x_4, x_5 and artificial variable x_6 and x_7, the standard form of the given LP problem is obtained as follows

$$\text{Maximise } z = 4x_1 + 5x_2 + 0x_3 + 0x_4 + 0x_5 - Mx_6 - Mx_7,$$

Subject to

$$x_1 + x_2 + x_3 = 4$$

$$2x_1 + 3x_2 - x_4 + x_6 = 12$$

$$4x_1 + 3x_2 - x_5 + x_7 = 18$$

and $$x_1, x_2, x_3, x_4, x_5, x_6, x_7 \geq 0$$

Note: The introduction of artificial variables x_6 and x_7 has become necessary, since the vectors corresponding to x_3, x_6 and x_7 alone provide the initial basis.

The initial and successive Simplex tables for the solution of the problem are given here.

		C_j	4	5	0	0	0	−M	−M		
B	C_B	X_B	A_1	A_2	A_3	A_4	A_5	A_6	A_7	RR1	RR2
A_3	0	4	1	1	1	0	0	0	0	4	1
A_6	−M	12	2	3	0	−1	0	1	0	4	0→
A_7	−M	18	4	3	0	0	−1	1	1	6	-
	$z = -30M$	$z_j - c_j$	(−6M − 4)	(−6M − 5) ↑	0	M	M	0	0	RR	
A_3	0	0	1/3	0	1	1/3	0	−1/3	0	0	→
A_2	5	4	2/3	1	0	−1/3	0	1/3	0	6	-
A_7	−M	6	2	0	0	1	−1	−1	1	3	-
	$z = 20 - 6M$	$z_j - c_j$	(−2M − 2/3) ↑	0	0	(−M − 5/3)	M	(2M + 5/3)	0		
A_1	4	0	1	0	3	1	0	−1	0	-	-
A_2	5	4	0	1	−2	−1	0	1/3	0	12	-
A_7	−M	6	0	0	−6	−1	−1	1	1	6	→
	$z = 20 - 6M$	$z_j - c_j$	0	0	(6M + 2)	(M − 1)	M	−7/3 ↑	0		
A_1	4	6	1	0	−3	0	−1	−0	1	-	-
A_2	5	2	0	1	0	−2/3	1/3	0	−1/3	-	-
A_6	−M	6	0	0	−6	−1	−1	1	1	-	-
	$z = 34 - 6M$	$z_j - c_j$	0	0	(6M − 12)	(M − 10/3)	(M − 7/3)	0	7/3	-	

We observe that all the entries in the index row are non-negative and so the optimality criterion is satisfied, but the artificial vector A_6 appears in the basis at the non-zero level (i.e., $x_6 = 6$).

This indicates that the LP problem has no feasible solution. When this sort of situation is encountered, we say that the LP problem has a pseudo-optimum BFS.

Example 8

Use the 'big M' method or penalty method to solve the following LP problem.

$$\text{Minimise } z = 25x_1 + 15x_2 + 5x_3,$$

Subject to
$$4x_1 + 2x_2 + x_3 \geq 10$$
$$2x_1 + 3x_2 \geq 6$$
$$3x_1 + x_3 \geq 8$$

and $$x_1, x_2, x_3 \geq 0$$

Solution

Introducing the surplus variable x_4, x_5 and x_6 and the artificial variable x_7, x_8 and x_9, the standard form of the given LP problem is obtained as given here

$$\text{Minimise } z = 25x_1 + 15x_2 + 5x_3 + 0x_4 + 0x_5 + 0x_6 + Mx_7 + Mx_8 + Mx_9,$$

Subject to
$$4x_1 + 2x_2 + x_3 - x_4 + x_7 = 10$$
$$2x_1 + 3x_2 - x_5 + x_8 = 6$$
$$3x_1 + x_3 - x_6 + x_9 = 8$$

and $$x_1, x_2, \ldots, x_9 \geq 0$$

The initial and successive Simplex tables for the solution of the problem are given here.

		C_j	25	15	5	0	0	0	M	M	M	
B	C_B	X_B	A_1	A_2	A_3	A_4	A_5	A_6	A_7	A_8	A_9	RR
A_1	M	10	4	2	1	−1	0	0	1	0	0	5/2 →
A_8	M	6	2	3	0	0	−1	0	0	1	0	3
A_9	M	8	3	0	1	0	0	1	0	0	1	8/3
	$z = 24M$	$z_j - c_j$	(9M − 25) ↑	5M − 15	24M − 5	−M	−M	−M	0	0	0	

Note: The vector to enter the basis corresponds to the most positive entry of the index row.

		C_j	25	15	5	0	0	0	M	M	M	
B	C_B	X_B	A_1	A_2	A_3	A_4	A_5	A_6	A_7	A_8	A_9	RR
A_1	25	5/2	1	1/2	1/4	−1/4	0	0	1/4	0	0	-
A_8	M	1	0	2	−1/2	1/2	−1	0	−1/2	1	0	2
A_9	M	1/2	0	−3/2	1/4	3/4	0	−1	−3/4	0	1	2/3 →
	$z = \frac{3M}{2} + \frac{125}{2}$	$z_j - c_j$	0	$\frac{M}{2} - \frac{5}{2}$	$\frac{-M}{4} + \frac{5}{4}$	$\frac{5M}{4} - \frac{25}{4}$ ↑	−M	−M	$\frac{-9M}{4} + \frac{25}{4}$	0	0	

		C_j	25	15	5	0	0	0	M	M	M	
B	C_B	X_B	A_1	A_2	A_3	A_4	A_5	A_6	A_7	A_8	A_9	RR
A_1	25	8/3	1	0	1/3	0	0	−1/3	0	0	1/3	-
A_8	M	2/3	0	3	−2/3	0	−1	2/3	0	1	−2/3	2/9 →

(Contd)

Table (*Contd*)

		c_j	25	15	5	0	0	0	M	M	M	
A_4	0	2/3	0	−2	1/3	1	0	−4/3	−1	0	4/3	-
	$z=\frac{2M}{2}+\frac{200}{3}$	z_j-c_j	0	$(3M-15)\uparrow$	$\frac{-2M}{3}+\frac{10}{3}$	0	$-M$	$\frac{2M}{3}-\frac{25}{3}$	$-M$	0	$\frac{-5M}{3}+\frac{25}{3}$	
A_1	25	8/3	1	0	1/3	0	0	−1/3	0	0	1/3	-
A_2	15	2/9	0	1	−2/9	0	−1/3	2/9	0	1/3	−2/9	-
A_4	0	10/9	0	0	−1/9	1	−2/3	−8/9	−1	2/3	8/9	-
	$z=70$	z_j-c_j	0	0	0	0	−5	−5	−M	(5 − M)	(5 − M)	

Since all the entries in the index row are non-positive, optimality has been obtained, as it is a minimisation problem.

The optimal solution is $x_1 = 8/3$, $x_2 = 2/9$ and $x_3 = 0$. (not in a basis)

The minimum value of $z = 70$.

Note: In the optimal table, we observe that there is an extra zero in the index row under A_3, in addition to the usual zeros under the basis vectors A_1, A_2, A_4. This means that the present LP problem has an alternative optimal solution. To get this alternative solution, we revise the final Simplex table by making A_3 enter the basis and A_1 leave the basis. The alternative optimal Simplex table is given here.

		c_j	**25**	**15**	**5**	**0**	**0**	**0**	**M**	**M**	**M**
B	C_B	X_B	A_1	A_2	A_3	A_4	A_5	A_6	A_7	A_8	A_9
A_3	5	8	3	0	1	0	0	−1	0	0	1
A_2	15	2	2/3	1	0	0	−1/3	0	0	1/3	0
A_4	0	2	1/3	0	0	1	−2/3	−1	−1	2/3	1
	$z=70$	z_j-c_j	0	0	0	0	−5	−5	−M	(5 − M)	(5 − M)

Thus, the alternative optimal solution is $x_1 = 0$ (not in the basis), $x_2 = 2$ and $x_3 = 8$ and the minimum value of z is the same.

Example 9

Solve the following LP problem by the Simplex method.

$$\text{Minimise } z = x_1 + x_2 - x_3,$$

Subject to

$$x_1 - 3x_2 - 4x_3 = 5$$

$$x_1 - 2x_2 \leq 3$$

$$x_2 + x_3 \geq 4$$

and $$x_1, x_2 \geq 0$$

and x_3 is unrestricted in sign.

SOLUTION

We put $x_3 = x_3' - x_3''$ where x_3' and $x_3'' \geq 0$.

Introducing the slack variable x_4, the surplus variable x_5 and artificial variables x_6 and x_7, the standard form of the LP problem becomes.

$$\text{Minimise } z = x_1 + x_2 - x_3' + x_3'' + 0x_4 + 0x_5 + Mx_6 + Mx_7,$$

Subject to
$$x_1 - 3x_2 - 4x_3' + 4x_3'' + x_6 = 5$$
$$x_1 - 2x_2 + x_4 = 3$$
$$2x_2 + x_3' - x_3'' - x_5 + x_7 = 4$$

and
$$x_1, x_2, x_3', x_3'' x_4, x_5, x_6, x_7 \geq 0$$

The initial and successive Simplex tables for the solution of the problem are given here.

		c_j	1	1	−1	1	0	0	M	M	
B	C_B	X_B	A_1	A_2	A_3'	A_3''	A_4	A_5	A_6	A_7	R.R
A_6	M	5	1	−3	−4	4	0	0	1	0	5/4 →
A_4	0	3	1	−2	0	0	1	0	0	0	-
A_7	M	4	0	2	1	−1	0	−1	0	1	-
	z = 9M	$z_j - c_j$	(M − 1)	(−M − 1)	(−3M + 1)	(3M − 1) ↑	0	−M	0	0	
A_3''	1	5/4	1/4	−3/4	−1	1	0	0	1/4	0	-
A_4	0	3	1	−2	0	0	1	0	0	0	-
A_7	M	21/4	1/4	5/4	0	0	0	−1	1/4	1	21/5 →
	z = 21M/4 + 5/4	$z_j - c_j$	(M/4 − 3/4)	(5M/4 − 7/4) ↑	0	0	0	−M	(−3M/4 + 1/4)	0	
A_3''	1	22/5	2/5	0	−1	1	0	−3/5	2/5	3/5	-
A_4	0	57/5	7/5	0	0	0	1	−8/5	2/5	8/5	-
A_2	1	21/5	1/5	1	0	0	0	−4/5	1/5	4/5	-
	z = 43/5	$z_j - c_j$	−2/5	0	0	0	0	−7/5	(−M + 3/5)	(−M + 7/5)	

Since all the entries in the index row are non-positive, optimality has been obtained. The optimal solution is $x_1 = 0,\ x_2 = 21/5, x_3' = 0$ and $x_3'' = 22/5$ or $x_1 = 0,\ x_2 = 21/5$ and $x_3 = -22/5$.

The minimum value of $z = 43/5$.

Example 10

Solve the following LP problem by the Simplex method.

$$\text{Minimise } z = x_1 - 3x_2 - 2x_3 - x_4,$$

Subject to
$$x_1 + x_2 + 2x_3 + x_4 = 10$$
$$3x_2 + 2x_3 + x_4 = 15$$
$$5x_2 + x_3 + 2x_4 = 20$$

and
$$x_1, x_2, x_3, x_4 \geq 0.$$

Solution

The constraints are already equalities. So it is enough to introduce artificial variables to get the initial basis. However, we note that the vector corresponding to the variable x_1 is a unit vector. So we need not introduce the artificial variable in the first constraint.

Introducing the artificial variables x_5 and x_6 in the second and third constraints, the standard form of the LP problem is obtained as.

$$\text{Minimise } z = x_1 - 3x_2 - 2x_3 - x_4 + Mx_5 + Mx_6,$$

Subject to
$$x_1 + x_2 + 2x_3 + x_4 = 10$$
$$3x_2 + 2x_3 + x_4 + x_5 = 15$$
$$5x_2 + x_3 + 2x_4 + x_6 = 20$$

and
$$x_i \geq 0;\ i = 1, 2, \ldots, 6$$

The initial and successive simplex tables for the solution of the problem are given as follows.

		c_j	1	−3	−2	−1	M	M	
B	C_B	X_B	A_1	A_2	A_3	A_4	A_5	A_6	RR
A_1	1	10	1	1	2	1	0	0	10
A_5	M	15	0	3	2	1	1	0	5
A_6	M	20	0	5	1	2	0	1	4 →
	z = 35M + 10	$z_j - c_j$	0	(8M + 4) ↑	(3M + 4)	(3M + 2)	0	0	

(Contd)

Table (*Contd*)

		C_j	1	−3	−2	−1	M	M	
A_1	1	6	1	0	9/5	3/5	0	−1/5	10/3
A_5	M	3	0	0	7/5	−1/5	1	−3/5	15/7 →
A_2	−3	4	0	1	1/5	2/5	0	1/5	20
	z = 35M − 60	$z_j - c_j$	0	0	(7M/5 + 16/5) ↑	(−M/5 + 2/5)	0	(−8M/5 − 4/5)	
A_1	1	15/7	1	0	0	6/7	−9/7	4/7	5/2 →
A_3	−2	15/7	0	0	1	−1/7	5/7	−3/7	-
A_2	−3	25/7	0	1	0	3/7	−1/7	2/7	25/3
	z = − 90/7	$z_j - c_j$	0	0	0	6/7 ↑	(−M − 16/7)	(−M + 4/7)	
A_4	−1	5/2	7/6	0	0	1	−3/2	2/3	-
A_3	−2	5/2	1/6	0	1	0	1/2	−1/3	-
A_2	−3	5/2	−1/2	1	0	0	1/2	0	-
	z = − 15	$z_j - c_j$	−1	0	0	0	(−M − 1)	−M	

Since all the entries in the index row are non-positive, optimality has been obtained. The optimal solution is $x_1 = 0$ and $x_2 = x_3 = x_4 = 5/2$.

The minimum value of $z = -15$.

Example 11

Solve the following LP problem by the Simplex method.

$$\text{Minimise } z = 4x_1 - 12x_2,$$

Subject to

$$2x_1 - x_2 \geq 3$$

$$x_1 - 2x_2 \geq 4$$

and

$$x_1, x_2 \geq 0.$$

SOLUTION

Introducing the surplus variable x_3, x_4 and artificial variables x_5, x_6, the standard form of the L.P. problem is got as.

$$\text{Minimise } z = 4x_1 - 12x_2 + 0x_3 + 0x_4 + Mx_5 + Mx_6,$$

Subject to

$$2x_1 - x_2 - x_3 + x_5 = 3$$

$$x_1 - 2x_2 - x_4 + x_6 = 4$$

and $x_i \geq 0$ $(i = 1, 2, \ldots, 6)$.

The initial and successive Simplex tables for the solution of the problem are given as follows.

		C_j	4	−12	0	0	M	M	
B	C_B	X_B	A_1	A_2	A_3	A_4	A_5	A_6	RR
A_5	M	3	2	−1	−1	0	1	0	3/2 →
A_6	M	4	1	−2	0	−1	0	1	4
	z = 7M	$z_j - c_j$	(3M − 4) ↑	(−3M + 12)	−M	−M	0	0	
A_1	4	3/2	1	−1/2	−1/2	0	1/2	0	–
A_6	M	5/2	0	−3/2	1/2	−1	−1/2	1	5 →
	z = 5M/2 + 6	$z_j - c_j$	0	(−3M/2 + 10)	(M/2 − 2) ↑	−M	(−3M/2 + 2)	0	
A_1	4	4	1	−2	0	−1	0	1	–
A_3	0	5	0	−3	1	−2	−1	2	–
	z = 16	$z_j - c_j$	0	4	0	−4	−M	(4 − M)	

At this stage, even though the optionality criterion is not satisfied, we cannot proceed further with the Simplex procedure. In other words, even though the vector to enter the basis has been identified as A_2, the vector to leave the basis could not be found out, as both the entries under A_2 column are negative. This is an indication that the LP problem has an unbounded solution.

EXERCISE 1(B)

Part A (Short answer questions)

1. Write the standard form of an LP problem. What are cost coefficients and structural coefficients?
2. Define slack and surplus variables with examples.
3. Define basic solution and basic feasible solution of an LP problem.
4. What is the difference between degenerate and non-degenerate basic feasible solutions of an LP problem?
5. While solving an LP problem by Simplex method, how will you conclude that the optimal solution has been attained?
6. What are the Simplex criteria for selecting the entering vector into and the leaving vector from the basis?
7. While solving an LP problem by the Simplex method, how will you find out that the problem has an unbounded solution?
8. While solving an LP problem by the Simplex method, how will you find out that the problem has an alternative solution?

9. Having found that an LP problem has an alternative solution, how will you find it in the Simplex procedure?
10. What is an artificial variable? What is the need for introducing it in the Simplex procedure?
11. How will you choose the initial basic feasible solution in the Simplex procedure?
12. Explain briefly how you will revise the current Simplex table.
13. What are degeneracy and cycling in LP problems?
14. What is Charne's perturbation technique that is used to resolve degeneracy in LP problems?

Part B

Use the Simplex method to solve the following LP problems, introducing 'big M' wherever required.

15. Maximise $z = 2x_1 + 3x_2$, Subject to $x_1 + x_2 \leq 1$, $3x_1 + x_2 \leq 4$; $x_1, x_2 \geq 0$
16. Maximise $z = 4x_1 + 3x_2$, Subject to $2x_1 + x_2 \leq 30$, $x_1 + 2x_2 \leq 24$; $x_1, x_2 \geq 0$
17. Maximise $z = 5x_1 + 3x_2$, Subject to $x_1 + x_2 \leq 15$, $2x_1 + x_2 \leq 20$, $x_1 \leq 6$; $x_1, x_2 \geq 0$
18. Maximise $z = 3x_1 + 5x_2$, Subject to $x_1 + x_2 \leq 2$, $2x_1 + 5x_2 \leq 10$, $8x_1 + 3x_2 \leq 12$; $x_1, x_2 \geq 0$
19. Maximise $z = 3x_1 + 5x_2 + 4x_3$,
    Subject to $2x_1 + 3x_2 \leq 8$, $2x_1 + 5x_3 \leq 10$, $3x_1 + 2x_2 + 4x_3 \leq 15$; $x_1, x_2, x_3 \geq 0$
20. Maximise $z = 4x_1 + 3x_2 + 2x_3$,
    Subject to $3x_1 + 2x_2 \leq 30$, $x_1 + x_2 + 2x_3 \leq 35$, $2x_1 + 3x_3 \leq 41$; $x_1, x_2, x_3 \geq 0$
21. Maximise $z = 2x_1 + 4x_2 + x_3$,
    Subject to $2x_1 + x_2 \leq 2$, $x_1 + x_2 + 2x_3 \leq 4$, $10x_1 + 8x_2 + 5x_3 \leq 40$; $x_1, x_2, x_3 \geq 0$
22. Maximise $z = 6x_1 + 4x_2 + 3x_3 + 4x_4$,
    Subject to $x_1 + 2x_2 + 2x_4 \leq 6$, $x_1 + x_2 + 3x_3 + 3x_4 \leq 8$, $4x_1 + 2x_2 + 2x_3 + x_4 \leq 8$; $x_1, x_2, x_3, x_4 \geq 0$
23. Maximise $z = 4x_1 + 5x_2 + 9x_3 + 11x_4$,
    Subject to $x_1 + x_2 + x_3 + x_4 \leq 3$, $3x_1 + 5x_2 + 10x_3 + 15x_4 \leq 20$, $7x_1 + 5x_2 + 3x_3 + 2x_4 \leq 24$; $x_i \geq 0$ $(i = 1, 2, 3, 4)$
24. Maximise $z = -x_2 + 3x_3 - 2x_5$,
    Subject to $x_1 + 3x_2 - x_3 + 2x_5 = 7$, $-2x_2 + 4x_3 + x_4 = 12$, $-4x_2 + 3x_3 + 8x_5 + x_6 = 10$ and $x_i \geq 0$ $(i = 1, 2, \ldots, 6)$ [Hint: Artificial variables need not be introduced]
25. Maximise $z = 5x_1 + 8x_2 + 3x_3$,
    Subject to $2x_1 + 5x_2 - x_3 \leq 1$, $-3x_1 - 8x_2 + 2x_3 \leq 4$, $-2x_1 - 12x_2 + 3x_3 \leq 9$ and x_1 is unrestricted in sign, $x_2, x_3 \geq 0$
26. Maximise $z = x_1 + x_2 + 3x_3$,
    Subject to $x_1 - x_2 + x_3 \leq 10$, $2x_1 - x_3 \leq 2$, $2x_1 + 2x_2 + 3x_3 \leq 0$, $x_1, x_3 \geq 0$ and x_2 is unrestricted in sign
27. Maximise $z = 2x_1 + 4x_2 + x_3$,
    Subject to $x_1 + 2x_2 \leq 4$, $2x_1 + x_2 \leq 3$, $x_2 + 4x_3 \leq 3$, $x_1, x_2 \geq 0$ and x_3 is unrestricted in sign
28. Maximise $z = 8x_1 + 5x_2$,
    Subject to $3x_1 + 2x_2 \leq 12$, $x_1 + 2x_2 \leq 18$, $3x_1 + x_2 \leq 12$; $x_1, x_2 \geq 0$. Use Charne's technique to resolve degeneracy.
29. Maximise $z = 25x_1 + 30x_2 + 22x_3$,
    Subject to $x_1 + x_2 \leq 5$, $x_1 + x_2 + 2x_3 \leq 10$, $2x_1 + 2x_2 + x_3 \leq 10$, $x_1, x_2, x_3 \geq 0$. Use Charne's technique to resolve degeneracy.

30. Maximise $z = 5x_1 - 2x_2 + 3x_3$,
Subject to $2x_1 + 2x_2 - x_3 \geq 2$, $3x_1 - 4x_2 \leq 3$, $x_2 + 3x_3 \leq 5$, $x_1, x_2, x_3 \geq 0$. Use Charne's technique to resolve degeneracy.
31. Maximise $z = 5x_1 + 2x_2$,
Subject to $x_1 + 2x_2 \leq 10$, $5x_1 + 2x_2 \leq 20$, $3x_1 + 2x_2 \leq 18$; $x_1, x_2 \geq 0$. Find also an alternative optimal solution, if it exists.
32. Maximise $z = 2x_1 + 2x_2 + 3x_3$,
Subject to $-x_1 - x_2 + 2x_3 \leq 4$, $x_1 + x_2 + x_3 \leq 6$, $x_1 + x_2 + 3x_3 \leq 9$ and $x_1, x_2, x_3 \geq 0$. Find also an alternative optimal solution, if it exists.
33. Show that the following LP problem has an unbounded solution.
Maximise $z = 2x_1 + x_2$,
Subject to $x_1 - x_2 \leq 1$, $2x_1 - x_2 \leq 4$; $x_1, x_2 \geq 0$.
34. Point out the peculiarity in the solution of the following LP problem, if any
Maximise $z = 2x_1 - x_2 + x_3$,
Subject to $x_1 + x_2 - 3x_3 \leq 8$, $4x_1 - x_2 + x_3 \geq 2$, $2x_1 + 3x_2 - x_3 \geq 4$ and $x_1, x_2, x_3 \geq 0$.
35. Maximise $z = x_1 + x_2$, Subject to $x_1 + x_2 \leq 2$, $-3x_1 + x_2 \geq 3$; $x_1, x_2 \geq 0$.
36. Maximise $z = 2x_1 + 3x_2 + 5x_3$,
Subject to $3x_1 + 10x_2 + 5x_3 \leq 15$, $33x_1 - 10x_2 + 9x_3 \leq 33$, $x_1 + 2x_2 + x_3 \geq 4$; $x_1, x_2, x_3 \geq 0$
37. Minimise $z = 8x_1 + 10x_2 + 15x_3$,
Subject to $2x_1 + 3x_3 \geq 3$, $-3x_1 + 2x_2 + 2x_3 \geq 5$, $5x_2 + 4x_3 \geq 4$; $x_1, x_2, x_3 \geq 0$.
38. Minimise $z = x_1 + 2x_2 + 3x_3$,
Subject to $x_1 - x_2 + x_3 \geq 4$, $x_1 + x_2 + 2x_3 \leq 8$, $x_2 - x_3 \geq 2$; $x_1, x_2, x_3 \geq 0$
39. Minimise $z = x_1 - x_2$,
Subject to $x_1 - 2x_2 \geq 1$, $2x_1 - x_2 \geq 1$, $4x_1 - 2x_2 \geq 1$; $x_1 \geq 0$ and x_2 unrestricted in sign.
40. Minimise $z = 2x_1 + 4x_2 + x_3$,
Subject to $x_1 + 2x_2 - x_3 \leq 5$, $2x_1 - x_2 + 2x_3 = 2$, $-x_1 + 2x_2 + 2x_3 \geq 1$ and $x_1, x_2, x_3 \geq 0$
41. Minimise $z = x_1 + x_2 + x_3$,
Subject to $x_1 - 3x_2 + 4x_3 = 5$, $x_1 - 2x_2 \leq 3$, $2x_2 - x_3 \geq 4$, $x_1, x_2 \geq 0$ and x_3 is unrestricted in sign.
42. Minimise $z = x_1 + x_2 + 2x_3$,
Subject to $x_1 + x_2 + x_3 \leq 9$, $2x_1 - 3x_2 + 3x_3 = 1$, $-3x_1 + 6x_2 - 4x_3 = 3$; $x_1, x_2, x_3 \geq 0$
43. Minimise $z = 2x_1 + x_2 - x_3 - x_4$,
Subject to $x_1 - x_2 + 2x_3 - x_4 = 2$, $2x_1 + x_2 - 3x_3 + x_4 = 6$, $x_1 + x_2 + x_3 + x_4 = 7$; $x_1, x_2, x_3, x_4 \geq 0$

ANSWERS

Exercise 1(A)

Part B

8. $x_1, x_2, x_3 \rightarrow$ Numbers of units of A, B, C respectively to be produced.
Maximise $z = 30x_1 + 20x_2 + 40x_3$. s.t $4x_1 + 3x_2 + 5x_3 \leq 2000$: $2x_1 + 2x_2 + 4x_3 \leq 2500$: $100 \leq x_1 \leq 150$: $x_2 \geq 200$ and $x_3 \geq 50$.

9. $x_1, x_2, x_3 \rightarrow$ Numbers of units of A, B, C respectively to be produced; Max $z = 15x_1 + 25x_2 + 20x_3$, s.t $3x_1 + 2x_3 \leq 8$: $5x_1 + 2x_2 \leq 10$: $4x_1 + 2x_2 + 3x_3 \leq 15$; $x_1, x_2, x_3 \geq 0$.
10. x_1 kgs of A and x_2 kgs of B to be used in a bag. Minimise $z = 20x_1 + 30x_2$, s.t $60x_1 + 30x_2 \geq 240$: $30x_1 + 60x_2 \geq 300$: $30x_1 + 180x_2 \geq 540$; $x_1, x_2 \geq 0$.
11. $x_1, x_2, x_3 \rightarrow$ Numbers of units of bread, butter and milk. To minimise $z = 5x_1 + 20x_2 + 6x_3$, s.t $4x_1 + 3x_2 + 3x_3 \geq 5$; $x_1 + 2x_2 + 2x_3 \geq 2$; $2x_1 + x_2 + x_3 \geq 3$; $x_i \geq 0$.
12. $x_1, x_2 \rightarrow$ Numbers of days in May when the plants P and Q operate.
    To minimise $z = 3000x_1 + 2000x_2$, s.t $3000x_1 + 1000x_2, \geq 24{,}000$: $1000x_1 + 1000x_2 \geq 16{,}000$, $2000x_1 + 6000x_2 \geq 48{,}000$; $x_1, x_2 \geq 0$.
13. $x_1, x_2, x_3, x_4 \rightarrow$ Numbers of units A, B, C, D respectively to be produced;
    Maximise $z = 100x_1 - 50x_4$. s.t $2x_1 + 3x_2 + 4x_3 + 2x_4 \leq 500$: $3x_1 + 2x_2 + x_3 + 2x_4 \leq 380$: $x_i \geq 0$.
14. $x_1, x_2 \rightarrow$ Numbers of production runs of process 1 and process 2. To maximise $z = 300x_1 + 400x_2$, s.t $5x_1 + 4x_2 \leq 200$; $3x_1 + 5x_2 \leq 150$; $5x_1 + 4x_2 \geq 100$; $8x_1 + 4x_2 \geq 80$; $x_i \geq 0$.
15. $x_1, x_2, x_3, x_4 \rightarrow$ Numbers of units A, B, C, D to be produced;
    To maximise $z = 12x_1 + 15x_2 + 16x_3 + 11x_4$, s.t $5x_1 + 5x_2 + 6x_3 + 7x_4 \leq 45$; $6x_1 + 6x_2 + 5x_3 + 7x_4 \leq 42$; $7x_1 + 5x_2 + 6x_3 + 6x_4 \leq 48$; $x_i \geq 0$.
16. $x_1, x_2 \rightarrow$ Numbers of days of operation of plants 1 and 2 respectively. To minimise $z = 2400x_1 + 3600x_2$, s.t $50x_1 + 60x_2 \geq 2400$; $100x_1 + 60x_2 \geq 3000$; $100x_1 + 200x_2 \geq 7000$; $x_i \geq 0$.
17. $x_1, x_2, x_3 \rightarrow$ Numbers of models M_1, M_2, M_3 to be produced. To maximise $z = 30x_1 + 40x_2 + 50x_3$, s.t $2x_1 + 3x_2 + 5x_3 \leq 2000$: $4x_1 + 4x_2 + 7x_3 \leq 3000$: $x_1 + \frac{1}{2}x_2 + \frac{1}{3}x_3 \leq 700$; $2x_1 = 3x_2$; $5x_2 = 2x_3$; $x_1 \geq 200$; $x_2 \geq 300$ and $x_3 \geq 250$
18. $x_1, x_2 \rightarrow$ Numbers of trucks to be sent from A to C and D respectively. To minimise $z = 5x_1 + 8x_2 + 32$, s.t $x_1 + x_2 \leq 6$; $0 \leq x_1 \leq 4$; $0 \leq x_2 \leq 7$; and $x_1 + x_2 \geq 3$.
19. $x_1, x_2 \rightarrow$ Numbers of units of A and B produced; $x_3 \rightarrow$ number of units of C to be sold; $x_4 \rightarrow$ number of units of C to be destroyed; To maximise $z = 3x_1 + 8x_2 + 2x_3 - x_4$, s.t $3x_1 + 4x_2 \leq 18$; $3x_1 + 5x_2 \leq 21$; $x_3 \leq 5$; $3x_2 - x_3 - x_4 = 0$; $x_i \geq 0$.
20. $x_1, x_2, x_3 \rightarrow$ Numbers of production runs for the departments 1, 2, 3 respectively. To maximise $z = x_4$, s.t $6x_1 + 5x_2 + 7x_3 - 4x_4 \geq 0$: $4x_1 + 8x_2 + 3x_3 - 3x_4 \geq 0$: $7x_1 + 4x_2 + 2x_3 \leq 100$; $5x_1 + 8x_2 + 7x_3 \leq 200$; $x_i \geq 0$. $x_4 = \text{Min}\left\{\frac{6x_1 + 5x_2 + 7x_3}{4}, \frac{4x_1 + 8x_2 + 3x_3}{3}\right\}$
21. 20 young hens only; maximum profit = Rs 70
22. 10,500 bottles of A and 34,500 bottles of B; Maximum profit = Rs 3,25,500/-
23. Mines A and B must work for 2 and 6 days respectively in a week; Minimum production cost = Rs 18,000/-
24. Plants 1 and 2 must operate for 12 and 30 days respectively in a month; Minimum total cost = Rs 1,36,800/-
25. $x_1 = 4$, $x_2 = 2$; Maximum $z = 10$;
26. $x_1 = 18$; $x_2 = 12$; Maximum $z = 72$
27. $x_1 = 30.7$, $x_2 = 11.5$; Maximum $z = 138.1$
28. $x_1 = 6$; $x_2 = 12$; Minimum $z = 240$
29. x = 4, y = 2; Minimum $z = 22$

30. $x_1 = 2; x_2 = 2$; Minimum $z = 18$
31. $x_1 = 4, x_2 = 2$; or alternative solution $x_1 = 2, x_2 = 4$; Maximum $z = 6$
32. Alternative solution $x_1 = 4, x_2 = 18$; or $x_1 = 6, x_2 = 12$; Minimum $z = 60$
33. Unbounded solution.
34. Unbounded solution.
35. Infeasible solution.

Exercise 1(B)

Part B

15. $x_1 = 0, x_2 = 1$; Max $z = 3$.
16. $x_1 = 12, x_2 = 6$; Max $z = 66$.
17. $x_1 = 5, x_2 = 10$; Max $z = 55$.
18. $x_1 = 0, x_2 = 2$; Max $z = 10$.
19. $x_1 = \frac{89}{41}, x_2 = \frac{50}{41}, x_3 = \frac{62}{41}$; Max $z = \frac{765}{41}$
20. $x_1 = 4, x_2 = 9, x_3 = 11$; Max $z = 65$.
21. $x_1 = 0, x_2 = 2, x_3 = 1$; Max $z = 9$.
22. $x_1 = \frac{18}{13}, x_2 = \frac{2}{13}, x_3 = 0, x_4 = \frac{28}{13}$; Max $z = \frac{228}{13}$
23. $x_1 = \frac{10}{7}, x_2 = 0, x_3 = \frac{11}{7}, x_4 = 0$; Max $z = \frac{139}{7}$
24. $x_1 = 0, x_2 = 4, x_3 = 5, x_4 = 0, x_5 = 0, x_6 = 11$; Max $z = 11$.
25. $x_1 = 0, x_2 = 3, x_3 = 14$; Max $z = 66$.
26. $x_1 = 0, x_2 = -6, x_3 = 4$; Max $z = 6$.
27. $x_1 = \frac{2}{3}, x_2 = \frac{5}{3}, x_3 = \frac{1}{3}$; Max $z = \frac{25}{3}$
28. $x_1 = 4, x_2 = 0$; Max $z = 32$.
29. $x_1 = \frac{5}{3}, x_2 = \frac{5}{3}, x_3 = \frac{10}{3}$; Max $z = 165$
30. $x_1 = \frac{23}{3}, x_2 = 5, x_3 = 0$; Max $z = \frac{85}{3}$
31. $x_1 = 4, x_2 = 0$; Max $z = 20$. Alternative optimal solution $x_1 = \frac{5}{2}, x_2 = \frac{15}{4}$
32. $x_1 = \frac{9}{2}, x_2 = 0, x_3 = \frac{3}{2}$; Max $z = \frac{27}{2}$. Alternative optimal solution $x_1 = 0, x_2 = \frac{9}{2}, x_3 = \frac{3}{2}$
33. From the optimal table, we conclude that the LPP has an unbounded solution.
34. The LP problem has an unbounded solution

35. No feasible solution

36. No feasible solution

37. $x_1 = 0,\ x_2 = \frac{3}{2},\ x_3 = 1$; Min $z = 30$.

38. $x_1 = 6, x_2 = 2, x_3 = 0$; Min $z = 10$.

39. $x_1 = \frac{1}{3},\ x_2 = \frac{-1}{3}$; Min $z = \frac{2}{3}$.

40. $x_1 = 0, x_2 = 0, x_3 = 1$; Min $z = 1$.

41. $x_1 = 0,\ x_2 = \frac{21}{5},\ x_3 = \frac{22}{5}$; Min $z = \frac{43}{5}$.

42. $x_1 = 0,\ x_2 = \frac{13}{6},\ x_3 = \frac{15}{6}$; Min $z = \frac{43}{6}$

43. $x_1 = 3, x_2 = 0, x_3 = 1, x_4 = 3$; Min $z = 2$.

Chapter 2

TWO-PHASE SIMPLEX METHOD

2.0 INTRODUCTION

The two-phase Simplex method is an alternative method to solve a linear programming problem in which one or more artificial variables are used. In the first phase, the usual Simplex method is applied to an auxiliary LPP (explained in this chapter) so as to get the basic feasible solution to the original problem. In the second phase, the usual Simplex method is applied to the basic feasible solution obtained at the end of the first phase so as to get an optimum basic feasible solution of the original problem.

2.1 COMPUTATIONAL PROCEDURE

Phase I

Step 1: Assign a Cost -1 to each artificial variable (if it is a maximisation problem): Cost 1 to each artificial variable (if it is a minimisation problem) and Cost 0 to all other variables (irrespective of their original costs). Then, the auxiliary LPP is to maximise $z^* = -$ (sum of the artificial variables) for a max. problem or to minimise $z^* =$ sum of the artificial variables for min. problem subject to the modified given constraints.

Step 2: Solve the auxiliary LPP by the Simplex method until either of the following three cases occur.

Case (i) When optimality has been attained, the optimum value of $z^* \# 0$ and at least one artificial variable (vector) appears in the basis at a positive level. In this case, the given LPP does not possess any feasible solution. We need not proceed to Phase II.

Case (ii) When optimality has been attained, the optimum value of $z^* = 0$ and at least one artificial variable (vector) appears in the basis at zero level.

Case (iii) When optimality has been attained, the optimum value of $z^* = 0$ and no artificial variable (vector) appears in the basis.

Phase II

If case (ii) or (iii) arises, we proceed to Phase II after modifying the final Simplex table of Phase I by deleting the non-basic artificial variables (vectors) from it and by assigning the actual costs to the variables in the objective function and a zero cost to every artificial variable that appears in the basis at the zero level [case (ii)]. We apply the Simplex method to the modified simplex table, until an optimum basic feasible solution is obtained. The solution thus obtained will be an optimal basic feasible solution to the original LPP.

WORKED EXAMPLES 2

Example 1

Use the two-phase Simplex method to solve the following linear programming problem:
Minimise $z = x_1 + x_2$, subject to $x_1 + 2x_2 \geq 2, -x_1 - x_2 \geq 1; x_1, x_2 \geq 0$.

SOLUTION

Introducing the surplus variables x_3, x_4 and artificial variables x_5 and x_6, the problem is converted as following:

Minimise $z = x_1 + x_2 + 0.x_3 + 0.x_4 + 1.x_5 + 1.x_6$, subject to $x_1 + 2x_2 - x_3 + x_5 = 2, -x_1 - x_2 - x_4 + x_6 = 1; x_i \geq 0$

Note: Since it is a minimisation problem, the cost coefficients corresponding to the artificial variables x_5 and x_6 have been taken as 1 each, corresponding to M of the ordinary Simplex method.

Phase I

The auxiliary LPP is to minimise $z^* = 0.x_1 + 0.x_2 + 0.x_3 + 0.x_4 + x_5 + x_6$, subject to the same constraints given earlier.

Starting Table

		c_j	0	0	0	0	1	1	
B	C_B	X_B	A_1	A_2	A_3	A_4	A_5	A_6	R.R
A_5	1	2	1	2	−1	0	1	0	1
A_6	1	1	−1	−1	0	−1	0	1	–
$z^* = 3$		$z_j^* - c_j$	0	1	−1	−1	0	0	–

Table after I iteration

		c_j:	0	0	0	0	1	1
B	C_B	X_B	A_1	A_2	A_3	A_4	A_5	A_6
A_2	0	1	$\frac{1}{2}$	1	$-\frac{1}{2}$	0	$\frac{1}{2}$	0
A_6	1	2	$-\frac{1}{2}$	0	$-\frac{1}{2}$	−1	$\frac{1}{2}$	1
	$z^* = 2$	$z_j^* - c_j$	$-\frac{1}{2}$	0	$-\frac{1}{2}$	−1	$-\frac{1}{2}$	0

Since all the entries of the index $(z_j^* - c_j)$ row are non-positive, optimality has been attained.

We note that the optimal value of z^* # 0, but one of the artificial variables (vectors), namely A_6 appears in the positive level, viz., $x_6 = 2$.

This indicates that the auxiliary LPP as well as the original LPP does not possess a feasible solution.

Example 2

Use two-phase Simplex method to solve the following linear programming problem:

Maximise $z = -x_1 - x_2$, subject to $x_1 + x_2 \leq 5$, $3x_1 + 2x_2 \geq 30$, $2x_1 - 3x_2 \geq 30$; $x_1, x_2, x_3 \geq 0$.

SOLUTION

Introducing the slack variable x_3, surplus variables x_4 and x_5 and artificial variables x_6 and x_7, the problem is converted as

Maximise $z = -x_1 - x_2, + 0.x_3 + 0.x_4 + 0.x_5 - 1.x_6 - 1.x_7$, subject to $x_1 + x_2 + x_3 = 5$, $3x_1 + 2x_2 - x_4 + x_6 = 30$, $2x_1 - 3x_2 - x_5 + x_7 = 0$; $x_i \geq 0$.

Note: The cost coefficients corresponding to the artificial variables x_6 and x_7 have been assumed as -1 each, corresponding to $-M$ of the ordinary Simplex method, as it is a maximisation problem.

Phase I

The auxiliary LPP is to maximise $z^* = 0.x_1 + 0.x_2 + 0.x_3 + 0.x_4 + 0.x_5 - 1.x_6 - 1.x_7$, subject to the same constraints given earlier.

Starting Table:

		c_j:	0	0	0	0	0	−1	−1	
B	C_B	X_B	A_1	A_2	A_3	A_4	A_5	A_6	A_7	RR
A_3	0	5	1	1	1	0	0	0	0	5
A_6	−1	30	3	2	0	−1	0	1	0	10
A_7	−1	30	2	−3	0	0	−1	0	1	15
$z^* = -60$		$z^* - c_j$	−5	−1	0	1	1	0	0	

Table after I iteration

		c_j:	0	0	0	0	0	−1	−1
B	C_B	X_B	A_1	A_2	A_3	A_4	A_5	A_6	A_7
A_1	0	5	1	1	1	0	0	0	0
A_6	−1	15	0	−1	−3	−1	0	1	0
A_7	−1	20	0	−5	−2	0	−1	0	1
$z^* = -35$		$z^* - c_j$	0	6	5	1	1	0	0

Optimality has been attained with optimal $z^* \# 0$, but both the artificial vectors (variables) A_6 and A_7 in the positive levels, viz., $x_6 = 15$ and $x_7 = 20$

$\therefore$ The given LPP does not possess a feasible solution.

Example 3

Use two-phase Simplex method to solve the following LPP:

Maximise $z = 5x_1 + 3x_2 + 2x_3$, subject to $x_1 + 2x_2 + x_3 \geq 4$, $5x_1 + 10x_2 + 3x_3 \leq 15$, $9x_1 - 10x_2 + 33x_3 \leq 33$; $x_1, x_2, x_3 \geq 0$.

SOLUTION

Converting the inequalities into equalities, the problem is re-stated as:

Maximise $z = 5x_1 + 3x_2 + 2x_3 + 0.x_4 + 0.x_5 + 0.x_6 - 1.x_7$, subject to $x_1 + 2x_2 + x_3 - x_4 + x_7 = 4$, $5x_1 + 10x_2 + 3x_3 + x_5 = 15$, $9x_1 - 10x_2 + 33x_3 + x_6 = 33$; $x_i \geq 0$, where x_4 is a surplus variable, x_5 and x_6 are slack variables and x_7 is an artificial variable.

Phase I

The auxiliary LPP is to maximise $z^* = 0.x_1 + 0.x_2 + 0.x_3 + 0.x_4 + 0.x_5 + 0.x_6 - 1.x_7$, subject to the same constraints given above.

Starting Table

		c_j:	0	0	0	0	0	0	−1		
B	C_B	X_B	A_1	A_2	A_3	A_4	A_5	A_6	A_7	RR	
A_7	−1	4	1	2	1	−1	0	0	1	2	
A_5	0	15	5	[10]	3	0	1	0	0	$\frac{3}{2}$	→
A_6	0	33	9	−10	33	0	0	1	0	–	
$z^* = -4$		$z_j^* - c_j$	−1	−2	−1	1	0	0	0	–	
				↑							

Tables after iterations

		c_j:	0	0	0	0	0	0	−1		
B	C_B	X_B	A_1	A_2	A_3	A_4	A_5	A_6	A_7	RR	
A_7	−1	1	0	0	$\frac{2}{5}$	−1	$-\frac{1}{5}$	0	1	$\frac{5}{2}$	
A_2	0	$\frac{3}{2}$	$\frac{1}{2}$	1	$\frac{3}{10}$	0	$\frac{1}{10}$	0	0	5	
A_6	0	48	14	0	[36]	0	1	1	0	$\frac{4}{3}$	→
$z^* = -1$		$z_j^* - c_j$	0	0	$-\frac{2}{5}$	1	$\frac{1}{5}$	0	0	–	
					↑						

(Contd)

Table (*Contd*)

		c_j:	0	0	0	0	0	0	−1
A_7	−1	$\frac{7}{15}$	$-\frac{7}{45}$	0	0	−1	$-\frac{19}{90}$	$-\frac{1}{90}$	1
A_2	0	$\frac{11}{10}$	$\frac{23}{60}$	1	0	0	$\frac{11}{120}$	$-\frac{1}{120}$	0
A_3	0	$\frac{4}{3}$	$\frac{7}{18}$	0	1	0	$\frac{1}{36}$	$\frac{1}{36}$	0
$z^* = -\frac{7}{15}$		$z_j^* - c_j$	$\frac{7}{45}$	0	0	1	$\frac{19}{90}$	$\frac{1}{90}$	0

After two iterations, optimality has been attained with $z^* \# 0$. However, the artificial variable (vector) A_7 appears in the basis at a positive level, viz., $x_7 = \frac{7}{15}$.

$\therefore$ The given LPP does not possess a feasible solution.

Example 4

Use two-phase Simplex method to solve the following LPP.:

Minimise $z = 6x_1 + 5x_2$, subject to the constraints $x_1 + 2x_2 \geq 2$, $3x_1 + 2x_2 \geq 5$, $x_1, x_2 \geq 0$.

SOLUTION

Converting the inequalities into equalities, the problem is re-stated as:

Minimise $z = 6x_1 + 5x_2 + 0.x_3 + 0.x_4 + 1.x_5 + 1.x_6$, subject to $x_1 + 2x_2 - x_3 + x_5 = 2$, $3x_1 + 2x_2 - x_4 + x_6 = 5$, $.x_i \geq 0$, where x_3 and x_4 are surplus variables and x_5 and x_6 are artificial variables.

Phase I

The auxiliary LPP is to minimise $z^* = 0.x_1 + 0.x_2 + 0.x_3 + 0.x_4 + 1.x_5 + 1.x_6$, subject to the same constraints given above.

Starting Table

		c_j:	0	0	0	0	1	1		
B	C_B	X_B	A_1	A_2	A_3	A_4	A_5	A_6	RR	
A_5	1	2	1	2	−1	0	1	0	2	
A_6	1	5	[3]	2	0	−1	0	1	$\frac{5}{3}$	→
$z^* = 7$		$z_j^* - c_j$	4	4	−1	−1	0	0		
			↑							

Tables after iterations

		c_j:	0	0	0	0	1	1	
B	C_B	X_B	A_1	A_2	A_3	A_4	A_5	A_6	RR
A_5	1	$\frac{1}{3}$	0	$\boxed{\frac{4}{3}}$	-1	$\frac{1}{3}$	1	$-\frac{1}{3}$	$\frac{1}{4}$ →
A_1	0	$\frac{5}{3}$	1	$\frac{2}{3}$	0	$-\frac{1}{3}$	0	$\frac{1}{3}$	$\frac{5}{2}$
$z^* = \frac{1}{3}$		$z_j^* - c_j$	0	$\frac{4}{3}$	-1	$\frac{1}{3}$	0	$-\frac{4}{3}$	–
A_2	0	$\frac{1}{4}$	0	1	$-\frac{3}{4}$	$\frac{1}{4}$	$\frac{3}{4}$	$-\frac{1}{4}$	
A_1	0	$\frac{3}{2}$	1	0	$\frac{1}{2}$	$-\frac{1}{2}$	$-\frac{1}{2}$	$\frac{1}{2}$	
$z^* = 0$		$z_j^* - c_j$	0	0	0	0	-1	-1	

The optimality criterion is satisfied with $z^* = 0$. Moreover, none of the artificial variables appears in the basis. Hence, we proceed to Phase II by using the final table of Phase I after deleting the two columns corresponding to A_5 and A_6 and restoring the actual costs to the variables.

Phase II: Starting Table

		c_j:	6	5	0	0
B	C_B	X_B	A_1	A_2	A_3	A_4
A_2	5	$\frac{1}{4}$	0	1	$-\frac{3}{4}$	$\frac{1}{4}$
A_1	6	$\frac{3}{2}$	1	0	$\frac{1}{2}$	$-\frac{1}{2}$
$z^* = \frac{41}{4}$		$z_j - c_j$	0	0	$-\frac{3}{4}$	$-\frac{7}{4}$

Since all the $(z_j - c_j)$ values are non-positive, the optimum basic feasible solution has been attained; viz., the optimal solution is given by $x_1 = \frac{3}{2}$ and $x_2 = \frac{1}{4}$ and the minimum value of $z = \frac{41}{4}$.

Example 5

Use two-phase Simplex method to solve the following LPP: Maximise $z = 3x_1 + 2x_2 + x_3$, subject to the constraints $2x_1 + x_2 + x_3 \leq 2$, $3x_1 + 4x_2 + 2x_3 \geq 8$; $x_1, x_2, x_3 \geq 0$.

SOLUTION

Converting the inequalities into equalities, the problem is re-stated as

Maximise $z = 3x_1 + 2x_2 + x_3 + 0.x_4 + 0.x_5 - 1.x_6$, subject to $2x_1 + x_2 + x_3 + x_4 = 2$, $3x_1 + 4x_2 + 2x_3 - x_5 + x_6 = 8$; $x_i \geq 0$, where x_4 is a slack variable, x_5 is a surplus variable and x_6 is an artificial variable.

Phase I

The auxiliary LPP is to maximise $z^* = 0.x_1 + 0.x_2 + 0.x_3 + 0.x_4 + 0.x_5 - 1.x_6$, subject to the same constraints given above.

Starting Table

		c_j:	0	0	0	0	0	−1		
B	C_B	X_B	A_1	A_2	A_3	A_4	A_5	A_6	RR	
A_4	0	2	2	[1]	1	1	0	0	2	→
A_6	−1	8	3	4	2	0	−1	1	2	
$z^* = -8$		$z_j^* - c_j$	−3	−4	−2	0	1	0	–	
				↑						
A_2	0	2	2	1	1	1	0	0		
A_6	−1	0	−5	0	−2	−4	−1	1		
$z^* = 0$		$z_j^* - c_j$	5	0	2	4	1	0		

The optimality criterion is satisfied with $z^* = 0$. Moreover, the artificial variable (vector) A_6 appears in the basis at zero level. Hence, we proceed to Phase II by using the final table of Phase I and by taking the cost coefficient corresponding to x_6 as zero and by restoring the actual costs to the other variables.

Phase II: Starting Table

		c_j:	3	2	1	0	0	0
B	C_B	X_B	A_1	A_2	A_3	A_4	A_5	A_6
A_2	2	2	2	1	1	1	0	0
A_6	0	0	−5	0	−2	−4	−1	1
$z = 4$		$z_j - c_j$	1	0	1	2	0	0

Since all the $(z_j - c_j)$ values are non-negative, the optimum basic feasible solution has been obtained. The optimal solution is given by $x_1 = 0$, $x_2 = 2$, and $x_3 = 0$ and the maximum value of $z = 4$.

Example 6

Use two-phase Simplex method to solve the following LPP: Minimise $z = 3x_1 + 2x_2$, subject to the constraints $x_1 + x_2 \geq 5$, $2x_1 + x_2 \geq 6$, $7x_1 + 2x_2 \geq 14$; $x_1, x_2 \geq 0$.

SOLUTION

Converting the inequalities into equalities, the problem becomes

Minimise $z = 3x_1 + 2x_2 + 0.x_3 + 0.x_4 + 0.x_5 + 1.x_6 + 1.x_7 + 1.x_8$, subject to $x_1 + x_2 - x_3 + x_6 = 5$, $2x_1 + x_2 - x_4 + x_7 = 6$, $7x_1 + 2x_2 - x_5 + x_8 = 14$; $x_i \geq 0$, where x_3, x_4, x_5 are surplus variables and x_6, x_7, x_8 are artificial variables.

Phase I

The auxiliary LPP is to minimise $z^* = 0.x_1 + 0.x_2 + 0.x_3 + 0.x_4 + 0.x_5 + 1.x_6 + 1.x_7 + 1.x_8$, subject to the same constraints given above.

Starting Table

		c_j:	0	0	0	0	0	1	1	1		
B	C_B	X_B	A_1	A_2	A_3	A_4	A_5	A_6	A_7	A_8	RR	
A_6	1	5	1	1	−1	0	0	1	0	0	5	
A_7	1	6	2	1	0	−1	0	0	1	0	3	
A_8	1	14	[7]	2	0	0	−1	0	0	1	2	→
$z^* = 25$		$z_j^* - c_j$	10	4	−1	−1	−1	0	0	0	–	
			↑									

B	C_B	X_B	A_1	A_2	A_3	A_4	A_5	A_6	A_7	A_8	RR	
A_6	1	3	0	[$\frac{5}{7}$]	−1	0	$\frac{1}{7}$	1	0	$-\frac{1}{7}$	$\frac{21}{5}$	→
A_7	1	2	0	$\frac{3}{7}$	0	−1	$\frac{2}{7}$	0	1	$-\frac{2}{7}$	$\frac{14}{3}$	
A_1	0	2	1	$\frac{2}{7}$	0	0	$-\frac{1}{7}$	0	0	$\frac{1}{7}$	7	
$z^* = 5$		$z_j^* - c_j$	0	$\frac{8}{7}$	−1	−1	$\frac{3}{7}$	0	0	$-\frac{10}{7}$	–	
				↑								

A_2	0	$\frac{21}{5}$	0	1	$-\frac{7}{5}$	0	$\frac{1}{5}$	$\frac{7}{5}$	0	$-\frac{1}{5}$	—	
A_7	1	$\frac{1}{5}$	0	0	$\boxed{\frac{3}{5}}$	−1	$\frac{1}{5}$	$-\frac{3}{5}$	1	$-\frac{1}{5}$	$\frac{1}{3}$	→
A_1	0	$\frac{4}{5}$	1	0	$\frac{2}{5}$	0	$-\frac{1}{5}$	$-\frac{2}{5}$	0	$\frac{1}{5}$	2	
$z^* = \frac{1}{5}$	$z_j^* - c_j$		0	0	$\frac{3}{5}$ ↑	−1	$\frac{1}{5}$	$\frac{8}{5}$	0	$-\frac{6}{5}$	—	

A_2	0	$\frac{14}{3}$	0	1	0	$-\frac{7}{3}$	$\frac{2}{3}$	0	$\frac{7}{3}$	$-\frac{2}{3}$
A_3	0	$\frac{1}{3}$	0	0	1	$-\frac{5}{3}$	$\frac{1}{3}$	−1	$\frac{5}{3}$	$-\frac{1}{3}$
A_1	0	$\frac{2}{3}$	1	0	0	$\frac{2}{3}$	$-\frac{1}{3}$	0	$-\frac{2}{3}$	$\frac{1}{3}$
$z^* = 0$	$z_j^* - c_j$		0	0	0	0	0	−1	−1	−1

Phase II: Starting Table

		c_j:	3	2	0	0	0		
B	C_B	X_B	A_1	A_2	A_3	A_4	A_5	RR	
A_2	2	$\frac{14}{3}$	0	1	0	$-\frac{7}{3}$	$\frac{2}{3}$	7	
A_3	0	$\frac{1}{3}$	0	0	1	$-\frac{5}{3}$	$\boxed{\frac{1}{3}}$	1	→
A_1	3	$\frac{2}{3}$	1	0	0	$\frac{2}{3}$	$-\frac{1}{3}$	—	
$z = \frac{34}{3}$	$z_j - c_j$		0	0	0	$-\frac{8}{3}$	$\frac{1}{3}$	—	

		c_j:	3	2	0	0	0
B	C_B	X_B	A_1	A_2	A_3	A_4	A_5
A_2	2	4	0	1	−2	1	0
A_3	0	1	0	0	3	−5	1
A_1	3	1	1	0	1	−1	0
$z = 11$	$z_j - c_j$		0	0	−1	−1	0

Since all $(z_j - c_j)$ values are non-positive, optimality has been attained.
The optimal solution is given by $x_1 = 1$ and $x_2 = 4$ and the minimum value $z = 11$.

Example 7

Use two-phase Simplex method to solve the following LPP:
Maximise $z = 3x_1 + 2x_2 + x_3$, subject to constraints $3x_1 + 4x_2 = 11$, $2x_1 + 5x_2 + x_3 = 12$, $x_1, x_2, x_3 \geq 0$

SOLUTION

Introducing the artificial variables x_4 and x_5, the problem is re-stated as:
Maximise $z = 3x_1 + 2x_2 + x_3 - 1.x_4 - 1.x_5$, subject to $3x_1 + 4x_2 + x_4 = 11$, $2x_1 + 5x_2 + x_3 + x_5 = 12$; $x_i \geq 0$

Phase I

The auxiliary LPP is to maximise $z^* = 0.x_1 + 0.x_2 + 0.x_3 - 1.x_4 - 1.x_5$, subject to the same constraints.

Starting Table

		c_j:	0	0	0	−1	−1		
B	C_B	X_B	A_1	A_2	A_3	A_4	A_5	RR	
A_4	−1	11	3	4	0	1	0	$\frac{11}{4}$	
A_5	−1	12	2	[5]	1	0	1	$\frac{12}{5}$	→
$z^* = -23$		$z_j^* - c_j$	−5	−9	0	0	0	–	
				↑					

A_4	−1	$\frac{7}{5}$	[$\frac{7}{5}$]	0	$-\frac{4}{5}$	1	$-\frac{4}{5}$	1	→
A_2	0	$\frac{12}{5}$	$\frac{2}{5}$	1	$\frac{1}{5}$	0	$\frac{1}{5}$	6	
$z^* = -\frac{7}{5}$		$z_j^* - c_j$	$-\frac{7}{5}$	0	$\frac{4}{5}$	0	$\frac{9}{5}$	–	
			↑						

A_1	0	1	1	0	$-\frac{4}{7}$	$\frac{5}{7}$	$-\frac{4}{7}$
A_2	0	2	0	1	$+\frac{3}{7}$	$-\frac{2}{7}$	$\frac{3}{7}$
$z^* = 0$		$z_j^* - c_j$	0	0	0	1	1

Phase II Starting Table

		c_j:	3	2	1	
B	C_B	X_B	A_1	A_2	A_3	RR
A_1	3	1	1	0	$-\frac{4}{7}$	–
A_2	2	2	0	1	$\boxed{\frac{3}{7}}$	$\frac{14}{3}$ →
$z = 7$		$z_j - c_j$	0	0	$-\frac{13}{7}$	–
					↑	

A_1	3	$\frac{11}{3}$	1	$\frac{4}{3}$	0
A_3	1	$\frac{14}{3}$	0	$\frac{7}{3}$	1
$z = \frac{47}{3}$		$z_j - c_j$	0	$\frac{13}{3}$	0

Since all $(z_j - c_j)$ values are non-negative, optimality has been attained.

The optimal solution is given by $x_1 = \frac{11}{3}$, $x_3 = \frac{14}{3}$, $x_2 = 0$ and the maximum value of $z = \frac{47}{3}$.

Example 8

Use two-phase Simplex method to solve the following LPP:

Minimise $z = x_1 + x_2 + x_3$ subject to the constraints $x_1 - 2x_2 \leq 3$. $x_1 - 3x_2 + 4x_3 = 5$, $2x_2 - x_3 \geq 4$; $x_1, x_2, x_3 \geq 0$.

SOLUTION

Converting the inequalities into equalities, the problem becomes:

Minimise $z = x_1 + x_2 + x_3 + 0.x_4 + 0.x_5 + 1.x_6 + 1.x_7$, subject to $x_1 - 2x_2 + x_4 = 3$, $x_1 - 3x_2 + 4x_3 + x_6 = 5$, $2x_2 - x_3 - x_5 + x_7 = 4$; $x_i \geq 0$, where x_4 is a slack variable, x_5 is a surplus variable and x_6 and x_7 are artificial variables.

Phase I

The auxiliary LPP is to minimise $z^* = 0.x_1 + 0.x_2 + 0.x_3 + 0.x_4 + 0.x_5 + 1.x_6 + 1.x_7$, subject to the same constraints given above.

Starting Table

		c_j:	0	0	0	0	0	1	1	
B	C_B	X_B	A_1	A_2	A_3	A_4	A_5	A_6	A_7	RR
A_4	0	3	1	–2	0	1	0	0	0	–

(Contd)

Table (*Contd*)

	c_j:		0	0	0	0	0	1	1		
A_6	1	5	1	−3	[4]	0	0	1	0	$\frac{5}{4}$	→
A_7	1	4	0	2	−1	0	−1	0	1	–	
$z^* = 9$		$z_j^* - c_j$	1	−1	3	0	−1	0	0	–	
					↑						

		c_j:	0	0	0	0	0	1	1		
B	C_B	X_B	A_1	A_2	A_3	A_4	A_5	A_6	A_7	RR	
A_4	0	3	1	−2	0	1	0	0	0	–	
A_3	0	$\frac{5}{4}$	$\frac{1}{4}$	$-\frac{3}{4}$	1	0	0	$\frac{1}{4}$	0	–	
A_7	1	$\frac{21}{4}$	$\frac{1}{4}$	[$\frac{5}{4}$]	0	0	−1	$\frac{1}{4}$	1	$\frac{21}{5}$	→
$z^* = \frac{21}{7}$		$z_j^* - c_j$	$\frac{1}{4}$	$\frac{5}{4}$	0	0	−1	$-\frac{3}{4}$	0	–	
				↑							

B	C_B	X_B	A_1	A_2	A_3	A_4	A_5	A_6	A_7
A_4	0	$\frac{57}{5}$	$\frac{7}{5}$	0	0	1	$-\frac{8}{5}$	$\frac{2}{5}$	$\frac{8}{5}$
A_3	0	$\frac{22}{5}$	$\frac{2}{5}$	0	1	0	$-\frac{3}{5}$	$\frac{2}{5}$	$\frac{3}{5}$
A_7	0	$\frac{21}{5}$	$\frac{1}{5}$	1	0	0	$-\frac{4}{5}$	$\frac{1}{5}$	$\frac{4}{5}$
$z^* = 0$		$z_j^* - c_j$	0	0	0	0	0	−1	−1

Phase II

		c_j:	1	1	1	0	0
B	C_B	X_B	A_1	A_2	A_3	A_4	A_5
A_4	0	$\frac{57}{5}$	$\frac{7}{5}$	0	0	1	$-\frac{8}{5}$
A_3	1	$\frac{22}{5}$	$\frac{2}{5}$	0	1	0	$-\frac{3}{5}$

(*Contd*)

Table (*Contd*)

		c_j:	1	1	1	0	0
A_2	1	$\frac{21}{5}$	$\frac{1}{5}$	1	0	0	$-\frac{4}{5}$
$z = \frac{43}{5}$		$z_j - c_j$	$-\frac{2}{5}$	0	0	0	$-\frac{7}{5}$

Since all $(z_j - c_j)$ values are non-positive, optimality has been attained.

The optimal solution is given by $x_1 = 0,\ x_2 = \frac{21}{5},\ x_3 = \frac{22}{5}$ and the minimum value of $z = \frac{43}{5}$.

EXERCISE 2

Part A (Short answer questions)

1. How will you write the auxiliary linear programming problem for a given LPP, while solving it by the two-phase Simplex method?
2. How will you identify that a given LPP possesses an infeasible solution, while solving it by the two-phase Simplex method?
3. Under what conditions do we proceed to Phase II of the two-phase Simplex method?
4. How do we write the starting table of Phase II using the final table of Phase I, when we solve an LPP by the two-phase Simplex method?

Part B

Solve the following linear programming problems by the two-phase Simplex method:

5. Maximise $z = 2x_1 + 3x_2$, subject to $x_1 + 2x_2 \leq 2$, $4x_1 + 3x_2 \geq 12$; $x_1, x_2 \geq 0$
6. Maximise $z = x_1 + x_2$, subject to $x_1 + x_2 \leq 2$, $-3x_1 + x_2 \geq 3$; $x_1, x_2 \geq 0$
7. Minimise $z = x_1 - 3x_2 - 2x_3$, subject to the constraints $2x_1 + 4x_2 + 3x_3 = 1$, $-2x_1 + 3x_2 + x_3 = 2$; $x_1, x_2, x_3 \geq 0$
8. Maximise $z = 2x_1 + x_2$, subject to the constraints $3x_1 + 2x_2 \leq 12$, $x_1 + x_2 \geq 8$, $-x_1 + x_2 \geq 4$, $x_1, x_2 \geq 0$
9. Maximise $z = 10x_1 + 3x_2 + 4x_3$, subject to the constraints $5x_1 + 5x_2 + 3x_3 \leq 15$, $9x_1 - 5x_2 + 33x_3 \leq 33$, $x_1 + x_2 + x_3 \geq 4$; $x_1, x_2, x_3 \geq 0$
10. Minimise $z = x_1 + x_2$, subject to the constraints $7x_1 + x_2 \geq 7$, $x_1 + 2x_2 \geq 4$; $x_1, x_2 \geq 0$
11. Minimise $z = 4x_1 + x_2$, subject to the constraints $x_1 + 5x_2 \geq 15$, $3x_1 + 4x_2 \geq 20$; $x_1, x_2 \geq 0$
12. Minimise $z = 4x_1 + x_2$, subject to the constraints $x_1 + 2x_2 \leq 3$, $3x_1 + x_2 = 3$, $4x_1 + 3x_2 \geq 6$; $x_1, x_2 \geq 0$
13. Maximise $z = -x_1 - 2x_2$, subject to the constraints $2x_1 + x_2 \leq 4$, $x_1 + 3x_2 = 3$, $3x_1 + 4x_2 \geq 6$; $x_1, x_2 \geq 0$
14. Minimise $z = 2x_1 + 9x_2 + x_3$, subject to the constraints $x_1 + 4x_2 + 2x_3 \geq 5$, $3x_1 + x_2 + 2x_3 \geq 4$; $x_1, x_2, x_3 \geq 0$

15. Minimise $z = x_1 + x_2 + 2x_3$, subject to the constraints $x_1 + x_2 + x_3 \le 9$, $2x_1 - 3x_2 + 3x_3 = 1$, $-3x_1 + 6x_2 - 4x_3 = 3$; $x_1, x_2, x_3 \ge 0$
16. Maximise $z = 2x_1 + x_2 + x_3$, subject to the constraints $3x_1 - 6x_2 - 4x_3 \le 1$, $4x_1 + 6x_2 + 3x_3 \le 8$, $2x_1 + 3x_2 - 5x_3 \ge 4$; $x_1, x_2, x_3 \ge 0$
17. Maximise $z = 5x_1 - 4x_2 + 3x_3$, subject to the constraints $2x_1 + x_2 - 6x_3 = 20$, $6x_1 + 5x_2 + 10x_3 \le 76$, $8x_1 - 3x_2 + 6x_3 \le 50$; $x_1, x_2, x_3 \ge 0$

ANSWERS

Exercise 2

Part B

5. No feasible solution.
6. No feasible solution.
7. No feasible solution.
8. No feasible solution.
9. No feasible solution.
10. $x_1 = \frac{10}{3}, x_2 = \frac{21}{13}$, Min. $z = \frac{31}{13}$
11. $x_1 = 0, x_2 = 5$; Min. $z = 5$
12. $x_1 = \frac{3}{5}, x_2 = \frac{6}{5}$, Min. $z = \frac{18}{5}$
13. $x_1 = \frac{6}{5}, x_2 = \frac{3}{5}$, Max. $z = -\frac{12}{5}$
14. $x_1 = 0, x_2 = 0, x_3 = \frac{5}{2}$; Min. $z = \frac{5}{2}$
15. $x_1 = 0, x_2 = \frac{13}{6}, x_3 = \frac{5}{2}$; Min. $z = \frac{43}{6}$
16. $x_1 = \frac{9}{7}, x_2 = \frac{10}{21}, x_3 = 0$; Max. $z = \frac{64}{21}$
17. $x_1 = \frac{55}{7}, x_2 = \frac{30}{7}, x_3 = 0$; Max. $z = \frac{155}{7}$

Chapter 3

REVISED SIMPLEX METHOD

3.0 INTRODUCTION

The revised Simplex method is a refined version of the usual Simplex method for solving a linear programming problem that provides an efficient computational procedure on digital computers. Use of the revised Simplex method enables us to obtain the solutions of large-sized and presumably unsolvable LPPs easily, decreases the computational time on large computers and hence reduces the cost of obtaining the solution.

While appling the usual Simplex method to solve an LPP, we have experienced that the major part of the computational work is due to the determination of all the A_j vectors in each iteration. Whenever a vector A_r is selected to enter the basis, its elements alone are required to find out the vector which is to leave the basis. During this process, the elements in the other columns of the basis remain unused until we proceed to the next iteration. It is not necessary to compute all the A_js every time; it is enough to find only one A_j at each iteration to compute the next basis.

3.1 COMPARISON OF USUAL AND REVISED SIMPLEX METHODS

The usual Simplex method is a straightforward algebraic procedure. However, it is not as efficient a computational procedure for electronic digital computers as the revised Simplex method.

The usual Simplex method computes and stores a lot of information that is not required for the current and also for subsequent iterations, whereas the revised Simplex method uses only the essential information for the current need in a more compact form. In the revised Simplex method, B^{-1}_{next} can be obtained directly from B^{-1}_{start}, without having to invert B^{-1}_{next}.

3.2 REVISED SIMPLEX ALGORITHM

Step 1: Convert the given LPP if necessary, into a problem of maximisation and re-write it in the standard form by introducing slack, surplus and artificial variables. Let the given standard problem be: Maximise $z = CX$, subject to $AX = b, X \geq 0$.

Step 2: Start with an initial basis $B = I_m$, where I_m is the unit matrix consisting of the unit vectors corresponding to the basis variables and form the auxiliary matrix $B^+ = \begin{pmatrix} B & 0 \\ C_B & 1 \end{pmatrix}$, where 0 is an $(m \times 1)$ null matrix and C_B is the $(1 \times m)$ matrix whose elements are the cost coefficients corresponding to the basic variables. Find the inverse of B^+ given by $(B^+)^{-1} = \begin{pmatrix} B^{-1} & 0 \\ C_B B^{-1} & 1 \end{pmatrix}$. It can be verified that $B^+ . (B^+)^{-1} = I_{m+1.}$ Find also $A^+ = \begin{pmatrix} A \\ -C \end{pmatrix}$ and $b^+ = \begin{pmatrix} b \\ 0 \end{pmatrix}$.

Step 3: Compute the net evaluations given by $z_j - c_j = (C_B B_S^{-1}, 1)\, A^+; j = 1, 2, 3..., n$, where $B_s^{-1} = B^{-1}_{\text{start}}$, viz., $z_j - c_j$ are obtained by multiplying the elements of the last row of $(B^+)_S^{-1}$ by the corresponding elements of the successive columns of A^+.
If all the values of $(z_j - c_j)$ are non-negative, the current (starting) basic solution itself is the optimal solution. Otherwise, viz., if at least one value of $z_j - c_j$ is negative, find the most negative value, say $(z_k - c_k)$ (Resolve the tie if occurs, by any standard method.). This indicates that the corresponding vector A_k^+ enters the basis.

Step 4: Compute $(A_k^+)_N = (B^+)_S^{-1} (A_k^+)_S\, X_B^+ = (B^+)_S^{-1} b^+$. If all the entries of $(A_k^+)_N$ are ≤ 0, the solution to the given problem is unbounded. If at least of one entry of $(A_k^+)_N$ is positive, find the replacement ratios by dividing the elements of the current X_B^+ by the corresponding entries of $(A_k^+)_N$, that are positive. The vector corresponding to the minimum RR is selected to leave the basis. If there is a tie, it is resolved as in the regular Simplex method.

Step 5: The revised Simplex table is displayed. The first m rows of the table are captioned by the vectors in the current basis and are put at the left-most end of the table. The first $(m + 1)$ columns are the same as those of $(B^+)_S^{-1}$, the next column is the same as the entering vector $(A_k^+)_N$ and the last column is the same as X_B^+. The replacement ratios are put at the right-most end of the table.

Step 6: The first $(m + 1)$ rows of $(B^+)_S^{-1}$ are subjected to elementary row operations that will convert $(A_k^+)_N$ into the unit vector in which the key element that corresponds to the least RR is first converted as 1. Thus, $(B^+)^{-1}$ is obtained and it is now taken as $(B^+)_S^{-1}$. We go to Step 3 and the optimality is tested. If optimal, we go to Step 7. If not, we go to Step 4 and continue the iteration, until the optimality condition is satisfied or an unbounded solution is indicated.

Step 7: Once the optimality is reached, the optimal solution is given by x_i (corresponding to A_i^+) = the cor. element of $(B^+)_S^{+1}, b^+$.

WORKED EXAMPLES 3

Example 1

Use revised Simplex method to solve the following linear programming problem:

Maximise $z = 6x_1 + 11x_2$, subject to $2x_1 + x_2 \leq 26$, $x_1 + 2x_2 \leq 19$; $x_1, x_2 \geq 0$.

SOLUTION

Step 1: The given problem is put in the standard form as:

Maximise $z = 6x_1 + 11x_2 + 0.x_3 + 0.x_4$, s.t. $2x_1 + x_2 + x_3 = 26$ and $x_1 + 2x_2 + x_4 = 19$: $x_i \geq 0$, where $\boldsymbol{x}_3$ and $\boldsymbol{x}_4$ are slack variables.

Step 2:

$$B = I_2 = \begin{pmatrix} 1 & 0 \\ 0 & 1 \end{pmatrix};$$

$$B^+ = \begin{pmatrix} B & 0 \\ -C_B & 1 \end{pmatrix} = \left(\begin{array}{cc|c} 1 & 0 & 0 \\ 0 & 1 & 0 \\ \hline 0 & 0 & 1 \end{array}\right);$$

$$(B^{+})^{-1} = \begin{pmatrix} B^{-1} & 0 \\ C_B B^{-1} & 1 \end{pmatrix}$$

viz.,
$$(B^{+})_S^{-1} = \left(\begin{array}{cc|c} 1 & 0 & 0 \\ 0 & 1 & 0 \\ \hline 0 & 0 & 1 \end{array}\right);$$

$$A^{+} = \begin{pmatrix} A \\ -C \end{pmatrix} = \begin{pmatrix} 2 & 1 & 1 & 0 \\ 1 & 2 & 0 & 1 \\ -6 & -11 & 0 & 0 \end{pmatrix};$$

$$b^{+} = \begin{pmatrix} 26 \\ 19 \\ 0 \end{pmatrix}$$

Step 3:
$$z_j - c_j = \text{last row of } (B^{+})_S^{-1} \times A^{+}$$

$$= (0,0,1)\begin{pmatrix} 2 & 1 & 1 & 0 \\ 1 & 2 & 0 & 1 \\ -6 & -11 & 0 & 0 \end{pmatrix} = (-6,-11,0,0)$$

Since the second entry is most negative, A_2^{+} enters the basis.

Step 4:
$$\left(A_2^{+}\right)_N = \left(B^{+}\right)_S^{-1}\left(A_2^{+}\right)_S$$

$$= \begin{pmatrix} 1 & 0 & 0 \\ 0 & 1 & 0 \\ 0 & 0 & 1 \end{pmatrix}\begin{pmatrix} 1 \\ 2 \\ -11 \end{pmatrix} = \begin{pmatrix} 1 \\ 2 \\ -11 \end{pmatrix}$$

$$X_B^{+} = \left(B^{+}\right)_S^{-1} b^{+}$$

$$= \begin{pmatrix} 1 & 0 & 0 \\ 0 & 1 & 0 \\ 0 & 0 & 1 \end{pmatrix}\begin{pmatrix} 26 \\ 19 \\ 0 \end{pmatrix} = \begin{pmatrix} 26 \\ 19 \\ 0 \end{pmatrix}$$

Step 5: Initial revised Simplex table

B_S^{+}		$(B^{+})_S^{-1}$		$(A_2^{+})_N$	X_B^{+}	RR
A_3^{+}	1	0	0	1	26	26
A_4^{+}	0	1	0	2*	19	19/2
-	0	1	0	−11	0	-

Since RR cor. to A_4^{+} is the least, A_4^{+} leaves the basis and A_2^{+} enters the basis.

Step 6: $(B^+)^{-1}_{\text{next}}$ or $(B^+)_N^{-1} = \begin{matrix} A_3^+ \\ A_2^+ \\ - \end{matrix}\begin{pmatrix} 1 & -\frac{1}{2} & 0 \\ 0 & \frac{1}{2} & 0 \\ 0 & \frac{11}{2} & 1 \end{pmatrix}\begin{matrix} 0 \\ 1 \\ 0 \end{matrix}$ which is taken as $(B^+)_S^{-1}$, for the computation of the next set of $(z_j - c_j)$ values and next iteration.

I iteration

Step 3:

$$z_j - c_j = \left(0, \ \frac{11}{2}, \ 1\right)\begin{pmatrix} 2 & 1 & 1 & 0 \\ 1 & 2 & 0 & 1 \\ -6 & -11 & 0 & 0 \end{pmatrix}$$

$$= \left(-\frac{1}{2}, \ 0, \ 0, \ \frac{11}{2}\right)$$

Since $-\frac{1}{2}$ is the only –ve entry in $z_j - c_j$, A_1^+ enters the basis.

Step 4:
$$\left(A_1^+\right)^N = \left(B^+\right)_S^{-1}\left(A_1^+\right)_S$$

$$= \begin{pmatrix} 1 & -\frac{1}{2} & 0 \\ 0 & \frac{1}{2} & 0 \\ 0 & \frac{11}{2} & 1 \end{pmatrix}\begin{pmatrix} 2 \\ 1 \\ -6 \end{pmatrix} = \begin{pmatrix} \frac{3}{2} \\ \frac{1}{2} \\ -\frac{1}{2} \end{pmatrix}$$

$$X_B^+ = \left(B^+\right)_S^{-1} b^+$$

$$= \begin{pmatrix} 1 & -\frac{1}{2} & 0 \\ 0 & \frac{1}{2} & 0 \\ 0 & \frac{11}{2} & 1 \end{pmatrix}\begin{pmatrix} 26 \\ 19 \\ 0 \end{pmatrix} = \begin{pmatrix} \frac{33}{2} \\ \frac{19}{2} \\ \frac{209}{2} \end{pmatrix}$$

Step 5: RS table after I iteration

B_S^+	$(B^+)_S^{-1}$			$(A_1^+)_N$	X_B^+	RR
A_3^+	1	$-\frac{1}{2}$	0	$\frac{3}{2}$ *	$\frac{33}{2}$	11✓
A_2^+	0	$\frac{1}{2}$	0	$\frac{1}{2}$	$\frac{19}{2}$	19
-	0	$\frac{11}{2}$	1	$-\frac{1}{2}$	$\frac{209}{2}$	-

Since RR cor. to A_3^+ is the least, A_3^+ leaves the basis and A_1^+ enters the basis.

$$\textit{Step 6:}\quad (B^+)_N^{-1} = \begin{matrix} A_1^+ \\ A_2^+ \\ - \end{matrix}\begin{pmatrix} \frac{2}{3} & -\frac{1}{3} & 0 & 1 \\ -\frac{1}{3} & \frac{2}{3} & 0 & 0 \\ \frac{1}{3} & \frac{16}{3} & 1 & 0 \end{pmatrix}$$

, which becomes $(B^+)_S^{-1}$ for the computation of the next set of values of $(z_j - c_j)$.

II iteration

Step 3:

$$z_j - c_j = \left(\frac{1}{3},\ \frac{16}{3},\ 1\right)\begin{pmatrix} 2 & 1 & 1 & 0 \\ 1 & 2 & 0 & 1 \\ -6 & -11 & 0 & 0 \end{pmatrix}$$

$$= \left(0,\ 0,\ \frac{1}{3},\ \frac{16}{3}\right)$$

Since all values of $(z_j - c_j)$ are non-negative, optimality has been obtained.

Step 7: The optimal solution is given by

$$\begin{pmatrix} A_1^+ \\ A_2^+ \\ z \end{pmatrix} = (B^+)_S^{-1}\, b^+$$

$$= \begin{pmatrix} \frac{2}{3} & -\frac{1}{3} & 0 \\ -\frac{1}{3} & \frac{2}{3} & 0 \\ \frac{1}{3} & \frac{16}{3} & 1 \end{pmatrix}\begin{pmatrix} 26 \\ 19 \\ 0 \end{pmatrix} = \begin{pmatrix} 11 \\ 4 \\ 110 \end{pmatrix}$$

Viz., the optimal solution is $x_1 = 11$ and $x_2 = 4$ and the maximum value of $z = 110$.

Example 2

Maximise $z = 5x_1 + 3x_2$, subject to $x_1 \leq 6, x_2 \leq 4, 2x_1 + 3x_2 \leq 18; x_1, x_2, \geq 0$, using the revised Simplex method.

Step 1: Introducing the slack variables x_3, x_4, x_5, the problem is re-stated as:

Maximise $z = 5x_1 + 3x_2 + 0.x_3 + 0.x_4 + 0.x_5$, subject to $x_1 + x_3 = 6, x_2 + x_4 = 4, 2x_1 + 3x_2 + x_5 = 18; x_i \geq 0$.

Viz., Maximise $z = CX$, $s.t\ AX = b$; $X \geq 0$,

where $C = (5, 3, 0, 0, 0)$;

$$A = \begin{pmatrix} 1 & 0 & 1 & 0 & 0 \\ 0 & 1 & 0 & 1 & 0 \\ 2 & 3 & 0 & 0 & 1 \end{pmatrix}$$

and
$$b = \begin{pmatrix} 6 \\ 4 \\ 18 \end{pmatrix}$$

Step 2:
$$B = \begin{pmatrix} 1 & 0 & 0 \\ 0 & 1 & 0 \\ 0 & 0 & 1 \end{pmatrix}; B^+ = \left(\begin{array}{ccc|c} 1 & 0 & 0 & 0 \\ 0 & 1 & 0 & 0 \\ 0 & 0 & 1 & 0 \\ \hline 0 & 0 & 0 & 1 \end{array}\right) = (B^+)_S^{-1};$$

$$A^+ = \begin{pmatrix} 1 & 0 & 1 & 0 & 0 \\ 0 & 1 & 0 & 1 & 0 \\ 2 & 3 & 0 & 0 & 1 \\ -5 & -3 & 0 & 0 & 0 \end{pmatrix}$$

and
$$b^+ = \begin{pmatrix} 6 \\ 4 \\ 18 \\ 0 \end{pmatrix}$$

Step 3:
$$z_j - c_j = (0, 0, 0, 1)\,A^+ = (-5, -3, 0, 0, 0)$$

Since $(z_1 - c_1)$ is the most negative, A_1^+ enters the basis.

Step 4:
$$(A_1^+)_N = (B^+)_S^{-1}(A_1^+)_S = \begin{pmatrix} 1 \\ 0 \\ 2 \\ -5 \end{pmatrix}$$

$$X_B^+ = (B^+)_S^{-1}\,b^+ = \begin{pmatrix} 6 \\ 4 \\ 18 \\ 0 \end{pmatrix}$$

Step 5: Initial revised Simplex table

B_S^+	$(B_S^+)^{-1}$				$(A_1^+)^N$	X_B^+	RR
A_3^+	1	0	0	0	1*	6	6✓
A_4^+	0	1	0	0	0	4	-
A_5^+	0	0	1	0	2	18	9
–	0	0	0	1	–5	0	-

Since RR cor. to A_3^+ is the least, A_3^+ leaves the basis and A_1^+ enters the basis.

Step 6: $$\left(B^+\right)_N^{-1} = \begin{matrix} A_1^+ \\ A_4^+ \\ A_5^+ \\ - \end{matrix}\left(\begin{array}{cccc|c} 1 & 0 & 0 & 0 & 1 \\ 0 & 1 & 0 & 0 & 0 \\ -2 & 0 & 1 & 0 & 0 \\ 5 & 0 & 0 & 1 & 0 \end{array}\right) \to \left(B^+\right)_S^{-1}$$ for the evaluation of next set of $(z_j - c_j)$values and next iteration.

I iteration

Step 3: $z_j - c_j = (5, 0, 0, 1)$. $A^+ = (0, -3, 5, 0, 0)$

Since $(z_2 - c_2)$ is the only –ve values, A_2^+ enters the basis.

Step 4:

$$\left(A_2^+\right)_N = \left(B^+\right)_S^{-1}\left(A_2^+\right)_S$$

$$= \begin{pmatrix} 1 & 0 & 0 & 0 \\ 0 & 1 & 0 & 0 \\ -2 & 0 & 1 & 0 \\ 5 & 0 & 0 & 1 \end{pmatrix}\begin{pmatrix} 0 \\ 1 \\ 3 \\ -3 \end{pmatrix} = \begin{pmatrix} 0 \\ 1 \\ 3 \\ -3 \end{pmatrix}$$

$$X_B^{\ +} = \left(B^+\right)_S^{-1} b^+$$

$$= \begin{pmatrix} 1 & 0 & 0 & 0 \\ 0 & 1 & 0 & 0 \\ -2 & 0 & 1 & 0 \\ 5 & 0 & 0 & 1 \end{pmatrix}\begin{pmatrix} 6 \\ 4 \\ 18 \\ 0 \end{pmatrix} = \begin{pmatrix} 6 \\ 4 \\ 6 \\ 30 \end{pmatrix}$$

Step 5: RS table after I iteration.

B_S^+	$(B^+)_S^{-1}$				$(A_2^+)_N$	X_B^+	RR
A_1^+	1	0	0	0	0	6	–
A_4^+	0	1	0	0	1	4	4
A_5^+	–2	0	1	0	3	6*	2✓
-	5	0	0	1	–3	30	–

Since RR cor.to A_5^+ is the least, A_5^+ leaves the basis and A_2^+ enters the basis.

Step 6: $$\left(B^+\right)_N^{-1} = \begin{matrix} A_1^+ \\ A_4^+ \\ A_2^+ \\ - \end{matrix}\left(\begin{array}{cccc|c} 1 & 0 & 0 & 0 & 0 \\ \frac{2}{3} & 1 & -\frac{1}{3} & 0 & 0 \\ -\frac{2}{3} & 0 & \frac{1}{3} & 0 & 1 \\ 3 & 0 & 1 & 1 & 0 \end{array}\right) \to \left(B^+\right)_S^{-1}$$

II iteration

Step 3: $z_j - c_j = (3, 0, 0, 1)\, A^+ = (0, 0, 3, 0, 1)$

Since all the values of $(z_j - c_j)$ are non-negative, optimality has been attained.

Step 7: The optimal solution is given by

$$\begin{pmatrix} A_1^+ \\ A_4^+ \\ A_2^+ \\ z \end{pmatrix} = (B^+)_S^{-1}\bar{b}$$

$$= \begin{pmatrix} 1 & 0 & 0 & 0 \\ \frac{2}{3} & 1 & -\frac{1}{3} & 0 \\ -\frac{2}{3} & 0 & \frac{1}{3} & 0 \\ 3 & 0 & 1 & 1 \end{pmatrix} \times \begin{pmatrix} 6 \\ 4 \\ 18 \\ 0 \end{pmatrix} = \begin{pmatrix} 6 \\ 2 \\ 2 \\ 36 \end{pmatrix}$$

Viz., the optimal solution is $x_1 = 6$ and $x_2 = 2$ and $(x_4 = 2)$ and the maximum value of $z = 36$.

Example 3

Maximise $z = x_1 + 3x_2 + x_3$, subject to $2x_1 + x_2 + 3x_3 \leq 3, x_1 + 2x_2 + 2x_3 \leq 2; x_1, x_2, x_3 \geq 0$, using revised Simplex method.

SOLUTION

Step 1: Introducing the slack variables x_4 and x_5, the problem becomes:

Maximise $z = x_1 + 3x_2 + x_3 + 0.x_4 + .0x_5$, such that $2x_1 + x_2 + 3x_3 + x_4 = 3, x_1 + 2x_2 + 2x_3 + x_5 = 2; x_i \geq 0$.

Viz., Maximise $z = CX$, s.t. $AX = b$; $x \geq 0$,

where $$C = (1,\ 3,\ 1,\ 0,\ 0);$$

$$A = \begin{pmatrix} 2 & 1 & 3 & 1 & 0 \\ 1 & 2 & 2 & 0 & 1 \end{pmatrix}; b = \begin{pmatrix} 3 \\ 2 \end{pmatrix}$$

Step 2: $$B = \begin{pmatrix} 1 & 0 \\ 0 & 1 \end{pmatrix}; B^+ = \left(\begin{array}{cc|c} 1 & 0 & 0 \\ 0 & 1 & 0 \\ \hline 0 & 0 & 1 \end{array}\right) = (B^+)_S^{-1};$$

$$A^+ = \begin{pmatrix} 2 & 1 & 3 & 1 & 0 \\ 1 & 2 & 2 & 0 & 1 \\ -1 & -3 & -1 & 0 & 0 \end{pmatrix}$$

and $$b^+ = \begin{pmatrix} 3 \\ 2 \\ 0 \end{pmatrix}$$

Step 3:

$$z_j - c_j = \begin{pmatrix} 0, & 0, & 1 \end{pmatrix} \begin{pmatrix} 2 & 1 & 3 & 1 & 0 \\ 1 & 2 & 2 & 0 & 1 \\ -1 & -3 & -1 & 0 & 0 \end{pmatrix}$$

$$= \begin{pmatrix} -1, & -3, & -1, & 0 & 0 \end{pmatrix}$$

Since $(z_2 - c_2)$ is most negative, A_2^+ enters the basis.

Step 4:

$$\left(A_2^+\right)_N = \left(B^+\right)_S^{-1} \left(A_2^+\right)_S = \begin{pmatrix} 1 \\ 2 \\ -3 \end{pmatrix};$$

$$X_B^+ = \left(B^+\right)_S^{-1} b^+ = \begin{pmatrix} 3 \\ 2 \\ 0 \end{pmatrix}$$

Step 5: Initial RS table.

B_S^+	$(B_S^+)^{-1}$			$(A_2)_N^+$	X_B^+	RR
A_4^+	1	0	0	1	3	3
A_5^+	0	1	0	2*	2	1✓
-	0	0	1	−3	0	-

Since RR cor. to A_5^+ is the least, A_5^+ leaves the basis and A_2^+ enters the basis.

Step 6:

$$\left(B^+\right)_N^{-1} = \begin{matrix} A_4^+ \\ A_2^+ \\ - \end{matrix} \begin{pmatrix} 0 & -\frac{1}{2} & 0 \\ 0 & \frac{1}{2} & 0 \\ 0 & \frac{3}{2} & 1 \end{pmatrix} \begin{matrix} 0 \\ 1 \\ 0 \end{matrix} \rightarrow \left(B^+\right)_S^{-1}$$

I iteration

Step 3:

$$z_j - c_j = \left(0, \ \frac{3}{2}, \ 1\right) \begin{pmatrix} 2 & 1 & 3 & 1 & 0 \\ 1 & 2 & 2 & 0 & 1 \\ -1 & -3 & -1 & 0 & 0 \end{pmatrix}$$

$$= \left(\frac{1}{2}, \ 0, \ 2, \ 0, \ \frac{3}{2}\right)$$

Since all the values of $(z_j - c_j)$ are non-negative, optimality has been obtained.

Step 7: The optimal solution is given by

$$\begin{pmatrix} A_4^+ \\ A_2^+ \\ z \end{pmatrix} = (B^+)_S^{-1} b^+$$

$$= \begin{pmatrix} 1 & -\frac{1}{2} & 0 \\ 0 & \frac{1}{2} & 0 \\ 0 & \frac{3}{2} & 1 \end{pmatrix} \begin{pmatrix} 3 \\ 2 \\ 0 \end{pmatrix} = \begin{pmatrix} 2 \\ 1 \\ 3 \end{pmatrix}$$

viz., the optimal solution is x_1 (not in the basis) $= 0$, $x_2 = 1$, x_3 (not in the basis) $= 0$ and $(x_4 = 2)$ and the maximum value of $z = 3$.

Example 4

Maximise $z = 2x_1 + 4x_2 + x_3$, subject to $x_1 + 2x_2 \leq 4$, $x_2 + 4x_3 \leq 3$, $5x_3 + 2x_1 \leq 3$; $x_1, x_2, x_3 \geq 0$, using revised Simplex method.

Step 1: Introducing the slack variables x_4, x_5, x_6, the problem becomes: maximise $z = 2x_1 + 4x_2 + x_3 + 0.x_4 + 0.x_5 + 0.x_6$, such that $x_1 + 2x_2 + x_4 = 4$, $x_2 + 4x_3 + x_5 = 3$, $2x_1 + 5x_3 + x_6 = 3$, $x_i \geq 0$.

viz., Maximise $z = CX$, s.t. $AX = b$, $x \geq 0$,

where
$$C = (2,\ 4,\ 1,\ 0,\ 0,\ 0), A = \begin{pmatrix} 1 & 2 & 0 & 1 & 0 & 0 \\ 0 & 1 & 4 & 0 & 1 & 0 \\ 2 & 0 & 5 & 0 & 0 & 1 \end{pmatrix}$$

and
$$b = \begin{pmatrix} 4 \\ 3 \\ 3 \end{pmatrix}$$

Step 2:
$$B = \begin{pmatrix} 1 & 0 & 0 \\ 0 & 1 & 0 \\ 0 & 0 & 1 \end{pmatrix}; B^+ = \left(\begin{array}{ccc|c} 1 & 0 & 0 & 0 \\ 0 & 1 & 0 & 0 \\ 0 & 0 & 1 & 0 \\ \hline 0 & 0 & 0 & 1 \end{array}\right) = (B^+)_S^{-1};$$

$$A^+ \begin{pmatrix} 1 & 2 & 0 & 1 & 0 & 0 \\ 0 & 1 & 4 & 0 & 1 & 0 \\ 2 & 0 & 5 & 0 & 0 & 1 \\ -2 & -4 & -1 & 0 & 0 & 0 \end{pmatrix};$$

$$b^+ = \begin{pmatrix} 4 \\ 3 \\ 3 \\ 0 \end{pmatrix}$$

Step 3:

$$z_j - c_j = (0, \ 0, \ 0, \ 1)A^+ = (-2, \ -4, \ -1, \ 0, \ 0, \ 0)$$

Since $(z_2 - c_2)$ is the most negative, A_2^+ enters the basis.

Step 4:

$$(A_2^+)_N = (B^+)_S^{-1}(A_2^+)_S = \begin{pmatrix} 2 \\ 1 \\ 0 \\ -4 \end{pmatrix};$$

$$X_B^+ = (B^+)_S^{-1} b^+ = \begin{pmatrix} 4 \\ 3 \\ 3 \\ 0 \end{pmatrix}$$

Step 5: Initial RS tables

B_S^+	$(B^+)_S^{-1}$				$(A_2^+)^N$	X_B^+	RR
A_4^+	1	0	0	0	2*	4	2✓
A_5^+	0	1	0	0	1	3	3
A_6^+	0	0	1	0	0	3	-
-	0	0	0	1	−4	0	-

Since RR cor. to A_4^+ is the least, A_4^+ leaves and A_2^+ enters the basis.

Step 6:

$$(B^+)_N^{-1} = \begin{matrix} A_2^+ \\ A_5^+ \\ A_6^+ \\ - \end{matrix}\begin{pmatrix} \frac{1}{2} & 0 & 0 & 0 \\ -\frac{1}{2} & 1 & 0 & 0 \\ 0 & 0 & 1 & 0 \\ 2 & 0 & 0 & 1 \end{pmatrix}\begin{matrix} 1 \\ 0 \\ 0 \\ 0 \end{matrix} \rightarrow (B^+)_S^{-1}.$$

I iteration

Step 3:

$$z_j - c_j = (2, \ 0, \ 0, \ 1)A^+ = (0, \ 0, \ -1, \ 2, \ 0, \ 0)$$

Since $(z_3 - c_3)$ is the most negative, A_3^+ enters the basis.

Step 4:

$$\left(A_3^+\right)_N = \left(B^+\right)_S^{-1}\left(A_3^+\right) = \begin{pmatrix} 0 \\ 4 \\ 5 \\ -1 \end{pmatrix};$$

$$X_B^+ = \left(B^+\right)_S^{-1} b^+ = \begin{pmatrix} 2 \\ 1 \\ 3 \\ 8 \end{pmatrix}$$

Step 5: RS table after I iteration

B_S^+	$(B^+)_S^{-1}$				$(A_3^+)_N$	X_B^+	RR
A_2^+	$\frac{1}{2}$	0	0	0	0	2	-
A_5^+	$-\frac{1}{2}$	1	0	0	4*	1	$\frac{1}{4}$ ✓
A_6^+	0	0	1	0	5	3	$\frac{3}{5}$
-	2	0	0	1	−1	8	-

∴ A_5^+ leaves the basis and A_3^+ enters the basis.

Step 6:

$$\left(B^+\right)_N^{-1} = \begin{matrix} A_2^+ \\ A_3^+ \\ A_6^+ \\ \\ \end{matrix}\begin{pmatrix} \frac{1}{2} & 0 & 0 & 0 \\ -\frac{1}{8} & \frac{1}{4} & 0 & 0 \\ \frac{5}{8} & -\frac{5}{4} & 1 & 0 \\ -\frac{15}{8} & \frac{1}{4} & 0 & 1 \end{pmatrix}\begin{matrix} 0 \\ 1 \\ 0 \\ 0 \end{matrix} \rightarrow \left(B^+\right)_S^{-1}$$

II iteration

Step 3:

$$z_j - c_j = \left(\frac{15}{8},\ \frac{1}{4},\ 0,\ 1\right)A^+ = \left(-\frac{1}{8},\ 0,\ 0,\ \frac{15}{8},\ \frac{1}{4},\ 0\right)$$

∴ A_1^+ enters.

Step 6:

$$(A_1)_N^+ = (B^+)_S^{-1}.(A_1^+)_S = \begin{pmatrix} \frac{1}{2} \\ -\frac{1}{8} \\ \frac{21}{8} \\ -\frac{1}{8} \end{pmatrix};$$

$$X_B^+ = (B^+)_S^{-1} b^+ = \begin{pmatrix} 2 \\ \frac{1}{4} \\ \frac{7}{4} \\ \frac{33}{4} \end{pmatrix}$$

Step 5: RS table after II iteration

B_S^+	$(B^+)_S^{-1}$				$(A_1^+)_N$	X_B^+	RR
A_2^+	$\frac{1}{2}$	0	0	0	$\frac{1}{2}$	2	4
A_3^+	$-\frac{1}{8}$	$\frac{1}{4}$	0	0	$-\frac{1}{8}$	$\frac{1}{4}$	-
A_6^+	$\frac{5}{8}$	$-\frac{5}{4}$	1	0	$\frac{21}{8}$ *	$\frac{7}{4}$	$\frac{2}{3}$ ✓
-	$\frac{15}{8}$	$\frac{1}{4}$	0	1	$-\frac{1}{8}$	$\frac{33}{4}$	-

$\therefore$ A_6^+ leaves and A_1^+ enters the basis.

Step 6:

$$(B^+)_N^{-1} = \begin{matrix} A_2^+ \\ A_3^+ \\ A_1^+ \\ - \end{matrix} \begin{pmatrix} \frac{8}{21} & \frac{5}{21} & -\frac{4}{21} & 0 \\ -\frac{2}{21} & \frac{4}{21} & \frac{1}{21} & 0 \\ \frac{5}{21} & -\frac{10}{21} & \frac{8}{21} & 0 \\ \frac{40}{21} & \frac{4}{21} & \frac{1}{21} & 1 \end{pmatrix} \begin{matrix} 0 \\ 0 \\ 1 \\ 0 \end{matrix} \rightarrow (B^+)_S^{-1}.$$

III iteration

Step 3:
$$z_j - c_j = \left(\frac{40}{21}, \frac{4}{21}, \frac{1}{21}, 1\right)A^+$$
$$= \left(0, 0, 0, \frac{40}{21}, \frac{4}{21}, \frac{1}{21}\right)$$

∴ Since all values of $(z_j - c_j)$ are non-negative, optimality has been obtained.

Step 7: The optimal solution is given by

$$\begin{pmatrix} A_2^+ \\ A_3^+ \\ A_1^+ \\ z \end{pmatrix} = (B^+)_S^{-1} b^+ = \begin{pmatrix} \frac{5}{3} \\ \frac{1}{3} \\ \frac{2}{3} \\ \frac{25}{3} \end{pmatrix}$$

viz., the optimal solution is $x_1 = \frac{2}{3}, x_2 = \frac{5}{3}, x_3 = \frac{1}{3}$ and the maximum value of $z = \frac{25}{3}$.

Example 5

Minimise $z = x_1 + x_2$, subject to $2x_1 + x_2 \geq 4$; $2x_1 + 7x_2 \geq 7$; $x_1, x_2 \geq 0$, using *R.S.* method.

Step1: Converting the problem as one of maximisation, we get

Maximise $z^* = -x_1 - x_2$, s.t. the same constraints, so that min $(z) = -$ max (z^*).

Introducing the surplus variables x_3, x_4 and artificial variables x_5, x_6, the problem becomes:

$$\text{Maximise } z^* = -x_1 - x_2 + 0.x_3 + 0.x_4 - M.x_5 - M.x_6 \text{ s.t.}$$
$$2x_1 + x_2 - x_3 + x_5 = 4,\ x_1 + 7x_2 - x_4 + x_6 = 7;\ x_i \geq 0.$$

viz, Maximise $z^* = Cx$, s.t $Ax = b, x \geq 0$.

$$C = (-1, -1, 0, 0, -M, -M),$$

$$A = \begin{pmatrix} 2 & 1 & -1 & 0 & 1 & 0 \\ 1 & 7 & 0 & -1 & 0 & 1 \end{pmatrix}$$

and
$$b = \begin{pmatrix} 4 \\ 7 \end{pmatrix}$$

Step 2:
$$B = \begin{pmatrix} 1 & 0 \\ 0 & 1 \end{pmatrix}; (B^+)^{-1} = \left(\begin{array}{cc|c} 1 & 0 & 0 \\ 0 & 1 & 0 \\ -M & -M & 1 \end{array}\right) = (B^+)_S^{-1};$$

$$A^+ \begin{pmatrix} 2 & 1 & -1 & 0 & 1 & 0 \\ 1 & 7 & 0 & -1 & 0 & 1 \\ 1 & 1 & 0 & 0 & M & M \end{pmatrix}; b^+ \begin{pmatrix} 4 \\ 7 \\ 0 \end{pmatrix}$$

Step 3:
$$z_j - c_j = (-M, -M, 1)\, A^+ = (-3M + 1, -8M + 1, M, M, 0, 0)$$

Since $(z_2 - c_2)$ is the most negative, A_2^+ enters the basis,

Step 4:
$$\left(A_2^+\right)_N = \left(B^+\right)_S^{-1}\left(A_2^+\right)_S$$

$$= \begin{pmatrix} 1 & 0 & 0 \\ 0 & 1 & 0 \\ -M & -M & 1 \end{pmatrix}\begin{pmatrix} 1 \\ 7 \\ 1 \end{pmatrix} = \begin{pmatrix} 1 \\ 7 \\ -8M+1 \end{pmatrix};$$

$$X_B{}^+ = \left(B^+\right)_S^{-1} b^+ = \begin{pmatrix} 4 \\ 7 \\ -11M \end{pmatrix}$$

Step 5: Initial RS table

B_S^+	$(B^+)_S^{-1}$			$(A_2^+)_N$	X_B^+	RR
A_5^+	1	0	0	1	4	4
A_6^+	0	1	0	7*	7	1✓
-	$-M$	$-M$	1	$(-8M+1)$	$-11M$	-

$\therefore$ A_6^+ leave the basis and A_2^+ enters the basis.

Step 6:
$$\left(B^+\right)_N^{-1} = \begin{matrix} A_5^+ \\ A_2^+ \\ - \end{matrix}\begin{vmatrix} 1 & -\frac{1}{7} & 0 \\ 0 & \frac{1}{7} & 0 \\ -M & \left(\frac{M}{7}-\frac{1}{7}\right) & 1 \end{vmatrix}\begin{matrix} 0 \\ 1 \\ 0 \end{matrix} \rightarrow \left(B^+\right)_S^{-1} \text{ for further iterations.}$$

Step 3:
$$z_j - c_j = \left(-M, \left(\frac{M}{7}-\frac{1}{7}\right), 1\right)$$

$$A^+ = \left(-\frac{13}{7}M + \frac{6}{7}, \ 0, \ M, \ -\frac{M}{7} + \frac{1}{7}, \ 0, \ \frac{8}{7}M - \frac{1}{7}\right)$$

Since $(z_1 - c_1)$ is the most negative, A_1^+ enters.

I iteration

Step 4:
$$\left(A_1^+\right)_N = \begin{pmatrix} 1 & -\frac{1}{7} & 0 \\ 0 & \frac{1}{7} & 0 \\ -M & \frac{M}{7}-\frac{1}{7} & 1 \end{pmatrix}\begin{pmatrix} 2 \\ 1 \\ 1 \end{pmatrix}$$

$$= \begin{pmatrix} \dfrac{13}{7} \\ \dfrac{1}{7} \\ -\dfrac{13M}{7}+\dfrac{6}{7} \end{pmatrix}; X_B{}^+ = \begin{pmatrix} 3 \\ 1 \\ -3M-1 \end{pmatrix}$$

Step 5: RS table after I iteration

$B_S{}^+$	$(B^+)_S{}^{-1}$			$(A_1{}^+)_N$	$X_B{}^+$	RR
$A_5{}^+$	1	$-\frac{1}{7}$	0	$\frac{13}{7}$ *	3	$\frac{21}{13}$ ✓
$A_2{}^+$	0	$\frac{1}{7}$	0	$\frac{1}{7}$	1	7
-	$-M$	$\left(\frac{M}{7}-\frac{1}{7}\right)$	1	$\left(-\frac{13M}{7}+\frac{6}{7}\right)$	$(-3M-1)$	-

$\therefore$ $A_5{}^+$ leaves the basis and $A_1{}^+$ enters

Step 6:
$$(B^+)_N{}^{-1} = \begin{matrix} A_1{}^+ \\ A_2{}^+ \\ - \end{matrix}\begin{pmatrix} \dfrac{7}{13} & -\dfrac{1}{13} & 0 \\ -\dfrac{1}{13} & \dfrac{2}{13} & 0 \\ -\dfrac{6}{13} & -\dfrac{1}{13} & 1 \end{pmatrix}\begin{matrix} 1 \\ 0 \\ 0 \end{matrix} \rightarrow (B^+)_S{}^{-1}$$

III Iteration

Step 3:
$$z_j - c_j = \left(-\frac{6}{13}, \ -\frac{1}{13}, \ 1\right)A^+ = \left(0, \ 0, \ \frac{6}{13}, \ \frac{1}{13}, \ M-\frac{6}{13}, \ M-\frac{1}{13}\right)$$

Since all values of $(z_j - c_j)$ are non-negative, the optimal solution has been obtained.

Step 7: Optimal solution is given by

$$\begin{pmatrix} A_1{}^+ \\ A_2{}^+ \\ z^* \end{pmatrix} = (B^+)_S^{-1}\, b^+$$

$$= \begin{pmatrix} \dfrac{7}{13} & -\dfrac{1}{13} & 0 \\ -\dfrac{1}{13} & \dfrac{2}{13} & 0 \\ -\dfrac{6}{13} & -\dfrac{1}{13} & 1 \end{pmatrix}\begin{pmatrix} 4 \\ 7 \\ 0 \end{pmatrix} = \begin{pmatrix} \dfrac{21}{13} \\ \dfrac{10}{13} \\ -\dfrac{31}{13} \end{pmatrix}$$

viz., $$x_1 = \frac{21}{13}, x_2 = \frac{10}{13}$$

and $$\max z^* = -\frac{31}{13} \text{ or } \min z = \frac{31}{13}.$$

Example 6

Using revised Simplex method, minimise $z = 2x_1 + x_2$, subject to $x_1 + 2x_2 \leq 40$, $3x_1 + x_2 \geq 30$; $4x_1 + 3x_2 \geq 60$; $x_1, x_2, x_3 \geq 0$.

SOLUTION

Step 1: Converting the problem to one of maximisation, we get:

Maximise $z^* = -2x_1 - x_2$, subject to the same constraints, so that $\min (z) = -\max (z^*)$.

Introducing the slack variable x_3 in the first constraint, the surplus variables x_4 and x_5 and the artificial variables x_6 and x_7 in the second and third constraints, the problem becomes:

Max $(z^*) = -2x_1 - x_2 + 0.x_3 + 0.x_4 + 0.x_5 - M.x_6 - M.x_7$, s.t. $x_1 + 2x_2 + x_3 = 40$, $3x_1 + x_2 - x_4 + x_6 = 30$, $4x_1 + 3x_2 - x_5 + x_7 = 60$; $x_i \geq 0$.

Viz., $$\max (z^*) = CX, \text{ s.t. } AX = b \text{ and } X \geq 0, \text{ where,}$$

$$C = (-2, \; -1, \; 0, \; 0, \; 0, \; -M, \; -M);$$

$$A = \begin{pmatrix} 1 & 2 & 1 & 0 & 0 & 0 & 0 \\ 3 & 1 & 0 & -1 & 0 & 1 & 0 \\ 4 & 3 & 0 & 0 & -1 & 0 & 1 \end{pmatrix}; b = \begin{pmatrix} 40 \\ 30 \\ 60 \end{pmatrix}$$

Step 2: $$A^+ = \begin{pmatrix} 1 & 2 & 1 & 0 & 0 & 0 & 0 \\ 3 & 1 & 0 & -1 & 0 & 1 & 0 \\ 4 & 3 & 0 & 0 & -1 & 0 & 1 \\ 2 & 1 & 0 & 0 & 0 & M & M \end{pmatrix};$$

$$(B^+)_S^{-1} = \begin{pmatrix} 1 & 0 & 0 & 0 \\ 0 & 1 & 0 & 0 \\ 0 & 0 & 1 & 0 \\ 0 & -M & -M & 1 \end{pmatrix}; b^+ = \begin{pmatrix} 40 \\ 30 \\ 60 \\ 0 \end{pmatrix}$$

Step 3: $$z_j - c_j = (0 - M - M\ 1)\, A^+ = \{(-7M + 2), (-4M+1), 0, M, M, 0, 0\}$$

Since $(z_1 - c_1)$ is the most negative, A_1^+ enters the basis

Step 4: $$(A_1^+)_N = (B^+)_S^{-1} (A_1^+)_S = \begin{pmatrix} 1 & 0 & 0 & 0 \\ 0 & 1 & 0 & 0 \\ 0 & 0 & 1 & 0 \\ 0 & -M & -M & 1 \end{pmatrix} \begin{pmatrix} 1 \\ 3 \\ 4 \\ 2 \end{pmatrix}$$

$$= \begin{pmatrix} 1 \\ 3 \\ 4 \\ -7M+2 \end{pmatrix}$$

$$X_B{}^+ = (B^+)_S^{-1}\, b^+ = (B^+)_S^{-1} \begin{pmatrix} 40 \\ 30 \\ 60 \\ 0 \end{pmatrix} = \begin{pmatrix} 40 \\ 30 \\ 60 \\ -90M \end{pmatrix}$$

Step 5: Initial RS table

$B_S{}^+$		$(B^+)_S{}^{-1}$			$(A_1{}^+)_N$	$X_B{}^+$	RR
$A_3{}^+$	1	0	0	0	1	40	40
$A_6{}^+$	0	1	0	0	3*	30	10✓
$A_7{}^+$	0	0	1	0	4	60	15
-	0	$-M$	$-M$	1	$-7M+2$	$-90M$	-

$\therefore$ $A_6{}^+$ leaves the basis (giving place to $A_1{}^+$)

Step 6:

$$(B^+)_N^{-1} = \begin{pmatrix} A_3{}^+ \\ A_6{}^+ \\ A_7{}^+ \\ - \end{pmatrix} \begin{pmatrix} 1 & -\frac{1}{3} & 0 & 0 \\ 0 & \frac{1}{3} & 0 & 0 \\ 0 & -\frac{4}{3} & 1 & 0 \\ 0 & \left(\frac{4M}{3} - \frac{2}{3}\right) & -M & 1 \end{pmatrix} \begin{matrix} 0 \\ 1 \\ 0 \\ 0 \end{matrix} \rightarrow (B^+)_S^{-1};$$

Iteration I

Step 3:

$$z_j - c_j = (B^+)_S^{-1} A^+ = \left(0, \frac{4M}{3}, \; -\frac{2}{3}, -M, \; 1\right) A^+$$

$$= \left\{\left(0, \; -\frac{5}{3}M + \; \frac{1}{3}, 0, \; -\frac{4M}{13} + \frac{2}{3}, M, \left(\frac{7}{3}M - \frac{2}{3}\right), 0\right)\right\}$$

Since $(z_2 - c_2)$ is the most negative, $A_2{}^+$ enters the basis

Step 4: $\quad (A_2{}^+)_N = (B^+)_S^{-1} (A_2{}^+)_S$

$$= \begin{pmatrix} 1 & -\frac{1}{3} & 0 & 0 \\ 0 & \frac{1}{3} & 0 & 0 \\ 0 & -\frac{4}{3} & 1 & 0 \\ 0 & \frac{4M}{3}-\frac{2}{3} & -M & 1 \end{pmatrix} \begin{pmatrix} 2 \\ 1 \\ 3 \\ 1 \end{pmatrix}$$

$$= \begin{pmatrix} \frac{5}{3} \\ \frac{1}{3} \\ \frac{5}{3} \\ -\frac{5}{3}M+\frac{1}{3} \end{pmatrix}$$

$$X_B{}^+ = (B^+)_S^{-1} b^+ = (B^+)_S^{-1} \begin{pmatrix} 40 \\ 30 \\ 60 \\ 0 \end{pmatrix} = \begin{pmatrix} 30 \\ 10 \\ 20 \\ -20M-20 \end{pmatrix}$$

Step 5: RS table after I iteration

$B_S{}^+$	$(B^+)_S{}^{-1}$				$(A_2{}^+)_N$	$X_B{}^+$	RR
$A_3{}^+$	1	$-\frac{1}{3}$	0	0	$\frac{5}{3}$	30	18
$A_1{}^+$	0	$\frac{1}{3}$	0	0	$\frac{1}{3}$	10	30
$A_7{}^+$	0	$-\frac{4}{3}$	1	0	$\frac{5}{3}$*	20	12✓
-	0	$\left(\frac{4}{3}M-\frac{2}{3}\right)$	−M	1	$-\frac{5}{3}M+\frac{1}{3}$	−20M −20	-

$\therefore$ $A_7{}^+$ leaves and is replaced by $A_2{}^+$

Step 6:

$$(B^+)_N^{-1} = \begin{pmatrix} A_3{}^+ \\ A_1{}^+ \\ A_2{}^+ \\ - \end{pmatrix} \begin{pmatrix} 1 & 1 & -1 & 0 \\ 0 & \frac{3}{5} & -\frac{1}{5} & 0 \\ 0 & -\frac{4}{5} & \frac{3}{5} & 0 \\ 0 & -\frac{2}{5} & -\frac{1}{5} & 1 \end{pmatrix} \begin{matrix} 0 \\ 0 \\ 1 \\ 0 \end{matrix} \rightarrow (B^+)_S^{-1}$$

Iteration II

Step 3:

$$z_j - c_j = (B^+)_S^{-1} A^+ = \left(0 \quad -\frac{2}{5} \quad -\frac{1}{5} \quad 1\right) A^+$$

$$= \left(0, \quad 0, \quad 0, \quad \frac{2}{5}, \quad \frac{1}{5}, \quad M - \frac{2}{5}, \quad M - \frac{1}{5}\right)$$

Since all the values of $(z_j - c_j)$ are non-negative, optimality has been reached.

Step 7: The optimal solution is given by.

$$\begin{pmatrix} A_3^+ \\ A_1^+ \\ A_2^+ \\ z^* \end{pmatrix} = \begin{pmatrix} 1 & 1 & -1 & 0 \\ 0 & \frac{3}{5} & -\frac{1}{5} & 0 \\ 0 & -\frac{4}{5} & \frac{3}{5} & 0 \\ 0 & -\frac{2}{5} & -\frac{1}{5} & 1 \end{pmatrix} \begin{pmatrix} 40 \\ 30 \\ 60 \\ 0 \end{pmatrix} = \begin{pmatrix} 10 \\ 6 \\ 12 \\ -24 \end{pmatrix}$$

viz., $x_1 = 6, x_2 = 12$ (and $x_3 = 10$), max $(z^*) = -24$ and min $(z) = 24$.

Example 7

Minimise $z = 2x_1 + 9x_2 + x_3$, subject to $x_1 + 4x_2 + 2x_3 \geq 5$, $3x_1 + x_2 + 2x_3 \geq 4$; $x_1, x_2, x_3 \geq 0$, using the revised Simplex method.

Solution

Step 1: Converting the problem to a maximisation problem and introducing the surplus variables x_4 and x_5 and artificial variables x_6 and x_7, the problem is re-stated as:

To maximise $z^* = -2x_1 - 9x_2 - x_3 + 0.x_4 + 0.x_5 - M.x_6 - M.x_7$, s.t. $x_1 + 4x_2 + 2x_3 - x_4 + x_6 = 5$, $3x_1 + x_2 + 2x_3 - x_5 + x_7 = 4$; $x_i \geq 0$, where min $(z) = -$ max (z^*).

viz., to maximise $z^* = CX$, s.t. $AX = b$ and $X \geq 0$, where

$$C = (-2, -9, -1, 0, 0, -M, -M);$$

$$A = \begin{pmatrix} 1 & 4 & 2 & -1 & 0 & 1 & 0 \\ 3 & 1 & 2 & 0 & -1 & 0 & 1 \end{pmatrix}; b = \begin{pmatrix} 5 \\ 4 \end{pmatrix}$$

Step 2:

$$A^+ = \begin{pmatrix} 1 & 4 & 2 & -1 & 0 & 1 & 0 \\ 3 & 1 & 2 & 0 & -1 & 0 & 1 \\ 2 & 9 & 1 & 0 & 0 & M & M \end{pmatrix};$$

$$(B^+)_S^{-1} = \begin{pmatrix} 1 & 0 & 0 \\ 0 & 1 & 0 \\ -M & -M & 1 \end{pmatrix}; b^+ = \begin{pmatrix} 5 \\ 4 \\ 0 \end{pmatrix}$$

Step 3: $$z_j - c_j = (-M, -M, 1)\, A^+ = (-4M + 2, -5M + 9, -4M + 1, M, M, 0, 0)$$

Since $(z_2 - c_2)$ is the most negative, A_2^+ enters the basis.

Step 4: $$\left(A_2^+\right)_N = \left(B^+\right)_S^{-1}\left(A_2^+\right)_S = \left(B^+\right)_S^{-1}\begin{pmatrix}4\\1\\9\end{pmatrix} = \begin{pmatrix}4\\1\\-5M+9\end{pmatrix};$$

$$X_B^+ = \left(B^+\right)_S^{-1} b^+ = \begin{pmatrix}5\\4\\-9M\end{pmatrix}$$

Step 5: Initial RS table.

B_S^+	$(B^+)_S^{-1}$			$(A_2^+)_N$	X_B^+	RR
A_6^+	1	0	0	4*	5	$\frac{5}{4}$ ✓
A_7^+	0	1	0	1	4	4
-	$-M$	$-M$	1	$-5M+9$	$-9M$	-

$\therefore$ A_6^+ is replaced by A_2^+.

Step 6:

$$\left(B^+\right)_N^{-1} = \begin{matrix}A_2^+\\A_7^+\\-\end{matrix}\begin{pmatrix}\frac{1}{4} & 0 & 0\\ -\frac{1}{4} & 1 & 0\\ \frac{M}{4}-\frac{9}{4} & -M & 1\end{pmatrix}\begin{matrix}1\\0\\0\end{matrix} \rightarrow \left(B^+\right)_S^{-1}$$

Iteration I

Step 3: $$z_j - c_j = \left\{\left(\frac{M}{4}-\frac{9}{4}\right), -M, 1\right\} A^+ = \left\{\left(-\frac{11}{4}M-\frac{1}{4}\right), 0, \left(-\frac{3M}{2}-\frac{7}{2}\right), \left(-\frac{M}{4}+\frac{9}{4}\right), M, \left(\frac{5M}{4}-\frac{9}{4}\right), 0\right\}$$

Since $(z_1 - c_1)$ is the most negative, A_1^+ enters the basis.

Step 4: $$\left(A_1^+\right)_N = \left(B^+\right)_S^{-1}\left(A_1^+\right)_S = \begin{pmatrix}\frac{1}{4} & 0 & 0\\ -\frac{1}{4} & 1 & 0\\ \frac{M}{4}-\frac{9}{4} & -M & 1\end{pmatrix}\begin{pmatrix}1\\3\\2\end{pmatrix} = \begin{pmatrix}\frac{1}{4}\\ \frac{11}{4}\\ -\frac{11}{4}M-\frac{1}{4}\end{pmatrix};$$

$$X_B^{+} = (B^{+})_S^{-1} b^{+} \begin{pmatrix} \frac{5}{4} \\ \frac{11}{4} \\ -\frac{11M}{4} - \frac{45}{4} \end{pmatrix}$$

Step 5: RS table after I iteration

B_S^{+}	$(B^{+})_S^{-1}$			$(A_1^{+})_N$	X_B^{+}	RR
A_2^{+}	$\frac{1}{4}$	0	0	$\frac{1}{4}$	$\frac{5}{4}$	5
A_7^{+}	$-\frac{1}{4}$	1	0	$\frac{11}{4}$*	$\frac{11}{4}$	1✓
-	$\frac{M}{4}-\frac{9}{4}$	$-M$	1	$-\frac{11}{4}M - \frac{1}{4}$	$-\frac{11}{4}M - \frac{45}{4}$	-

$\therefore$ A_7^{+} is replaced by A_1^{+}.

Step 6:

$$(B^{+})_N^{-1} = \begin{matrix} A_2^{+} \\ A_1^{+} \\ - \end{matrix} \begin{pmatrix} \frac{3}{11} & -\frac{1}{11} & 0 \\ -\frac{1}{11} & \frac{4}{11} & 0 \\ -\frac{25}{11} & \frac{1}{11} & 1 \end{pmatrix} \begin{matrix} 0 \\ 1 \\ 0 \end{matrix} \rightarrow (B^{+})_S^{-1}$$

Iteration II

Step 3:

$$z_j - c_j = \left(-\frac{25}{11} \quad \frac{1}{11} \quad 1\right) A^{+} = \left\{0,\ 0,\ -\frac{37}{11},\ \frac{25}{11},\ -\frac{1}{11},\ M-\frac{25}{11},\ M+\frac{1}{11}\right\}$$

Since $(z_3 - c_3)$ is the most negative, A_3^{+} enters the basis.

Step 4:

$$(A_3^{+})_N = (B^{+})_S^{-1} (A_3^{+})_S = \begin{pmatrix} \frac{3}{11} & -\frac{1}{11} & 0 \\ -\frac{1}{11} & \frac{4}{11} & 0 \\ -\frac{25}{11} & \frac{1}{11} & 1 \end{pmatrix} \begin{pmatrix} 2 \\ 2 \\ 1 \end{pmatrix}$$

$$= \begin{pmatrix} \frac{4}{11} \\ \frac{6}{11} \\ -\frac{37}{11} \end{pmatrix}; X_B^+ = (B^+)_S^{-1} b^+ = \begin{pmatrix} 1 \\ 1 \\ -11 \end{pmatrix}$$

Step 5: RS table after II iteration

B_S^+	$(B^+)_S^{-1}$			$(A_3^+)_N$	X_B^+	RR
A_2^+	$\frac{3}{11}$	$-\frac{1}{11}$	0	$\frac{4}{11}$	1	$\frac{11}{4}$
A_1^+	$-\frac{1}{11}$	$\frac{4}{11}$	0	$\frac{6}{11}$ *	1	$\frac{11}{6}$ ✓
	$-\frac{25}{11}$	$\frac{1}{11}$	1	$-\frac{37}{11}$	−1	-

$\therefore$ A_1^+ leaves the basis; it is replaced by A_3^+.

Step 6:

$$(B^+)_N^{-1} = \begin{matrix} A_2^+ \\ A_3^+ \\ - \end{matrix} \begin{pmatrix} \frac{1}{3} & -\frac{1}{3} & 0 \\ -\frac{1}{6} & \frac{2}{3} & 0 \\ -\frac{17}{6} & \frac{7}{3} & 1 \end{pmatrix} \begin{matrix} 0 \\ 1 \\ 0 \end{matrix} \rightarrow (B^+)_S^{-1}$$

Iteration III

Step 3:

$$z_j - c_j = \left(-\frac{17}{6}, \frac{7}{3}, 1\right) A^+ = \left(\frac{37}{6}, 0, 0, \frac{17}{6}, -\frac{7}{3}, M - \frac{17}{6}, M + \frac{7}{3}\right)$$

Since $z_5 - c_5$ is the most negative, A_5^+ enters the basis.

Step 4:

$$(A_5^+)_N = \begin{pmatrix} \frac{1}{3} & -\frac{1}{3} & 0 \\ -\frac{1}{6} & \frac{2}{3} & 0 \\ -\frac{7}{6} & \frac{7}{3} & 1 \end{pmatrix} \begin{pmatrix} 0 \\ -1 \\ 0 \end{pmatrix} = \begin{pmatrix} \frac{1}{3} \\ -\frac{2}{3} \\ -\frac{7}{3} \end{pmatrix};$$

$$X_B^+ = (B^+)_S^{-1} \begin{pmatrix} 5 \\ 4 \\ 0 \end{pmatrix} = \begin{pmatrix} \frac{1}{3} \\ \frac{11}{6} \\ -\frac{29}{6} \end{pmatrix}$$

Step 5: RS table after III iteration

B_S^+	$(B^+)_S^{-1}$			$(A_5^+)_N$	X_B^+	RR
A_2^+	$\frac{1}{3}$	$-\frac{1}{3}$	0	$\frac{1}{3}$*	$\frac{1}{3}$	1✓
A_3^+	$-\frac{1}{6}$	$\frac{2}{3}$	0	$-\frac{2}{3}$	$\frac{11}{6}$	-
-	$-\frac{17}{6}$	$\frac{7}{3}$	1	$-\frac{7}{3}$	$-\frac{29}{6}$	-

A_2^+ is replaced by A_5^+.

Step 6:

$$(B^+)_N^{-1} = \begin{matrix} A_5^+ \\ A_3^+ \\ - \end{matrix}\left(\begin{matrix} 1 & -1 & 0 \\ \frac{1}{2} & 0 & 0 \\ -\frac{1}{2} & 0 & 1 \end{matrix}\right)\begin{matrix} 1 \\ 0 \\ 0 \end{matrix} \rightarrow (B^+)_S^{-1}$$

Iteration IV

Step 3:

$$z_j - c_j = \left(-\frac{1}{2}, \ 0, \ 1\right)A^+ = \left(\frac{3}{2}, \ 7, \ 0, \ \frac{1}{2}, \ 0, \ M - \frac{1}{2}, \ M\right)$$

Since all the values of $(z_j - c_j)$ are non-negative, optimal solution is got.

Step 7: The optimal solution is given by $\begin{pmatrix} A_5^+ \\ A_3^+ \\ z* \end{pmatrix}\begin{pmatrix} 1 & -1 & 0 \\ \frac{1}{2} & 0 & 0 \\ -\frac{1}{2} & 0 & 1 \end{pmatrix}\begin{pmatrix} 5 \\ 4 \\ 0 \end{pmatrix} = \begin{pmatrix} 1 \\ \frac{5}{2} \\ -\frac{5}{2} \end{pmatrix}$

viz., the solution required is $x_1 = x_2 = 0, x_3 = \frac{5}{2}$ (and $x_4 = 0, x_5 = 1, x_6 = 0, x_7 = 0$) max $(z^*) = -\frac{5}{2}$ and min $(z) = \frac{5}{2}$.

Example 8

Minimise $z = x_1 + x_2 + x_3$, subject to the constraints $x_1 - 2x_2 \leq 3, x_1 - 3x_2 + 4x_3 = 5, 2x_2 - x_3 \geq 4; x_1, x_2, x_3 \geq 0$, using the revised Simplex method.

SOLUTION

Step 1: Converting the problem to a maximisation problem and introducing the slack variable x_4, the surplus variable x_5 and artificial variables x_6 and x_7, the problem is re-stated as:

To maximise $z^* = -x_1 - x_2 - x_3 + 0.x_4 + 0.x_5 - M.x_6 - M.x_7$ s.t. $x_1 - 2x_2 + x_4 = 3$, $x_1 - 3x_2 + 4x_3 + x_6 = 5$; $2x_2 - x_3 - x_5 + x_7 = 4$; $x_i \geq 0$, where min $(z) = -$ max (z^*).

viz., to maximise $z^* = CX$, s.t. $AX = b$ and $x \geq 0$, where

$$C = (-1, -1, -1, 0, 0, -M, -M);$$

$$A = \begin{pmatrix} 1 & -2 & 0 & 1 & 0 & 0 & 0 \\ 1 & -3 & 4 & 0 & 0 & 1 & 0 \\ 0 & 2 & -1 & 0 & -1 & 0 & 1 \end{pmatrix}; b = \begin{pmatrix} 3 \\ 5 \\ 4 \end{pmatrix}$$

Step 2:

$$A^+ = \begin{pmatrix} 1 & -2 & 0 & 1 & 0 & 0 & 0 \\ 1 & -3 & 4 & 0 & 0 & 1 & 0 \\ 0 & 2 & -1 & 0 & -1 & 0 & 1 \\ 1 & 1 & 1 & 0 & 0 & M & M \end{pmatrix};$$

$$(B^+)_S^{-1} = \left(\begin{array}{ccc|c} 1 & 0 & 0 & 0 \\ 0 & 1 & 0 & 0 \\ 0 & 0 & 1 & 0 \\ \hline 0 & -M & -M & 1 \end{array}\right); b^+ = \begin{pmatrix} 3 \\ 5 \\ 4 \\ 0 \end{pmatrix}$$

Step 3:

$$z_j - c_j \ (0, -M, -M, 1)\, A^+ = \{(-M+1), (M+1), (-3M+1), 0, M, 0, 0\}$$

As $(z_3 - c_3)$ is the most negative, A_3^+ enters the basis.

$$(A_3^+)_N = (B^+)_S^{-1} (A_3^+)_S$$

Step 4:

$$= \begin{pmatrix} 1 & 0 & 0 & 0 \\ 0 & 1 & 0 & 0 \\ 0 & 0 & 1 & 0 \\ 0 & -M & -M & 1 \end{pmatrix} \begin{pmatrix} 0 \\ 4 \\ -1 \\ 1 \end{pmatrix}$$

$$= \begin{pmatrix} 0 \\ 4 \\ -1 \\ -3M+1 \end{pmatrix};$$

$$X_B^+ = (B^+)_S^{-1}\, b^+$$

$$= \begin{pmatrix} 0 \\ 4 \\ -1 \\ -9M \end{pmatrix}$$

Step 5: Initial RS table

B_S^+	$(B^+)_S^{-1}$				$(A_3^+)_N$	X_B^+	RR
A_4^+	1	0	0	0	0	3	-
A_6^+	0	1	0	0	4*	5	$\frac{5}{4}$
A_7^+	0	0	1	0	−1	4	-
-	0	−M	−M	1	−3M+1	−9M	-

A_6^+ leaves the basis, giving place to A_3^+.

Step 6:
$$(B^+)_N^{-1} = \begin{matrix} A_4^+ \\ A_3^+ \\ A_7^+ \\ - \end{matrix}\begin{pmatrix} 1 & 0 & 0 & 0 \\ 0 & \frac{1}{4} & 0 & 0 \\ 0 & \frac{1}{4} & 1 & 0 \\ 0 & \left(-\frac{M}{4}-\frac{1}{4}\right) & -M & 1 \end{pmatrix}\begin{matrix} 0 \\ 1 \\ 0 \\ 0 \end{matrix} \to (B^+)_S^{-1}$$

Iteration I

Step 3:
$$z_j - c_j = \left\{0, \left(-\frac{M}{4}-\frac{1}{4}\right), -M, 1\right\} A^+ = \left\{\left(-\frac{M}{4}+\frac{3}{4}\right), \left(-\frac{5M}{4}+\frac{7}{4}\right), 0, 0, M, \left(\frac{3M}{4}-\frac{1}{4}\right)\right\},$$

Since $(z_2 - c_2)$ is the most negative, A_2^+ enters.

Step 4:
$$(A_2^+)_N = (B^+)_S^{-1}(A_2^+)_S$$

viz.,
$$\left(A_2^+\right)_N = \begin{pmatrix} 1 & 0 & 0 & 0 \\ 0 & \frac{1}{4} & 0 & 0 \\ 0 & \frac{1}{4} & 1 & 0 \\ 0 & \left(-\frac{M}{4}-\frac{1}{4}\right) & -M & 1 \end{pmatrix}\begin{pmatrix} -2 \\ -3 \\ 2 \\ 1 \end{pmatrix}$$

$$= \begin{pmatrix} -2 \\ -\frac{3}{4} \\ \frac{5}{4} \\ -\frac{5}{4}M+\frac{7}{4} \end{pmatrix}; X_B^+ = \left(B^+\right)_S^{-1}\begin{pmatrix} 3 \\ 5 \\ 4 \\ 0 \end{pmatrix}$$

$$= \begin{pmatrix} 3 \\ \frac{5}{4} \\ \frac{21}{4} \\ -\frac{21}{4}M - \frac{5}{4} \end{pmatrix}$$

Step 5: RS table after I iteration

B_S^+	$(B^+)_S^{-1}$				$(A_2^+)_N$	X_B^+	RR
A_4^+	1	0	0	0	-2	3	-
A_3^+	0	$\frac{1}{4}$	0	0	$-\frac{3}{4}$	$\frac{5}{4}$	-
A_7^+	0	$\frac{1}{4}$	1	0	$\frac{5}{4}$ *	$\frac{21}{4}$	$\frac{21}{5}$ ✓
-	0	$\left(\frac{-M}{4}-\frac{1}{4}\right)$	$-M$	1	$-\frac{5}{4}M+\frac{7}{4}$	$-\frac{21M}{4}-\frac{5}{4}$	-

$\therefore$ A_7^+ leaves the basis, giving place to A_2^+.

Step 6:

$$(B^+)_N^{-1} = \begin{array}{c} A_4^+ \\ A_3^+ \\ A_2^+ \\ - \end{array} \left| \begin{array}{cccc|c} 1 & \frac{2}{5} & \frac{8}{5} & 0 & 0 \\ 0 & \frac{2}{5} & \frac{3}{5} & 0 & 0 \\ 0 & \frac{1}{5} & \frac{4}{5} & 0 & 1 \\ 0 & -\frac{3}{5} & -\frac{7}{5} & 1 & 0 \end{array} \right| \rightarrow (B^+)_S^{-1}$$

Iteration II

Step 3:

$$z_j - c_j = \left(0, -\frac{3}{5}, -\frac{7}{5}, 1\right) A^+ = \left\{\frac{2}{5}, 0, 0, 0, \frac{7}{5}, M - \frac{3}{5}, M - \frac{7}{5}\right\}$$

Since all values of $(z_j - c_j)$ are non-negative, optimality is reached.

Step 7: The optimal solution is given by

$$\begin{pmatrix} A_4^+ \\ A_3^+ \\ A_2^+ \\ z^* \end{pmatrix} = \begin{pmatrix} 1 & \frac{2}{5} & \frac{8}{5} & 0 \\ 0 & \frac{2}{5} & \frac{3}{5} & 0 \\ 0 & \frac{1}{5} & \frac{4}{5} & 0 \\ 0 & -\frac{3}{5} & -\frac{7}{5} & 1 \end{pmatrix} \begin{pmatrix} 3 \\ 5 \\ 4 \\ 0 \end{pmatrix} = \begin{pmatrix} \frac{57}{5} \\ \frac{22}{5} \\ \frac{21}{5} \\ -\frac{43}{5} \end{pmatrix}$$

viz., the optimal solution is $x_1 = 0, x_2 = \frac{21}{5}, x_3 = \frac{22}{5}$

and $$\max (z^*) = -\frac{43}{5},$$

viz., $$\min (z) = \frac{43}{5}.$$

EXERCISE 3

Part A (Short answer questions)

1. What is the advantage of using the revised Simplex method instead of the usual Simplex method?
2. Compare briefly the usual and revised simplex methods.
3. If $B = I_m$ is the initial basis of the Simplex table, write down the formulas for B^+ and $(B^+)^{-1}$, where B^+ is the modified basis used in the revised Simplex procedure.
4. What is the optimality criterion used in the revised Simplex procedure?
5. When the optimality has been reached in the revised Simplex procedure, explain how to get the optimal solution and the optimal value of the maximised function.

Part B

Solve the following linear programing problems using the revised Simplex method:

6. Maximise $z = 2x_1 + 3x_2$, subject to $x_1 + 2x_2 \leq 10$, $3x_1 + x_2 \leq 6$; $x_1, x_2 \geq 0$.
7. Maximise $z = 3x_1 + 4x_2$, subject $x_1 + x_2 \leq 40$, $x_1 + 5x_2 \leq 180$; $x_1, x_2 \geq 0$.
8. Maximise $z = 5x_1 + 3x_2$, subject to $3x_1 + 5x_2 \leq 15$, $5x_1 + 2x_2 \leq 10$; $x_1, x_2 \geq 0$.
9. Maximise $z = 5x_1 + 7x_2$, subject to $x_1 + x_2 \leq 4$, $3x_1 + 8x_2 \leq 24$, $10x_1 + 7x_2 \leq 35$; $x_1, x_2 \geq 0$.
10. Maximise $z = 3x_1 + 5x_2$, subject to $x_1 + 2x_2 \leq 20$, $x_1 + x_2 \leq 15$, $x_2 \leq 6$; $x_1, x_2 \geq 0$.
11. Maximise $z = 3x_1 + 2x_2$, subject to $-2x_1 + x_2 \leq 1$, $x_1 \leq 2$, $x_1 + x_2 \leq 3$; $x_1, x_2 \geq 0$.
12. Maximise $z = 5x_1 + 12x_2 + 4x_3$, subject to $x_1 + 2x_2 + x_3 \leq 5$, $2x_1 - x_2 + 3x_3 \leq 2$; $x_1, x_2, x_3 \geq 0$.
13. Maximise $z = x_1 + x_2 + x_3$, subject to $2x_1 + x_2 + 2x_3 \leq 2$, $4x_1 + 2x_2 + x_3 \leq 2$; $x_1, x_2, x_3 \geq 0$.
14. Maximise $z = 2x_1 + 2x_2 + 3x_3$, subject to $x_1 + 2x_2 + 4x_3 \leq 8$, $2x_1 + 2x_2 + x_3 \leq 6$, $3x_1 + x_2 + x_3 \leq 8$; $x_1, x_2, x_3 \geq 0$.
15. Maximise $z = 3x_1 + 5x_2 + 4x_3$, subject to $2x_1 + 3x_2 \leq 8$, $2x_2 + 5x_3 \leq 10$, $3x_1 + 2x_2 + 4x_3 \leq 15$; $x_1, x_2, x_3 \geq 0$.
16. Minimise $z = 5x_1 + 6x_2$, subject to $2x_1 + 5x_2 \geq 15$, $3x_1 + x_2 \geq 12$; $x_1, x_2 \geq 0$.
17. Minimise $z = 2x_1 + x_2$, subject to $5x_1 + 2x_2 \geq 6$, $x_1 + x_2 \geq 2$; $x_1, x_2 \geq 0$.
18. Minimise $z = x_1 + x_2$, subject to $2x_1 + x_2 \geq 7$, $x_1 + 4x_2 \geq 6$; $x_1, x_2 \geq 0$.
19. Minimise $z = 4x_1 + 2x_2$, subject to $3x_1 + x_2 \geq 27$, $x_1 + x_2 \geq 21$, $x_1 + 2x_2 \geq 30$; $x_1, x_2 \geq 0$.

20. Minimise $z = x_1 + 2x_2$, subject to $6x_1 + x_2 \geq 18, x_1 + 4x_2 \geq 12, 2x_1 + x_2 \geq 10; x_1, x_2 \geq 0$.
21. Minimise $z = 3x_1 + 2x_2$, subject to $5x_1 + x_2 \geq 10, x_1 + x_2 \geq 6, x_1 + 4x_2 \geq 12; x_1, x_2 \geq 0$.
22. Minimise $z = 4x_1 + 2x_2 + 3x_3$, subject to $2x_1 + 4x_3 \geq 5, 2x_1 + 3x_2 + x_3 \geq 4; x_1, x_2, x_3 \geq 0$.
23. Minimise $z = 4x_1 + 6x_2 + 9x_3$, subject to $-x_1 + x_2 + x_3 \geq 2, 2x_1 + x_2 + 3x_3 \geq 2; x_1, x_2, x_3\ 0$.
24. Minimise $z = x_1 + 2x_2 + 3x_3$, subject to $x_1 - x_2 + x_3 \geq 4, x_1 + x_2 + 2x_3 \leq 8, x_2 - x_3 \geq 2; x_1, x_2, x_3 \geq 0$.
25. Minimise $z = 3x_1 + 6x_2 + x_3$, subject to $x_1 + x_2 + x_3 \geq 6, x_1 - 5x_2 - x_3 \geq 4, x_1 + 5x_2 + x_3 \geq 24; x_1, x_2, x_3 \geq 0$.

ANSWERS

Exercise 3

Part B

6. $x_1 = \frac{2}{5}, x_2 = \frac{24}{5}$; max $z = \frac{76}{5}$.
7. $x_1 = 5, x_2 = 35$; max $z = 155$.
8. $x_1 = \frac{20}{19}, x_2 = \frac{45}{19}$; max $z = \frac{235}{19}$.
9. $x_1 = 1.6, x_2 = 2.4$; max $z = 24.8$.
10. $x_1 = 10, x_2 = 5$; max $z = 55$.
11. $x_1 = 2, x_2 = 1$; max $z = 8$.
12. $x_1 = \frac{9}{5}, x_2 = \frac{8}{5}, x_3 = 0$; max $z = \frac{141}{5}$.
13. $x_1 = 0, x_2 = \frac{2}{3}, x_3 = \frac{2}{3}$; max $z = \frac{4}{3}$.
14. $x_1 = \frac{28}{13}, x_2 = \frac{2}{13}, x_3 = \frac{18}{13}$; max $z = \frac{114}{13}$.
15. $x_1 = \frac{89}{41}, x_2 = \frac{50}{41}, x_3 = \frac{62}{41}$; max $z = \frac{765}{41}$.
16. $x_1 = \frac{45}{13}, x_2 = \frac{21}{13}$; min $z = 27$.
17. $x_1 = \frac{2}{3}, x_2 = \frac{4}{3}$; min $z = \frac{8}{3}$.
18. $x_1 = \frac{22}{7}, x_2 = \frac{5}{7}$; min $z = \frac{27}{7}$.
19. $x_1 = 12, x_2 = 9$; min $z = 66$.
20. $x_1 = 4, x_2 = 2$; min $z = 8$.
21. $x_1 = 1, x_2 = 5$; min $z = 13$.
22. $x_1 = 0, x_2 = \frac{11}{12}, x_3 = \frac{5}{4}$; min $z = \frac{67}{12}$.
23. $x_1 = 0, x_2 = \frac{3}{2}, x_3 = \frac{1}{2}$; min $z = \frac{27}{2}$.
24. $x_1 = 6, x_2 = 2, x_3 = 0$; min $z = 10$.
25. $x_1 = 14, x_2 = 0, x_3 = 10$; min $z = 52$.

4 DUALITY AND DUAL SIMPLEX PROGRAMMING

4.0 DUALITY THEORY – INTRODUCTION

Corresponding to every linear programing problem, there exists another unique linear programming problem such that they involve the same data and possess the same optimal value for their objective functions. The latter is called the *dual*, while the original problem is called the *primal*. In fact, either problem can be considered as the primal and the other as the dual. The optimal Simplex table of either problem gives information about the optimal solutions of the other. If the given LP problem contains a large number of constraint conditions and less number of variables, it will be advantageous to solve the dual problem and deduce from this solution, the solutions of the original problem.

4.1 DEFINITION OF THE DUAL

If the given LP problem is assumed to be in the canonical form as

Maximise $Z = CX$, subject to $AX \le b$ and $X \ge 0$, where C is a row vector, X and b are column vectors and A is an $(m \times n)$ matrix,

Then the dual problem is defined as

Minimise $w = b^T Y$, subject to $A^T Y \le C^T$ and $Y \ge 0$.

On the other hand, if the given problem is:

Minimise $z = CX$, subject to $AX \ge b$ and $X \ge 0$, then the primal problem is:

Maximise $w = b^T Y$, subject to $A^T Y \le C^T$ and $Y \ge 0$.

The definition simply means that if we assume the primal problem in the practical non-vector form to be

Maximise (or minimise) $z = c_1 x_1 + c_2 x_2 + \ldots + c_n x_n$, such that

$$a_{11}x_1 + a_{12}x_2 + \cdots + a_{1n}x_n \le (\text{or} \ge)\, b_1$$
$$a_{21}x_1 + a_{22}x_2 + \cdots + a_{2n}x_n \le (\text{or} \ge)\, b_2$$
$$\ldots$$
$$a_{m1}x_1 + a_{m2}x_2 + \cdots + a_{mn}x_n \le (\text{ or } \ge) b_m$$

and $\quad x_1 x_2 \ldots x_n \ge 0,$

then the dual problem will be

Minimise (or maximise) $w = b_1 y_1 + b_2 y_2 + \ldots + b_m y_m$, such that

$$a_{11}y_1 + a_{21}y_2 + \ldots + a_{m1}y_m \geq (\text{or} \leq)\ c_1$$

$$a_{12}y_1 + a_{22}y_2 + \ldots + a_{m2}y_m \geq (\text{or} \leq)\ c_2$$

$$\ldots$$

$$a_{1n}y_1 + a_{2n}y_2 + \ldots + a_{mn}y_m \geq (\text{or} \leq)\ c_n$$

and $$y_1, y_2, \ldots, y_m \geq 0 .$$

Notes:

1. The primal and dual problems written in the manner given earlier are called *symmetric dual problems*.
2. The variables $x_1, x_2, \ldots, x_n$ used in the primal problem are called *primal variables* and the variables $y_1, y_2, \ldots, y_m$ used in the dual problem are called *dual variables*.
3. If the primal problem is one of maximisation, all the constraints should be of the $\leq$ type. The minimisation problem should be of the $\geq$ type.
4. If a constraint in a primal maximisation problem has a $\geq$, it is multiplied by -1 throughout so that the sign is changed to $\leq$.)
5. For getting the dual, the entries of b vector in the primal are needed not be positive. Some important axioms of the duality theory are stated now without proof.

4.2 AXIOMS OF DUALITY THEORY

1. Dual of the dual of a given primal problem is the primal problem itself.
2. If the primal problem has a finite optimal solution, then the dual problem also has a finite optimal solution with the same optimal value of the objective function.
3. If the primal problem has an unbounded solution, the dual problem has either no feasible solution or an unbounded solution.

Note: *Axioms* (2) and (3) are sometimes called *duality theorems.*

4. If the ith primal constraint of a maximisation (minimisation) LP problem is a strict equality, then the dual can be constructed treating it as an $\leq$ ($\geq$) inequality, but the ith dual variables will be unrestricted in sign while all other dual variables will be non-negative.
5. Conversely if the ith primal variable is unrestricted in sign, then the ith dual constraint will be a strict equality, while all other dual constraints will be of the $\leq$ or $\geq$ type according as the dual problem is a maximisation or minimisation problem.
6. If the optimal solution for the dual problem has been obtained by Simplex method, the optimal solution of the primal problem can be obtained by the following rule:

The optimal values of the primal variables are the entries in the net evaluation (index) row of the optimal Simplex table for the solution of the dual problem that occur under the initial basis vectors in order, subject to the conditions

(i) $\pm M$ in the entries are deleted and

(ii) the entry under the ith initial basis vector is changed in sign, if the ith dual constraint has been multiplied by -1 for the application of the Simplex method.

Dual Simplex Method

This is another method for solving LP problems. The regular Simplex method and the dual Simplex method are differentiated in the following table.

	Regular Simplex method	Dual Simplex method
1.	We start with a (basic) feasible solution that is not optimal. By the iterative procedure, we keep the solution feasible but move gradually towards optimality.	We start with a solution that satisfies the optimality criterion but is infeasible, viz., the values of some or all the basic variables are negative. By the iterative procedure, we keep the solution to satisfy the optimality criterion but move gradually towards feasibility.
2.	The vector that enters the basis is first determined and then the leaving vector is found out.	The vector that leaves the basis is first determined and then the entering vector is found out.
3.	The replacement ratios are obtained by dividing the elements of the X_B column by the corresponding elements of the entering vector that are strictly positive.	The replacement ratios are obtained by dividing the index row elements by the corresponding elements of the leaving vector that are strictly negative.
4.	The leaving vector corresponds to the minimum (positive) replacement ratio, irrespective of whether the problem is one of maximisation or minimisation.	The entering vector corresponds to the maximum of the negative replacement ratios for a maximisation problem and to the minimum of the positive replacement ratios for a minimisation problem.
5.	Artificial and surplus variables are used.	Use of artificial and surplus variables is not required since all the constraints are expressed as $\leq$ type constraints.

4.3 DUAL SIMPLEX ALGORITHM

Step 1: The constraints of the LP problem are converted to $\leq$ type constraints, irrespective of whether the problem is one of maximisation or minimisation.

Step 2: If all the b_is (the constants in the right-hand side of the constraints) are positive, the dual Simplex method cannot be applied. If not, the initial Simplex table is constructed after introducing the slack variables.

Step 3: If at least one index row entry is negative for maximisation problems or positive for minimisation problems, the dual Simplex method fails. If not, we proceed to Step 4.

Step 4: The most negative entry in the X_B column is selected and the basis vector corresponding to this entry is the vector to leave the basis.

Step 5: If all the row entries corresponding to this leaving vector are positive, it indicates that the LP problem has no feasible solution. If not, we proceed to Step 6.

Step 6: The replacement ratios are obtained by dividing the index row elements by the corresponding row elements of the leaving vector that are negative.

Step 7: The vector corresponding to the maximum of the negative replacement ratios for maximisation problems and the minimum of the positive replacement ratios for the minimisation problems is selected as the entering vector.

Step 8: The Simplex table is reversed as usual and iteration continued until the feasibility is achieved or the infeasibility is established.

WORKED EXAMPLES 4

Example 1

Solve the dual of the following LP problem by Simplex method and hence deduce the solution:

$$\text{Maximise } z = 2x_1 + x_2$$

Subject to $\quad x_1 + 2x_2 \leq 10, x_1 + x_2 \leq 6, x_1 - x_2 \leq 2, x_1 - 2x_2 \leq 1$ and $x_1, x_2 \geq 0$.

SOLUTION

Since the given problem is in the canonical form, the dual is written directly as:

$$\text{Minimise } w = 10y_1 + 6y_2 + 2y_3 + y_4,$$

Subject to
$$y_1 + y_2 + y_3 + y_4 \geq 2,$$
$$2y_1 + y_2 - y_3 - 2y_4 \geq 1$$

and
$$y_1, y_2, y_3, y_4 \geq 0.$$

Note: The number of primal variables has become the number of dual constraints and the number of primal constraints has become the number of dual variables. As it is easy to solve the dual problem, this approach is preferred whenever the number of primal constraints is more than the number of primal variables.

Introducing the surplus variables y_5 and y_6 and artificial variables y_7 and y_8, the standard form of the dual problem becomes:

$$\text{Minimise } w = 10y_1 + 6y_2 + 2y_3 + y_4 + 0 \cdot y_5 + 0 \cdot y_6 + M \cdot y_7 + M \cdot y_8$$

Subject to
$$y_1 + y_2 + y_3 + y_4 - y_5 + y_7 = 2$$
$$2y_1 + y_2 - y_3 - 2y_4 - y_6 + y_8 = 1$$

and
$$y_i \geq 0.$$

Now we proceed with the Simplex method.

		c_j:	10	6	2	1	0	0	M	M	
B	C_B	Y_B	A_1	A_2	A_3	A_4	A_5	A_6	A_7	A_8	R.R
A_7	M	2	1	1	1	1	−1	0	1	0	2
A_8	M	1	2	1	−1	−2	0	−1	0	1	$\frac{1}{2}$ →
$w = 3M$		$w_j - c_j$	$3M - 10$	$2M - 6$	−2	$-M - 1$	$-M$	$-M$	0	0	
A_7	M	$\frac{3}{2}$	0	$\frac{1}{2}$	$\frac{3}{2}$	2	−1	$\frac{1}{2}$	1	$-\frac{1}{2}$	$\frac{3}{4}$ →
A_1	10	$\frac{1}{2}$	1	$\frac{1}{2}$	$-\frac{1}{2}$	−1	0	$-\frac{1}{2}$	0	$\frac{1}{2}$	–
$w = \frac{3}{2}M + 5$		$w_j - c_j$	0	$\frac{M}{2} - 1$	$\frac{3M}{2} - 7$	$2M - 11$	$-M$	$\frac{M}{2} - 5$	0	$-\frac{M}{2} + 5$	

(Contd)

Table (*Contd*)

		c_j:	10	6	2	1	0	0	M	M	
A_4	1	$\frac{3}{4}$	0	$\frac{1}{4}$	$\frac{3}{4}$	1	$-\frac{1}{2}$	$\frac{1}{4}$	$\frac{1}{2}$	$-\frac{1}{4}$	3
A_1	10	$\frac{5}{4}$	1	$\frac{3}{4}$	$\frac{1}{4}$	0	$-\frac{1}{2}$	$-\frac{1}{4}$	$\frac{1}{2}$	$\frac{1}{4}$	$\frac{5}{3}$
$w=\frac{53}{4}$		$w_j - c_j$	0	$\frac{7}{4}$	$\frac{5}{4}$	0	$-\frac{11}{2}$	$-\frac{9}{4}$	$\frac{11}{2}-M$	$\frac{9}{4}-M$	
A_4	1	$\frac{1}{3}$	$-\frac{1}{3}$	0	$\frac{2}{3}$	1	$-\frac{1}{3}$	$\frac{1}{3}$	$\frac{1}{3}$	$-\frac{1}{3}$	$\frac{1}{2}$
A_2	6	$\frac{5}{3}$	$\frac{4}{3}$	1	$\frac{1}{3}$	0	$-\frac{2}{3}$	$-\frac{1}{3}$	$\frac{2}{3}$	$\frac{1}{3}$	5
$w=\frac{31}{3}$		$w_j - c_j$	$-\frac{7}{3}$	0	$\frac{2}{3}$	0	-3	$-\frac{5}{3}$	$\frac{13}{3}-M$	$\frac{5}{3}-M$	
A_3	2	$\frac{1}{2}$	$-\frac{1}{2}$	0	1	$\frac{3}{2}$	$-\frac{1}{2}$	$\frac{1}{2}$	$\frac{1}{2}$	$-\frac{1}{2}$	
A_2	6	$\frac{3}{2}$	$\frac{3}{2}$	1	0	$-\frac{1}{2}$	$-\frac{1}{2}$	$-\frac{1}{2}$	$\frac{1}{2}$	$\frac{1}{2}$	
$w = 10$		$w_j - c_j$	-2	0	0	-1	-4	-2	$4-M$	$2-M$	

From the optimal table, we get the optimal solution of the dual problem as
$y_1 = 0, y_2 = 6, y_3 = 2$ and $y_4 = 0$ and minimum value of $w = 10$.

The optimal solution of the original (primal) problem is obtained as follows:

In the solution of the dual problem, A_7 and A_8 were chosen as the initial basis vectors. The entries in the index row of the optimal table for the solution of the dual problem under A_7 and A_8 are $4 - M$ and $2 - M$.

Deleting $-M$, the optimal solutions of the primal problem is $x_1 = 4$ and $x_2 = 2$ and the maximum value of $z =$ the minimum value of $w = 10$.

Example 2

Find the solution of the following LP problem by solving its dual:

Minimise $z = 2x_1 + 2x_2$, subject to the constraints $2x_1 + 4x_2 \geq 1$, $x_1 + 2x_2 \geq 1$, $2x_1 + x_2 \geq 1$ and x_1, $x_2 \geq 0$.

Solution

Since the given problem is in the canonical form, the dual is written directly as:

$$\text{Maximise } w = y_1 + y_2 + y_3,$$

Subject to

$$2y_1 + y_2 + 2y_3 \leq 2$$

$$4y_1 + 2y_2 + y_3 \leq 2$$

and $\qquad y_1, y_2, y_3 \geq 0.$

Introducing the slack variables y_4 and y_5, the standard form of the dual problem becomes

$$\text{Maximise } w = y_1 + y_2 + y_3 + 0 \cdot y_4 + 0 \cdot y_5,$$

Subject to
$$2y_1 + y_2 + 2y_3 + y_4 = 2$$
$$4y_1 + 2y_2 + y_3 + y_5 = 2$$

and $\qquad y_1, y_2, \ldots, y_5 \geq 0.$

Solution of the problem by the Simplex procedure is given here:

		c_j:	1	1	1	0	0	
B	C_B	Y_B	A_1	A_2	A_3	A_4	A_5	R.R
A_4	0	2	2	1	2	1	0	1
A_5	0	2	4	2	1	0	1	$\frac{1}{2}$
$w = 0$		$w_j - c_j$	−1	−1	−1	0	0	
A_4	0	1	0	0	$\frac{3}{2}$	1	$-\frac{1}{2}$	$\frac{2}{3}$
A_1	1	$\frac{1}{2}$	1	$\frac{1}{2}$	$\frac{1}{4}$	0	$\frac{1}{4}$	2
$w = \frac{1}{2}$		$w_j - c_j$	0	$-\frac{1}{2}$	$-\frac{3}{4}$	0	$\frac{1}{4}$	
A_3	1	$\frac{2}{3}$	0	0	1	$\frac{2}{3}$	$-\frac{1}{3}$	–
A_1	1	$\frac{1}{3}$	1	$\frac{1}{2}$	0	$-\frac{1}{6}$	$\frac{1}{3}$	$\frac{2}{3}$
$w = 1$		$w_j - c_j$	0	$-\frac{1}{2}$	0	$\frac{1}{2}$	0	
A_3	1	$\frac{2}{3}$	0	0	1	$\frac{2}{3}$	$-\frac{1}{3}$	
A_2	1	$\frac{2}{3}$	2	1	0	$-\frac{1}{3}$	$\frac{2}{3}$	
$w = \frac{4}{3}$		$w_j - c_j$	1	0	0	$\frac{1}{3}$	$\frac{1}{3}$	

The optimal solution of the dual problem is $y_3 = \frac{2}{3}$, $y_2 = \frac{2}{3}$, $y_1 = 0$ and the maximum value of $w = \frac{4}{3}$. Since the entries of the index row of the optimal table under A_4 and A_5 (the initial basis vectors) are $\frac{1}{3}$ each, the optimal solution of the primal problem is $x_1 = \frac{1}{3}$, $x_2 = \frac{1}{3}$ and the minimum $z = \frac{4}{3}$.

Example 3

Find the solution of the following LP problem by solving its dual:

$$\text{Maximise } z = 5x_1 - 2x_2 + 3x_3,$$

Subject to

$$2x_1 + 2x_2 - x_3 \geq 2$$
$$3x_1 - 4x_2 \leq 3$$
$$x_2 + 3x_3 \leq 5$$

and

$$x_1, x_2, x_3 \geq 0$$

Solution

To reduce the problem to the canonical form, we multiply the first constraint by -1 and then write the dual as follows.

$$\text{Minimise } w = -2y_1 + 3y_2 + 5y_3,$$

Subject to

$$-2y_1 + 3y_2 \geq 5$$
$$-2y_1 - 4y_2 + y_3 \geq -2$$
$$y_1 + 3y_3 \geq 3$$

and

$$y_1, y_2, y_3 \geq 0.$$

In order to apply the Simplex method to solve the dual problem, the constant in the right-hand side of all the constraints should be positive. Hence, we multiply the second dual constraint by -1 and get $2y_1 + 4y_2 - y_3 \leq 2$.

Now we solve the dual problem by Simplex methods, after introducing the surplus variables y_4 and y_6, slack variable y_5 and artificial variables y_7 and y_8.

B	C_B	Y_B	A_1	A_2	A_3	A_4	A_5	A_6	A_7	A_8	RR
A_7	M	5	-2	3	0	-1	0	0	1	0	$\frac{5}{3}$
A_5	0	2	2	4	-1	0	1	0	0	0	$\frac{1}{2}$
A_8	M	3	1	0	3	0	0	-1	0	1	–
$w = 8M$		$w_j - c_j$	$-M+2$	$3M - 3$	$3M-5$	$-M$	0	$-M$	0	0	
A_7	M	$\frac{7}{2}$	$-\frac{7}{2}$	0	$\frac{3}{4}$	-1	$-\frac{3}{4}$	0	1	0	$\frac{14}{3}$
A_2	3	$\frac{1}{2}$	$\frac{1}{2}$	1	$-\frac{1}{4}$	0	$\frac{1}{4}$	0	0	0	–
A_8	M	3	1	0	3	0	0	-1	0	1	1
$w = \frac{13}{2}M + \frac{3}{2}$		$w_j - c_j$	$-\frac{5}{2}M + \frac{7}{2}$	0	$\frac{15}{4}M - \frac{23}{4}$	$-M$	$-\frac{3}{4}M + \frac{3}{4}$	$-M$	0	0	

(Contd)

Table (*Contd*)

B	C_B	Y_B	A_1	A_2	A_3	A_4	A_5	A_6	A_7	A_8	RR
A_7	M	$\frac{11}{4}$	$-\frac{15}{4}$	0	0	−1	$-\frac{3}{4}$	$\frac{1}{4}$	1	$-\frac{1}{4}$	11
A_2	3	$\frac{3}{4}$	$\frac{7}{12}$	1	0	0	$\frac{1}{4}$	$-\frac{1}{12}$	0	$\frac{1}{12}$	–
A_3	5	1	$\frac{1}{3}$	0	1	0	0	$-\frac{1}{3}$	0	$\frac{1}{3}$	–
$w=\frac{11}{4}M+\frac{29}{4}$		w_j-c_j	$-\frac{15}{4}M+\frac{65}{12}$	0	0	$-M$	$-\frac{3}{4}M+\frac{3}{4}$	$\frac{1}{4}M-\frac{23}{12}$	0	$-\frac{5}{4}M+\frac{23}{12}$	
A_6	0	11	−15	0	0	−4	−3	1	4	−1	
A_2	3	$\frac{5}{3}$	$-\frac{2}{3}$	1	0	$-\frac{1}{3}$	0	0	$\frac{1}{3}$	0	
A_3	5	$\frac{14}{3}$	$-\frac{14}{3}$	0	1	$-\frac{4}{3}$	−1	0	$\frac{4}{3}$	0	
$w=\frac{85}{5}$		w_j-c_j	$-\frac{76}{3}$	0	0	$-\frac{23}{3}$	−5	0	$\frac{23}{3}-M$	$-M$	

The optimal solution of the dual problem is $y_1 = 0,\ y_2 = \frac{5}{3},\ y_3 = \frac{14}{3}$ and the minimum value of $w = \frac{85}{3}$.

The entries of the index row of the optimal table under A_7, A_5, A_8 (viz., the order of the unit vectors in the initial basis) are $\frac{23}{3} - M$, -5 and $-M$.

We have multiplied the second dual constraint by -1 in order to use the Simplex procedure. Hence, we change the sign of the second entry (under A_5) and get 5. Moreover, deleting $-M$ from the first and third entries (under A_7 and A_8 respectively), we get $\frac{23}{3}$ and 0.

Hence, the optimal solution of the primal problem is $x_1 = \frac{23}{3}$, $x_2 = 5$ and $x_3 = 0$ and the maximum value of $z = \frac{85}{3}$.

Example 4

Find the solution of the following LP problem by solving its dual:

$$\text{Minimise } z = x_1 + 2x_2 + 3x_3,$$

Subject to

$$x_1 + x_2 + 2x_3 \leq 8$$

$$x_1 - x_2 + x_3 \geq 4$$

$$x_2 - x_3 \geq 2$$

and $$x_1, x_2, x_3 \geq 0.$$

SOLUTION

To reduce the given problem to the canonical form, we multiply the constraint by -1. Then we write the dual as follows.

$$\text{Maximise } w = -8y_1 + 4y_2 + 2y_3,$$

Subject to
$$-y_1 + y_2 \leq 1$$
$$-y_1 - y_2 + y_3 \leq 2$$
$$-2y_1 + y_2 - y_3 \leq 3$$

and $$y_1, y_2, y_3 \geq 0.$$

After introducing the slack variables y_4, y_5 and y_6, we proceed with the Simplex method to solve the dual problem.

	c_j:		−8	4	2	0	0	0	
B	C_B	Y_B	A_1	A_2	A_3	A_4	A_5	A_6	
A_4	0	1	−1	1	0	1	0	0	1
A_5	0	2	−1	−1	1	0	1	0	–
A_6	0	3	−2	1	−1	0	0	1	3
	$w = 0$	$w_j - c_j$	8	−4	−2	0	0	0	
A_2	4	1	−1	1	0	1	0	0	–
A_5	0	3	−2	0	1	1	1	0	3
A_6	0	2	−1	0	−1	−1	0	1	–
	$w = 4$	$w_j - c_j$	4	0	−2	4	0	0	
A_2	4	1	−1	1	0	1	0	0	
A_3	2	3	−2	0	1	1	1	0	
A_6	0	5	−3	0	0	0	1	1	
	$w = 10$	$w_j - c_j$	0	0	0	6	2	0	

The optimal solution of the dual problem is $y_1 = 0, y_2 = 1, y_3 = 3$ and the maximum value of $w = 10$.

The entries of the index row of the final table under the initial basis vectors A_4, A_5 and A_6 are 6, 2 and 0 respectively.

Hence, the optimal solution of the primal problem is $x_1 = 6$, $x_2 = 2$ and $x_3 = 0$ and the minimum value of $z = 10$.

Example 5

Find the solution of the following LP problem by solving its dual:

$$\text{Maximise } z = 6x_1 + 5x_2 + 3x_3,$$

Subject to $x_1 + x_2 + x_3 \le 5$

$$x_2 + x_3 \ge \frac{3}{2}$$

$$x_1 + x_2 - x_3 = -1$$

and $x_1, x_2, x_3 \ge 0.$

SOLUTION

The second constraint is multiplied by -1 to reduce the problem to the canonical form. Even though the third constraint is a strict equality, it is regarded as a $\le$ type inequality. The dual of the given problem is:

$$\text{Minimise } w = 5y_1 - 3y_2 - 1y_3,$$

Subject to $y_1 + y_3 \ge 6$

$$y_1 - 2y_2 + y_3 \ge 5$$

$$y_1 - 2y_2 - y_3 \ge 3$$

and $y_1, y_2 \ge 0$ and y_3 is unrestricted in sign

After putting $y_3 = y_3' - y_3''$, where $y_3', y_3'' \ge 0$ and introducing the surplus variables y_4, y_5, y_6 and artificial variables y_7, y_8, y_9, we solve the dual problem by the Simplex procedure.

		c_j:	5	−3	−1	1	0	0	0	M	M	M	
B	C_B	Y_B	A_1	A_2	A_3'	A_3''	A_4	A_5	A_6	A_7	A_8	A_9	R.R.
A_7	M	6	1	0	1	−1	−1	0	0	1	0	0	6
A_8	M	5	1	−2	1	−1	0	−1	0	0	1	0	5
A_9	M	3	1	−2	−1	1	0	0	−1	0	0	1	3
$w = 14M$	$w_j - c_j$	$3M - 5$	$-4M + 3$	$M + 1$	$-M - 1$	$-M$	$-M$	$-M$	0	0	0		
A_7	M	3	0	2	2	−2	−1	0	1	1	0	−1	$\frac{3}{2}$
A_8	M	2	0	0	2	−2	0	−1	1	0	1	−1	1
A_1	5	3	1	−2	−1	1	0	0	−1	0	0	1	–
$w = 5M + 15$	$w_j - c_j$	0	$2M - 7$	$4M - 4$	$-4M + 4$	$-M$	$-M$	$2M - 5$	0	0	$-3M + 5$		

(Contd)

Table (*Contd*)

		c_j:	5	−3	−1	1	0	0	0	*M*	*M*	*M*	
A_7	*M*	1	0	2	0	0	−1	1	0	1	−1	0	$\frac{1}{2}$
A_3'	−1	1	0	0	1	−1	0	$-\frac{1}{2}$	$\frac{1}{2}$	0	$\frac{1}{2}$	$-\frac{1}{2}$	–
A_1	5	4	1	−2	0	0	0	$-\frac{1}{2}$	$-\frac{1}{2}$	0	$\frac{1}{2}$	$\frac{1}{2}$	–
$w = M + 19$		$w_j - c_j$	0	$2M - 7$	0	0	$-M$	$M - 2$	−3	0	$-2M + 2$	$-M + 3$	
A_2	−3	$\frac{1}{2}$	0	1	0	0	$-\frac{1}{2}$	$\frac{1}{2}$	0	$\frac{1}{2}$	$-\frac{1}{2}$	0	1
A_3'	−1	1	0	0	1	−1	0	$-\frac{1}{2}$	$\frac{1}{2}$	0	$\frac{1}{2}$	$-\frac{1}{2}$	–
A_1	5	5	1	0	0	0	−1	$\frac{1}{2}$	$-\frac{1}{2}$	1	$-\frac{1}{2}$	$\frac{1}{2}$	10
$w = \frac{45}{2}$		$w_j - c_j$	0	0	0	0	$-\frac{7}{2}$	$\frac{3}{2}$	−3	$\frac{7}{2} - M$	$-\frac{3}{2} - M$	$3 - M$	
A_5	0	1	0	2	0	0	−1	1	0	1	−1	0	
A_3'	−1	$\frac{3}{2}$	0	1	1	−1	$-\frac{1}{2}$	0	$\frac{1}{2}$	$\frac{1}{2}$	0	$-\frac{1}{2}$	
A_1	5	$\frac{9}{2}$	1	−1	0	0	$-\frac{1}{2}$	0	$-\frac{1}{2}$	$\frac{1}{2}$	0	$\frac{1}{2}$	
$w = 21$		$w_j - c_j$	0	−3	0	0	−2	0	−3	$2 - M$	$-M$	$3 - M$	

The optimal solution of the dual problem is $y_1 = \frac{9}{2}$, $y_2 = 0$, $y_3 = y_3' - y_3'' = \frac{3}{2} - 0 = +\frac{3}{2}$ and the minimum value of $w = 21$.

The entries of the index row of the final table under the initial basis vectors A_7, A_8 and A_9 are $2 - M$, $-M$, $3 - M$.

Deleting $-M$, we get the optimal solution of the primal problem as $x_1 = 2$, $x_2 = 0$ and $x_3 = 3$ and the maximum value of $z = 21$.

Example 6

Find the solution of the following LP problem by solving its dual:

$$\text{Minimise } z = 4x_1 + x_2,$$

Subject to $3x_1 + x_2 = 3,\ 4x_1 + 3x_2 \geq 6,\ x_1 + 2x_2 \leq 3$ and $x_1, x_2 \geq 0$.

Solution

Converting the third constraint into a $\geq$ inequality and noting that the first constraint is an equality, we write down the dual problem as given here:

$$\text{Maximise } w = 3y_1 + 6y_2 - 3y_3,$$

Subject to $3y_1 + 4y_2 - y_3 \leq 4$

$$y_1 + 3y_2 - 2y_3 \leq 1$$

and $y_2, y_3 \geq 0$ and y_1 is unrestricted in sign.

The solution of the dual problem is given in the following Simplex tables:

		c_j:	3	−3	6	−3	0	0	
B	C_B	Y_B	A_1'	A_1''	A_2	A_3	A_4	A_5	R.R.
A_4	0	4	3	−3	4	−1	1	0	1
A_5	0	1	1	−1	3	−2	0	1	$\frac{1}{3}$
$w = 0$		$w_j - c_j$	−3	3	−6	3	0	0	
A_4	0	$\frac{8}{3}$	$\frac{5}{3}$	$-\frac{5}{3}$	0	$\frac{5}{3}$	1	$-\frac{4}{3}$	$\frac{8}{5}$
A_2	6	$\frac{1}{3}$	$\frac{1}{3}$	$-\frac{1}{3}$	1	$-\frac{2}{3}$	0	$\frac{1}{3}$	1
$w = 2$		$w_j - c_j$	−1	1	0	−1	0	2	
A_4	0	1	0	0	−5	5	1	−3	$\frac{1}{5}$
A_1'	3	1	1	−1	3	−2	0	1	–
$w = 3$		$w_j - c_j$	0	0	3	−3	0	3	
A_3	−3	$\frac{1}{5}$	0	0	−1	1	$\frac{1}{5}$	$-\frac{3}{5}$	
A_1'	3	$\frac{7}{5}$	1	−1	1	0	$\frac{2}{5}$	$-\frac{1}{5}$	
$w = \frac{18}{5}$		$w_j - c_j$	0	0	0	0	$\frac{3}{5}$	$\frac{6}{5}$	

The optimal solution of the dual problem is $y_1 = y_1' - y_1'' = \frac{7}{5}$, $y_2 = 0$, $y_3 = \frac{1}{5}$ and the maximum value of $w = \frac{18}{5}$.

The optimal solution of the primal problem is $x_1 = \frac{3}{5}$, $x_2 = \frac{6}{5}$ and the minimum value of $z = \frac{18}{5}$.

Example 7

Solve the following LP problem by solving its dual:

$$\text{Maximise } z = 5x_1 + 2x_2,$$

subject to $\quad x_1 \le 3, x_2 \le 4, 2x_1 + x_2 \le 8$ and $x_1 \ge 0$ and x_2 is unrestricted.

SOLUTION

The given primal problem is in the canonical form and so the dual is written directly. Since the second primal variable is unrestricted, the second dual constraint is a strict equality. The dual of the given problem is as follows.

$$\text{Minimise } w = 3y_1 + 4y_2 + 8y_3,$$

Subject to
$$y_1 + 2y_3 \ge 5$$
$$y_2 + y_3 = 2$$

and
$$y_1, y_2, y_3 \ge 0$$

The solution of the dual problem is given in the following simplex tables, where y_4 is a surplus variable and y_5, y_6 are artificial variables.

		c_j:	3	4	8	0	M	M	
B	C_B	Y_B	A_1	A_2	A_3	A_4	A_5	A_6	R.R.
A_5	M	5	1	0	2	–1	1	0	$\frac{5}{2}$
A_6	M	2	0	1	1	0	0	1	2
$w = 7M$		$w_j - c_j$	$M-3$	$M-4$	$3M-8$	$-M$	0	0	
A_5	M	1	1	–2	0	–1	1	–2	1
A_3	8	2	0	1	1	0	0	1	–
$w = M + 16$		$w_j - c_j$	$M-3$	$-2M+4$	0	$-M$	0	$-3M+8$	
A_1	3	1	1	–2	0	–1	1	–2	
A_3	8	2	0	1	1	0	0	1	
$w = 19$		$w_j - c_j$	0	–2	0	–3	$3-M$	$2-M$	

The optimal solution of the dual problem is $y_1 = 1, y_2 = 0, y_3 = 2$ and the minimum value of $w = 19$. The primal problem has the optimal solution $x_1 = 3, x_2 = 2$ and the maximum value of $z = 19$.

Note: Since the vectors corresponding to y_1 and y_2 form the initial basis, we could have proceeded without introducing artificial variables. However, by this approach we can only get the optimal solution of the dual problem; we cannot deduce that of the primal problem. As we are interested in the primal problem, we have to introduce the artificial variables.

Example 8

Solve the following LP problem by solving its dual:

$$\text{Minimise } z = 4x_1 + 3x_2 - 6x_3,$$

Subject to $x_1 - x_3 \geq 2$, $x_2 - x_3 \geq 5$ and $x_1, x_2 \geq 0$ and x_3 is unrestricted in sign.

SOLUTION

Noting that the given problem is in the canonical form and the third primal variable is unrestricted in sign, the dual is written as given here:

$$\text{Maximise } w = 2y_1 + 5y_2,$$

Subject to
$$y_1 \leq 4$$
$$y_2 \leq 3$$
$$y_1 + y_2 = 6$$

and
$$y_1, y_2, y_3 \geq 0.$$

After introducing the slack variables y_3, y_4 and artificial variables y_5, we proceed to solve the dual problem by the Simplex method. The solution is given in the following simplex tables.

		c_j:	2	5	0	0	$-M$	
B	C_B	Y_B	A_1	A_2	A_3	A_4	A_5	R.R.
A_3	0	4	1	0	1	0	0	–
A_4	0	3	0	1	0	1	0	3
A_5	$-M$	6	1	1	0	0	1	6
$w = -6M$		$w_j - c_j$	$-M-2$	$-M-5$	0	0	0	
A_3	0	4	1	0	1	0	0	4
A_2	5	3	0	1	0	1	0	–
A_5	$-M$	3	1	0	0	-1	1	3
$w = -3M + 15$		$w_j - c_j$	$-M-2$	0	0	$M+5$	0	
A_3	0	1	0	0	1	1	-1	
A_2	5	3	0	1	0	1	0	
A_1	2	3	1	0	0	-1	–	
$w = 21$		$w_j - c_j$	0	0	0	3	$M+2$	

The optimal solution of the dual problem is $y_1 = 3$, $y_2 = 3$ and the maximum value of $w = 21$.

The entries in the index row of the dual optimal table under the initial basis vectors A_3, A_4 and A_5 are 0, 3 and $M + 2$ respectively.

Since we have changed the sign throughout the third dual constraint for applying the Simplex method, the optimal solution of the primal problem is

$x_1 = 0, x_2 = 3$ and $x_3 = -2$ (after deleting $-M$) and the minimum value of $z = 21$.

Example 9

Solve the dual of the following LP problem by the Simplex method:

$$\text{Maximise } z = 3x_1 + 2x_2,$$

Subject to
$$x_1 - x_2 \leq 1$$
$$-x_1 - x_2 \leq -3$$

and
$$x_1, x_2 \geq 0.$$

SOLUTION

The dual of the given problem is

$$\text{Minimise } w = y_1 - 3y_2,$$

Subject to
$$y_1 - y_2 \geq 3$$
$$-y_1 - y_2 \geq 2$$

and
$$y_1, y_2 \geq 0$$

After introducing the surplus variables y_3, y_4 and artificial variables y_5, y_6, we proceed to solve the dual problem by the Simplex method. The solution is given in the following tables:

		c_j:	1	−3	0	0	M	M
B	C_B	Y_B	A_1	A_2	A_3	A_4	A_5	A_6
A_5	M	2	1	−1	−1	0	1	0
A_6	M	2	−1	−1	0	−1	0	1
w = 5M		$w_j - c_j$	−1	−2M + 3	−M	−M	0	0

Even though the optimality condition is satisfied, the artificial variables are present in the basis at the non-zero level. This indicates that the dual problem does not have a feasible solution. Hence, we may infer that the primal problem has an unbounded solution. This can be verified by the reader.

Example 10

Solve the dual of the following LP problem by the Simplex method.

$$\text{Minimise } z = 2x_1 - 3x_2,$$

Subject to
$$x_1 + 3x_2 \geq 1$$
$$x_1 - x_2 \geq 1$$

and
$$x_1, x_2 \geq 0.$$

SOLUTION

The dual of the given problem is

$$\text{Maximise } w = y_1 + y_2,$$

Subject to $$y_1 + y_2 \le 2$$

$$3y_1 - y_2 \le -3$$

and $$y_1, y_2 \ge 0.$$

The solution of the dual problem, obtained after introducing the slack variables y_3, the surplus variable y_4 and the artificial variable y_5, is given in the following Simplex tables.

		c_j:	1	1	0	0	$-M$	
B	C_B	Y_B	A_1	A_2	A_3	A_4	A_5	R.R.
A_3	0	2	1	1	1	-1	0	2
A_5	$-M$	3	-3	1	0	0	1	3
$w = -3M$		$w_j - c_j$	$3M - 1$	$-M - 1$	0	0	0	
A_2	1	2	1	1	1	-1	0	–
A_5	$-M$	1	-4	0	-1	1	1	1
$w = -M + 2$		$w_j - c_j$	$4M$	0	$M + 1$	$-M - 1$	0	
A_2	1	3	-3	1	0	0	1	
A_4	0	1	-4	0	-1	1	1	
$w = 3$		$w_j - c_j$	-4	0	0	0	$M + 1$	

From the final table, we observe that the dual problem has an unbounded solution. We can verify that the primal problem too has an unbounded solution.

Example 11

Use the dual Simplex method to solve the following LP problem.

$$\text{Maximise } z = -x_1 - x_2 - 2x_3,$$

Subject to $$3x_1 + 6x_2 + 4x_3 \le 8,$$

$$x_1 - 9x_2 + x_3 \le -3,$$

$$5x_1 - 3x_2 - 2x_3 \le -4,\ x_1, x_2, x_3 \ge 0.$$

Since all the constraints are of the $\le$ type, we can apply the dual Simplex method after introducing the slack variables x_4, x_5 and x_6. The dual Simplex tables leading to the solution are as follows.

		c_j:	-1	-1	-2	0	0	0
B	C_B	X_B	A_1	A_2	A_3	A_4	A_5	A_6
A_4	0	8	3	6	4	1	0	0
A_5	0	-3	1	-9	1	0	1	0
A_6	0	-4	5	-3	-2	0	0	1

(Contd)

Table (*Contd*)

		c_j:	−1	−1	−2	0	0	0
$z = 0$		$z_j - c_j$	1	1	2	0	0	0
		R.R	–	$-\frac{1}{3}$	−1	–	–	–
A_4	0	0	−13	0	0	1	0	2
A_5	0	9	−14	0	7	0	1	−3
A_2	−1	$\frac{4}{3}$	$-\frac{5}{3}$	1	$\frac{2}{3}$	0	0	$-\frac{1}{3}$
$z = \frac{-4}{3}$		$z_j - c_j$	$\frac{8}{3}$	0	$\frac{4}{3}$	0	0	$\frac{1}{3}$

Since all the index row elements are non-negative, the optimality continues to be satisfied. Moreover, in the final table, the entries in the X_B column are non-negative, indicating that the optimal solution is also (basic) feasible.

Hence, the required optimal solution is $x_1 = 0,\ x_2 = \frac{4}{3},\ x_3 = 0$ and the maximum value of $z = -\frac{4}{3}$.

Example 12

Solve the following LP problem by the dual Simplex method:

$$\text{Minimise } z = x_1 + 2x_2 + 3x_3,$$

subject to $x_1 - x_2 + x_3 \geq 4,$

$$x_1 + x_2 + 2x_3 \leq 8,$$

$$x_2 - x_3 \geq 2$$

and $x_1, x_2, x_3 \geq 0.$

Solution

Converting the first and third constraints as $\leq$ type constraints and introducing the slack variables x_4, x_5 and x_6, we proceed with the dual Simplex method. The solution is given in the following dual Simplex tables.

		c_j	1	2	3	0	0	0
B	C_B	X_B	A_1	A_2	A_3	A_4	A_5	A_6
A_4	0	−4	−1	1	−1	1	0	0
A_5	0	8	1	1	2	0	1	0
A_6	0	−2	0	−1	1	0	0	1
$z = 0$		$z_j - c_j$	−1	−2	−3	0	0	0
		R.R.	1	–	3	–	–	–

(*Contd*)

Table (*Contd*)

		C_j	1	2	3	0	0	0
A_1	1	4	1	−1	1	−1	0	0
A_5	0	4	0	2	1	1	1	0
A_6	0	−2	0	−1	1	0	0	1
$z = 4$		$z_j - c_j$	0	−3	−2	−1	0	0
		R.R.	–	3	–	–	–	–
A_1	1	6	1	0	0	−1	0	−1
A_5	0	0	0	0	3	1	1	2
A_2	2	2	0	1	−1	0	0	−1
$z = 10$		$z_j - c_j$	0	0	−5	−1	0	−3

The optimal solution is $x_1 = 6, x_2 = 2, x_3 = 0$ and the minimum value of $z = 10$.

Example 13

Solve the following LP problem by the dual Simplex method:

$$\text{Maximise } z = -3x_1 - 2x_2 - x_3 - 4x_4,$$

Subject to

$$2x_1 + 4x_2 + 5x_3 + x_4 \geq 10,$$
$$3x_1 - x_2 + 7x_3 - 2x_4 \geq 2,$$
$$5x_1 + 2x_2 + x_3 + 6x_4 \geq 15,$$

and

$$x_1, x_2, x_3, x_4 \geq 0.$$

SOLUTION

After converting the constraints into $\leq$ type constraints and introducing the slack variables x_5, x_6, x_7, we proceed with the dual Simplex method. The dual Simplex tables are given here:

		C_j	−3	−2	−1	−4	0	0	0
B	C_B	X_B	A_1	A_2	A_3	A_4	A_5	A_6	A_7
A_5	0	−10	−2	−4	−5	−1	1	0	0
A_6	0	−2	−3	1	−7	2	0	1	0
A_7	0	−15	−5	−2	−1	−6	0	0	1
$z = 0$		$z_j - c_j$	3	2	1	4	0	0	0
		R.R.	$-\frac{3}{5}$	−1	−1	$-\frac{2}{3}$	–	–	–

(*Contd*)

Table (*Contd*)

		C_j	−3	−2	−1	−4	0	0	0
A_5	0	−4	0	$-\frac{16}{5}$	$-\frac{23}{5}$	$\frac{7}{5}$	1	0	$-\frac{2}{5}$
A_6	0	7	0	$\frac{11}{5}$	$-\frac{32}{5}$	$\frac{28}{5}$	0	1	$-\frac{3}{5}$
A_1	−3	3	1	$\frac{2}{5}$	$\frac{1}{5}$	$\frac{6}{5}$	0	0	$-\frac{1}{5}$
$z=-9$		$z_j - c_j$	0	$\frac{4}{5}$	$\frac{2}{5}$	$\frac{2}{5}$	0	0	$\frac{3}{5}$
		R.R.	–	$-\frac{1}{4}$	$-\frac{2}{23}$	–	–	–	$-\frac{3}{2}$
A_3	−1	$\frac{20}{23}$	0	$\frac{16}{23}$	1	$-\frac{7}{23}$	$-\frac{5}{23}$	0	$\frac{2}{23}$
A_6	0	$\frac{289}{23}$	0	$\frac{153}{23}$	0	$\frac{84}{23}$	$-\frac{32}{23}$	1	$-\frac{1}{23}$
A_1	−3	$\frac{65}{23}$	1	$\frac{6}{23}$	0	$\frac{29}{23}$	$\frac{1}{23}$	0	$-\frac{5}{23}$
$z=-\frac{215}{23}$		$z_j - c_j$	0	$\frac{12}{23}$	0	$\frac{12}{23}$	$\frac{2}{23}$	0	$\frac{13}{25}$

The optimal solution is $x_1 = \frac{65}{23}$, $x_2 = 0$, $x_3 = \frac{20}{23}$, $x_4 = 0$ and the maximum value of $z = \frac{-215}{23}$.

Example 14

Solve the following LP problem by the dual Simplex method:

$$\text{Minimise } z = 3x_1 + 5x_2 + 6x_3 + 7x_4,$$

subject to

$$x_1 + x_2 + 2x_3 + 5x_4 \geq 8,$$
$$5x_1 - 6x_2 + x_4 \geq 10,$$
$$-3x_1 + 4x_2 + 5x_3 + 6x_4 \geq 12,$$

and

$$x_1, x_2, x_3, x_4 \geq 0.$$

Solution

After converting the constraints into $\leq$ type constraints and introducing the slack variables x_5, x_6, x_7, we proceed with the dual Simplex tables given here:

		C_j:	3	5	6	7	0	0	0
B	C_B	X_B	A_1	A_2	A_3	A_4	A_5	A_6	A_7
A_5	0	−8	−1	−1	−2	−5	1	0	0
A_6	0	−10	−5	6	0	−1	0	1	0
A_7	0	−12	3	−4	−5	−6	0	0	1
$z=0$		z_j-c_j	−3	−5	−6	−7	0	0	0
		R.R.	–	$\frac{5}{4}$	$\frac{6}{5}$	$\frac{7}{6}$	–	–	–
A_5	0	2	$-\frac{7}{2}$	$\frac{7}{3}$	$\frac{13}{6}$	0	1	0	$-\frac{5}{6}$
A_6	0	−8	$-\frac{11}{2}$	$\frac{20}{3}$	$\frac{5}{6}$	0	0	1	$-\frac{1}{6}$
A_4	7	2	$-\frac{1}{2}$	$\frac{2}{3}$	$\frac{5}{6}$	1	0	0	$-\frac{1}{6}$
$z=14$		z_j-c_j	$-\frac{13}{2}$	$-\frac{1}{3}$	$-\frac{1}{6}$	0	0	0	$-\frac{7}{6}$
		R.R.	$\frac{13}{11}$	–	–	–	–	–	7
A_5	0	$\frac{78}{11}$	0	$-\frac{21}{11}$	$\frac{18}{11}$	0	1	$-\frac{7}{11}$	$-\frac{8}{11}$
A_1	3	$\frac{16}{11}$	1	$-\frac{40}{33}$	$-\frac{5}{33}$	0	0	$-\frac{2}{11}$	$\frac{1}{33}$
A_4	7	$\frac{30}{11}$	0	$\frac{2}{33}$	$\frac{25}{33}$	1	0	$\frac{1}{11}$	$-\frac{5}{33}$
$z=\frac{258}{11}$		z_j-c_j	0	$-\frac{106}{33}$	$-\frac{38}{33}$	0	0	$-\frac{13}{11}$	$-\frac{32}{33}$

The optimal solution is $x_1=\frac{16}{11}, x_2=0, x_3=0$ and $x_4=\frac{30}{11}$ and the minimum value of $z=\frac{258}{11}$.

Example 15

Use the dual Simplex method to solve the following LP problem:

$$\text{Minimise } z = x_1 + x_2$$

Such that $\quad x_1 + 2x_2 \geq 2,$

$$-x_1 - x_2 \geq 1$$

and $$x_1, x_2 \geq 0.$$

Solution

After converting the constraints into $\leq$ type constrains and introducing the slack variables x_3, x_4, we proceed with the dual Simplex method. The solution is given in the following dual Simplex tables:

		c_j	1	1	0	0
B	C_B	X_B	A_1	A_2	A_3	A_4
A_3	0	−2	−1	−2	1	0
A_4	0	−1	1	1	0	1
	$z = 0$	$z_j - c_j$	−1	−1	0	0
		R.R.	1	$\frac{1}{2}$	–	–
A_2	1	1	$\frac{1}{2}$	1	$-\frac{1}{2}$	0
A_4	0	−2	$\frac{1}{2}$	0	$\frac{1}{2}$	1
	$z = 1$	$z_j - c_j$	$-\frac{1}{2}$	0	$-\frac{1}{2}$	0

Though the optimality condition is satisfied and A_4 is identified as the leaving vectors, all the entries in the row corresponding to A_4 are positive. This is an indication that the LP problem has no feasible solution.

EXERCISE 4

Part A (Short answer questions)

1. Define the dual of a given linear programming problem.
2. What do you mean by symmetric dual problems?
3. State duality theorems.
4. State the important postulates of the duality theory.
5. State the differences between regular Simplex and the dual Simplex method.
6. Give the steps of the dual Simplex algorithm.
7. Why are surplus and artificial variables not required in the dual Simplex method?
8. When does the dual simplex method fail in solving an LP problem?
9. How will you select the leaving vector and entering vector when solving an LP problem by the dual Simplex method?
10. How will you identify that an LP problem has no feasible solution, when it is solved by the dual Simplex method?

Write down the duals of the following LP problems:

11. Maximise $z = 8x_1 + 7x_2 + 6x_3$, subject to $7x_1 + 2x_2 + x_3 \leq 60$, $3x_1 + 2x_3 \leq 20$, $5x_1 + 4x_2 \leq 45$, $x_1 \leq 15$ and $x_1, x_2 \geq 0$ and x_3 is unrestricted in sign.
12. Minimise $z = 5x_1 + 2x_2$, subject to $6x_1 - 2x_2 + 2x_3 \geq 20$, $2x_1 + 4x_2 + x_3 \geq 40$ and x_1, x_2 are unrestricted in sign and $x_3 \geq 0$.
13. Minimise $z = -6x_1 - 8x_2 + 10x_3$, subject to $3x_1 + x_2 - x_3 = 5$, $2x_1 + 4x_2 + x_3 = 8$ and $x_1, x_2, x_3 \geq 0$.
14. Maximise $z = 2x_1 + x_2 + 13x_3 - 5x_4$, subject to $10x_1 - 8x_2 + 2x_3 + 4x_4 \leq 25$, $-x_1 + x_2 + x_3 x_4 = 15$, $6x_1 + 5x_2 - 7x_3 + 2x_4 \geq 18$ and $x_1, x_2, x_3, x_4 \geq 0$.
15. Minimise $z = x_1 + x_2 + x_3$, subject to $x_1 - 3x_2 + 4x_3 = 5$, $x_1 - 2x_2 \leq 3$, $2x_2 - x_3 \geq 4$ and $x_1, x_2 \geq 0$, x_3 is unrestricted in sign.

Part B

Solve the dual of each of the following LP problems by the Simplex method and hence deduce the solution of the given problem:

16. Maximise $z = 22x_1 + 30x_2 + 25x_3$, subject to $x_1 + x_2 \leq 5$, $2x_1 + x_2 + x_3 \leq 10$, $x_1 + 2x_2 + 2x_3 \leq 10$ and $x_1, x_2, x_3 \geq 0$.
17. Maximise $z = 7x_1 + 12x_2 + 16x_3$, subject to $2x_1 + x_2 + x_3 \leq 1$, $x_1 + 2x_2 + 4x_3 \leq 2$ and $x_1, x_2, x_3 \geq 0$.
18. Minimise $z = 10x_1 + 2x_2 - 12x_3$, subject to $x_1 - 2x_2 - 4x_3 \geq 2$, $x_1 + 3x_2 - x_3 \geq -1$ and $x_1, x_2, x_3 \geq 0$.
19. Minimise $z = 5x_1 + 2x_2 - 9x_3$, subject to $x_1 + x_2 - 3x_3 \geq 1$, $x_1 + 2x_2 + x_3 \geq -1$, $3x_1 - 4x_2 - x_3 \geq 2$ and $x_1, x_2, x_3 \geq 0$.
20. Maximise $z = x_1 + 2x_2 + x_3$, subject to $2x_1 + x_2 - x_3 \leq 2$, $-2x_1 + x_2 - 5x_3 \geq -6$, $4x_1 + x_2 + x_3 \leq 6$ and $x_1, x_2, x_3 \geq 0$.
21. Minimise $z = 6x_1 + 4x_2$, subject to $2x_1 + 3x_2 \leq 30$, $3x_1 + 2x_2 \leq 24$, $x_1 + x_2 \geq 3$ and $x_1, x_2, \geq 0$.
22. Minimise $z = x_1 + 2x_2$, subject to $2x_1 + x_2 \leq 40$, $x_1 + 3x_2 \geq 30$, $3x_1 + 4x_2 \geq 60$ and $x_1, x_2 \geq 0$.
23. Minimise $z = 3x_1 + 4x_2$, subject to $x_1 + 2x_2 \geq 10$, $2x_1 - 3x_2 \leq 6$, $x_1 + x_2 \geq 6$and $x_1, x_2 \geq 0$.
24. Minimise $z = x_1 + x_2 + x_3$, subject to $x_1 - 3x_2 + 4x_3 = 5$, $x_1 - 2x_2 \leq 3$, $2x_2 - x_3 \geq 4$ and $x_1, x_2, x_3 \geq 0$.
25. Minimise $z = 8x_1 + 10x_2 + 11x_3$, subject to $x_1 + x_2 + x_3 = 20$, $x_1 \leq 6$, $x_2 \geq 3$, $x_3 \geq 4$ and $x_1 \geq 0$.
26. Maximise $z = 6x_1 + 5x_2 + 3x_3$, subject to $x_1 + x_2 + x_3 \leq 50$, $x_1 + x_2 \geq 15$, $x_1 + x_2 - x_3 = -10$ and $x_1, x_2, x_3 \geq 0$.
27. Maximise $z = x_1 + 2x_2 + 3x_3 - x_4$, subject to $x_1 + 2x_2 + 3x_3 = 15$, $2x_1 + x_2 + 5x_3 = 20$, $x_1 + 2x_2 + x_3 + x_4 = 10$ and $x_1, x_2, x_3, x_4 \geq 0$
28. Maximise $z = 4x_1 + 3x_2 + 10x_3$, subject to $3x_1 + 2x_2 + 8x_3 \leq 5$, $3x_2 + 4x_3 \leq 3$, $x_1 + 2x_3 \leq 1$and x_2, $x_3 \geq 0$ and x_1 is unrestricted.
29. Minimise $z = 8x_1 - 10x_2$, subject to $8x_1 - 6x_2 \geq 150$ and $2x_1 - 9x_2 \geq 120$ and $x_1 \geq 0$, x_2 is unrestricted.
30. Maximise $z = 3x_1 + 2x_2$, subject to $2x_1 + x_2 \leq 2$, $3x_1 + 4x_2 \geq 12$ and $x_1, x_2 \geq 0$.
31. Minimise $z = 2x_1 - 12x_2$, subject to $2x_1 - 3x_2 \geq 3$, $x_1 - 4x_2 \geq 2$ and $x_1, x_2 \geq 0$.

Solve the following linear programing problems by the dual Simplex method:

32. Maximise $z=-2x_1 - 2x_2 - 4x_3$, subject to $2x_1 + 3x_2 + 5x_3 \geq 2$, $3x_1 + x_2 + 7x_3 \leq 3$, $x_1 + 4x_2 + 6x_3 \leq 5$ and $x_1, x_2, x_3 \geq 0$.
33. Miximise $z = -3x - 2x_1$, subject to $x_1 + x_2 \geq 1$, $x_1 + x_2 \leq 7$, $x_1 + 2x_2 \geq 10$, $x_2 \leq 3$ and $x_1, x_2 \geq 0$.
34. Maximise $z = -4x_1 - 6x_2 - 18x_3$, subject to $x_1 + 3x_3 \geq 3$, $x_2 + 2x_3 \geq 5$ and $x_1, x_2, x_3 \geq 0$.
35. Maximise $z = -3x_1 - x_2$, subject to $x_1 + x_2 \geq 1$, $2x_1 + 3x_2 \geq 2$, and $x_1, x_2 \geq 0$.
36. Minimise $z = 3x_1 + x_2 + 2x_3$, subject to $x_1 + x_2 - 2x_3 \geq 1$, $x_1 - 2x_2 + x_3 \geq -1$ and $x_1, x_2, x_3 \geq 0$.
37. Minimise $z = 4x_1 + 5x_2 + 10x_3$, subject to $-3x_1 + 2x_2 + 3x_3 \geq 3$, $2x_1 + 4x_3 \geq 10$ and $x_1, x_2, x_3 \geq 0$
38. Minimise $z = 4x_1 + 6x_2 + 3x_3$, subject to $3x_1 + 4x_2 + x_3 \geq 10$, $-2x_1 - 3x_2 + 2x_3 \leq -5$, $x_1 - 2x_2 - 3x_3 \leq -1$, $3x_1 + 2x_2 + 2x_3 \geq 5$ and $x_1, x_2, x_3 \geq 0$.
39. Minimise $z = 6x_1 + 7x_2 + 3x_3 + 5x_4$ subject to $5x_1 + 6x_2 - 3x_3 + 4x_4 \geq 12$, $x_2 + 5x_3 - 6x_4 \geq 10$, $2x_1 + 5x_2 + x_3 + x_4 \geq 8$ and $x_1, x_2, x_3, x_4 \geq 0$.
40. Maximise $z = -3x_1 - 2x_2$, subject to $2x_1 + x_2 \leq 2$, $3x_1 + 4x_2 \geq 12$ and $x_1, x_2 \geq 0$.

ANSWERS

Exercise 4

Part A

11. Minimise $w = 60y_1 + 20y_2 + 45y_3 + 15y_4$, s.t $7y_1+3y_2+5y_3+ y_4 \geq 8$, $2y_1 + 4y_3 \geq 7$, $y_1 + 2y_2 = 6$; $y_i \geq 0$ $(i = 1, 2, 3, 4)$.
12. Maximise $w = 20y_1 + 40y_2$, s.t $6y_1+2y_2 = 5$, $-2y_1+4y_2 = 2$, $2y_1+ y_2 \geq 0$.
13. Maximise $w = 5y_1 + 8y_2$, s.t $3y_1 + 2y_2 \leq -6$, $y_1 + 4y_2 \leq -8$, $- y_1 - y_2 \leq 10$; y_1, y_2 are unrestricted in sign.
14. Minimise $w = 25y_1 + 15y_2 - 18y_3$, s.t $10y_1 - y_2 - 6y_3 \geq 2$, $-8y_1 + y_2 - 5y_3 \geq 1$, $2y_1 + y_2 + 7y_3 \geq 13$, $4y_1 - y_2 - 2y_3 \geq -5$; $y_1, y_3 \geq 0$ and y_2 is unrestricted in sign.
15. Maximise $w = 5y_1 + 3y_2 - 4y_3$, s.t $y_1 + y_2 \leq 1$, $-3y_1 - 2y_2 - 2y_3 \leq 1$, $4y_1 + y_3 = 1$; y_1 is restricted; $y_2, y_3 \geq 0$.

Part B

16. $x_1 = \frac{10}{3}$, $x_2 = \frac{5}{3}$, $x_3 = \frac{5}{3}$; Max $z = 165 =$ min w; $y_1 = 5$, $y_2 = 3$, $y_3 = 11$.
17. $x_1 = 0$, $x_2 = 1$, $x_3 = 0$; Max $z = 12 =$ min w; $y_1 = 8$, $y_2 = 2$.
18. Min $z = 20 =$ max w; $x_1 = 2$, $x_2 = 0$, $x_3 = 0$; $y_1 = 10$, $y_2 = 0$.
19. Min $z = \frac{32}{7} =$ max w; $x_1 = \frac{6}{7}$, $x_2 = \frac{1}{7}$, $x_3 = 0$; $y_1 = \frac{26}{7}$, $y_2 = 0$, $y_3 = \frac{3}{7}$.
20. Max $z = 10 =$ min w; $x_1 = 0$, $x_2 = 4$, $x_3 = 2$; $y_1 = \frac{1}{2}$, $y_2 = 0$, $y_3 = \frac{3}{2}$.
21. Max $z = 48 =$ min w; $x_1 = 8$, $x_2 = 0$; $y_1 = 0$, $y_2 = 2$, $y_3 = 0$.
22. Min $z = 24 =$ max w; $x_1 = 12$, $x_2 = 6$; $y_1 = 10$, $y_2 = \frac{2}{5}$, $y_3 = \frac{1}{5}$.
23. Min $z = 22 =$ max w; $x_1 = 2$, $x_2 = 4$; $y_1 = 10$, $y_2 = 0$, $y_3 = 6$.

24. Min $z=\frac{43}{3}=\max w; x_1=0,\ x_2=\frac{21}{5},\ x_3=\frac{22}{5};\ y_1=\frac{3}{5}, y_2=0,\ y_3=\frac{7}{5}$.

25. Min $z=192=\max w; x_1=6, x_2=10, x_3=4; y_1=0, y_2=2, y_3=0, y_4=1$.

26. Max $z=210=\min w; x_1=20, x_2=0, x_3=30;\ y_1=\frac{9}{2}, y_2=0,\ y_3=\frac{3}{2}$.

27. Max $z=15=\min w;\ x_1=\frac{5}{2},\ x_2=\frac{5}{2},\ x_3=\frac{5}{2}, x_4=0; y_1=1, y_2=0, y_3=0$.

28. Max $z=7=\min w; x_1=1, x_2=1, x_3=0; y_1=0, y_2=1, y_3=4$.

29. Min $z=194=\max w;\ x_1=\frac{21}{2}, x_2=-11;\ y_1=\frac{13}{15},\ y_2=\frac{8}{15}$.

30. Dual has an unbounded solution. Hence, the principal problem has no feasible solutions.

31. Dual has an infeasible solution. Hence, the principal problem has unbounded solutions.

32. $x_1=0,\ x_2=\frac{2}{3}, x_3=0$; Max $z=\frac{-4}{3}$.

33. $x_1=4, x_2=3$; Max $z=-18$.

34. $x_1=0, x_2=3, x_3=1$; Max $z=-36$.

35. $x_1=0, x_2=1$; max $z=-1$.

36. $x_1=\frac{1}{3},\ x_2=\frac{2}{3}, x_3=0$; min $z=\frac{5}{3}$.

37. $x_1=1, x_2=0, x_3=2$; min $z=24$.

38. $x_1=\frac{8}{5},\ x_2=\frac{13}{10}, x_3=0, x_4=0$; min $z=\frac{71}{5}$.

39. $x_1=0,\ x_2=\frac{30}{11},\ x_3=\frac{16}{11}, x_4=0$; min $z=\frac{258}{11}$.

40. No feasible solutions.

5 DYNAMIC PROGRAMMING

5.0 INTRODUCTION

Dynamic programing is mathematical technique that involves a dynamic multi-stage decision process, viz., it is a procedure used for determining the combination of inter-related decisions taken at various stages of the problem for optimising the objective function. In other words, the dynamic programming technique decomposes the original problem in n variables into n sub-problems (segments), each sub-problem being a decision-making problem in one variable only. The solution to a DPP (dynamic programming problem) is obtained sequentially starting from one (initial) stage and moving to the next till the final stage is reached.

5.1 BELLMAN'S PRINCIPLE OF OPTIMALITY

The concept of dynamic programming is based on the principle of optimality due to Bellman which is stated as follows.

An optimal policy (set of decisions) is one that does not depend on the initial state and initial decision to enter a particular state and that subsequent decisions constitute an optimal policy for leaving that state.

Solution procedure for a multi-(finite) stage problem using the dynamic programming approach

1. A recurrence relation (functional equation) connecting the optimal decision function for the n-stage problem with the optimal decision function for the $(n - 1)$-stage sub-problem $(n = 1, 2, \ldots, n)$ is found out.
2. To solve the functional equation for determining the optimal policy, we first write the optimal decision function for the one-stage sub-problem and solve it. Then, we solve the optimal decision function for 2-stage, 3-stage, …, $(n - 1)$ stage and n stage problem.

Note: Generally, the solution of a recurrence relation involves two types of computations, according as the system is continuous or discrete. In the first case, the optimal decisions at each stage are obtained by the calculus method of optimisation. In the second case, a tabular computational scheme is followed. The number of rows in each table is equal to the number of corresponding feasible state values and the number of columns is equal to the number of possible decisions.

WORKED EXAMPLES 5

Example 1

Use dynamic programming to maximise $y = x_1x_2x_3$, subject to the constraint

$$x_1 + x_2 + x_3 = 6; x_1, x_2, x_3 \geq 0.$$

SOLUTION

Let $f_n(6)$ represent the maximum value of y when 6 is divided into n parts.

When $n = 1, f_1(6) = 6$, since 6 is divided into one part only, viz., stage is 1 and state in this stage is 6.

When $n = 2$, viz., when 6 is divided into two parts, let the two parts be z and $6 - z$.

$$f_2(6) = \max_{0\leq z\leq 6}\{z.(6-z)\}$$

$$\therefore \quad = \max_{0\leq z\leq 6}\{z. f_1(6-z)\}, \text{ since } f_1(6) = 6.$$

Now the maximum of $\phi(z) = z(6 - z)$ is given by $\phi'(z) = 6-2z = 0$, i.e., $z = 3$, since $\phi''(3) < 0$.

$$\therefore \quad f_2(6) = 3^2 \text{ or } \left(\frac{6}{2}\right)^2, \text{ viz., now stage is 2 and state is 3.}$$

When $n = 3$, viz., when 6 is divided into 3 parts, let the first part be z and the sum of the other two parts be $6 - z$.

$$\therefore \quad f_3(6) = \max_{0\leq z\leq 6}\left\{z\cdot\left(\frac{6-z}{2}\right)^2\right\}$$

$$= \max_{0\leq z\leq 6}\{z\cdot f_2(6-z)\}, \text{ since } f_2(6) = \left(\frac{6}{2}\right)^2$$

Now the maximum of $\chi(z) = \dfrac{z(6-z)^2}{4}$ is given by $\chi'(z) = (6 - z)^2 - 2z(6 - z) = 0$. viz., $z = 2$, since $\chi''(2) < 0$

$\therefore$ The required optimal policy is $\{x_1 = 2, x_2 = 2 \text{ and } x_3 = 2\}$ and the maximum value of $y = 2^3 = 8$.

Note: Generalising this problem, if a positive quantity C is to be divided into n parts so that the product of the n parts may be a maximum, the required recurrence relation will be $f_n(C) = \max_{0\leq z\leq C}\{z\, f_{n-1}(C-z)\}$ and the optimal sub-division will be $\left\{\frac{C}{n}, \frac{C}{n}, \frac{C}{n}, \ldots, \frac{C}{n}\right\}$.

Example 2

Use the principle of optimality to minimise $y = x_1 + x_2 + x_3 + \ldots + x_n$, when $x_1\, x_2\, x_3 \ldots x_n = p$ (a constant) and $x_1, x_2, x_3, \ldots, x_n \geq 0$.

SOLUTION

Let $f_n(p)$ denote the minimum sum attainable for $\sum x_n$, when p is factorised into n factors.

When $n = 1$, viz., if p is factorised into only one factor say z, then $z = p$ or

$$f_1(p) = p \tag{1}$$

When $n = 2$, viz., if p is factorised into two factors z and p/z, then

$$f_2(p) = \min_{0 \le z \le p}\left(z + \frac{p}{z}\right) = \min_{0 \le z \le p}\left\{z + f_1\left(\frac{p}{z}\right)\right\}, \text{ since } f_1(p) = p \tag{2}$$

When $n = 3$, viz., when p is factorised into 3 factors, the first of which is z and the product of the other two is $\frac{p}{z}$, then

$$f_3(p) = \min_{0 \le z \le p}\left\{z + f_2\left(\frac{p}{z}\right)\right\} \tag{3}$$

Proceeding in this way, the recurrence relation, in general, will be obtained as

$$f_j(p) = \min_{0 \le z \le p}\left\{z + f_{j-1}\left(\frac{p}{z}\right)\right\} \quad j = 1, 2, \ldots, n \tag{4}$$

Let us now solve equation (4), treating it as an equation in z.

$$f_1(p) = p = 1.p^{1/1} \tag{1)$'$}$$

From (2),

$$f_2(p) = \min_{0 \le z \le p}\left\{z + f_1\left(\frac{p}{z}\right)\right\} = \min_{0 \le z \le p}\left(z + \frac{p}{z}\right) = \min_{0 \le z \le p}\{\phi(z)\}, \text{ say}$$

Min of $\phi(z) = z + \frac{p}{z}$ is given by $\phi'(z) = 0$, i.e., $1 - \frac{p}{z^2} = 0$, i.e.,

when $z = \pm p^{1/2}$, since $f''(z) = \frac{2p}{z^3} > 0$, when $z = p^{1/2}$

$$\therefore \quad f_2(p) = \min\left\{z + \frac{p}{z}\right\} = 2p^{1/2} \tag{5}$$

From (3)

$$f_3(p) = \min_{0 \le z \le p}\left\{z + f_2\left(\frac{p}{z}\right)\right\}$$

$$= \min_{0 \le z \le p}\left\{z + 2\left(\frac{p}{z}\right)^{1/2}\right\}, \text{ from (5)}$$

$$= \min_{0 \le z \le p}\{\chi(z)\}, \text{ say}$$

Now min $\chi(z)$ is given by $\chi'(z) = 0$, i.e., $1 + \left(\frac{p}{z}\right)^{-1/2} \times \frac{-p}{z^2} = 0$

i.e., when $1 - \frac{z^{1/2}}{p^{1/2}} \cdot \frac{p}{z^2} = 0$, i.e., when $z = p^{1/3}$

Now $\chi''(z) = \frac{3}{2} \cdot \frac{p^{1/2}}{z^{5/2}} > 0$, when $z = p^{1/3}$.

$$\therefore \qquad f_3(p) = p^{1/3} + 2\left(p^{2/3}\right)^{1/2} = 3p^{1/3} \qquad (6)$$

Results in steps (1)′, (5) and (6) suggest that for the n-stage problem,

$$f_n(p) = np^{1/n} \qquad (7)$$

which can be established by induction as follows:
Let us assume that (7) is true for $n = m$.

Then, $$f_m(p) = mp^{1/m} \qquad (8)$$

Now $$f_{m+1}(p) = \min_{0 \le z \le p}\left\{z + f_m\left(\frac{p}{z}\right)\right\} \text{ by (4)}$$

$$= \min_{0 \le z \le p}\left\{z + m\left(\frac{p}{z}\right)^{1/m}\right\} \text{ by (8)}$$

$$= (m+1)p^{1/m+1}$$, in the same way in which (6) was derived from (5)

Thus, result (7) is true for $n = m + 1$.
Hence, the optimal (factorisation) policy is $(p^{1/n}, p^{1/n}, \ldots, p^{1/n})$ and the minimum value of $y = np^{1/n}$.

Example 3

Use Bellman's principle of optimality to find the maximum value of $y = a_1x_1 + a_2x_2 + a_3x_3 + \ldots + a_nx_n$ when $x_1 + x_2 + x_3 + \ldots + x_n = s$, where $s > 0$ and $0 < a_1 < a_2 < a_3 < \ldots a_n$ are constants and $x_i \geq 0$; $i = 1, 2, 3, \ldots n$.

Solution

Let $f_n(s)$ denote the maximum possible value of $\sum a_i x_i$, when s is divided into n parts.

When $n = 1$, i.e., when s is divided into one part, say z, then $z = s$ and

$$f_1(s) = \max(y) = a_1 s \qquad (1)$$

When $n = 2$, i.e., when s is divided into two parts, say $s - z$ and z,

$$f_2(s) = \max_{0 \le z \le s}\{a_1(s-z) + a_2 z\}$$

$$= \max_{0 \le z \le s}\{a_2 z + f_1(s-z)\} \qquad (2)$$

since $$f_1(s) = a_1 s$$

$$= \max_{0 \le z \le s}\{(a_2 - a_1)z + a_1 s\}$$

Since $a_2 - a_1 > 0$ and $a_1 s > 0$, $f_2(s)$ is maximum when $z = s$

$$f_2(s) = a_2 s \qquad (3)$$

The optimal division at this stage is $(0, s)$
When $n = 3$,

$$f_3(s) = \max_{0 \le z \le s}\{a_3 z + f_2(s-z)\}$$

$$= \max_{0 \le z \le s}\{a_3 z + a_2(s-z)\}, \text{ from (3)}$$

$$= \max_{0 \le z \le s} \{(a_3 - a_z)z + a_2 s\}$$

Since $a_3 - a_2 > 0$ and $a_2 s > 0$, $f_3(s)$ is maximum when $z = s$

$\therefore$
$$f_3(s) = a_3 s \tag{4}$$

Steps (1), (2), (3) and (4) suggest that for the n-stage problem,

$$f_n(s) = a_n s \tag{5}$$

Step (5) is verified by induction as follows:
Let (5) be true for $n = m$; i.e., $z = a_m(s)$

Now
$$f_{m+1}(s) = \max_{0 \le z \le s} \{(a_{m+1} - a_m)z + a_m s\}$$

$$= a_{m+1}s, \text{ as in previous steps}$$

i.e., step (5) is true for $n = m + 1$.

Hence, the optimal division policy is $\{0, 0, 0, \ldots, a_n s\}$ and the maximum value of $y = a_n s$.

Example 4

Use dynamic programming to find the minimum value of $y = x_1 \log x_1 + x_2 \log x_2 + x_3 \log x_3 + \ldots + x_n \log x_n$, subject to the constraint $x_1 + x_2 + x_3 + \ldots + x_n = 1$.

SOLUTION

Let $f_n(1)$ be the minimum possible value of $\sum_{i=1}^{n} x_i \log x_i$.

When $n = 1$, the sum is divided into one part only. Then,

$$x_1 = 1 \text{ and } f_1(1) = 1 \log 1, \tag{1}$$

which is true trivially.
When n = 2, the sum 1 is divided into two parts. Let them be z and $1 - z$.

$\therefore$
$$f_2(1) = \min_{0 \le z \le 1} \{z \log z + (1 - z)\log(1 - z)\} \tag{2}$$

$$= \min_{0 \le z \le 1} \{z \log z + f_1(1 - z)\}$$

When $n = 3$,

$$f_3(1) = \min_{0 \le z \le 1} \{z \log z + f_2(1 - z)\} \tag{3}$$

Proceeding like this, the recurrence equation for the n-stage problem,

$$f_n(1) = \min_{0 \le z \le 1} \{z \log z + f_{n-1}(1 - z)\}$$

To find $f_2(1)$, we minimise $\phi(z) = z \log z + (1 - z)\log(1 - z)$.
$\phi(z)$ is minimum, when $\phi'(z) = 0$.

i.e., when $\log z - \log(1 - z) = 0$ or $\frac{z}{1-z} = 1$ or $z = \frac{1}{2}$

$$\phi''(z) = \frac{1}{z} + \frac{1}{1-z}$$

$\therefore$
$$\phi''\left(\frac{1}{2}\right) = 4 > 0$$

$$\therefore \qquad f_2(1) = \frac{1}{2}\log\frac{1}{2} + \frac{1}{2}\log\frac{1}{2} \text{ or } 2\times\left(\frac{1}{2}\log\frac{1}{2}\right)$$

$\therefore$ Similarly, we can prove that $f_3(1) = 3\times\left(\frac{1}{3}\log\frac{1}{3}\right)$

From the values of $f_2(1)$ and $f_3(1)$, we can guess that

$f_n(1) = n\times\left(\frac{1}{n}\log\frac{1}{n}\right)$, which can be verified by induction.

$\therefore$ The optimal solution is $x_1 = x_2 = \ldots = x_n = \frac{1}{n}$ and the optimal value of $y = \log\left(\frac{1}{n}\right)$.

Example 5

By dynamic programming technique, find the minimum value of $y = x_1^2 + x_2^2 + x_3^2$, subject to the constraints $x_1 + x_2 + x_3 \geq 18$; $x_1, x_2, x_3 \geq 0$.

Solution

Taking x_1, x_2, x_3 as the policy or decision variables and assuming that $s_1 = x_1$, $s_2 = x_1 + x_2$ and $s_3 = x_1 + x_2 + x_3$, the problem becomes a three-stage problem with the constraints.

$$s_3 \geq 18,\ s_2 = x_1 + x_2 = s_3 - x_3 \text{ and } s_1 = x_1 = s_2 - x_2.$$

Let $f_i(s_i)$ be the minimum value of y at the ith stage. Then the functional equations for the stages 1, 2 and 3 are

$$f_1(s_1) = \min_{0\leq x_1\leq s_1}(x_1^2) \text{ or } \min_{0\leq x_1\leq s_1}(s_2 - x_2)^2 \qquad (1)$$

$$f_2(s_2) = \min_{0\leq x_2\leq s_2}\left(x_1^2 + x_2^2\right) \text{ or } \min_{0\leq x_2\leq s_2}\left\{f_1(s_1) + x_2^2\right\} \qquad (2)$$

and

$$f_3(s_3) = \min_{0\leq x_3\leq s_3}\left(x_1^2 + x_2^2 + x_3^2\right) \text{ or } \min_{0\leq x_3\leq s_3}\left\{f_2(s_2) + x_3^2\right\} \qquad (3)$$

To find $f_2(s_2)$, let us find the minimum of $u = f_1(s_1) + x_2^2$, i.e., $u = (s_2 - x_2)^2 + x_2^2$

Now
$$\frac{du}{dx_2} = -2(s_2 - x_2) + 2x_2 = 0,$$

when
$$x_2 = \frac{1}{2}s_2;\ \frac{d^2u}{dx^2} = 4 > 0.$$

$$\therefore \qquad f_2(s_2) = \left(\frac{1}{2}s_2\right)^2 + \left(\frac{1}{2}s_2\right)^2 = \frac{1}{2}s_2^2.$$

Similarly, we can get $f_3(s_3) = \frac{1}{3}s_3^2$. This occurs when $x_3 = \frac{1}{3}s_3$.

$\therefore$ Minimum value of $y = \frac{1}{3}s_3^2$.

i.e., since the minimum value of $s_3 = 18$, the minimum value of $y = \frac{1}{3}(18)^2 = 108$.

This occurs when $x_3 = \frac{1}{3}(18) = 6$, $x_2 = \frac{1}{2}s_2 = \frac{1}{2}(s_3 - x_3) = 6$ and $x_1 = 6$.

Example 6

Use dynamic programming technique, to find the maximum value of $y = -x_1^2 - 2x_2^2 + 3x_2 + x_3$, subject to the constraints $x_1 + x_2 + x_3 \le 1; x_1, x_2, x_3 \ge 0$.

SOLUTION

Taking x_1, x_2, x_3 as the decision variables, the problem can be considered as a three-stage problem with $s_3 = x_1 + x_2 + x_3 \le 1$ (for stage 3), $s_2 = x_1 + x_2 = s_3 - x_3$ (for stage 2) and $s_1 = x_1 = s_2 - x_2$ (for stage 1).

Let $f_i(s_i)$ be the maximum value of y at the ith stage; $i = 1, 2, 3$.

∴ The functional equations are

$$f_1(s_1) = \max_{0 \le x_1 \le s_1} \left(-x_1^2\right)$$

$$\max_{0 \le x_1 \le s_1} \left(-x_1^2\right) = 0; \text{ viz., } f_1(s_1) = 0 \qquad (1)$$

$$f_2(s_2) = \max_{0 \le x_2 \le s_2} \left\{-x_1^2 - 2x_2^2 + 3x_2\right\}$$

$$= \max_{0 \le x_2 \le s_2} \left\{f_1(s_1) - 2x_2^2 + 3x_2\right\} \text{ or } \max_{0 \le x_2 \le s_2} \left\{-2x_2^2 + 3x_2\right\}$$

By calculus, $f_2(s_2)$ occurs at $x_2 = \frac{3}{4}$.

Now two cases will arise, depending on the position of x_2.

If $$s_2 \ge \frac{3}{4}, f_2(s_2) = f_2(3/4) = \frac{9}{8}$$

If $s_2 < \frac{3}{4}, f_2(s_2) = -2s_2^2 + 3s_2$, as x_2 can take the largest possible value satisfying the condition $0 \le x_2 \le s_2$.

Case (i):

$$x_2 = \frac{3}{4}, s_2 \ge \frac{3}{4} \text{ and } f_2(s_2) = \frac{9}{8}$$

$$f_3(s_3) = \max_{0 \le x_3 \le s_3} \left\{f_2(s_2) + x_3\right\}$$

Now $$x_3 = s_3 - s_2$$

∴ x_3 is maximum, when s_3 is maximum (= 1) and s_2 is minimum (= 3/4).

∴ $$\text{Maximum of } x_3 = 1 - \frac{3}{4} = \frac{1}{4}$$

∴ $$\text{Max } f_3(s_3) = \frac{9}{8} + \frac{1}{4} = \frac{11}{8}$$

Case (ii):

$$x_2 = s_2, s_2 < \frac{3}{4} \text{ and } f_2(s_2) = -2s_2^2 + 3s_2$$

$$f_3(s_3) = \max_{0\le x_3\le s_3}\{f_2(s_2)+x_3\}$$

$$= \max_{0\le x_3\le s_3}\{(-2s_2^2+3s_2)+x_3\}$$

$$= \max_{0\le x_3\le s_3}\left\{-2(s_3-x_3)^2+3(s_3-x_3)+x_3\right\},\ \text{since } s_2 = s_3 - x_3$$

Let us consider $f(x_3) = -2(s_3 - x_3)^2 + 3(s_3 - x_3) + x_3$

$$f'(x_3) = 4(s_3 - x_3) - 3 + 1 = 0, \text{ when } s_3 - x_3 = \frac{1}{2} \text{ or } x_3 = s_3 - \frac{1}{2}$$

$$f''(x_3) = -4 < 0$$

$\therefore$ $f(x_3)$ is maximum when $x_3 = s_3 - \frac{1}{2}$

$\therefore$ Maximum of $f(x_3)$ occurs when $x_3 = 1 - \frac{1}{2} = \frac{1}{2}$, since the maximum of $s_3 = 1$.

$$\therefore \quad [f_3(s_3)]_{x_3=s_3-1/2} = [f_3(s_3)]_{s_3-x_3=1/2\ \&\ x_3=1/2} = -2\times\frac{1}{4}+3\times\frac{1}{2}+\frac{1}{2}=\frac{3}{2}$$

Comparing the values of $f_3(s_3)$ in the two cases, $\left(\text{since}\,\frac{3}{2} > \frac{11}{8}\right)$, $x_3^{(0)} = \frac{1}{2}$ and $f_3^{(0)}(s_3) = \frac{3}{2}$

Since $\quad x_2 = s_3 - x_3,\ x_2^{(0)} = 1 - \frac{1}{2} = \frac{1}{2}$

Since $\quad x_1 = s_2 - x_2,\ x_1^{(0)} = \frac{1}{2} - \frac{1}{2} = 0$

$\therefore$ The optimal policy is $\left\{x_1 = 0,\ x_2 = \frac{1}{2} \text{ and } x_3 = \frac{1}{2}\right\}$ and the optimal value of $y = \frac{3}{2}$.

Solution of a Linear Programming Problem as a Dynamic Programming Problem

Let us consider the following LPP with n decision variables and constraints:

$$\text{To maximise } Z = \sum_{j=1}^{n} c_j x_j,$$

subject to $\quad \sum_{j=1}^{n} a_{ij} x_j \le b_i; i = 1, 2, \ldots, m$ and $x_j \ge 0; j = 1, 2, \ldots, n$

This problem can be considered as an n-stage decision problem where the value of the decision variable x_j must be found at stage j. The RHS of the constraints, namely, $b_i (i = 1, 2, \ldots, m)$ can be treated as m type of resources to be allocated among the different kinds of activities x_j. Hence, when the value of the decision variable x_j is determined at the jth stage, $a_{1j}x_j$ units of resource 1, $a_{2j}x_j$ units of resource 2,…, $a_{mj}x_j$ units of resource m will be allocated to the jth activity, if sufficient unused resources exist. Thus, the amounts of the available resources must be determined before allocating them to any particular activity. In other words, the states of the system, viz., the amounts of resources remaining for allocation, must be known before making a decision about allocation at any stage of the n-stage system. In this problem, since there are m resources, we have m state variables or parameters constituting the state vector.

Let $B_{1j}, B_{2j}, \ldots, B_{mj}$ be the state vector at stage j, viz., the amounts of resources 1, 2, …, m respectively, available for allocation at the current and subsequent stages. Let $f_j(B_{1j}, B_{2j}, \ldots B_{mj})$ be the optimal value of the objective function for the stage j.

The optimal solution may be found by forward recursion procedure or backward recursion procedure. In the forward procedure, the computation starts at the first stage and advances towards the last stage, whereas in the backward procedure, the computation starts at the last stage and proceeds towards the first stage.

5.1.1 Recursive Equations

If forward recursion procedure is adopted, the recursive equations to be used are

$$f_1(B_{11}, B_{21}, \ldots, B_{m1}) = \max_{\substack{0 \le a_{i1}x_1 \le B_{i1} \\ [i=1,2,\ldots,m]}} (c_1 x_1) \qquad (1)$$

where B_{i1} is the resource available for allocation at stage 1 and hence equal to b_i.

$$f_j(B_{1j}, B_{2j}, \ldots B_{mj}) = \max_{\substack{0 \le a_{ij}x_j \le B_{ij} \\ [i=1,2,\ldots,m]}} \left[c_j x_j + f_{j-1}(B_{1j} - a_{1j}x_j, B_{2j} - a_{2j}x_j, \ldots, B_{mj} - a_{mj}x_j)\right]$$

where $\qquad j = 2, 3, \ldots, n \qquad (2)$

If backward recursion procedure is adopted, the recursive equations to be used are

$$f_n(B_{1n}, B_{2n}, \ldots B_{mn}) = \max_{\substack{0 \le a_{in}x_n \le B_{in} \\ [i=1,2,\ldots,m]}} (c_n x_n) \qquad (3)$$

$$f_j(B_{1j}, B_{2j}, \ldots B_{mj}) = \max_{\substack{0 \le a_{ij}x_j \le B_{ij} \\ [i=1,2,\ldots,m]}} \left[c_j x_j + f_{j+1}(B_{1j} - a_{1j}x_j, B_{2j} - a_{2j}x_j, \ldots B_{mj} - a_{mj}x_j)\right]$$

where $\qquad j = 1, 2, \ldots, (n-1)$, and $0 \le B_{ij} \le b_i$ for all i and j.

Example 7

Solve the following linear programming problem using dynamic programming algorithm with (i) forward recursive procedure and (ii) backward recursive procedure.

Maximise $Z = 4x_1 + 3x_2$, subject to constraints $2x_1 + x_2 \le 30$, $x_1 + 2x_2 \le 24$ and $x_1, x_2 \ge 0$.

Solution

In this problem, there are two resources (structural constraints) and two decision variables (x_1 and x_2).

$\therefore$ The states of the equivalent dynamic programming are B_{1j} and B_{2j} ($j = 1,2$)

Forward Recursive Procedure

Stage (i):

$$f_1(B_{11}, B_{21}) = 4 \max_{\substack{0 \le 2x_1 \le 30 \\ \& 0 \le x_1 \le 24}} (x_1) = 4 \max_{0 \le x_1 \le 15} (x_1)$$

Also

$$f_1(B_{11}, B_{21}) = 4 \times \min\left\{\frac{30 - x_2}{2}, \frac{24 - 2x_2}{1}\right\} \qquad (1)$$

Since $x_1 \geq 0$ (non-negativity constraint), we see that $\frac{30-x_2}{2} \geq 0$ and $\frac{24-2x_2}{1}$ or $x_2 \leq 30$ and $x_2 \leq 12$.

Maximum value of $x_1 = 15$ and that of $x_2 = 12$.

Stage (ii): $f_2(B_{12}, B_{22}) = \max[3x_2 + f_1(B_{11}, B_{21})]$

$$= \max\left[3x_2 + 4\min\left\{\frac{30-x_2}{2}, \frac{24-2x_2}{1}\right\}\right] \text{ from (1)}$$

$$= \max\left[3x_2 + 4\begin{cases}\frac{30-x_2}{2}, \text{if } 0 \leq x_2 \leq 6\\ \frac{24-2x_2}{1}, \text{if } 6 \leq x_2 \leq 16\end{cases}\right]$$

$$= \max\begin{bmatrix}x_2 + 60, \text{if } 0 \leq x_2 \leq 6\\ 96 - 5x_2, \text{if } 6 \leq x_2 \leq 12\end{bmatrix}$$

i.e., max $Z = f_2^{(0)}(B_{12}, B_{22}) = 66$, corresponding to $x_2^{(0)} = 6$.

When $$x_2^{(0)} = 6, x_1^{(0)} = \min\left\{\frac{30-x_2^{(0)}}{2}, \frac{24-2x_2^{(0)}}{1}\right\} = \min(12,12) = 12$$

$\therefore$ Z is maximum when $x_1 = 12$ and $x_2 = 6$ and the maximum value of Z = 66.

Backward Recursive Procedure

Stage (ii): $f_2(B_{12}, B_{22}) = 3 \max_{\substack{0 \leq x_2 \leq 30 \\ \& 0 \leq 2x_2 \leq 24}} (x_2) = 3 \max_{0 \leq x_2 \leq 12} (x_2)$

Also $$f_2(B_{12}, B_{22}) = 3 \times \min\left\{30 - 2x_1, \frac{24-x_1}{2}\right\} \quad (2)$$

Since $x_2 \geq 0$ (non-negativity constraint), we see that $30 - 2x_1 \geq 0$ and $\frac{24-x_1}{2}$ or $x_1 \leq 15$ and $x_1 \leq 24$

$\therefore$ Maximum value of $x_1 = 15$ and that of $x_2 = 12$.

Stage (i): $f_1(B_{11}, B_{12}) = \max[4x_1 + f_2(B_{12}, B_{22})]$

$$= \max\left[4x_1 + 3\min\left\{30 - 2x_1, \frac{24-x_1}{2}\right\}\right]$$

$$= \max\left[4x_1 + 3\begin{cases}\frac{24-x_1}{2}, \text{ if } 0 \leq x_1 \leq 12\\ 30 - 2x_1, \text{ if } 12 \leq x_1 \leq 15\end{cases}\right]$$

$$= \max\begin{bmatrix}\frac{5}{2}x_1 + 36, \text{if } 0 \leq x_1 \leq 12\\ 90 - 2x_1, \text{if } 12 \leq x_1 \leq 15\end{bmatrix}$$

i.e., $$\text{Max } Z = f_1^{(0)}(B_{11}, B_{12}) = 66, \text{ corresponding to } x_1^{(0)} = 12.$$

when $$x_1^{(0)} = 12, x_2^{(0)} = \min\left\{30 - 2x_1, \frac{24 - x_1}{2}\right\} = 6$$

$\therefore$ Z is maximum, when $x_1 = 12$ and $x_2 = 6$ and the maximum value of $Z = 66$.

Example 8

Solve the following linear programming problem using dynamic programming algorithm with (i) forward recursive procedure and (ii) backward recursive procedure.

Maximise $Z = 3x_1 - 2x_2$, subject to constraints $x_1 - x_2 \leq 2$, $x_1 - 2x_2 \leq 1$ and $x_1, x_2 \geq 0$.

SOLUTION

This problem is a two-resources and two-decision variable problem. Hence, the states of the equivalent dynamic programming problem are B_{ij}; $i = 1, 2$ and $j = 1, 2$.

Forward recursive procedure

Stage (i) or ($j = 1$):

$$f_1(B_{11}, B_{21}) = 3 \times \max(x_1) = 3 \max_{0 \leq x_1 \leq 3}(x_1)$$

$$= 3 \times \max(x_2 + 2, 2x_2 + 1), \text{ using constraints} \qquad (1)$$

Since $x_1 \geq 0$, by non-negativity constraint,

$$x_2 + 2 \geq 0 \text{ and } 2x_2 + 1 \geq 0$$

i.e., $x_2 \geq -2$ and $x_2 \geq \dfrac{-1}{2}$, which are true,

since $x_2 \geq 0$, $\max(x_1) = 3$ and $\min(x_2) = 0$.

Stage (ii) or ($j = 2$):

$$f_2(B_{12}, B_{22}) = \max[-2x_2 + f_1(B_{11}, B_{12})]$$

$$= \max[-2x_2 + 3\max(x_2 + 2, 2x_2 + 1)], \text{ from (1)}$$

$$= \max\left[-2x_2 + \begin{cases} 3(x_2 + 2), \text{ if } x_2 \leq 1 \\ 3(2x_2 + 1), \text{ if } x_2 \geq 1 \end{cases}\right]$$

Since $x_2 + 2 \geq 2x_2 + 1$, when $x_2 \leq 1$ and $2x_2 + 1 \geq x_2 + 2$, when $x_2 \geq 1$.

$$f_2(B_{12}, B_{22}) = \max\begin{cases} x_2 + 6, \text{ if } x_2 \leq 1 \\ 4x_2 + 3, \text{ if } x_2 \geq 1 \end{cases}$$

Note: $x_2 \geq 1, f_2(B_{12}, B_{22})$ becomes unbounded and so $x_2^{(0)} = 1$

$\therefore$ $$\max Z = f_2^{(0)}(B_{12}, B_{22}) = 7, \text{ corresponding to } x_2^{(0)} = 1.$$

when $$x_2^{(0)}=1,\ x_1^{(0)}=\max\left\{x_2^{(0)}+2,\ 2x_2^{(0)}+1\right\}=3$$

$\therefore$ Maximum of $Z = 7$ at $x_1 = 3$ and $x_2 = 1$.

Backward recursive procedure

Stage (ii) or ($j = s$):

$$f_2\left(B_{12}, B_{22}\right)=\max\left(-2x_2\right), \text{ s.t } x_2 \geq -2 \text{ and } x_2 \geq \frac{-1}{2}$$
$$=\max\left(-2x_2\right), \text{ s.t } x_2 \geq 0$$
$$=\min\left\{4-2x_1, 1-x_1\right\}$$

Since x_2 is a non-negativity constraint, $4 - 2x_1 \leq 0$ and $1 - x_1 \leq 0$

i.e., $x_1 \geq 2$ and $x_1 \geq 1$; i.e., $x_1 \geq 2$

$\therefore$ Minimum values of x_1 and x_2 are 2 and 0 respectively.

Stage (i) or ($j = 1$):

$$f_1\left(B_{11}, B_{12}\right)=\max\left[3x_1+f_2\left(B_{12}, B_{22}\right)\right]$$
$$=\max\left[3x_1+\min\left(4-2x_1, 1-x_1\right)\right]$$
$$=\max\left[3x_1+\begin{cases}4-2x_1, \text{ if } x_1 \geq 3\\ 1-x_1, \text{ if } 0 \leq x_1 \leq 3\end{cases}\right]$$
$$=\max\begin{cases}x_1+4, \text{ if } x_1 \geq 3\\ 2x_1+1, \text{ if } 0 \leq x_1 \leq 3\end{cases}$$

Note: $x_1 \geq 3, f_1(B_{11}, B_{21})$ becomes unbounded and so $x_1^{(0)}=3$.

$\therefore$ $$\text{Max } Z=f_1^{(0)}\left(B_{11}, B_{21}\right)=7, \text{ corresponding to } x_1^{(0)}=3.$$

When $$x_1^{(0)}=3,\ x_2^{(0)}=\max\left\{x_1^{(0)}-2, \frac{x_1^{(0)}-1}{2}\right\}=1$$

$\therefore$ Maximum of $Z = 7$ at $x_1 = 3$ and $x_2 = 1$.

Example 9

Solve the following linear programming problem using dynamic programming algorithm with forward recursive procedure.

Maximise $Z = 5x_1 + 7x_2$, subject to constraints $x_1 + x_2 \leq 4$, $3x_1 + 8x_2 \leq 24$, $10x_1 + 7x_2 \leq 35$ and x_1, $x_2 \geq 0$.

Solution

There are three resources ($i = 1, 2, 3$) and two decision variables ($j = 1, 2$) in this problem. Hence, the states of the equivalent dynamic programming problem are B_{ij}; $i = 1, 2, 3$ and $j = 1, 2$.

Stage (i); i.e., (j = 1):

$$f_1(B_{11}, B_{21}, B_{31}) = 5 \times \max(x_1), \text{ s.t } 0 \le x_1 \le 4, 0 \le 3x_1 \le 24, 0 \le 10x_1 \le 35$$

Also $$f_1(B_{11}, B_{21}, B_{31}) = 5 \times \min\left\{4 - x_2, \frac{24-8x_2}{3}, \frac{35-7x_2}{10}\right\}$$

Since $x_1 \ge 0, x_2 \le 4, x_2 \le 3$ and $x_2 \le 5$.

∴ Maximum of $x_1 = 3.5$ and that of $x_2 = 3$.

Stage (ii); i.e., (j = 2):

$$f_2(B_{12}, B_{22}, B_{32}) = \max\left[7x_2 + 5\min\left\{4 - x_2, \frac{24-8x_2}{3}, \frac{35-7x_2}{10}\right\}\right]$$

Let us compare the three quantities $4 - x_2$, $\frac{24-8x_2}{3}$ and $\frac{35-7x_2}{10}$.

$4 - x_2$ is the least of the three, if $4 - x_2 \le \frac{24-8x_2}{3}$ and $4 - x_2 \le \frac{35-7x_2}{10}$

viz., if $5x_2 \le 12$ and $3x_2 \ge 5$

viz., if $x_2 \le \frac{12}{5}$ and $x_2 \ge \frac{5}{3}$ or $\frac{5}{3} \le x_2 \le \frac{12}{5}$

$\frac{24-8x_2}{3}$ is the least of the three, if $\frac{24-8x_2}{3} \le 4 - x_2$ and $\frac{24-8x_2}{3} \le \frac{35-7x_2}{10}$

viz., if $5x_2 \le 12$ and $59x_2 \ge 135$

viz., if $x_2 \le \frac{12}{5}$ and $x_2 \ge \frac{135}{39}$ or $x_2 \ge \frac{12}{5}$

$\frac{35-7x_2}{10} \le 4 - x_2$ and $\frac{35-7x_2}{10} \le \frac{24-8x_2}{3}$

viz., if $3x_2 \le 5$ and $59x_2 \le 135$

viz., if $x_2 \le \frac{5}{3}$ and $x_2 \le \frac{135}{39}$ or $x_2 \le \frac{5}{3}$

$$f_2(B_{12}, B_{22}, B_{32}) = \max\left[\begin{array}{l} 7x_2 + 5\left(\frac{35-7x_2}{10}\right), \text{if } 0 \le x_2 \le \frac{5}{3} \\ 7x_2 + 5(4 - x_2), \text{if } \frac{5}{3} \le x_2 \le \frac{12}{5} \\ 7x_2 + 5\left(\frac{24-8x_2}{3}\right), \text{if } \frac{12}{5} \le x_2 \le 3 \end{array}\right.$$

$$= \max \begin{bmatrix} \frac{1}{2}(7x_2 + 35), \text{if } 0 \le x_2 \le \frac{5}{3} \\ 2x_2 + 20, \text{if } \frac{5}{3} \le x_2 \le \frac{12}{5} \\ \frac{1}{3}(120 - 19x_2), \text{if } \frac{12}{5} \le x_2 \le 3 \end{bmatrix}$$

$$= \max \begin{bmatrix} \frac{70}{3}, \text{at } x_2 = \frac{5}{3} \\ \frac{124}{5}, \text{at } x_2 = \frac{12}{5} \end{bmatrix}$$

$\therefore$ Max $Z = f_2^{(0)}(B_{12}, B_{22}, B_{32}) = \frac{124}{5}$, corresponding to $x_2^{(0)} = \frac{12}{5}$

When $x_2^{(0)} = \frac{12}{5}$, $x_1^{(0)} = \min\left\{4 - x_2^{(0)}, \frac{24 - 8x_2^{(0)}}{3}, \frac{35 - 7x_2^{(0)}}{10}\right\}$

$$= \min\left\{\frac{8}{5}, \frac{8}{5}, \frac{91}{10}\right\} = \frac{8}{5}$$

$\therefore$ Z is maximum when $x_1 = \frac{8}{5}$ and $x_2 = \frac{12}{5}$ and the maximum value of $Z = \frac{124}{5}$

Example 10

Solve the following linear programming problem using dynamic programming algorithm with backward recursive procedure.

Maximise $Z = 2x_1 + x_2$ subject to constraints $x_1 + 2x_2 \le 500$, $5x_1 + 2x_2 \le 1000$, $3x_1 + 2x_2 \le 900$ and $x_1, x_2 \ge 0$.

SOLUTION

The states of the equivalent dynamic programming problem are B_{ij}; $i = 1, 2, 3$ and $j = 1, 2$ as there are 3 resources and 2 decision variables.

Stage (ii); ($j = 2$):

$$f_2(B_{12}, B_{22}, B_{32}) = \max(x_2), \text{ s.t } 0 \le x_2 \le 250, 0 \le x_2 \le 500, \text{ and } 0 \le x_2 \le 450$$

Also $f_2(B_{12}, B_{22}, B_{32}) = \min\left\{\frac{500 - x_1}{2}, \frac{1000 - 5x_1}{2}, \frac{900 - 3x_1}{2}\right\}$

Since $x_2 \ge 0$, $x_1 \le 500$, $x_1 \le 200$ and $x_1 \le 300$.

$\therefore$ Maximum of $x_1 = 200$ and that of $x_2 = 250$.

Stage (i); ($j = 1$):

$$f_1(B_{11}, B_{21}, B_{31}) = \max\left[2x_1 + \min\left\{\frac{500 - x_1}{2}, \frac{1000 - 5x_1}{2}, \frac{900 - 3x_1}{2}\right\}\right]$$

Let us compare the three quantities $\frac{500-x_1}{2}, \frac{1000-5x_1}{2}$ and $\frac{900-3x_1}{2}$.

$\frac{500-x_1}{2}$ is the minimum of the three quantities, if $500 - x_1 \leq 1000 - 5x_1$ and $500 - x_1 \leq 900 - 3x_1$

i.e., $4x_1 \leq 500$ or $x_1 \leq 125$ and $2x_1 \leq 400$ or $x_1 \leq 200$

i.e., $0 \leq x_1 \leq 125$

$\frac{1000-5x_1}{2}$ is the minimum, if $1000 - 5x_1 \leq 500 - x_1$ and $1000 - 5x_1 \leq 900 - 3x_1$

i.e., $4x_1 \geq 500$ or $x_1 \geq 125$ and $2x_1 \geq 100$ or $x_1 \geq 50$

i.e., $x_1 \geq 125$

$\frac{900-3x_1}{2}$ is the minimum, if $900-3x_1 \leq 500-x_1$ and $900-3x_1 \leq 1000-5x_1$

i.e., $2x_1 \geq 400$ or $x_1 \geq 200$ and $2x_1 \leq 100$; $2x_1 \leq 100$ or $x_1 \leq 50$

$x_1 \leq 50$ and $x_1 \geq 200$ are inconsistent.

$$f_1(B_{11}, B_{21}, B_{31}) = \max \begin{bmatrix} 2x_1 + \frac{1}{2}(500 - x_1), \text{ if } 0 \leq x_1 \leq 125 \\ 2x_1 + \frac{1}{2}(1000 - 5x_1), \text{ if } x_1 \geq 125 \end{bmatrix}$$

$$= \max \begin{bmatrix} \frac{1}{2}(3x_1 + 500), \text{ if } 0 \leq x_1 \leq 125 \\ \frac{1}{2}(1000 - x_1), \text{ if } 125 \leq x_1 \leq 200 \end{bmatrix}$$

$$= \max \left(\frac{875}{2}, \frac{875}{2} \right) \text{ corresponding to } x_1 = 125$$

$\therefore$ $$f_1^{(0)}(B_{11}, B_{21}, B_{31}) = 437.5 \text{ at } x_1^{(0)} = 125$$

When $$x_1^{(0)} = 125,\ x_2^{(0)} = \min \left\{ \frac{500 - x_1^{(0)}}{2}, \frac{1000 - 5x_1^{(0)}}{2}, \frac{900 - 3x_1^{(0)}}{2} \right\}$$

$$= \min (187.5, 187.5, 262.5)$$

$$= 187.5$$

$\therefore$ max $(Z) = 437.5$, corresponding to $x_1 = 125$ and $x_2 = 187.5$.

Example 11

Max $Z = 3x_1 + 2x_2$, subject to constraints $2x_1 - x_2 \leq 4$, $x_1 + x_2 \leq 6$, $3x_1 + x_2 \leq 9$ and $x_1, x_2 \geq 0$, using dynamic programming algorithm with backward recursive procedure.

SOLUTION

Stage (ii); ($j = 2$):

$$f_2(B_{12}, B_{22}, B_{32}) = 2\max(x_2), \text{ s.t } x_2 \geq 0, 0 \leq x_2 \leq 6, \text{ and } 0 \leq x_2 \leq 9$$

Note: Since one of the coefficients in the constraint inequalities is negative, we find the bounds for x_1 and x_2 as follows:

$x_1 = \dfrac{10}{3}$ and $x_2 = \dfrac{8}{3}$ satisfy the inequalities $2x_1 - x_2 \leq 4$ and $x_1 + x_2 \leq 6$

$x_1 = \dfrac{3}{2}$ and $x_2 = \dfrac{9}{2}$ satisfy $x_1 + x_2 \leq 6$ and $3x_1 + x_2 \leq 9$

$x_1 = \dfrac{13}{5}$ and $x_2 = \dfrac{6}{5}$ satisfy $3x_1 + x_2 \leq 9$ and $2x_1 - x_2 \leq 4$

Of these, only $\left(x_1 = \dfrac{3}{2} \text{ and } x_2 = \dfrac{9}{2}\right)$ and $\left(x_1 = \dfrac{13}{5} \text{ and } x_2 = \dfrac{6}{5}\right)$ satisfy all the three inequalities.

The maximum of $\left(x_1 = \dfrac{3}{2} \text{ and } x_1 = \dfrac{13}{5}\right)$ is the upper bound for x_1; i.e., $x_1 \leq \dfrac{13}{5}$

Similarly, the upper bound for x_2 is $\dfrac{9}{2}$, i.e., $x_2 \leq \dfrac{9}{2}$

Also $f_2(B_{12}, B_{22}, B_{32}) = 2\min\{\infty, 6 - x_1, 9 - 3x_1\}$, from the given constraints.

Stage (i);($j = 1$):

$$f_1(B_{11}, B_{21}, B_{31}) = \max[3x_1 + 2 \times \min\{\infty, 6 - x_1, 9 - 3x_1\}]$$

$$= \max\left[3x_1 + 2 \times \begin{cases} 6 - x_1, \text{ if } x_1 \leq \dfrac{3}{2} \\ 9 - 3x_1, \text{ if } x_1 \geq \dfrac{3}{2} \end{cases}\right]$$

$$= \max\begin{bmatrix} x_1 + 12, \text{ if } x_1 \leq \dfrac{3}{2} \\ 18 - 3x_1, \text{ if } x_1 \geq \dfrac{3}{2} \end{bmatrix}$$

$$= \max\left(\frac{27}{2}, \frac{27}{2}\right) \text{ at } x_1^{(0)} = \frac{3}{2}$$

$\therefore$ $$\max z = f_1^{(0)}(B_{11}, B_{21}, B_{31}) = \frac{27}{2} \text{ at } x_1^{(0)} = \frac{3}{2}$$

When $$x_1^{(0)} = \frac{3}{2}, x_2^{(0)} = \min\left\{\infty, \frac{9}{2}, \frac{9}{2}\right\} = \frac{9}{2}$$

$\therefore$ The optimal solution is $x_1^{(0)} = \dfrac{3}{2}$ and $x_2^{(0)} = \dfrac{9}{2}$ and the maximum value of $Z = Z^{(0)} = \dfrac{27}{2}$.

Example 12

Minimise $Z = x_1 + 2x_2$, subject to constraints $6x_1 + x_2 \geq 18, x_1 + 4x_2 \geq 12, 2x_1 + x_2 \geq 10$ and $x_1, x_2 \geq 0$, using dynamic programming algorithm with backward recursive procedure.

SOLUTION

Stage (ii); (j = 2):

$$f_2(B_{12}, B_{22}, B_{32}) = 2\min(x_2), \text{ s.t } x_2 \geq 18, x_2 \geq 3, \text{ and } x_2 \geq 10$$

Better lower bounds for x_1 and x_2 are found out as in worked EXAMPLE 11. They are found from ($x_1 \geq 4$ and $x_1 \geq 2$) and from ($x_2 \geq 2$ and $x_2 \geq 6$). Thus, the lower bounds are given by $x_1 \geq 2$ and $x_2 \geq 2$.

Also $$f_2(B_{12}, B_{22}, B_{32}) = 2\max\left\{18 - 6x_1, \frac{1}{4}(12 - x_1), 10 - 2x_1\right\}$$

Now $18 - 6x_1 \geq \frac{1}{4}(12 - x_1)$ if $x_1 \leq \frac{60}{23}$ and $18 - 6x_1 \geq 10 - 2x_1$, $18 - 6x_1 \geq 10 - 2x_1$, if $x_1 \leq 2$

$\therefore$ $18 - 6x_1$ is the maximum of the three quantities, if $x_1 \leq 2$

$$\frac{1}{4}(12 - x_1) \geq 18 - 6x_1,$$

if $$x_1 \geq \frac{60}{23}$$

and $$\frac{1}{4}(12 - x_1) \geq 10 - 2x_1, \text{ if } x_1 \geq 4$$

$\therefore$ $\frac{1}{4}(12 - x_1)$ is the maximum of the three, if $x_1 \geq 4$

$10 - 2x_1 \geq 18 - 6x_1$, if $x_1 \geq 2$ and $10 - 2x_1 \geq \frac{1}{4}(12 - x_1)$, if $x_1 \leq 4$

$10 - 2x_1$ is the maximum of three, if $2 \leq x_1 \leq 4$

Stage (i); (j = 1):

$$f_1(B_{11}, B_{21}, B_{31}) = \min\left\{x_1 + 2\max\left[18 - 6x_1, \frac{1}{4}(12 - x_1), 10 - 2x_1\right]\right\}$$

$$= \min\left\{x_1 + 2\begin{bmatrix} 18 - 6x_1, \text{ if } x_1 \leq 2 \\ 10 - 2x_1, \text{ if } 2 \leq x_1 \leq 4 \\ \frac{1}{4}(12 - x_1), \text{ if } x_1 \geq 4 \end{bmatrix}\right\}$$

$$= \min\begin{cases} 36 - 11x_1, \text{ if } x_1 \leq 2 \\ 20 - 3x_1, \text{ if } 2 \leq x_1 \leq 4 \\ \frac{1}{2}(x_1 + 12), \text{ if } x_1 \geq 4 \end{cases}$$

$$= \min(14, 8, 8)$$

$$f_1^{(0)}(B_{11}, B_{21}, B_{31}) = 8 \text{, which corresponds to } x_1^{(0)} = 4$$

When $x_1^{(0)} = 4$, $x_2^{(0)} = \max\left[18 - 6x_1^{(0)}, \frac{1}{4}\left(12 - x_1^{(0)}\right), 10 - 2x_1^{(0)}\right]$

$$= \max[-6, 2, 2] = 2$$

$\therefore$ min (Z) = 8, corresponding to $x_1^{(0)} = 4$ and $x_2^{(0)} = 2$

EXERCISE 5

Part A (Short answer questions)

1. Explain briefly what you mean by dynamic programming.
2. State Bellman's principle of optimality.
3. Explain the solution procedure of a multi-stage problem using the dynamic programming approach.
4. Explain briefly how a linear programming problem can be considered as a dynamic programming problem.
5. State the recursive equations used when the forward recursive procedure is adopted to solve an LPP.
6. State the recursive equations used when an LPP is solved using the backward recursive procedure.

Part B

7. Use the dynamic programming method to maximise $Z = x_1x_2x_3$, subject to $x_1 + x_2 + x_3 = 5$; x_1, x_2, $x_3 \geq 0$.
8. A positive quantity C is to be divided into n parts in such a way that the product of the n parts is to be a maximum. Obtain the optimal division using dynamic programming.
9. Use Bellman's principle of optimality to find the minimum of $Z = x_1 + x_2 + x_3$ subject to the constraint $x_1x_2x_3 = 64$; $x_i \geq 0$.
10. Use dynamic programming method to find the maximum of $Z = -\sum_{i=1}^{n} p_i \log p_i$, subject to the constraint $\sum_{i=1}^{n} p_i = 1$.
11. Use dynamic programming method to find the minimum value of $Z = x_1^2 + x_2^2 + x_3^2 + \ldots + x_n^2$, subject to the constraint $x_1x_2 \ldots x_n = C$; $x_i \geq 0$ for $i = 1, 2, \ldots, n$.
12. Use dynamic programming method to find the maximum value of $Z = x_1^2 + 2x_2^2 + 4x_3$, subject to the constraints $x_1 + 2x_2 + x_3 \leq 8$; $x_1, x_2, x_3 \geq 0$.
13. Use dynamic programming method to find the minimum value of $Z = x_1^2 + 2x_2^2 + 4x_3$, subject to the constraints $x_1 + 2x_2 + x_3 \geq 8$; $x_1, x_2, x_3 \geq 0$.

Solve the following LP problems using dynamic programming algorithm with both forward and backward recursive procedures.

14. Maximise $Z = 12x_1 + 9x_2$, s.t. $2x_1 + x_2 \leq 5$, $x_1 + x_2 \leq 4$; $x_1, x_2 \geq 0$.
15. Maximise $Z = 2x_1 + 3x_2$, s.t. $x_1 + x_2 \leq 6$, $x_1 + 3x_2 \leq 9$; $x_1, x_2 \geq 0$.
16. Maximise $Z = 5x_1 + 3x_2$, s.t. $3x_1 + 5x_2 \leq 15$, $5x_1 + 2x_2 \leq 10$; $x_1, x_2 \geq 0$.

Solve the following LP problems using the dynamic programming algorithm with forward recursive procedure.

17. Maximise $Z = 2x_1 + 5x_2$, s.t. $x_1 + 4x_2 \le 24$, $3x_1 + x_2 \le 21$, $x_1 + x_2 \le 9$; $x_1, x_2, x_3 \ge 0$.
18. Maximise $Z = 5x_1 + 3x_2$, s.t. $x_1 + x_2 \le 15$, $2x_1 + x_2 \le 20$, $x_1 \le 6$; $x_1, x_2, x_3 \ge 0$.
19. Maximise $Z = 50x_1 + 100x_2$, s.t. $10x_1 + 5x_2 \le 2500$, $4x_1 + 10x_2 \le 2000$, $2x_1 + 3x_2 \le 450$; $x_1, x_2, x_3 \ge 0$.

Solve the following LP problems using the dynamic programming algorithm with backward recursive procedure.

20. Maximise $Z = 5x_1 + 4x_2$, s.t. $4x_1 + 5x_2 \le 10$, $3x_1 + 2x_2 \le 6$, $8x_1 + 3x_2 \le 12$; $x_1, x_2 \ge 0$.
21. Maximise $Z = 10x_1 + 7x_2$ s.t. $4x_1 + x_2 \le 8$, $7x_1 + 3x_2 \le 27$, $x_1 + 2x_2 \le 6$; $x_1, x_2 \ge 0$..
22. Maximise $Z = 3x_1 + 5x_2$, s.t. $x_1 + x_2 \le 2$, $2x_1 + 5x_2 \le 10$, $8x_1 + 3x_2 \le 12$; $x_1, x_2 \ge 0$.
23. Maximise $Z = 3x_1 + x_2$, s.t. $6x_1 + x_2 \le 8$, $2x_1 + x_2 \le 4$, $6x_1 - x_2 \le 8$; $x_1, x_2 \ge 0$.

Solve the following LP problems, using the dynamic programming algorithm with backward recursive procedure.

24. Minimise $Z = 3x_1 + 2x_2$, s.t. $5x_1 + x_2 \ge 10$, $x_1 + x_2 \ge 6$, $x_1 + 4x_2 \ge 12$; $x_1, x_2 \ge 0$.
25. Minimise $Z = 4x_1 + 2x_2$, s.t. $x_1 + 2x_2 \ge 2$, $3x_1 + x_2 \ge 3$, $4x_1 + 3x_2 \ge 6$; $x_1, x_2 \ge 0$.

ANSWERS

Exercise 5

Part B

7. $x_1 = x_2 = x_3 = \frac{5}{3}$; max $z = \frac{125}{27}$.
8. Optimal division is $\left\{\frac{C}{n}, \frac{C}{n}, \frac{C}{n}, \ldots, \frac{C}{n}\right\}$.
9. Minimum $Z = 12$, cor. to $x_1 = x_2 = x_3 = 4$.
10. Optimal solution is $p_1 = p_2 = \ldots = p_n = \frac{1}{n}$; Optimal $z = \log n$.
11. Minimum value of $Z = n.C^{2/n}$, corresponding to $x_1 = x_2 = \ldots = x_n = C^{1/n}$.
12. Maximum $Z = 64$; cor. to $x_1 = 8, x_2 = 0, x_3 = 0$.
13. Minimum $Z = 20$; cor. to $x_1 = 2, x_2 = 0, x_3 = 0$
14. Maximum $Z = 39$; cor. to $x_1 = 1$ and $x_2 = 3$.
15. Maximum $Z = \frac{27}{2}$; cor. to $x_1 = \frac{9}{2}$ and $x_2 = \frac{3}{2}$
16. Maximum $Z = \frac{235}{19}$; cor. to $x_1 = \frac{20}{19}$ and $x_2 = \frac{45}{19}$.
17. Maximum $Z = 33$; cor. to $x_1 = 4$ and $x_2 = 5$.
18. Maximum $Z = 55$; cor. to $x_1 = 5$ and $x_2 = 12$
19. Maximum $Z = 21{,}875$; cor. to $x_1 = 187.5$ and $x_2 = 125$.
20. Maximum $Z = \frac{139}{14}$; cor. to $x_1 = \frac{15}{14}$ and $x_2 = \frac{8}{7}$.
21. Maximum $Z = \frac{212}{7}$; cor. to $x_1 = \frac{10}{7}$ and $x_2 = \frac{16}{7}$.
22. Maximum $Z = 10$; cor. to $x_1 = 0$ and $x_2 = 2$.
23. Maximum $Z = 5$; cor. to $x_1 = 1$ and $x_2 = 2$.
24. Minimum $Z = 13$; cor. to $x_1 = 1$ and $x_2 = 5$.
25. Minimum $Z = \frac{24}{5}$; cor. to $x_1 = \frac{3}{5}$ and $x_2 = \frac{6}{5}$.

Chapter 6

INTEGER PROGRAMMING

6.0 INTRODUCTION

The integer programming problem (abbreviated as IPP) is a special case of the linear programming problem in which all or some of the decision variables are restricted to be non-negative integer values. In many decision-making situations, we come across certain practical problems in which fractional values of the decision variables are physically meaningless. For example, the number of cars manufactured by a company, the number of lathes in a workshop, etc., cannot be fractions; they can only be non-negative integers.

One may think that the IPP can be treated as an ordinary LPP which can be solved by using the regular Simplex method and then rounding off the fractional values in the optimal solution, but in reality, it is difficult to round off the solution without violating any of the given constraints. Moreover, the rounded off solution may give a value of the objective function that is very much different from the original (integer) objective function value. These difficulties can be overcome if we use a systematic procedure for obtaining the exact optimum integer solution to such problems.

There are two types of integer programming problems:

1. *All integer programming problem* (All IPP): An IPP is said to be an *all IPP* (or pure IPP), if all the decision variables are restricted to assume non-negative integer values in the optional solution.
2. *Mixed integer programming problem* (Mixed IPP): An IPP is said to be a *mixed IPP* if only some decision variables are restricted to assume non-negative integer values, while the remaining decision variables may be just non-negative values in the optimal solution.

Note: The discussion of the solution of Type II IPPs is beyond the scope of this book.

6.1 GOMORY'S METHOD FOR ALL IPPS

In this method, the given IPP is first solved as an ordinary LPP by ignoring the integer constraints. After an optimum basic feasible solution has been obtained, if all the decision variables assume integer values, the solution obtained is the required optimum integer solution. On the other hand, if a basic variable has a fractional value, a new constraint known as *Gomory's constraint* (explained later) is introduced into the LPP. The optimum solution to this modified LPP is obtained using any standard algorithm. If all the variables in this solution are integers, then the optimum solution of the given IPP is obtained. Otherwise, another Gomory's constraint is introduced into the LPP and the procedure is repeated iteratively, until the required integer-valued optimum solution is obtained.

6.2 GOMORY'S CONSTRAINT

We will learn about Gomory's constraint by considering the following example. Assume that we are trying to find the integer solution for the following LPP: To maximise $z = c_1x_1 + c_2x_2$, subject to the constraints $a_{11}x_1 + a_{12}x_2 \leq b_1$, $a_{21}x_1 + a_{22}x_2 \leq b_2$, $x_1, x_2 \geq 0$ and are integers. After introducing the slack variables x_3 and x_4 and solving the problem by the usual Simplex method, let the final Simplex table giving the optimum basic feasible non-integer solution be

	c_j	C_1	C_2	0	0
B	X_B	A_1	A_2	A_3	A_4
A_2	$\bar{b}_1$	$\bar{a}_{11}$	$\bar{a}_{12}$	$\bar{a}_{13}$	$\bar{a}_{14}$
A_4	$\bar{b}_2$	$\bar{a}_{21}$	$\bar{a}_{22}$	$\bar{a}_{23}$	$\bar{a}_{24}$

Note: Since A_2 is the first unit vector in the basis, $\bar{a}_{12} = 1$ and $\bar{a}_{22} = 0$ Similarly, $\bar{a}_{14} = 0$ and $\bar{a}_{24} = 1$. Let us assume that $\bar{b}_1$ (and hence), $\bar{a}_{11}$ and $\bar{a}_{13}$ are fractional.

The corresponding constraint equation is

$$\bar{a}_{11}x_1 + x_2 + \bar{a}_{13}x_3 = \bar{b}_1 \tag{1}$$

viz., $x_2 = \bar{b}_1 - \bar{a}_{11}x_1 - \bar{a}_{13}x_3$. The RHS of this equation is a linear combination of non-basic variables x_1 and x_3. Since $\bar{b}_1 \geq 0$, the fractional part of $\bar{b}_1$ will also be non-negative.

Let $\bar{a}_{10}\ (= \bar{b}_1)$, $\bar{a}_{11}$ and $\bar{a}_{13}$ each be expressed as the sum of an integral part and a non-negative fractional part.

viz., let

$$\bar{a}_{ij} = I_{1j} + f_{1j};\ j = 0,1 \text{ and } 3.$$

Then (1) may be written as

$$(I_{10} + f_{10}) = (I_{11} + f_{11})x_1 + x_2 + (I_{13} + f_{13})x_3$$

viz.,

$$f_{10} - f_{11}x_1 - f_{13}x_3 = x_2 + I_{11}x_1 + I_{13}x_3 - I_{10} \tag{2}$$

Comparing (1) and (2), we can guess that if we add an additional constraint so that the LHS is an integer, the non-integer $\bar{a}_{10}(= \bar{b}_1)$ will become an integer and hence the corresponding non-integer solution will become an integer solution.

The additional constraint required (called the Gomory's constraint) should be $f_{10} - f_{11}x_1 - f_{13}x_3 \leq 0$. Otherwise, if $f_{10} - f_{11}x_1 - f_{13}x_3 = h$, where h is a positive integer, then $f_{10} = h + f_{11}x_1 + f_{13}x_3$ is obviously greater than 1, which contradicts our assumption that $0 < f_{10} < 1$.

Thus, the Gomory's constraint is $f_{11}x_1 + f_{13}x_3 \geq f_{10}$

viz.,

$$-f_{11}x_1 - f_{13}x_3 \leq -f_{10} \tag{3}$$

Or $-f_{11}x_1 - f_{13}x_3 + g_1 = -f_{10}$, where g_1 is a slack variable in the first Gomory's constraint.

After introducing the additional (Gomory's) constraint to the given LPP, the optimum Simplex table will become:

	c_j	c_1	c_2	0	0	0
B	X_B	A_1	A_2	A_3	A_4	G_1
A_2	$\bar{b}_1$	$\bar{a}_{11}$	$\bar{a}_{12}$	$\bar{a}_{13}$	$\bar{a}_{14}$	0
A_4	$\bar{b}_2$	$\bar{a}_{21}$	$\bar{a}_{22}$	$\bar{a}_{23}$	$\bar{a}_{24}$	0
G_1	$-f_{10}$	$-f_{11}$	0	$-f_{13}$	0	1

Note: $\bar{a}_{12} = 1, \bar{a}_{22} = 0$ and $\bar{a}_{14} = 0, \bar{a}_{24} = 1$.

Since $-f_{10}$ is negative, the optimum solution is infeasible and so the dual Simplex method is applied for obtaining a feasible solution. If the resulting solution is still a non-integer one, we construct the second Gomory's constraint from the resulting Simplex table and introduce it in that table and solve by the dual Simplex method. The procedure is repeated until an integer solution is obtained.

6.3 ALL IPP ALGORITHM BY GOMORY'S METHOD

Step 1: If the given IPP is one of minimisation, it is converted to a maximisation problem.

Step 2: Slack/surplus variables are introduced to convert the constraints into equations.

Step 3: The optimum solution of the given LPP is obtained by using the regular Simplex algorithm.

Step 4: The integrality of the optimum solution is examined.

Two possibilities may occur as given here:

(a) The optimum solution contains all integer values; then the required solution has been obtained.

(b) The optimum solution does not contain all integer values; then we proceed to the next step.

Step 5: If only one decision variable (in the optional basis) assumes fractional value, then we form the Gomory's constraint of the form (3) given in the previous discussion, corresponding to the row in which the fractional variable lies in the optimum Simplex table.

If more than one decision variable in the optimal basis are fractional, then the Gomory's constraint is formed corresponding to that decision variable with the largest fractional part.

Step 6: The Gomory's constraint equation is appended to the optimum Simplex table obtained in Step 3.

Step 7: Using this new set of constraint equations, the new optimum solution is obtained by the dual Simplex algorithm (so that the Gomory's slack variable g_1 may be the first basic variable to leave the table)

Step 8: If this new optimum solution for the modified LPP is an integer solution, it is also feasible and optimum for the given IPP. Otherwise, we return to Step 4(b) and repeat the process until an optimum integer solution is obtained.

Algorithm for the branch and bound technique for solving all IPP

Note: Though the branch and bound technique can be applied to solve mixed integer programming problems also, we are concerned with solving all IPPs only)

Step 1: The given IPP is solved by the regular Simplex method ignoring the integer value constraint.

Step 2: The integrality of the optimum solution is examined. Two possibilities may arise as given here.

(a) The optimum solution contains all integer values; then the required solution has been obtained.

(b) The optimum solution does not contain all integer values; then we proceed to the next step.

Step 3: If the optimum value x_i^* of the variable x_i is fractional, then two sub-problems are formed.

Sub-problem 1 is the given problem with the given constraints and one or more constraint $x_i \le [x_i^*]$, where $[x_i^*]$ is the integral part of x_i^*.

Sub-problem 2 is the given problem with the given constraints and one more constraint $x_i \ge [x_i^*] + 1$.

Step 4: The two sub-problems obtained in Step 3 are solved by the regular Simplex method. Now three possibilities may arise:

(a) The optimal solutions of both the sub-problems are integer valued. In this case, the required solution is that which gives the larger value of the objective function z.

(b) The optimal solution of one sub-problem is integer valued, while the other sub-problem has no feasible optimal solution. In this case, the required solution is the optimal solution of the former.

(c) The optimal solution of one sub-problem is integer valued, while that of the other sub-problem is fraction valued. In this case, the integer-valued solution is recorded. Steps (3) and (4) are repeated for the fraction-valued sub-problems. They are continued iteratively till all integer-valued solutions are recorded.

Step 5: Of all the integer-valued solutions recorded, that solution which gives the largest value of z is the required optimal solution of the problem.

WORKED EXAMPLES 6

Example 1

Find the optimum solution to the following IPP using Gomory's technique.

Maximise $z = 4x_1 + 5x_2$, subject to the constraint $3x_1 + 2x_2 \le 7$; $x_1, x_2 \ge 0$ and are integers.

SOLUTION

Introducing the slack variable x_3 in the given constraint, the problem is re-stated as:

$$\text{Maximise } z = 4x_1 + 5x_2 + 0.x_3, \text{ s.t. } 3x_1 + 2x_2 + x_3 = 7;\ x_1, x_2, x_3 \ge 0$$

Initial Simplex table:

		c_j	4	5	0
B	C_B	X_B	A_1	A_2	A_3
A_3	0	7	3	[2]	1
$z = 0$		$z_j - c_j$	-4	-5	0
				↑	

→

I iteration

		c_j	4	5	0
B	C_B	X_B	A_1	A_2	A_3
A_2	5	$\frac{7}{2}$	$\frac{3}{2}$	1	$\frac{1}{2}$
$z = \frac{35}{2}$		$z_j - c_j$	$\frac{7}{2}$	0	$\frac{5}{2}$

Since all $(z_j - c_j)$ values are ≥ 0 and the entry in the X_B column is also ≥ 0, an optimum basic feasible solution has been obtained, but the optimal solution is not an integer solution. To obtain the optimum integer solution, we append Gomory's fractional cut constraint to the optimal table, which is got as follows:

The row in the optimal Simplex table means

$$\frac{3}{2}x_1 + x_2 + \frac{1}{2}x_3 = \frac{7}{2}$$

i.e., $$\left(1+\frac{1}{2}\right)x_1 + x_2 + \left(0+\frac{1}{2}\right)x_3 = \left(3+\frac{1}{2}\right)$$

$\therefore$ Gomory's cut constraint is $-\frac{1}{2}x_1 - \frac{1}{2}x_3 \le -\frac{1}{2}$

Introducing Gomory's slack variable g_1, this inequality becomes $-\frac{1}{2}x_1 - \frac{1}{2}x_3 + g_1 = -\frac{1}{2}$ and this cut constraint equation is appended to the optimum table already obtained as given in the following table.

The modified problem with the new constraint is solved by the dual Simplex method.

Initial dual Simplex table:

		c_j	4	5	0	0	
B	C_B	X_B	A_1	A_2	A_3	G_1	
A_2	5	$\frac{7}{2}$	$\frac{3}{2}$	1	$\frac{1}{2}$	0	
G_1	0	$-\frac{1}{2}$	$-\frac{1}{2}$	0	$\boxed{-\frac{1}{2}}$	1	→
$z=\frac{35}{2}$		(z_j-c_j)	$\frac{7}{2}$	0	$\frac{5}{2}$	0	
		RR	−7	–	−5	–	
					↑		

I iteration

		c_j	4	5	0	0
B	C_B	X_B	A_1	A_2	A_3	G_1
A_2	5	3	1	1	0	1
A_3	0	1	1	0	1	−2
$z=15$		(z_j-c_j)	1	0	0	5

The final table shows that the optimum feasible solution has been obtained. The solution is also integer valued.

$\therefore$ The required solution is $x_1 = 0$ and $x_2 = 3$ and the maximum value of $z = 15$.

Example 2

Find the optimum solution to the following IPP using Gomory's technique:

Maximise $z = 2x_1 + x_2$, subject to the constraint $8x_1 + 5x_2 \le 15$; $x_1, x_2 \ge 0$ and are integers.

SOLUTION

Introducing the slack variable x_3 in the given constraint, the problem is re-stated as:

$$\text{Maximise } z = 2x_1 + x_2 + 0.x_3, \text{ s.t. } 8x_1 + 5x_2 + x_3 = 15;\ x_1, x_2, x_3 \ge 0$$

Initial Simplex table:

B	C_B	X_B	c_j: 2 A_1	1 A_2	0 A_3
A_3	0	15	[8]	5	1
$z=0$	z_j-c_j		−2	−1	0
			↑		

I iteration

B	C_B	X_B	c_j: 2 A_1	1 A_2	0 A_3
A_1	2	$\frac{15}{8}$	1	$\frac{5}{8}$	$\frac{1}{8}$
$z=\frac{15}{4}$	z_j-c_j		0	$\frac{1}{4}$	$\frac{1}{4}$

Though the solution obtained after the I iteration is optimum feasible, it is not integral. So we introduce Gomory's cut constraint, obtained as explained here:

The row in the optimum table means

$$x_1+\frac{5}{8}x_2+\frac{1}{8}x_3=\frac{15}{8}$$

i.e.,
$$x_1+\left(0+\frac{5}{8}\right)x_2+\left(0+\frac{1}{8}\right)x_3=\left(1+\frac{7}{8}\right)$$

$\therefore$ Gomory's cut constraint is $-\frac{5}{8}x_2-\frac{1}{8}x_3\le-\frac{7}{8}$

Introducing Gomory's slack variable g_1,

we get
$$-\frac{5}{8}x_2-\frac{1}{8}x_3+g_1=-\frac{7}{8}$$

Initial dual Simplex table

B	C_B	X_B	c_j: 2 A_1	1 A_2	0 A_3	0 G_1
A_1	2	$\frac{15}{8}$	1	$\frac{5}{8}$	$\frac{1}{8}$	0
G_1	0	$-\frac{7}{8}$	0	[$-\frac{5}{8}$]	$-\frac{1}{8}$	1 →
$z=\frac{15}{4}$	(z_j-c_j)		0	$\frac{1}{4}$	$\frac{1}{4}$	0
RR			–	$-\frac{2}{5}$	−2	–
				↑		

I iteration

B	C_B	X_B	c_j: 2 A_1	1 A_2	0 A_3	0 G_1
A_1	2	1	1	0	0	1
A_2	1	$\frac{7}{5}$	0	1	$\frac{1}{5}$	$-\frac{8}{5}$
$z=\frac{17}{5}$	(z_j-c_j)		0	1	$\frac{1}{5}$	$\frac{2}{5}$

Though the solution is optimum feasible, it is still not integral. So we introduce a second Gomory's cut constraint.

From the second row of the optimal table, we have

$$x_2 + \frac{1}{5}x_3 - \frac{8}{5}g_1 = \frac{7}{5}$$

i.e.,
$$x_2 + \left(0 + \frac{1}{5}\right)x_3 + \left(-2 + \frac{2}{5}\right)g_1 = 1 + \frac{2}{5}$$

$\therefore$ The Gomory's cut constraint is

$$-\frac{1}{5}x_3 - \frac{2}{5}g_1 \le -\frac{2}{5}$$

i.e., $-\frac{1}{5}x_3 - \frac{2}{5}g_1 + g_2 = -\frac{2}{5}$, where g_2 is the second Gomory's second slack variable.

Stage II/Initial D.S.T

		c_j	2	1	0	0	0	
B	C_B	X_B	A_1	A_2	A_3	G_1	G_2	
A_1	2	1	1	0	0	1	0	
A_2	1	$\frac{7}{5}$	0	1	$\frac{1}{5}$	$-\frac{8}{5}$	0	
G_2	0	$-\frac{2}{5}$	0	0	$\boxed{-\frac{1}{5}}$	$-\frac{2}{5}$	1	→
$z = \frac{17}{5}$		$(z_j - c_j)$	0	0	$\frac{1}{5}$	$\frac{2}{5}$	0	
		RR	–	–	–1	–1	–	
					↑			

Stage II/I iteration

		c_j	2	1	0	0	0
B	C_B	X_B	A_1	A_2	A_3	G_1	G_2
A_1	2	1	1	0	0	1	0
A_2	1	1	0	1	0	–2	1
A_3	0	2	0	0	1	2	–5
$z = 3$		$(z_j - c_j)$	0	0	0	0	1

The optimum feasible integral solution has been obtained. The required solution is $x_1 = 1, x_2 = 1$ and the maximum value of $z = 3$.

Note: The replacement ratios corresponding to A_3 and G_1 are equal in the stage II/initial dual Simplex table. We could have introduced G_1 in the basis instead of A_3. This would have given the following final table:

		c_j	2	1	0	0	0
B	C_B	X_B	A_1	A_2	A_3	G_1	G_2
A_1	2	0	1	0	$-\frac{1}{2}$	0	$\frac{5}{2}$
A_2	1	3	0	1	1	0	–4
G_1	0	1	0	0	$\frac{1}{2}$	1	$-\frac{5}{2}$
$z = 3$		$(z_j - c_j)$	0	0	0	0	1

∴ The alternative optimum feasible integral solution is $x_1 = 0$, $x_2 = 3$ and the maximum value of $z = 3$.

Example 3

Find the optimum solution to the following IPP using Gomory's technique:

Maximise $z = 2x_1 + 3x_2$, subject to the constraints $4x_1 + 5x_2 \leq 40$, $x_1 + 2x_2 \leq 12$; $x_1, x_2 \geq 0$ and integers.

SOLUTION

Introducing the slack variables x_3 and x_4 in the given constraints, the problem is re-stated as:

Maximise $z = 2x_1 + 3x_2 + 0.x_3 + 0.x_4$, s.t. $4x_1 + 5x_2 + x_3 = 40$, $x_1 + 2x_2 + x_4 = 12$; $x_i \geq 0$.

Initial Simplex table

		c_j	2	3	0	0	
B	C_B	X_B	A_1	A_2	A_3	A_4	
A_3	0	40	4	5	1	0	
A_4	0	12	1	[2]	0	1	→
$z=0$		$z_j - c_j$	−2	−3	0	0	
				↑			

I iteration

		c_j	2	3	0	0	
B	C_B	X_B	A_1	A_2	A_3	A_4	
A_3	0	10	$\frac{3}{2}$	0	1	$-\frac{5}{2}$	→
A_2	3	6	$\frac{1}{2}$	1	0	$\frac{1}{2}$	
$z = 18$		$z_j - c_j$	$-\frac{1}{2}$	0	0	$\frac{3}{2}$	
			↑				

II iteration

		c_j	2	3	0	0
B	C_B	X_B	A_1	A_2	A_3	A_4
A_1	2	$\frac{20}{3}$	1	0	$\frac{2}{3}$	$-\frac{5}{3}$
A_2	3	$\frac{8}{3}$	0	1	$-\frac{1}{3}$	$\frac{4}{3}$
$z = \frac{64}{3}$		$z_j - c_j$	0	0	$\frac{1}{3}$	$\frac{2}{3}$

Though the solution obtained after two iterations is optimum feasible, it is not integral. So we introduce the Gomory's cut constraint got from the first row in which the x_B entry is greater.

Note: The fractional parts of x_B entries are $\frac{2}{3}$ each. We have selected the first row arbitrarily for Gomory's cut constraint.

The first row in the optimum table means

$$x_1 + \frac{2}{3}x_3 - \frac{5}{3}x_4 = \frac{20}{3}$$

i.e.,
$$x_1 + \left(0+\frac{2}{3}\right)x_3 + \left(-2+\frac{1}{3}\right)x_4 = \left(6+\frac{2}{3}\right)$$

$\therefore$ Gomory's cut constant is $-\frac{2}{3}x_3 - \frac{1}{3}x_4 + g_1 = \left(6+\frac{2}{3}\right)$

Introducing this constraint in the optimum table, we proceed:

Initial dual Simplex table

		c_j	2	3	0	0	0
B	C_B	X_B	A_1	A_2	A_3	A_4	G_1
A_1	2	$\frac{20}{3}$	1	0	$\frac{2}{3}$	$-\frac{5}{3}$	0
A_2	3	$\frac{8}{3}$	0	1	$-\frac{1}{3}$	$\frac{4}{3}$	0
G_1	0	$-\frac{2}{3}$	0	0	$\boxed{-\frac{2}{3}}$	$-\frac{1}{3}$	1 →
$z=\frac{64}{3}$		z_j-c_j	0	0	$\frac{1}{3}$	$\frac{2}{3}$	0
		RR:	-	-	$\frac{1}{2}$ ↑	-2	-

I iteration

		c_j	2	3	0	0	0
B	C_B	X_B	A_1	A_2	A_3	A_4	G_1
A_1	2	6	1	0	0	-2	1
A_2	3	3	0	1	0	$\frac{3}{2}$	$-\frac{1}{2}$
A_3	0	1	0	0	1	$\frac{1}{2}$	$-\frac{3}{2}$
$z=21$		z_j-c_j	0	0	0	$\frac{1}{2}$	$\frac{1}{2}$

The optimum feasible integral solution has been obtained.
The required solution is $x_1 = 6$, $x_2 = 3$ and the maximum value of $z = 21$.

Example 4

Find the optimal solution to the following IPP using Gomory's technique:
Maximise $z = 5x_1 + 7x_2$, subject to the constraints $x_1 \leq 2$, $x_1 + x_2 \leq 4$, $3x_1 + 8x_2 \leq 24$; $x_1, x_2 \geq 0$ and integers.

SOLUTION

Introducing the slack variable x_3, x_4, x_5 in the given constraints, the problem is re-stated as:
Maximise $z = 5x_1 + 7x_2 + 0{\cdot}x_3 + 0{\cdot}x_4 + 0{\cdot}x_5$, s.t. $x_1 + x_3 = 2$, $x_1 + x_2 + x_4 = 4$, $3x_1 + 8x_2 + x_5 = 24$; $x_i \geq 0$.

Initial Simplex table

		c_j	5	7	0	0	0
B	C_B	X_B	A_1	A_2	A_3	A_4	A_5
A_3	0	2	1	0	1	0	0
A_4	0	4	1	1	0	1	0
A_5	0	24	3	$\boxed{8}$	0	0	1 →
$z=0$		z_j-c_j	-5	-7 ↑	0	0	0

I iteration

A_3	0	2	1	0	1	0	0
A_4	0	1	$\boxed{\frac{5}{8}}$	0	0	1	$-\frac{1}{8}$ →
A_2	7	3	$\frac{3}{8}$	1	0	0	$\frac{1}{8}$
$z = 21$		$z_j - c_j$	$-\frac{19}{8}$	0	0	0	$\frac{7}{8}$
			↑				

II iteration

A_3	0	$\frac{2}{5}$	0	0	1	$-\frac{8}{5}$	$\frac{1}{5}$
A_1	5	$\frac{8}{5}$	1	0	0	$\frac{8}{5}$	$-\frac{1}{5}$
A_2	7	$\frac{12}{5}$	0	1	0	$-\frac{3}{5}$	$\frac{1}{5}$
$z = \frac{124}{5}$		$z_j - c_j$	0	0	0	$\frac{19}{5}$	$\frac{2}{5}$

The final table indicates that the solution is optimum feasible, but not integral. So we introduce the Gomory's cut constraint corresponding to the third row for which the X_B entry is maximum. The third row in the optimal table represents the equation $x_2 - \frac{3}{5}x_4 + \frac{1}{5}x_5 = \frac{12}{5}$.

Viz.,
$$x_2 + \left(-1 + \frac{2}{5}\right)x_4 + \left(0 + \frac{1}{5}\right)x_5 = \left(2 + \frac{2}{5}\right)$$

∴ Gomory's cut constraint equation is $-\frac{2}{5}x_4 - \frac{1}{5}x_5 + g_1 = -\frac{2}{5}$

Introducing this constraint in the optimal table, we proceed further:

		c_j	5	7	0	0	0	0	
B	C_B	X_B	A_1	A_2	A_3	A_4	A_5	G_1	
A_3	0	$\frac{2}{5}$	0	0	1	$-\frac{8}{5}$	$\frac{1}{5}$	0	
A_1	5	$\frac{8}{5}$	1	0	0	$\frac{8}{5}$	$-\frac{1}{5}$	0	
A_2	7	$\frac{12}{5}$	0	1	0	$-\frac{3}{5}$	$\frac{1}{5}$	0	
G_1	0	$-\frac{2}{5}$	0	0	0	$-\frac{2}{5}$	$\boxed{-\frac{1}{5}}$	1	→

(Contd)

Table (*Contd*)

	c_j	5	7	0	0	0	0
$z=\frac{124}{5}$	z_j-c_j	0	0	0	$\frac{19}{5}$	$\frac{2}{5}$	0
	RR:	–	–	–	$-\frac{19}{2}$	-2 ↑	–

I iteration

		c_j	5	7	0	0	0	0
B	C_B	X_B	A_1	A_2	A_3	A_4	A_5	G_1
A_3	0	0	0	0	1	−2	0	1
A_1	5	2	1	0	0	2	0	−1
A_2	7	2	0	1	0	−1	0	1
A_5	0	2	0	0	0	2	1	−5
$z=24$		z_j-c_j	0	0	0	3	0	2

The optimum feasible integral solution has been obtained.

The required solution is $x_1 = 2$, $x_2 = 2$ and the maximum value of $z = 24$.

Note: According to Step 5 of the IPP algorithm, we ought to have used the Gomory's constraint equation corresponding to the second row of the optimal Simplex table, since the maximum fractional part of the entries in the X_B column corresponds to this row. However, we have used the third row for the construction of Gomory's constraint equation, because our procedure gives the final solution after using the dual Simplex algorithm only once.

Example 5

Find the optimal solution to the following IPP using Gomory's technique:

Maximise $z = 2x_1 + 2x_2 + 3x_3$, subject to the constraints $x_1 + 2x_2 + 4x_3 \leq 8$, $3x_1 + x_2 + x_3 \leq 8$; x_1, x_2, $x_3 \geq 0$.

SOLUTION

Introducing the slack variables x_4 and x_5 in the given constraints, the problem is re-stated as:

Maximise $z = 2x_1 + 2x_2 + 3x_3 + 0.x_4 + 0.x_5$, s. t. $x_1 + 2x_2 + 4x_3 + x_4 = 8$, $3x_1 + x_2 + x_3 + x_5 = 8$; $x_i \geq 0$.

Initial Simplex table

		c_j	2	2	3	0	0	
B	C_B	X_B	A_1	A_2	A_3	A_4	A_5	
A_4	0	8	1	2	[4]	1	0	→

(*Contd*)

Table (*Contd*)

		c_j	2	2	3	0	0	
A_5	0	8	3	1	1	0	1	
$z=0$		(z_j-c_j)	-2	-2	-3	0	0	
					↑			

I iteration

A_3	3	2	$\frac{1}{4}$	$\frac{1}{2}$	1	$\frac{1}{4}$	0	
A_5	0	6	$\boxed{\frac{11}{4}}$	$\frac{1}{2}$	0	$-\frac{1}{4}$	1	→
$z=6$		(z_j-c_j)	$-\frac{5}{4}$	$-\frac{1}{2}$	0	$\frac{3}{4}$	0	
			↑					

II iteration

A_3	3	$\frac{16}{11}$	0	$\boxed{\frac{5}{11}}$	1	$\frac{3}{11}$	$-\frac{1}{11}$	→
A_1	2	$\frac{24}{11}$	1	$\frac{2}{11}$	0	$-\frac{1}{11}$	$\frac{4}{11}$	
$z=\frac{96}{11}$		(z_j-c_j)	0	$-\frac{3}{11}$	0	$\frac{7}{11}$	$\frac{5}{11}$	
				↑				

III iteration

A_2	2	$\frac{16}{5}$	0	1	$\frac{11}{5}$	$\frac{3}{5}$	$-\frac{1}{5}$
A_1	2	$\frac{8}{5}$	1	0	$-\frac{2}{5}$	$-\frac{1}{5}$	$\frac{2}{5}$
$z=\frac{48}{5}$		(z_j-c_j)	0	0	$\frac{3}{5}$	$\frac{4}{5}$	$\frac{2}{5}$

The optimal solution is feasible but not integral.

To get the integer-valued solution, a Gomory's cut constraint is to be introduced.

The fractional parts of X_B column entries are $\frac{1}{5}$ and $\frac{3}{5}$ respectively and the larger of these two is $\frac{3}{5}$, which corresponds to the second row of the optimal table.

From the second row, we get

$$x_1+\left(-1+\frac{3}{5}\right)x_3+\left(-1+\frac{4}{5}\right)x_4+\left(0+\frac{2}{5}\right)x_5=\left(1+\frac{3}{5}\right)$$

$\therefore$ The fractional cost is $-\frac{3}{5}x_3 - \frac{4}{5}x_4 - \frac{2}{5}x_5 + g_1 = -\frac{3}{5}$

Introducing this constraint in the optimal table, we proceed further.

Initial dual Simplex table

		c_j	2	2	3	0	0	0	
B	C_B	X_B	A_1	A_2	A_3	A_4	A_5	G_1	
A_2	2	$\frac{16}{5}$	0	1	$\frac{11}{5}$	$\frac{3}{5}$	$-\frac{1}{5}$	0	
A_1	2	$\frac{8}{5}$	1	0	$-\frac{2}{5}$	$-\frac{1}{5}$	$\frac{2}{5}$	0	
G_1	0	$-\frac{3}{5}$	0	0	$\boxed{-\frac{3}{5}}$	$-\frac{4}{5}$	$-\frac{2}{5}$	1	$\rightarrow$
$Z = \frac{48}{5}$		$(z_j - c_j)$	0	0	$\frac{3}{5}$	$\frac{4}{5}$	$\frac{2}{5}$	0	
		RR	-	-	-1	-1	-1	-	
					$\uparrow$				

II stage/I iteration

		c_j	2	2	3	0	0	0
B	C_B	X_B	A_1	A_2	A_3	A_4	A_5	G_1
A_2	2	1	0	1	0	$-\frac{7}{3}$	$-\frac{5}{3}$	$\frac{11}{3}$
A_1	2	2	1	0	0	$\frac{1}{3}$	$\frac{2}{3}$	$-\frac{2}{3}$
A_3	3	1	0	0	1	$\frac{4}{3}$	$\frac{2}{3}$	$-\frac{5}{3}$
$z = 9$		$(z_j - c_j)$	0	0	0	0	0	1

The optimum feasible integral solution has been obtained.
The required solution is $x_1 = 2, x_2 = 1, x_3 = 1$ and the maximum value of $z = 9$.

Example 6

Find the optimal solution to the following IPP using Gomory's technique:

Maximise $z = 3x_1 + 5x_2 + 4x_3$, subject to the constraints $2x_1 - 3x_2 \leq 8$, $2x_2 + 5x_3 \leq 10$, $3x_1 + 2x_2 + 4x_3 \leq 15$; $x_1, x_2, x_3 \geq 0$

Solution

Introducing the slack variables x_4, x_5, x_6 in the given constraints, the problem is re-stated as

Maximise $z = 3x_1 + 5x_2 + 4x_3 + 0.x_4 + 0.x_5 + 0.x_6$, s. t. $2x_1 - 3x_2 + x_4 = 8$, $2x_2 + 5x_3 + x_5 = 10$, $3x_1 + 2x_2 + 4x_3 + x_6 = 15$; $x_i \geq 0$.

Initial Simplex table:

		c_j	3	5	4	0	0	0	
B	C_B	X_B	A_1	A_2	A_3	A_4	A_5	A_6	
A_4	0	8	2	−3	0	1	0	0	
A_5	0	10	0	2	5	0	1	0	→
A_6	0	15	3	2	4	0	0	1	
$z=0$		$(z_j - c_j)$	−3	−5	−4	0	0	0	
				↑					

I iteration

		c_j	3	5	4	0	0	0	
B	C_B	X_B	A_1	A_2	A_3	A_4	A_5	A_6	
A_4	0	23	2	0	$\frac{15}{2}$	1	$\frac{3}{2}$	0	
A_2	5	5	0	1	$\frac{5}{2}$	0	$\frac{1}{2}$	0	
A_6	0	5	[3]	0	−1	0	−1	1	→
$z = 25$		$(z_j - c_j)$	−3	0	$\frac{17}{2}$	0	$\frac{5}{2}$	0	
			↑						

II iteration

A_4	0	$\frac{59}{3}$	0	0	$\frac{49}{6}$	1	$\frac{13}{6}$	$-\frac{2}{3}$
A_2	5	5	0	1	$\frac{5}{2}$	0	$\frac{1}{2}$	0
A_1	3	$\frac{5}{3}$	1	0	$-\frac{1}{3}$	0	$-\frac{1}{3}$	$\frac{1}{3}$
$z = 30$		$(z_j - c_j)$	0	0	$\frac{15}{2}$	0	$\frac{3}{2}$	1

This final solution is optimal and feasible, but not integer valued. To get the integer-valued solution, a Gomory's cut constraint is to be introduced.

The fractional parts of $\frac{59}{3}$ and $\frac{5}{3}$ that occur in the X_B column are $\frac{2}{3}$ each. So we arbitrarily choose the first row of the final table in which $\frac{59}{3}$ occurs for the construction of Gomory's cut constraint.

The first row in the final table represents the equation

$$\frac{49}{6}x_3 + x_4 + \frac{13}{6}x_5 - \frac{2}{3}x_6 = \frac{59}{3}$$

Viz., $$\left(8+\frac{1}{6}\right)x_3 + x_4 + \left(2+\frac{1}{6}\right)x_5 + \left(-1+\frac{1}{3}\right)x_6 = \left(19+\frac{2}{3}\right)$$

$\therefore$ Gomory's cut constraint equation is

$$-\frac{1}{6}x_3 - \frac{1}{6}x_5 - \frac{1}{3}x_6 + g_1 = -\frac{2}{3}$$

Introducing this constraint in the final table, we proceed further:

Initial dual Simplex table

		c_j	3	5	4	0	0	0	0	
B	C_B	X_B	A_1	A_2	A_3	A_4	A_5	A_6	G_1	
A_4	0	$\frac{59}{3}$	0	0	$\frac{49}{6}$	1	$\frac{13}{6}$	$-\frac{2}{3}$	0	
A_2	5	5	0	1	$\frac{5}{2}$	0	$\frac{1}{2}$	0	0	
A_1	3	$\frac{5}{3}$	1	0	$-\frac{1}{3}$	0	$-\frac{1}{3}$	$\frac{1}{3}$	0	
G_1	0	$-\frac{2}{3}$	0	0	$-\frac{1}{6}$	0	$-\frac{1}{6}$	$\boxed{-\frac{1}{3}}$	1	→
$z = 30$	$(z_j - c_j)$		0	0	$\frac{15}{2}$	0	$\frac{3}{2}$	1	0	
		RR	-	-	−45	-	−9	−3	-	
								↑		

II stage/I iteration

		c_j	3	5	4	0	0	0	0
B	C_B	X_B	A_1	A_2	A_3	A_4	A_5	A_6	G_1
A_4	0	21	0	0	$\frac{17}{2}$	1	$\frac{5}{2}$	0	−2
A_2	5	5	0	1	$\frac{5}{2}$	0	$\frac{1}{2}$	0	0
A_1	3	1	1	0	$-\frac{1}{2}$	0	$-\frac{1}{2}$	0	1
A_6	0	2	0	0	$\frac{1}{2}$	0	$\frac{1}{2}$	1	−3
$z = 28$		$(z_j - c_j)$	0	0	7	0	1	0	3

The optimum feasible integral solution has been obtained.
The required solution is $x_1 = 1, x_2 = 5, x_3 = 0$ and the maximum *m* value of $z = 28$.

Example 7

Find the optimum integer solution to the following IPP using the branch and bound technique:

Maximise $z = -2x_1 + x_2$, subject to the constraint $2x_1 + 4x_2 \leq 15$; $x_1, x_2 \geq 0$ and are integers.

Solution

After introducing the slack variable x_3, the solution is given as follows

Initial Simplex table

		c_j	-2	1	0	
B	C_B	X_B	A_1	A_2	A_3	
A_3	0	15	2	4	1	→
$z = 0$		$(z_j - c_j)$	-2	-1	0	
				↑		

I iteration

		c_j	-2	1	0
B	C_B	X_B	A_1	A_2	A_3
A_2	1	$\frac{15}{4}$	$\frac{1}{2}$	1	$\frac{1}{4}$
$z = \frac{15}{4}$		$(z_j - c_j)$	$\frac{5}{2}$	0	$\frac{1}{4}$

Note: The final solution after I iteration is optimal and feasible; it is not integer valued.

The integral part of $x_2 = \frac{15}{4}$ is given by $[x_2] = 3$.

This gives rise to two sub-problems, namely,

Sub-problem 1: Maximise $z = -2x_1 + x_2$, s.t. $2x_1 + 4x_2 \leq 15$, $x_2 \leq 3$; $x_1, x_2 \geq 0$ and integers.

Sub-problem 2: Maximise $z = -2x_1 + x_2$, s.t. $2x_1 + 4x_2 \leq 15$, $x_2 \geq 4$; $x_1, x_2 \geq 0$ and integers.

Solution of Sub-problem 1

After introducing the slack variables x_3 and x_4 and solving the LP problem by the Simplex method, the following tables giving the required solution are obtained:

		c_j	-2	1	0	0
B	C_B	X_B	A_1	A_2	A_3	A_4
A_3	0	15	2	4	1	0
A_4	0	3	0	1	0	1
$z = 0$		$z_j - c_j$	2	-1	0	0

		c_j	-2	1	0	0
B	C_B	X_B	A_1	A_2	A_3	A_4
A_3	0	3	2	0	1	-4
A_2	1	3	0	1	0	1
$z = 3$		$z_j - c_j$	2	0	0	1

The optimal solution is given by $x_1 = 0$, $x_2 = 3$ and the maximum value of $z = 3$.

Solution of Sub-problem 2

After introducing the slack variable x_3, surplus variable x_4 and artificial variable x_5 and solving the problem by the Simplex method, the following table is obtained

		c_j	-2	1	0	0	$-M$
B	C_B	X_B	A_1	A_2	A_3	A_4	A_5
A_3	0	15	2	4	1	0	0
A_5	$-M$	4	0	1	0	-1	1
$z=-4M$		z_j-c_j	2	$-M-1$	0	M	0

		c_j	-2	1	0	0	$-M$
B	C_B	X_B	A_1	A_2	A_3	A_4	A_5
A_2	1	$\frac{15}{4}$	$\frac{1}{2}$	1	$\frac{1}{4}$	0	0
A_5	$-M$	$\frac{1}{4}$	$-\frac{1}{2}$	0	$-\frac{1}{4}$	-1	1
$Z=-\frac{M}{4}$		z_j-c_j	$\frac{M}{2}+\frac{5}{2}$	0	$\frac{M}{4}+\frac{1}{4}$	M	0

The final table indicates that the problem has no optimal BFS.

$\therefore$ The optimal integer solution is given by $x_1 = 0, x_2 = 3$ and the maximum value of $z = 3$.

Example 8

Find the optimal integer solution to the following IPP using the branch and bound technique:

Maximise $z = x_1 + x_2$, subject to the constraints $2x_1+x_2 \leq 4$, $3x_1 + 5x_2 \leq 8$; $x_1, x_2 \geq 0$ and integers.

SOLUTION

After introducing the slack variables x_3 and x_4 and solving the LP problem, the final simplex table is obtained after two iterations as:

		c_j	1	1	0	0
B	C_B	X_B	A_1	A_2	A_3	A_4
A_1	1	$\frac{12}{7}$	1	0	$\frac{5}{7}$	$-\frac{1}{7}$
A_2	1	$\frac{4}{7}$	0	1	$-\frac{3}{7}$	$\frac{2}{7}$
$z=\frac{16}{7}$		z_j-c_j	0	0	$\frac{2}{7}$	$\frac{1}{7}$

Though the final solution is optimal and feasible, it is not integer-valued.

The integral part of $x_1 = \frac{12}{7}$ is given by$[x_1] = 1$.

This gives rise to two sub-problems, namely:

Sub-problem 1: Maximise $z = x_1 + x_2$, s. t. $2x_1+x_2 \leq 4$, $3x_1 + 5x_2 \leq 8$, $x_1 \leq 1$; $x_1, x_2 \geq 0$ and integers.

Sub-problem 2: Maximise $z = x_1 + x_2$, s. t. $2x_1+x_2 \leq 4$, $3x_1 + 5x_2 \leq 8$, $x_1 \geq 2$; $x_1, x_2 \geq 0$ and integers.

The integral part of $x_2 = \frac{4}{7}$ is given by $[x_2] = 0$.

According to the branch and bound technique, the additional constraints are given by $x_2 \leq 0$ and $x_2 \geq 1$. Of these, $x_2 \leq 0$ is ignored, since as per the non-negatively constraint, $x_2 \geq 0$. So this gives rise to one more sub-problem, namely,

Sub-problem 3: Maximise $z = x_1 + x_2$, s. t. $2x_1+x_2 \leq 4$, $3x_1 + 5x_2 \leq 8$, $x_2 \geq 1$; $x_1, x_2 \geq 0$ and integers.

Solution of sub-problem 1

After introducing the slack variables x_3, x_4 and x_5 and solving the LP problem by the Simplex method, the final Simplex table obtained after two iterations is as follows:

		c_j	1	1	0	0	0
B	C_B	X_B	A_1	A_2	A_3	A_4	A_5
A_3	0	1	0	0	1	$-\frac{1}{5}$	$-\frac{7}{5}$
A_2	1	1	0	1	0	$\frac{1}{5}$	$-\frac{3}{5}$
A_1	1	1	1	0	0	0	1
$z = 2$		$z_j - c_j$	0	0	0	$\frac{1}{5}$	$\frac{2}{5}$

The final solution is optimal, feasible and integer-valued. The solution is given by $x_1 = 1, x_2 = 1$ and the maximum value of $z = 2$.

Solution of sub-problem 2

After introducing the slack variables x_3, x_4, surplus variable x_5 and artificial variable x_6 and solving the LP problem by the Simplex method, the final Simplex table obtained after two iterations is as follows:

		c_j	1	1	0	0	0	$-M$
B	C_B	X_B	A_1	A_2	A_3	A_4	A_5	A_6
A_2	1	0	0	1	1	0	2	−2
A_4	0	2	0	0	−5	1	−7	7
A_1	1	2	1	0	0	0	−1	1
$z = 2$		$z_j - c_j$	0	0	1	0	1	$M-1$

The final solution is optimal, feasible and integer-valued.
The solution is given by $x_1 = 2, x_2 = 0$ and the maximum value of $z = 2$.

Solution of sub-problem 3

After introducing the slack variables x_3, x_4, surplus variable x_5 and artificial variable x_6 and solving the LP problem by the Simplex method, the final Simplex table obtained after two iterations is as follows:

		c_j	1	1	0	0	0	$-M$
B	C_B	X_B	A_1	A_2	A_3	A_4	A_5	A_6
A_3	0	1	0	0	1	$-\frac{2}{3}$	$-\frac{7}{3}$	$\frac{7}{3}$
A_1	1	1	1	0	0	$\frac{1}{3}$	$\frac{5}{3}$	$-\frac{5}{3}$
A_2	1	1	0	1	0	0	−1	1
$z = 2$		$z_j - c_j$	0	0	0	$\frac{1}{3}$	$\frac{2}{3}$	$-\frac{2}{3}+M$

The final solution is optimal, feasible and integer valued.

The solution is given by $x_1 = 1, x_2 = 1$ and the maximum value of $z = 2$.

In all the three sub-problems, the maximum value of z is the same. Hence, the maximum value of $z = 2$.

Example 9

Find the optimal integer solution to the following IPP using the branch and bound technique:

Maximise $z = 5x_1 + 7x_2$, subject to the constraints $x_1 + 2x_2 \leq 5, 2x_1 + x_2 \leq 5, x_1 + 3x_2 \leq 10; x_1, x_2, x_3 \geq 0$ and integers.

SOLUTION

After introducing the slack variables x_3, x_4, x_5 and solving the LP problem, the final Simplex table obtained after two iterations is as follows:

		c_j	5	7	0	0	0
B	C_B	X_B	A_1	A_2	A_3	A_4	A_5
A_2	7	$\frac{5}{3}$	0	1	$\frac{2}{3}$	$-\frac{1}{3}$	0
A_1	5	$\frac{5}{3}$	1	0	$-\frac{1}{3}$	$\frac{2}{3}$	0
A_5	0	$\frac{10}{3}$	0	0	$-\frac{5}{3}$	$\frac{1}{3}$	1
$z = 20$		$z_j - c_j$	0	0	3	1	0

Though the final solution is optimal feasible, it is not integer valued. The optimal solution is given by $x_1 = \frac{5}{3}$ and $x_2 = \frac{5}{3}$. $x_1 = \frac{5}{3}$ contributes two sub-problems and $x_2 = \frac{5}{3}$ gives rise to two more sub-problems as detailed here:

Solution of sub-problem 1, (corresponding to $x_1 \leq 1$) given below:

Maximise $z = 5x_1 + 7x_2 + 0.x_3 + 0.x_4 + 0.x_5 + 0.x_6$, s. t. the constraints $x_1 + 2x_2 + x_3 = 5$, $2x_1 + x_2 + x_4 = 5$, $x_1 + 3x_2 + x_5 = 10$, $x_1 + x_6 = 1$, $x_i \geq 0$ and integers, where x_3, x_4, x_5, x_6 are slack variables

Solving this LP problem, the final optimal Simplex table obtained after two iterations is as follows:

		c_j	5	7	0	0	0	0
B	C_B	X_B	A_1	A_2	A_3	A_4	A_5	A_6
A_2	7	2	0	1	$+\frac{1}{2}$	0	0	$-\frac{1}{2}$
A_4	0	1	0	0	$-\frac{1}{2}$	1	0	1
A_5	0	3	0	0	$-\frac{3}{2}$	0	1	$\frac{1}{2}$
A_1	5	1	1	0	0	0	0	1
$z = 19$		$z_j - c_j$	0	0	$\frac{7}{2}$	0	0	$\frac{3}{2}$

The optimal feasible and integer-valued solution obtained is given by $x_1 = 1$ and $x_2 = 2$ and the maximum value of $z = 19$.

Solution of sub-problem 2 (corresponding to $x_1 \geq 2$)

Maximise $z = 5x_1 + 7x_2 + 0.x_3 + 0.x_4 + 0.x_5 + 0.x_6 - M.x_7$, s. t. the constraints $x_1 + 2x_2 + x_3 = 5$, $2x_1 + x_2 + x_4 = 5$, $x_1 + 3x_2 + x_5 = 10$, $x_1 - x_6 + x_7 = 2$, $x_i \geq 0$ and integers, where x_3, x_4, x_5, are slack variables, x_6 is a surplus variable and x_7 is an artificial variable.

Solving this LP problem, the final optimal Simplex table obtained after two iterations is as follows:

		c_j	5	7	0	0	0	0	$-M$
B	C_B	X_B	A_1	A_2	A_3	A_4	A_5	A_6	A_7
A_3	0	1	0	0	1	−2	0	−3	3
A_2	7	1	0	1	0	1	0	2	−2
A_5	0	5	0	0	0	−3	1	−5	5
A_1	5	2	1	0	0	0	0	−1	1
$z = 17$		$z_j - c_j$	0	0	0	7	0	9	$M - 9$

The optimal feasible and integer-valued solution obtained is given by $x_1 = 2$ and $x_2 = 1$ and the maximum value of $z = 17$.

Solution of sub-problem 3 (corresponding to $x_2 \leq 1$)

Maximise $z = 5x_1 + 7x_2 + 0.x_3 + 0.x_4 + 0.x_5 + 0.x_6$, s. t. the constraints $x_1 + 2x_2 + x_3 = 5$, $2x_1 + x_2 + x_4 = 5$, $x_1 + 3x_2 + x_5 = 10$, $x_2 + x_6 = 1$; $x_i \geq 0$ and integers, where x_3, x_4, x_5, x_6 are slack variables

Solving this LP problem, the final optimal Simplex table obtained after two iterations is as follows:

		c_j	5	7	0	0	0	0
B	C_B	X_B	A_1	A_2	A_3	A_4	A_5	A_6
A_3	0	1	0	0	1	$-\frac{1}{2}$	0	$-\frac{3}{2}$
A_1	5	2	1	0	0	$\frac{1}{2}$	0	$-\frac{1}{2}$
A_5	0	5	0	0	0	$-\frac{1}{2}$	1	$-\frac{5}{2}$
A_2	7	1	0	1	0	0	0	1
$z = 17$		$z_j - c_j$	0	0	0	$\frac{5}{2}$	0	$\frac{9}{2}$

The optimal feasible and integer-valued solution obtained is given by $x_1 = 2$ and $x_2 = 1$ and the maximum value of $z = 17$.

Solution of sub-problem 3 (corresponding to $x_2 \geq 2$)

Maximise $z = 5x_1 + 7x_2 + 0.x_3 + 0.x_4 + 0.x_5 + 0.x_6 - M.x_7$, s. t. the constraints $x_1 + 2x_2 + x_3 = 5$, $2x_1 + x_2 + x_4 = 5$, $x_1 + 3x_2 + x_5 = 10$, $x_2 - x_6 + x_7 = 2$, $x_i \geq 0$ and integers, where x_3, x_4, x_5, are slack variables, x_6 is a surplus variable and x_7 is an artificial variable.

Solving this LP problem, the final optimal Simplex table obtained after three iterations is as follows:

		c_j	5	7	0	0	0	0	$-M$
B	C_B	X_B	A_1	A_2	A_3	A_4	A_5	A_6	A_7
A_1	5	1	1	0	1	0	0	2	-2
A_4	0	1	0	0	-2	1	0	-3	3
A_5	0	3	0	0	0	0	1	1	-1
A_2	7	2	0	1	0	0	0	-1	1
$z = 19$		$z_j - c_j$	0	0	5	0	0	3	$M - 3$

The optimal feasible and integer-valued solution obtained is given by $x_1 = 1$ and $x_2 = 2$ and the maximum value of $z = 19$.

Among the maximum values corresponding to the four sub-problems, the larger value is $z = 19$, corresponding to $x_1 = 1$ and $x_2 = 2$, given by sub-problems (1) and (4), which is the required solution.

Example 10

Find the optimal integer solution to the following IPP using the branch and bound technique:

Maximise $z = 2x_1 + 2x_2 + 3x_3$ subject to the constraints $x_1 + 2x_2 + 4x_3 \leq 8$, $3x_1 + x_2 + x_3 \leq 8$, x_i, $x_2 \geq 0$ and integers.

Solution

After introducing the slack variables x_4, x_5 and solving the LP problem, the final optimal Simplex table obtained after three iterations is as follows:

		c_j	2	2	3	0	0
B	C_B	X_B	A_1	A_2	A_3	A_4	A_5
A_2	2	$\frac{16}{5}$	0	1	$\frac{11}{5}$	$\frac{3}{5}$	$-\frac{1}{5}$
A_1	2	$\frac{8}{5}$	1	0	$-\frac{2}{5}$	$-\frac{1}{5}$	$\frac{2}{5}$
$Z = \frac{48}{5}$		$z_j - c_j$	0	0	$\frac{3}{5}$	$\frac{4}{5}$	$\frac{2}{5}$

Though the final solution is optimal and feasible, it is not integer valued. The optimal solution is given by $x_1 = \frac{8}{5}$ and $x_2 = \frac{16}{5}$ and $x_3 = 0$. $x_1 = \frac{8}{5}$ gives rise to two sub-problems and $x_2 = \frac{16}{5}$ gives rise to two more sub-problems as detailed here:

Solution of sub-problem 1 (corresponding to $x_1 \leq 1$)

Maximise $z = 2x_1 + 2x_2 + 3x_3 + 0.x_4 + 0.x_5 + 0.x_6$, s. t. the constraints $x_1 + 2x_2 + 4x_3 + x_4 = 8$, $3x_1 + x_2 + x_3 + x_5 = 8$, $x_1 + x_6 = 1$; $x_i \geq 0$ and integers, where x_4, x_5, x_6 are slack variables.

Solving this LP problem, the final optimal Simplex table obtained after three iterations is as follows:

		c_j	2	2	3	0	0	0
B	C_B	X_B	A_1	A_2	A_3	A_4	A_5	A_6
A_2	2	$\frac{7}{2}$	0	1	2	$\frac{1}{2}$	0	$-\frac{1}{2}$
A_5	0	$\frac{3}{2}$	0	0	−1	$-\frac{1}{2}$	1	$-\frac{5}{2}$
A_1	2	1	1	0	0	0	0	1
$z = 9$		$z_j - c_j$	0	0	1	1	0	1

Though the final solution is optimal and feasible, it is not integer valued. The optimal solution is given by $x_1 = 1$ and $x_2 = \frac{7}{2}$, and $x_3 = 0$ gives rise to sub-problems, which are the same as the two sub-problems corresponding to $x_2 = \frac{16}{5}$.

Solution of sub-problem 2 (corresponding to $x_1 \geq 2$)

Maximise $z = 2x_1 + 2x_2 + 3x_3 + 0.x_4 + 0.x_5 + 0.x_6 - M.x_7$, s. t. the constraints $x_1 + 2x_2 + 4x_3 + x_4 = 8$, $3x_1 + x_2 + x_3 + x_5 = 8$, $x_1 - x_6 + x_7 = 2$, $x_i \geq 0$ and integers, where x_4, x_5, are slack variables, x_6 is a surplus variable and x_7 is an artificial variable.

Solving this LP problem, the final optimal Simplex table obtained after four iterations is as follows:

		c_j	2	2	3	0	0	0	$-M$
B	C_B	X_B	A_1	A_2	A_3	A_4	A_5	A_6	A_7
A_6	0	$\frac{6}{5}$	0	0	$\frac{11}{5}$	$\frac{3}{5}$	$-\frac{1}{5}$	1	−1
A_1	2	$\frac{8}{5}$	1	0	$-\frac{2}{5}$	$-\frac{1}{5}$	$\frac{2}{5}$	0	0

(Contd)

Table (*Contd*)

		c_j	2	2	3	0	0	0	$-M$
A_2	2	$\frac{16}{5}$	0	1	$\frac{11}{5}$	$\frac{3}{5}$	$-\frac{1}{5}$	0	0
$z=\frac{48}{5}$		z_j-c_j	0	0	$\frac{3}{5}$	$\frac{4}{5}$	$\frac{2}{5}$	0	M

Though the final solution is optimal and feasible, it is not integer valued. The optimal solution is given by $x_1=\frac{8}{5}$, $x_2=\frac{16}{5}$ and $x_3=0$. These values give rise to four sub-problems, which have been already enlisted.

Solution of sub-problem 3 (corresponding to $x_2 \le 3$)

Maximise $z = 2x_1 + 2x_2 + 3x_3 + 0.x_4 + 0.x_5 + 0.x_6$ s. t. the constraints $x_1 + 2x_2 + 4x_3 + x_4 = 8$, $3x_1 + x_2 + x_3 + x_5 = 8$, $x_2 + x_6 = 3$, $x_i \ge 0$ and integers, where x_4, x_5, x_6 are slack variables.

Solving this LP problem, the final optimal Simplex table obtained after three iterations is as follows:

		c_j	2	2	3	0	0	0
B	C_B	X_B	A_1	A_2	A_3	A_4	A_5	A_6
A_3	3	$\frac{1}{11}$	0	0	1	$\frac{3}{11}$	$-\frac{1}{11}$	$-\frac{5}{11}$
A_1	2	$\frac{18}{11}$	1	0	1	$-\frac{1}{11}$	$\frac{4}{11}$	$-\frac{2}{11}$
A_2	2	3	0	1	0	0	0	1
$z=\frac{105}{11}$		z_j-c_j	0	0	0	$\frac{7}{11}$	$\frac{5}{11}$	$\frac{3}{11}$

Though the final solution is optimal and feasible, it is not integer valued. The optimal solution is given by $x_1=\frac{18}{11}$, $x_2 = 3$. and $x_3=\frac{1}{11}$. The value $x_1=\frac{18}{11}$ gives rise to two sub-problems which have been already enlisted. The value $x_3=\frac{1}{11}$ gives rise to a new sub-problem, which will be considered later:

Solution of sub-problem 4 (corresponding to $x_2 \ge 4$)

Maximise $z = 2x_1 + 2x_2 + 3x_3 + 0.x_4 + 0.x_5 + 0.x_6 - M.x_7$, s. t. the constraints $x_1 + 2x_2 + 3x_3 + x_4 = 8$, $3x_1 + x_2 + x_3 + x_5 = 8$, $x_2 - x_6 + x_7 = 4$, $x_i \ge 0$ and integers, where x_4, x_5, are slack variables, x_6 is a surplus variable and x_7 is an artificial variable.

Solving this LP problem, the final optimal Simplex table obtained after three iterations is as follows:

		c_j	2	2	3	0	0	0	$-M$
B	C_B	X_B	A_1	A_2	A_3	A_4	A_5	A_6	A_7
A_1	2	0	1	0	4	1	0	2	-2
A_5	0	4	0	0	-11	-3	1	-5	5
A_2	2	4	0	1	0	0	0	-1	1
$z = 8$		$z_j - c_j$	0	0	5	2	0	2	$M-2$

The final solution is optimal, feasible and integer valued.
The optimal solution is given by $x_1 = 0, x_2 = 4$ and $x_3 = 0$ and the maximum value of $z = 8$.

Solution of sub-problem 5 (corresponding to $x_3 \geq 1$)
Maximise $z = 2x_1 + 2x_2 + 3x_3 + 0.x_4 + 0.x_5 + 0.x_6 - M.x_7$, s. t. the constraints $x_1 + 2x_2 + 3x_3 + x_4 = 8$, $3x_1 + x_2 + x_3 + x_5 = 8$, $x_3 - x_6 + x_7 = 1$, $x_i \geq 0$ and integers, where x_4, x_5, are slack variables, x_6 is a surplus variable and x_7 is an artificial variable.

Solving this LP problem, the final optimal Simplex table, obtained after four iterations, is as follows:

		c_j	2	2	3	0	0	0	$-M$
B	C_B	X_B	A_1	A_2	A_3	A_4	A_5	A_6	A_7
A_2	2	1	0	1	0	$\frac{3}{5}$	$-\frac{1}{5}$	$\frac{11}{5}$	$-\frac{11}{5}$
A_1	2	2	1	0	0	$-\frac{1}{5}$	$\frac{2}{5}$	$-\frac{2}{5}$	$\frac{2}{5}$
A_3	3	1	0	0	1	0	0	-1	1
$z = 9$		$z_j - c_j$	0	0	0	$\frac{4}{5}$	$\frac{2}{5}$	$\frac{3}{5}$	$M-\frac{3}{5}$

The final solution is optimal, feasible and integer valued.
The optimal solution is given by $x_1 = 2, x_2 = 1$ and $x_3 = 1$ and the maximum value of $z = 9$.

Solutions of sub-problems (4) and (5) are eligible integer solutions of the given main problem. Since the maximum value of z obtained from sub-problem (5) is greater than that from sub-problem (4), the required solution of the given main problem is given by $x_1 = 2, x_2 = 1$ and $x_3 = 1$ and the maximum value of $z = 9$.

EXERCISE 6

Part A (Short answer questions)

1. Explain how to construct Gomory's constraint using an example.
2. If more than one decision variable in the optimal basis are fractional, which one is preferred for forming Gomory's constraint and why?
3. When applying the branch and bound technique, if the optimum value x_i^* of the variable x_i is fractional, how will you get the corresponding sub-problems?

Part B

Find the optimal solutions of the following integer programming problems using Gomory's technique:

4. Maximise $z = x_1 + x_2$, s. t. $3x_1 + 2x_2 \leq 5$; $x_1, x_2 \geq 0$ and integers.
5. Maximise $z = x_1 + 5x_2$, s. t. $2x_1 + 3x_2 \leq 8$; $x_1, x_2 \geq 0$ and integers.
6. Maximise $z = 2x_1 + x_2$, s. t. $5x_1 + 4x_2 \leq 12$; $x_1, x_2 \geq 0$ and integers.
7. Maximise $z = 3x_1$, s. t. $2x_1 + 3x_2 \leq 7$; $x_1 - x_2 \leq 2$ and integers.
8. Maximise $z = 9x_1 + 7x_2$, s. t. $3x_1 - x_2 \leq 6$; $x_1 + 7x_2 \leq 35$; $x_1, x_2 \geq 0$ and integers.
9. Maximise $z = x_1 + x_2$, s. t. $2x_1 + x_2 \leq 4$, $3x_1 + 5x_2 \leq 8$; $x_1, x_2 \geq 0$ and integers.
10. Maximise $z = 4x_1 + x_2$, s. t. $4x_1 + 2x_2 \leq 7$, $3x_1 + 5x_2 \leq 15$; $x_1, x_2 \geq 0$ and integers.
11. Maximise $z = x_1 + x_2$, s. t. $5x_1 + 2x_2 \leq 16$, $5x_1 + 6x_2 \leq 30$; $x_1, x_2 \geq 0$ and integers.
12. Maximise $z = x_1 + x_2$, s. t. $2x_1 + x_2 \leq 4$, $3x_1 + 5x_2 \leq 8$, $x_1 \geq 1$; $x_1, x_2 \geq 0$ and integers.
13. Maximise $z = 2x_1 + x_2$, s. t. $2x_1 \leq 7$, $2x_2 \leq 11$, $x_1 + x_2 \leq 7$; $x_1, x_2 \geq 0$ and integers.
14. Maximise $z = 5x_1 + 7x_2$, s. t. $x_1 + x_2 \leq 4$, $3x_1 + 8x_2 \leq 24$, $10x_1 + 7x_2 \leq 35$; $x_1, x_2 \geq 0$ and integers.
15. Maximise $z = 3x_1 + 5x_2 - 2x_3$, s. t. $x_1 + 2x_2 + 2x_3 \leq 10$, $2x_1 + 4x_2 + 3x_3 \leq 15$; $x_1, x_2, x_3 \geq 0$ and integers.
16. Maximise $z = 30x_1 + 23x_2 + 29x_3$, s. t. $6x_1 + 5x_2 + 3x_3 \leq 26$, $4x_1 + 2x_2 + 5x_3 \leq 7$; $x_1, x_2, x_3 \geq 0$ and integers.

Find the optimal solutions of the following integer programming problems using the branch and bound technique:

17. Maximise $z = x_1 + 5x_2$, s. t. $3x_1 + 4x_2 \leq 5$, $x_1, x_2, \geq 0$ and integers.
18. Maximise $z = 5x_1 + x_2$, s. t. $4x_1 + x_2 \leq 5$, $x_1, x_2, \geq 0$ and integers
19. Maximise $z = 4x_1 + 5x_2$, s. t. $3x_1 + 2x_2 \leq 7$, $x_1 \geq 0$, $x_2 \leq 3$, $x_1, x_2 \geq 0$ and integers.
20. Maximise $z = 13x_1 + 3x_2 + 3x_3$, s. t. $7x_1 + 6x_2 - 3x_2 \leq 8$, $7x_1 - 3x_2 + 6x_3 \leq 8$, $x_1, x_2, x_3 \geq 0$ and integers

ANSWERS

Exercise 6

Part B

4. $x_1 = x_2 = 1$; Max $z = 2$
5. $x_1 = 1, x_2 = 2$; Max $z = 11$
6. $x_1 = 2, x_2 = 0$; Max $z = 4$
7. $x_1 = 2, x_2 = 1$ (or) $x_1 = 2, x_2 = 0$; Max $z = 6$
8. $x_1 = 3, x_2 = 4$; Max $z = 55$
9. $x_1 = 1, x_2 = 1$; Max $z = 2$
10. $x_1 = 1, 2 = 3$; Max $z = 5$
11. $x_1 = 2, x_2 = 3$; Max $z = 5$
12. $x_1 = 1, x_2 = 1$; Max $z = 2$
13. $x_1 = 3, x_2 = 4$; Max $z = 10$.
14. $x_1 = 0, x_2 = 3$; Max $z = 21$
15. $x_1 = 7, x_2 = 0, x_3 = 0$; Max $z = 21$
16. $x_1 = 0, x_2 = 3, x_3 = 0$; Max $z = 69$
17. $x_1 = 0, x_2 = 1$; Max $z = 5$.
18. $x_1 = 1, x_2 = 1$; Max $z = 6$.
19. $x_1 = 1, x_2 = 2$, Max $z = 14$.
20. $x_1 = 1, x_2 = 0, x_3 = 0$; Max $z = 13$.

Chapter 7

NON-LINEAR PROGRAMMING

7.0 INTRODUCTION

In linear programming problems, we find the values of a finite number of decision variables $x_1, x_2, \ldots x_n$ that maximise (or minimise) the value of an objective function which is linear in $x_1, x_2, \ldots, x_n$ subject to certain constraints that are also linear in $x_1, x_2, \ldots, x_n$.

If either the objective function or at least one constraint or both are non-linear in $x_1, x_2, \ldots, x_n$, then the problem is a non-linear programming problem (NLPP)

A general non-linear programming problem (GNPP) can be stated mathematically as follows:

To find the values of $x_1, x_2, \ldots, x_n$, so as to optimise

$z = f(x_1, x_2, \ldots, x_n)$, subject to $g_i(x_1, x_2, \ldots, x_n)$ $(\leq, =$ or $\geq)$ b_i; $i = 1, 2, \ldots, m$ and $m < n$ and $x_j \geq 0$; $j = 1, 2, \ldots, n$, where either f or one of g_i's is non-linear.

Note: If both $f(X)$ and $g_i(X)$, where $X = (x_1, x_2, \ldots, x_n)^T$ are convex functions, then the NLPP is called a *convex programming problem.*

If $f(X)$ is a quadratic expression and all g_is are linear then the NLPP is called a *quadratic programming problem.*

7.1 DEFINITIONS

A function $f(X)$ is said to be a *convex function* if for any pair of points $X_r = (x_1^{(r)}, x_2^{(r)}, \ldots, x_n^{(r)})$; $r = 1, 2$ and for all λ, $0 \leq \lambda \leq 1$.

$$f\{\lambda X_1 + (1-\lambda)X_2\} \leq \lambda f(X_1) + (1-\lambda) f(X_2).$$

If $\leq$ is replaced by $<$, then $f(X)$ is said to be *strictly convex*. Alternatively, the function $f(X)$ is convex or strictly convex, according as the Hessian matrix $H(X) = \left[\dfrac{\partial^2 f(X)}{\partial x_i, \partial x_j}\right]$ is positive semi-definite or positive definite for all real values of X. A function $f(X)$ is a *concave function* if $-f(X)$ is a convex function, viz., if $f\{\lambda X_1 + (1-\lambda) X_2\} \geq \lambda f(X_1) + (1-\lambda) f(X_2)$ for all λ, $0 \leq \lambda \leq 1$. Alternatively, the function $f(X)$ is concave or strictly concave, according as the Hessian matrix $H(X)$ is negative semi-definite or negative definite for all real values of X.

7.2 NLPP WITH EQUALITY CONSTRAINTS (CLASSICAL OPTIMISATION PROBLEM)

The problem is stated as follows:

To find $x_1, x_2, \ldots, x_n$, which optimise $z = f(x_1, x_2, \ldots, x_n)$ subject to $g_i(x_1, x_2, \ldots, x_n) = b_i$ or $h_i(x_1, x_2, \ldots, x_n) = 0$; $i = 1, 2, \ldots, m$ and $m < n$, where $h_i = g_i - b_i$ $(i = 1, 2, \ldots, m)$.

Necessary conditions for the maximum (minimum) of an NLPP with equality constraints are provided by the *Lagrange multiplier method* explained in the following discussion.

We consider the Lagrangian function

$$L(x_1, x_2, \ldots, x_n; \lambda_1, \lambda_2, \ldots, \lambda_m) = f(x_1, x_2, \ldots, x_n) - \sum_{i=1}^{m} \lambda_i h_i(x_1, x_2, \ldots, x_n),$$

where $\lambda_1, \lambda_2, \ldots, \lambda_m$ are the Lagrange multipliers.

According to the Lagrange multiplier method, the necessary conditions for the maximum (minimum) points of f are given by $\dfrac{\partial L}{\partial x_j} = 0$, for $j = 1, 2, \ldots, n$

and $$\frac{\partial L}{\partial \lambda_i} = 0,$$

for $i = 1, 2, \ldots, m$

Note: 1. These necessary conditions also become sufficient conditions for a maximum (minimum), if the objective function f is concave (convex) and the constraints are equalities.

2. If the concavity (convexity) of the objective function is not known, the following sufficient conditions can be used.

The *bordered Hessian matrix* $H^{\text{B}} = \left[\begin{array}{c|c} O & U \\ \hline U^{\text{T}} & V \end{array}\right]$, which is an $(m + n, m + n)$ square matrix is formed

where O is an $(m \times m)$ null matrix,

$$U = \left[\frac{\partial h_i}{\partial x_j}(x_1, x_2, \ldots, x_n)\right];$$

$i = 1, 2, \ldots, m$ and $j = 1, 2, \ldots, n$ is an $(m \times n)$ matrix and

$$V = \left[\frac{\partial^2 L}{\partial x_i, \partial x_j}(x_1, x_2, \ldots, x_n; \lambda_1, \lambda_2, \ldots, \lambda_m)\right] \text{ is an } (n \times n) \text{ matrix.}$$

If the elements of H^B involve variables, they are evaluated at the stationary point (x_0, y_0) and the resulting H^B is denoted as H^{B_0}

Sufficient conditions for the stationary point (x_0, y_0) to be a maximum (minimum) point

(i) *When $m < n - 1$*, if the $(n - m)$ principal minors of H^{B_0} of orders $2m + 1, 2m + 2, \ldots, m + n$ are of alternating signs, starting with $(-1)^{m+1}$, then the stationary point (x_0, y_0) is a maximum point of the objective function z.

If they are of the same sign, namely, $(-1)^m$, then (x_0, y_0) is a minimum point of the objective function z.

(ii) *When $m = n - 1$*, the only principal minor of H^{B_0} is itself. Hence, (x_0, y_0) is a maximum point of z, if the sign of $|H^{B_0}|$ is $(-1)^{m+1}$ and a minimum point of z, if the sign of $|H^{B_0}|$ is $(-1)^m$.

Note The case corresponding to $m > n - 1$ does not arise as it is inconsistent with $m < n$, since m is a positive integer.

(iii) *In the rare case when $m = n$*, the principal minors of H^{B_0} of orders greater than or equal to $2m + 1$(or $2n + 1$) are not possible, since H^{B_0} itself is of order $2m$ (or $2n$).

In this case, we consider the Hessian matrix of $f(X)$, namely $H_f(X) = \left[\dfrac{\partial^2 f(X)}{\partial x_i, \partial x_j}\right]$. If $H_f(X)$ is positive definite or positive semi-definite, $f(X)$ is a convex function and the stationary point (x_0, y_0) is a minimum point of z.

If $H_f(X)$ is negative definite or negative semi-definite, $f(X)$ is a concave function and the stationary point (x_0, y_0) is a maximum point of z.

NLPP with inequality constraints

To maximise $z = f(x_1, x_2, \ldots, x_n)$ subject to the constraints $g_i = f(x_1, x_2, \ldots, x_n) \le b_i$ or $h_i(x_1, x_2, \ldots, x_n) \le 0$; $i = 1, 2, \ldots, m$ and $x_j \ge 0$; $j = 1, 2, \ldots, n$.

The inequality constraints are converted into equality constraints by introducing the slack variables s_i^2 or x_{n+i}^2 $(i = 1, 2, \ldots, m)$

Note: x_{n+i} is unrestricted in sign, as $x_{n+i}^2 \ge 0$)

We consider the Lagrangian function

$$L(x_1, x_2, \ldots, x_n; x_{n+1}, x_{n+2}, \ldots, x_{n+m}; \lambda_1, \lambda_2, \ldots, \lambda_m) = f(x_1, x_2, \ldots, x_n)$$
$$-\sum_{i=1}^{m} \lambda_i \left[h_i(x_1, x_2, \ldots, x_n) + x_{n+i}^2\right].$$

By the Lagrange multiplier method, the necessary conditions for the maximum (or minimum) of the objective function z are $\dfrac{\partial L}{\partial x_j} = 0$,

viz.,
$$\frac{\partial f}{\partial x_j} - \sum_{i=1}^{m} \lambda_i \frac{\partial h_i}{\partial x_j} = 0;$$
$$j = 1, 2, \ldots, n \tag{1}$$
$$\frac{\partial L}{\partial \lambda_i} = 0$$

viz.,
$$h_i + x_{n+i}^2 = 0;$$
$$i = 1, 2, \ldots, m \tag{2}$$

and
$$\frac{\partial L}{\partial x_{n+i}} = 0$$

viz.,
$$-2\lambda_i x_{n+i} = 0; i = 1, 2, \ldots, m \tag{3}$$

From (3), we get either $\lambda_i = 0$ or $x_{n+i} = 0$ (4)

If $x_{n+i} = 0$, then from (2), we have $h_i = 0$ (5)

From (4) and (5), we get either $\lambda_i = 0$ or $h_i = 0$

$$\therefore \lambda_i h_i = 0; i = 1, 2, m \tag{6}$$

In (4), if we assume that $\lambda_i = 0$, then from (6), $h_i \ne 0$ and from (2),

$$h_i = -x_{n+i}^2 \le 0 \tag{7}$$

Note: Whenever $h_i < 0$, from (6), we get $\lambda_i = 0$

and whenever $\lambda_i > 0$, from (6), we get $h_i = 0$

When $h_i = 0$ (viz., when the constraints are equality constraints), from (6), we get that λ_is are unrestricted in sign.

Summarising these discussions, the necessary conditions for the maximum (or minimum) of $z = f(X) - \sum_{i=1}^{m} \lambda_i h_i(X)$; $h_i(X) \le 0$, in the simplified form, called *Kuhn–Tucker conditions* are given here:

$$\frac{\partial f}{\partial x_j} - \sum_{i=1}^{m} \lambda_i \cdot \frac{\partial h_i}{\partial x_j} = 0;\ j = 1, 2, \ldots, n$$

and

$$\left.\begin{aligned} \lambda_i h_i &= 0 \\ h_i &\le 0 \\ \lambda_i &\ge 0 \end{aligned}\right\}$$

for $i = 1, 2, \ldots, m$

Note: The Kuhn–Tucker conditions for the maximum (or minimum) of $z = f(X) - \sum_{i=1}^{m} \lambda_i h_i(X)$, subject to $h_i(x) \ge 0$ are as follows:

$$\frac{\partial f}{\partial x_j} - \sum_{i=1}^{m} \lambda_i \frac{\partial h_i}{\partial x_j} = 0;\ j = 1, 2, \ldots, n$$

and

$$\left.\begin{aligned} \lambda_i h_i &= 0 \\ h_i &\ge 0 \\ \lambda_i &\ge 0 \end{aligned}\right\} \text{ for } i = 1, 2, \ldots, m.$$

Sufficient conditions

1. For the standard maximisation problem, viz., in the problem in which we maximise $f(X)$ subject to the constraints $h_i(X) \le 0$ and $X \ge 0$, the Kunh–Tucker conditions are also sufficient conditions for the maximum of $z = f(X)$, provided $f(X)$ is concave and each $h_i(X)$ is convex.
2. For the standard minimisation problem, viz., in the problem in which we minimise $f(X)$ subject to the constraint $h_i(X) \ge 0$ and $X \ge 0$, the Kunh–Tucker conditions are also sufficient conditions for the minimum of $z = f(X)$, provided $f(X)$ is convex and each $h_i(X)$ is concave.

WORKED EXAMPLES 7(A)

Example 1

Maximise $z = 2x_1^2 - 3x_2^2 + 18x_2$, subject to the constraint $2x_1 + x_2 = 8$; $x_i \ge 0$.

SOLUTION

The problem is an NLPP with an equality constraint. So we use the Lagrange multiplier method to find the stationary point of z.

$$L(x_1, x_2, \lambda) = 2x_1^2 - 3x_2^2 + 18x_2 - \lambda(2x_1 + x_2 - 8).$$

The necessary conditions for the optimum of z are

$$\frac{\partial L}{\partial x_1} = 0,$$

i.e., $$4x_1 - 2\lambda = 0 \tag{1}$$

$$\frac{\partial L}{\partial x_2} = 0,$$

i.e., $$-6x_2 + 18 - \lambda = 0 \tag{2}$$

$$\frac{\partial L}{\partial \lambda} = 0,$$

i.e., $$2x_1 + x_2 - 8 = 0 \tag{3}$$

From (1), $$x_1 = \frac{\lambda}{2};$$

From (2), $$x_2 = \frac{18-\lambda}{6}$$

Using these values in (3), we have

$$\lambda + \frac{18-\lambda}{6} = 8,$$

i.e., $$5\lambda = 30$$

$\therefore$ $$\lambda = 6$$

$\therefore$ $$x_1 = 3 \text{ and } x_2 = 2$$

i.e., the optimum point of z is (3, 2).

To determine whether the optimum point of z is a maximum point or a minimum point, we consider

$H^B = \left(\begin{array}{c|c} O & U \\ \hline U^T & V \end{array}\right)$, where $O \equiv 0$; $U \equiv \left(\frac{\partial h}{\partial x_1}, \frac{\partial h}{\partial x_2}\right) = (2,1)$ and $V \equiv \begin{pmatrix} Lx_1x_1 & Lx_1x_2 \\ Lx_2x_1 & Lx_2x_2 \end{pmatrix} = \begin{pmatrix} 4 & 0 \\ 0 & -6 \end{pmatrix}$ i.e.,

$H^B = \begin{pmatrix} 0 & 2 & 1 \\ 2 & 4 & 0 \\ 1 & 0 & -6 \end{pmatrix}$. In this problem, $m = n - 1$, as $m = 1$ and $n = 2$. Hence, we consider the sign of the minor of order $2m + 1$ $(= 3)$, viz., that of $|H^B|$ itself.

Now $$|H^B| = \begin{vmatrix} 0 & 2 & 1 \\ 2 & 4 & 0 \\ 1 & 0 & -6 \end{vmatrix} = -2(-12) + 1(-4) = +20$$

i.e., the sign of $$|H^B| = (-1)^{1+1};$$

$\therefore$ The optimum point (3, 2) is a maximum point of z and the maximum values of $z = 2 \times 9 - 3 \times 4 + 18 \times 2 = 42$.

Example 2

Minimise $z = 2x_1^2 - 24x_1 + 2x_2^2 - 8x_2 + 2x_3^2 - 12x_3 + 200$, subject to the constraint $x_1 + x_2 + x_3 = 11$; $x_2 \geq 0$.

SOLUTION

This problem is an NLPP with an equality constraint in which $m = 1$ and $n = 3$.

We consider

$$L(x_1, x_2 x_3, \lambda) = 2x_1^2 - 24x_1 + 2x_2^2 - 8x_2 + 2x_3^2 - 12x_3 + 200 - \lambda(x_1 + x_2 + x_3 - 11).$$

The necessary conditions for the optimum of z are

$$\frac{\partial L}{\partial x_1} = 0,$$

i.e.,
$$4x_1 - 24 - \lambda = 0 \tag{1}$$

$$\frac{\partial L}{\partial x_2} = 0,$$

i.e.,
$$4x_2 - 8 - \lambda = 0 \tag{2}$$

$$\frac{\partial L}{\partial x_3} = 0,$$

i.e.,
$$4x_3 - 12 - \lambda = 0 \tag{3}$$

$$\frac{\partial L}{\partial \lambda} = 0,$$

i.e.,
$$x_1 + x_2 + x_3 = 11 \tag{4}$$

From (1), (2) and (3), we get

$$x_1 = \frac{\lambda + 24}{4},\ x_2 = \frac{\lambda + 8}{4},\ x_3 = \frac{\lambda + 12}{4} \tag{5}$$

Using (5) in (4), we get $3\lambda + 44 = 44$

$$\therefore \quad \lambda = 0$$

From (5), using $\lambda = 0$, we get $x_1 = 6, x_2 = 2, x_3 = 3$

$\therefore$ The optimum point of z is (6, 2, 3)

To decide the nature of the optimum point, we consider $H^B = \left(\begin{array}{c|c} O & U \\ \hline U^T & V \end{array}\right)$.

where
$$O \equiv 0;\ U = (h_{x_1}, h_{x_2}, h_{x_3}) = (1, 1, 1)$$

and
$$V \equiv \begin{pmatrix} L_{x_1x_1} & L_{x_1x_2} & L_{x_1x_3} \\ L_{x_2x_1} & L_{x_2x_2} & L_{x_2x_3} \\ L_{x_3x_1} & L_{x_3x_2} & L_{x_3x_3} \end{pmatrix} = \begin{pmatrix} 4 & 0 & 0 \\ 0 & 4 & 0 \\ 0 & 0 & 4 \end{pmatrix};$$

i.e.,
$$H^B = \begin{pmatrix} 0 & 1 & 1 & 1 \\ 1 & 4 & 0 & 0 \\ 1 & 0 & 4 & 0 \\ 1 & 0 & 0 & 4 \end{pmatrix}$$

Now we consider the values of the principal minors of H^B of order $(2m + 1)$ (= 3) and 4.

$$D_3 = \begin{vmatrix} 0 & 1 & 1 \\ 1 & 4 & 0 \\ 1 & 0 & 4 \end{vmatrix} = -4 - 4 = -8;$$

$$D_4 = \begin{vmatrix} 0 & 1 & 1 & 1 \\ 1 & 4 & 0 & 0 \\ 1 & 0 & 4 & 0 \\ 1 & 0 & 0 & 4 \end{vmatrix} = -48$$

As D_3 and D_4 are of the same sign, namely $(-1)^m = (-1)^1$, the optimum point (6, 2, 3) is a minimum point of z.

The minimum value of $z = 72 - 144 + 8 - 16 + 18 - 36 + 200 = 102$

Example 3

Minimise $z = x_1^2 + x_2^2 + 2x_1 - 4x_2$, subject to the constraints $3x_1 + 2x_2 = 8$ and $2x_1 + 3x_2 = 7$; $x_1, x_2 \geq 0$

Solution

This problem is an NLPP with equality constraints in which $m = 2$ and $n = 2$.

If we consider the Lagrangian function $L(x_1, x_2, \lambda_1, \lambda_2) = x_1^2 + x_2^2 + 2x_1 - 4x_2 - \lambda_1(3x_1 + 2x_2 - 8) - \lambda_2(2x_1 + 3x_2 - 7)$, the stationary points of z are given by

$$\frac{\partial L}{\partial x_1} = 0 \qquad (1);$$

$$\frac{\partial L}{\partial x_2} = 0 \qquad (2);$$

$$\frac{\partial L}{\partial \lambda_1} = 0 \qquad (3);$$

$$\frac{\partial L}{\partial \lambda_2} = 0 \qquad (4)$$

Solving (1) and (2), we will get

$$x_1 = -\frac{1}{2}(3\lambda_1 + 2\lambda_2 + 2) \qquad (5)$$

and

$$x_2 = -\frac{1}{2}(2\lambda_1 + 3\lambda_2 - 4) \qquad (6)$$

Using (5) and (6) in (3) and (4), we will get

$$13\lambda_1 + 12\lambda_2 = -14 \qquad (7)$$

and

$$12\lambda_1 + 13\lambda_2 = -6 \qquad (8)$$

Solving (7) and (8), we will get $\lambda_1 = -\frac{22}{5}$ and $\lambda_2 = \frac{18}{5}$

Using these values of λ_1 and λ_2 in (5) and (6), we can get $x_1 = 2$ and $x_2 = 1$.

∴ The stationary point of z is (2, 1).

Note: Instead of finding the stationary point in this circuitous way, we can also find it by solving the constraint equations, as the stationary point has to satisfy the two constraint equations.
To find the nature of the optimal point (2, 1), we cannot use H^B which is of order 4 and hence, we cannot find the values of the principal minors of H^B of order 5, 6, … (viz., $2m + 1$, $2m + 2$, …)

So we consider $$H_f \equiv \begin{pmatrix} fx_1x_1 & fx_1x_2 \\ fx_2x_1 & fx_2x_2 \end{pmatrix} = \begin{pmatrix} 2 & 0 \\ 0 & 2 \end{pmatrix}$$

H_f is positive definite, as $D_1 = 2 > 0$ and $D_2 = 4 > 0$.
∴ $z = f(x_1, x_2)$ is a convex function and hence, the optimal point (2, 1) is a minimum point of z.
The minimum value of $z = 4 + 1 + 4 - 4 = 5$.

Example 4

Minimise $z = x_1^2 + x_2^2 + x_3^2$, subject to the constraints $x_1 + x_2 + 3x_3 = 2$, $5x_1 + 2x_2 + x_3 = 5$; $x_i \geq 0$.

Solution

This problem is an NLPP with equality constraints in which $m = 2$ and $n = 3$.
We consider $L(x_1, x_2, x_3, \lambda_1, \lambda_2) = x_1^2 + x_2^2 + x_3^2 - \lambda_1(x_1 + x_2 + 3x_3 - 2) - \lambda_2(5x_1 + 2x_2 + x_3 - 5)$.
The stationary points of z are given by

$$\frac{\partial L}{\partial x_1} \equiv 2x_1 - \lambda_1 - 5\lambda_2 = 0 \quad (1);$$

$$\frac{\partial L}{\partial x_2} \equiv 2x_2 - \lambda_1 - 2\lambda_2 = 0 \quad (2);$$

$$\frac{\partial L}{\partial x_3} \equiv 2x_3 - 3\lambda_1 - \lambda_2 = 0 \quad (3);$$

$$\frac{\partial L}{\partial \lambda_1} \equiv -(x_1 + x_2 + 3x_3 - 2) = 0 \quad (4);$$

$$\frac{\partial L}{\partial \lambda_2} \equiv -(5x_1 + 2x_2 + x_3 - 5) = 0 \quad (5)$$

Using the values of x_1, x_2, x_3 obtained from (1), (2) and (3) in (4) and (5), we have

$$\frac{1}{2}(\lambda_1 + 5\lambda_2) + \frac{1}{2}(\lambda_1 + 2\lambda_2) + \frac{3}{2}(3\lambda_1 + \lambda_2) = 2$$

and $$\frac{5}{2}(\lambda_1 + 5\lambda_2) + \frac{2}{2}(\lambda_1 + 2\lambda_2) + \frac{1}{2}(3\lambda_1 + \lambda_2) = 5$$

i.e., $$11\lambda_1 + 10\lambda_2 = 4$$

and $$\lambda_1 + 3\lambda_2 = 1$$

Solving, we get $$\lambda_1 = \frac{2}{23}$$

and $$\lambda_2 = \frac{7}{23}$$

∴ $$x_1 = \frac{1}{2}(\lambda_1 + 5\lambda_2) = \frac{37}{46};$$

$$x_2 = \frac{1}{2}(\lambda_1 + 2\lambda_2) = \frac{16}{46};$$

$$x_3 = \frac{1}{2}(3\lambda_1 + \lambda_2) = \frac{13}{46}.$$

$\therefore$ The stationary point of $z \equiv \left(\frac{37}{46}, \frac{16}{46}, \frac{13}{46}\right)$

To find the nature of the optimal point, we consider H^B, which is the following (5 × 5) matrix:

$$H^B = \left(\begin{array}{cc|ccc} 0 & 0 & 1 & 1 & 3 \\ 0 & 0 & 5 & 2 & 1 \\ \hline 1 & 5 & 2 & 0 & 0 \\ 1 & 2 & 0 & 2 & 0 \\ 3 & 1 & 0 & 0 & 2 \end{array}\right)$$

The only principal minor of H^B of order $(2m + 1) = 5$ is H^B itself.

$|H^B| = 460$; i.e., the sign of $|H^B| = (-1)^2$.

$\therefore$ The optimal point is a minimum point of z.

$\therefore$ The minimum value of $z = \frac{897}{1058} \simeq 0.8478$.

Example 5

Maximise $z = 10x_1 + 8x_2 - x_1^2 - x_2^2$, subject to the constraint $2x_1 + 3x_2 \leq 6$; $x_1, x_2 \geq 0$.

Solution

$$|H_f| = \begin{vmatrix} fx_1x_1 & fx_1x_2 \\ fx_2x_1 & fx_2x_2 \end{vmatrix} = \begin{vmatrix} -2 & 0 \\ 0 & -2 \end{vmatrix};$$

$$D_1 = -2 < 0 \text{ and } D_2 = 4 > 0.$$

$\therefore$ $z = f(x_1, x_2)$ is negative definite and hence concave.

$h \equiv (2x_1 + 3x_2 - 6)$, being linear, can be regarded as a convex function. Moreover, the given constraint is of the form $h \leq 0$. Hence, the Kunh–Tucker conditions that are the necessary conditions for the maximum of z are also the sufficient conditions.

The Kunh–Tucker conditions are

$$\frac{\partial f}{\partial x_1} - \lambda \frac{\partial h}{\partial x_1} = 0;$$

i.e.,

$$10 - 2x_1 - 2\lambda = 0 \tag{1}$$

$$\frac{\partial f}{\partial x_2} - \lambda \frac{\partial h}{\partial x_2} = 0;$$

i.e.,

$$8 - 2x_2 - 3\lambda = 0 \tag{2}$$

$$\lambda h = 0;$$

i.e.,

$$\lambda(2x_1 + 3x_2 - 6) = 0 \tag{3}$$

$$h \leq 0,$$

i.e., $$2x_1 + 3x_2 - 6 \leq 0 \quad (4)$$

$$\lambda \geq 0,$$

i.e., $$\lambda \geq 0 \quad (5)$$

From (3), either $\lambda = 0$ or $2x_1 + 3x_2 - 6 = 0$

If $\lambda = 0$, from (1) and (2), we get $x_1 = 5$ and $x_2 = 4$.

For these values of x_1 and x_2, condition (4) is not satisfied.

$\therefore$ From (3), $$2x_1 + 3x_2 - 6 = 0 \quad (6)$$

From (1) and (2), $$\frac{10 - 2x_1}{8 - 2x_2} = \frac{2}{3}$$

or $$3x_1 - 2x_2 = 7 \quad (7)$$

Solving (6) and (7), we get $$x_1 = \frac{33}{13}$$

and $$x_2 = \frac{4}{13}$$

and hence, from (1) and (2), $$\lambda = \frac{32}{13}$$

Thus, all the Kunh–Tucker conditions are satisfied by $\left(\frac{33}{13}, \frac{4}{13}\right)$.

$\therefore$ The maximal point of z is $\left(\frac{33}{13}, \frac{4}{13}\right)$ and the maximum value of $z = \frac{3361}{169}$ or 19.89.

Example 6

Maximise $z = 2x_1^2 - 7x_2^2 + 12x_1x_2$, subject to the condition $2x_1 + 5x_2 \leq 98$; $x_1, x_2 \geq 0$.

SOLUTION

$$|H_f| = \begin{vmatrix} 4 & 12 \\ 12 & -14 \end{vmatrix};$$

$$D_1 = 4 > 0;$$

$$D_2 = -200 < 0.$$

$\therefore$ f is indefinite. Hence, the sufficiency condition of the optimum of $z = f(x_1, x_2)$ cannot be tested. If we find the two optimum points using Kunh–Tucker (necessary) conditions, we can determine which optimum point corresponds to the maximum of z.

Kunh–Tucker conditions are

$$\frac{\partial f}{\partial x_1} - \lambda \frac{\partial h}{\partial x_1} = 0,$$

i.e., $$4x_1 + 12x_2 - 2\lambda = 0 \quad (1)$$

$$\frac{\partial f}{\partial x_2} - \lambda \frac{\partial h}{\partial x_2} = 0,$$

i.e., $$12x_1 - 14x_2 - 5\lambda = 0 \quad (2)$$

$$\lambda h = 0,$$

i.e., $$\lambda(2x_1 + 5x_2 - 98) = 0 \quad (3)$$

$$h \le 0,$$

i.e., $$2x_1 + 5x_2 - 98 \le 0 \quad (4)$$

$$\lambda \ge 0,$$

i.e., $$\lambda \ge 0 \quad (5)$$

From (3), either $\lambda = 0$ or $2x_1 + 5x_2 - 98 = 0$
If $\lambda = 0$, from (1) and (2), we get $4x_1 + 12x_2 = 0$ and $12x_1 - 14x_2 = 0$
Solving these equations, we get $x_1 = 0$ and $x_2 = 0$.
For these values of x_1 and x_2, (4) is satisfied viz., all the Kunh–Tucker conditions are satisfied by the point (0, 0).
∴ (0, 0) is an optimum point of $z = f(x_1, x_2)$.
In (3), if $2x_1 + 5x_2 - 98 = 0$ (and $\lambda > 0$), from (1) and (2), we get $\frac{2x_1 + 6x_2}{12x_1 - 14x_2} = \frac{1}{5}$, viz., $x_1 + 8x_2 = 0$.
Solving the equations $2x_1 + 5x_2 = 98$ and $x_1 - 22x_2 = 0$, we get $x_1 = 44$ and $x_2 = 2$.
For these values of x_1 and x_2, from (1) or (2), we get $\lambda = 100$.
∴ All the Kunh–Tucker conditions are satisfied by the point (44, 2)
∴ (44, 2) is an optimum point of $z = f(x_1, x_2)$
Corresponding to the point (0, 0), the optimal value of $z = 0$.
Corresponding to the point (44, 2), the optimal value of $z = 4900$.
∴ (44, 2) is the maximum point of z and the maximum value of $z = 4900$.

Example 7

Minimise $z = 5x_1^2 + 6x_2^2$, subject to the constraint $5x_1 + x_2 \ge 3$; $x_1, x_2 \ge 0$.

Solution

$$|H_f| = \begin{vmatrix} f_{x_1x_1} & f_{x_1x_2} \\ f_{x_2x_1} & f_{x_2x_2} \end{vmatrix},$$

where $$z \equiv f = 5x_1^2 + 6x_2^2$$

$$= \begin{vmatrix} 10 & 0 \\ 0 & 12 \end{vmatrix};$$

$$D_1 = 10 > 0$$

and $$D_2 = 120 > 0$$

∴ f is positive definite and so $f(x_1, x_2)$ is a convex function $h \equiv (5x_1 + x_2 - 3)$ is a linear function in x_1, x_2 and can be considered as a concave function. Moreover, the given constraint is of the form $h \ge 0$.

Hence, the Kunh–Tucker conditions are both necessary and sufficient conditions for the minimum of $z = f(x_1, x_2)$.

They are

$$\frac{\partial f}{\partial x_1} - \lambda \frac{\partial h}{\partial x_1} = 0;$$

i.e., $$10x_1 - 5\lambda = 0 \quad (1)$$

$$\frac{\partial f}{\partial x_2} - \lambda \frac{\partial h}{\partial x_2} = 0;$$

i.e., $$12x_2 - \lambda = 0 \quad (2)$$

$$\lambda h = 0;$$

i.e., $$\lambda(5x_1 + x_2 - 3) = 0 \quad (3)$$

$$h \geq 0;$$

i.e., $$5x_1 + x_2 - 3 \geq 0 \quad (4)$$

and $$\lambda \geq 0;$$

i.e., $$\lambda \geq 0 \quad (5)$$

From (3), either $$\lambda = 0$$

or $$5x_1 + x_2 - 3 = 0$$

If $\lambda = 0$, from (1) and (2), we get $x_1 = 0$ and $x_2 = 0$

For these values of x_1 and x_2, condition (4) is not satisfied.

$\therefore$ $$5x_1 + x_2 - 3 = 0 \quad (6)$$

From (1) and (2), $$\frac{10x_1}{12x_2} = 5$$

or $$x_1 - 6x_2 = 0 \quad (7)$$

Solving (6) and (7), we get $$x_1 = \frac{18}{31}$$

and $$x_2 = \frac{3}{31}$$

For these values of x_1 and x_2, $$\lambda = \frac{36}{31} > 0$$

Thus, all the Kunh–Tucker conditions are satisfied by the point $\left(\frac{18}{31}, \frac{3}{31}\right)$

$\therefore$ The point $\left(\frac{18}{31}, \frac{3}{31}\right)$ is the minimum point of z and the minimum value of $z = \frac{1674}{961} \simeq 1.742$.

Example 8

Maximise $z = 4x_1 + 6x_2 - x_1^2 - x_2^2 - x_3^2$, subject to the constraints $x_1 + x_2 \leq 2$, $2x_1 + 3x_2 \leq 12$; $x_1, x_2, x_3 \geq 0$.

SOLUTION

$$z = f(x_1, x_2, x_3) = 4x_1 + 6x_2 - x_1^2 - x_2^2 - x_3^2$$

$$|H_f| = \begin{vmatrix} fx_1x_1 & fx_1x_2 & fx_1x_3 \\ fx_2x_1 & fx_2x_2 & fx_2x_3 \\ fx_3x_1 & fx_3x_2 & fx_3x_3 \end{vmatrix} = \begin{vmatrix} -2 & 0 & 0 \\ 0 & -2 & 0 \\ 0 & 0 & -2 \end{vmatrix};$$

$$D_1 < 0, D_2 > 0$$

and $$D_3 < 0$$

$\therefore$ f is a negative definite and hence concave.

$h_1 \equiv x_1 + x_2 - 2$ and $h_2 \equiv 2x_1 + 3x_2 - 12$ are linear functions in x_1, x_2, x_3 and can be regarded as convex functions

Hence, the optimal point that satisfies the Kunh–Tucker (KT) conditions will also satisfy the sufficiency for maximum of z, since $h_1 \leq 0$ and $h_2 \leq 0$.

The Kunh–Tucker conditions for this problem are

$$\frac{\partial f}{\partial x_1} - \lambda_1 \frac{\partial h_1}{\partial x_1} - \lambda_2 \frac{\partial h_2}{\partial x_1} = 0$$

viz., $$4 - 2x_1 - \lambda_1 - 2\lambda_2 = 0 \tag{1}$$

$$\frac{\partial f}{\partial x_2} - \lambda_1 \frac{\partial h_1}{\partial x_2} - \lambda_2 \frac{\partial h_2}{\partial x_2} = 0,$$

viz., $$6 - 2x_2 - \lambda_1 - 3\lambda_2 = 0 \tag{2}$$

$\Rightarrow$ $$\frac{\partial f}{\partial x_3} - \lambda_1 \frac{\partial h_1}{\partial x_3} - \lambda_2 \frac{\partial h_2}{\partial x_3} = 0,$$

viz., $$-2x_3 = 0 \tag{3}$$

$$\lambda_1 h_1 = 0,$$

viz., $$\lambda_1(x_1 + x_2 - 2) = 0 \tag{4}$$

$$\lambda_2 h_2 = 0,$$

viz., $$\lambda_2(2x_1 + 3x_2 - 12) = 0 \tag{5}$$

$$h_1 \leq 0,$$

viz., $$x_1 + x_2 - 2 \leq 0 \tag{6}$$

$$h_2 \leq 0,$$

viz., $$2x_1 + 3x_2 - 12 \leq 0 \tag{7}$$

viz., $$\lambda_1 \geq 0 \tag{8}$$

and $$\lambda_2 \geq 0 \tag{9}$$

Case (i): $\lambda_1 = 0$ and $\lambda_2 = 0$

From (1), (2) and (3), we get $x_1 = 2$, $x_2 = 3$, and $x_3 = 0$.
These values do not satisfy (6) and (7).
$\therefore$ $\lambda_1 = 0$ and $\lambda_2 = 0$ do not satisfy the KT conditions

Case (ii): $\lambda_1 \neq 0$ and $\lambda_2 \neq 0$

From (4) and (5), we have $x_1 + x_2 = 2$, and $2x_1 + 3x_2 = 12$
Solving these equations, we get $x_1 = -6$, $x_2 = 8$, $x_3 = 0$
These values do not satisfy the non-negativity constraints.

Case (iii): $\lambda_1 = 0$ and $\lambda_2 \neq 0$

From (1) and (2), we have $x_1 = 2 - \lambda_2$

and $$x_2 = \frac{1}{2}(6 - 3\lambda_2)$$

Solving these equations, we get $x_1 = -6$, $x_2 = 8$, $x_3 = 0$
These values do not satisfy the non-negativity constraints.
Using these values in (5),

we get $$2(2-\lambda_2) + \frac{3}{2}(6 - 3\lambda_2) - 12 = 0$$

i.e., $$-13\lambda_2 + 2 = 0$$

or $$\lambda_2 = \frac{2}{13}$$

Using this value in (1) and (2), $$x_1 = 2 - \frac{2}{13} = \frac{24}{13}$$

and $$x_2 = \frac{1}{2}\left(6 - \frac{6}{13}\right) = \frac{36}{13}$$

Though these values of x_1, x_2 satisfy (7), they do not satisfy (6).
$\therefore$ $\lambda_1 = 0$ and $\lambda_2 \neq 0$, do not satisfy the KT conditions.

Case (iv): $\lambda_1 \neq 0$ and $\lambda_2 = 0$.

From (1) and (2), we have $$x_1 = \frac{1}{2}(4 - \lambda_1)$$

and $$x_2 = \frac{1}{2}(6 - \lambda_1)$$

Using these values in (4),

$$\frac{1}{2}(4 - \lambda_1) + \frac{1}{2}(6 - \lambda_1) - 2 = 0$$

i.e., $$6 - 2\lambda_1 = 0$$

or $$\lambda_1 = 3$$

Using this value in (1) and (2), $x_1 = \frac{1}{2}$

and $x_2 = \frac{3}{2}$

These values of x_1, x_2 satisfy (6) and (7).
Thus, all the KT conditions are satisfied by the point $\left(\frac{1}{2}, \frac{3}{2}, 0\right)$

Maximum point of z is $\left(\frac{1}{2}, \frac{3}{2}, 0\right)$ and the maximum value of $z = 4\times\frac{1}{2} + 6\times\frac{3}{2} - \frac{1}{4} - \frac{9}{4} - 0 = \frac{17}{2}$.

Example 9

Minimise $z = x_1^2 + x_2^2$, subject to the constraints $x_1 + x_2 \geq 4$, $x_1 + 2x_2 \geq 5$; $x_1, x_2 \geq 0$.

SOLUTION

The given problem is a standard minimisation problem.

$$z = f(x_1, x_2) = x_1^2 + x_2^2$$

$$|H_f| = \begin{vmatrix} fx_1x_1 & fx_1x_2 \\ fx_2x_1 & fx_2x_2 \end{vmatrix} = \begin{vmatrix} 2 & 0 \\ 0 & 2 \end{vmatrix}; D_1 = 2 > 0$$

and $D_2 = 4 > 0$.

$f(x_1, x_2)$ is a positive definite function and hence, convex. Also $h_1 = x_1 + x_2 - 4$ and $h_2 = x_1 + 2x_2 - 5$ are linear functions in x_1, x_2 and can be regarded as concave functions. Hence, the optimal point that satisfies the Kunh–Tucker conditions will also satisfy the sufficiency for minimum of z.

The KT conditions for the problem are

$$\frac{\partial f}{\partial x_1} - \lambda_1 \frac{\partial h_1}{\partial x_1} - \lambda_2 \frac{\partial h_2}{\partial x_1} = 0,$$

viz., $$2x_1 - \lambda_1 - \lambda_2 = 0 \qquad (1)$$

$$\frac{\partial f}{\partial x_2} - \lambda_1 \frac{\partial h_1}{\partial x_2} - \lambda_2 \frac{\partial h_2}{\partial x_2} = 0,$$

viz., $$2x_2 - \lambda_1 - 2\lambda_2 = 0 \qquad (2)$$

$$\lambda_1 h_1 = 0,$$

viz., $$\lambda_1(x_1 + x_2 - 4) = 0 \qquad (3)$$

$$\lambda_2 h_2 = 0,$$

viz., $$\lambda_2(x_1 + 2x_2 - 5) = 0 \qquad (4)$$

$$h_1 \geq 0,$$

viz., $$x_1 + x_2 - 4 \geq 0 \qquad (5)$$

$$h_2 \geq 0,$$

viz., $$x_1 + 2x_2 - 5 \geq 0 \qquad (6)$$

and $$\lambda_1 \geq 0 \tag{7}$$

$$\lambda_2 \geq 0 \tag{8}$$

It can be verified that the pairs of values (i) $\lambda_1 = 0$ and $\lambda_2 = 0$, (ii) $\lambda_1 = 0$ and $\lambda_2 \neq 0$ and (iii) $\lambda_1 \neq 0$ and $\lambda_2 \neq 0$ do not satisfy the KT conditions.

Let us now consider the values $\lambda_1 \neq 0$ and $\lambda_2 = 0$.

For these values of λ_1 and λ_2 from (1) and (2),

we get $$x_1 = \frac{\lambda_1}{2}$$

and $$x_2 = \frac{\lambda_1}{2}.$$

From (3), $$x_1 + x_2 - 4 = 0,$$

since $$\lambda_1 \neq 0$$

i.e., $$\lambda_1 = 4$$

$\therefore$ $x_1 = 2$ and $x_2 = 2$.

These values satisfy all the K T conditions

$\therefore$ The required minimal point of z is (2, 2) and the minimum values of $z = 8$.

Example 10

Minimise $z = (x_1 - 2)^2 + (x_2 - 1)^2$, subject to the constraints $x_1^2 - x_2 \leq 0$, $x_1 + x_2 - 2 \leq 0$; $x_1, x_2 \geq 0$.

SOLUTION

In order to convert the problem into a standard minimisation problem, we rewrite the constraints as $-x_1^2 + x_2 \geq 0$ and $-x_1 - x_2 + 2 \geq 0$

$$z = f(x_1, x_2) = (x_1 - 2)^2 + (x_2 - 1)^2$$

$$|H_f| = \begin{vmatrix} 2 & 0 \\ 0 & 2 \end{vmatrix}; D_1 > 0$$

and $$D_2 > 0$$

$\therefore$ $f(x_1, x_2)$ is a positive definite function and hence, convex. Also $h_1 = -x_1^2 + x_2$ is concave, since $|H_{h_1}| = \begin{vmatrix} -2 & 0 \\ 0 & 0 \end{vmatrix}$; $D_1 < 0$ and $D_2 = 0$ and hence negative semi definite.

$h_2(x_1, x_2)$, being a linear function in (x_1, x_2), can be regarded as a concave function.

Hence, the optimal point that satisfies the KT conditions will also satisfy the sufficiency for minimum of z.

The KT conditions for this problem are

$$2(x_1 - 2) + \lambda_1 2x_1 + \lambda_2 = 0 \quad (1);$$

$$2(x_2 - 1) - \lambda_1 + \lambda_2 = 0 \tag{2}$$

$$\lambda_1(-x_1^2 + x_2) = 0 \quad (3);$$

$$\lambda_2(-x_1 - x_2 + 2) = 0 \tag{4}$$

$$-x_1^2 + x_2 \geq 0 \qquad (5);$$

$$-x_1 - x_2 + 2 \geq 0 \qquad (6);$$

$$\lambda_1 \geq 0 \qquad (7);$$

$$\lambda_2 \geq 0 \qquad (8)$$

In the three cases corresponding to (i) $\lambda_1 = 0$ and $\lambda_2 = 0$, (ii) $\lambda_1 = 0$ and $\lambda_2 \neq 0$ and (iii) $\lambda_1 \neq 0$ and $\lambda_2 = 0$, KT conditions are not satisfied and the optimal values of x_1 and x_2 cannot be determined uniquely. Let us now consider the case when $\lambda_1 \neq 0$ and $\lambda_2 \neq 0$.

In this case, from (3) and (4), we have

$$x_1^2 = x_2$$

and $$x_1 + x_2 = 2$$

i.e., $$x_1^2 + x_1 - 2 = 0$$

or $$(x_1 + 2)(x_1 - 1) = 0$$

$\therefore$ $$x_1 = 1$$

or $$x_1 = -2$$

$\therefore$ Since, $x_1 \geq 0$, the value $x_1 = -2$ is discarded

$\therefore$ $x_1 = 1$ and so $x_2 = 1$

For these values of x_1 and x_2, we get from (1) and (2),

$$2\lambda_1 + \lambda_2 = 2$$

and $$\lambda_1 - \lambda_2 = 0$$

Solving these equations, we get $$\lambda_1 = \lambda_2 = \frac{2}{3}.$$

$\therefore$ Equations (7) and (8) are satisfied.

Thus, $x_1 = 1 = x_2$ satisfy all the Kunh–Tucker conditions

$\therefore$ The required minimal point of z is (1, 1) and the minimum value of $z = 1$.

EXERCISE 7(A)

Part A (Short answer questions)

1. Give the mathematical form of a general non-linear programming problem.
2. What is the difference between a convex programming problem and a quadratic programming problem?
3. How will you find whether a function $f(X) = f(x_1, x_2, \ldots, x_n)$ is convex or concave, using the Hessian matrix of $f(X)$?
4. State the necessary conditions for the optimum of an NLPP with equality constraints.
5. When will the necessary conditions for the optimum of an NLPP with equality constraints become sufficient conditions?

6. Define the bordered Hessian matrix for a classical optimisation problem.
7. State the sufficient conditions for the optimal point (x_0, y_0) to be a maximum point of a classical optimisation problem when the number m of equality constraints is less than the number n of variables.
8. State the same for (x_0, y_0) to be a minimum point when $m = n - 1$.
9. State the Kunh–Tucker conditions for the optimum of a standard maximisation problem. When will they become sufficient conditions?
10. State the Kunh–Tucker conditions for the optimum of a standard minimisation problem. When will they become sufficient conditions?

Part B

Solve the following NLPPs, using the method of Lagrange's multipliers:

11. Maxmise $z = 4x_1 - 0.02x_1^2 + x_2 - 0.02x_2^2$, subject to the constraint $x_1 + 2x_2 = 120; x_1, x_2 \geq 0$.
12. Minimise $z = 6x_1^2 + 5x_2^2$ subject to the constraint $x_1 + 5x_2 = 3$; $x_1, x_2 \geq 0$
13. Minimise $z = 2x_1^2 + x_2^2 + 3x_3^2 + 10x_1 + 8x_2 + 6x_3 - 100$, subject to the constraint $x_1 + x_2 + x_3 = 20, x_i \geq 0$
14. Minimise $z = x_1^2 + 2x_2^2 + x_3^2$, subject to the constraint $2x_1 + x_2 + 2x_3 = 30; x_i \geq 0$.
15. Optimise $z = x_1^2 + x_2^2 + x_3^2$, subject to the constraint $4x_1 + x_2^2 + 2x_3 = 14$; $x_i \geq 0$. [Hint: $\frac{\partial L}{\partial x_2} = 0$ leads to the equation $(1-\lambda)x_2 = 0; \lambda = 1$ corresponds to the minimum of z and $x_2 = 0$ corresponds to the maximum of z.]
16. Maximise $z = 8x_1 + 6x_2 - x_1^2 - x_2^2$, subject to the constraints $3x_1 + 4x_2 = 16$, $5x_1 + 3x_2 = 15$; $x_1, x_2 \geq 0$
17. Minimise $z = 2x_1^2 + x_2^2 + 3x_3^2$, subject to the constraints $2x_1 + x_2 + 3x_3 = 30$ and $3x_1 - 2x_2 + 4x_3 = 25; x_1, x_2, x_3 \geq 0$
18. Minimise $z = 4x_1^2 + 2x_2^2 + x_3^2 - 4x_1x_2$, subject to the constraints $x_1 + x_2 + x_3 = 15$ and $2x_1 - x_2 + 2x_3 = 20$; $x_1, x_2, x_3 \geq 0$

Solve the following NLPPs using Kunh–Tucker conditions:

19. Maximise $z = 3x_1^2 + 14x_1x_2 - 8x_2^2$, subject to the constraint $x_1 + 2x_2 \leq 24; x_1, x_2 \geq 0$
20. Maximise $z = x_1^2 + x_1x_2 - 2x_2^2$, subject to the constraint $2x_1 + x_2 \leq 12; x_1, x_2 \geq 0$
21. Maximise $z = 2x_1 + x_2 - x_1^2$, subject to the constraints $x_1 + x_2 \leq 6; 2x_1 + x_2 \leq 4; x_1, x_2 \geq 0$
22. Minimise $z = 7x_1^2 + 6x_1 + 5x_2$, subject to the constraint $x_1 + 2x_2 \leq 10; x_1 - 3x_2 \leq 9; x_1, x_2 \geq 0$ [Hint: Convert the problem into a standard minimisation problem, by changing $\leq$ into $\geq$ in the constraints]
23. Minimise $z = x_1^2 + x_2^2 - 2x_1 - 4x_2$, subject to the constraint $x_1 + 4x_2 - 5 \leq 0, 2x_1 + 3x_2 - 6 \leq 0; x_1, x_2 \geq 0$
[Hint: Convert the problem into a standard minimisation problem]
24. Minimise $z = (x_1 - 1)^2 + (x_2 - 2)^2$, subject to the constraints $0 \leq x_1 \leq 2, 0 \leq x_2 \leq 1, x_1, x_2 \geq 0$.
25. Minimise $z = (x_1 - 1)^2 + (x_2 - 2)^2$, subject to the constraints $x_1 - x_2^2 \geq 0, x_1 + x_2 \leq 2; x_1, x_2 \geq 0$

7.3 QUADRATIC PROGRAMMING

As was pointed out in the beginning, a quadratic programming problem is a particular case of non-linear programming problem (NLPP) in which we have to maximise (or minimise) an objective function that is a quadratic expression subject to a set of linear inequality constraints.

An algorithm given by Wolfe to solve quadratic programming problems is based on fairly simple modifications of the Simplex method and converges to a finite number of iterations.

Note: *Wolfe's algorithm* is applied to a standard maximisation problem of the following form:

$$\text{Maximise } z = f(x_1, x_2, \ldots x_n) = \sum_{j=1}^{n} c_j x_j + \frac{1}{2}\sum_{j=1}^{n}\sum_{k=1}^{n} c_{jk} x_j x_k \text{ subject to the constraints } \sum_{j=1}^{n} a_{ij} x_j \le b_i,$$

for $i = 1, 2, \ldots m$ and $x_j \ge 0$, for $j = 1, 2, \ldots, n$, where $\sum_{j=1}^{n}\sum_{k=1}^{n} c_{jk} x_j x_k$ is a negative semi-definite quadratic form in which $c_{jk} = c_{kj}$ for all j and k.

If the quadratic programming problem is one of minimisation, it should be converted into a standard maximisation problem. Only then can Wolfe's algorithm be applied to solve it.

Step-by-step procedure of Wolfe's modified simplex algorithm

Step 1: Convert the inequality structural constraints into equations by adding the non-negative slack variables x_{n+i} ($i = 1, 2, \ldots, m$) and add the non-negative slack variables x_{n+m+j} ($j = 1, 2, \ldots, n$) to the non-negative constraints after converting them to $-x_j \le 0$($j = 1, 2, \ldots, n$)

Step 2: Construct the Lagrangian function

$$L = f(x_1, x_2, \ldots, x_n) - \sum_{i=1}^{m} \lambda_i \left\{\left[\sum_{j=1}^{n} a_{ij} x_j\right] - b_i + x_{n+i}\right\} - \sum_{j=1}^{n} \lambda_{m+j}\left(-x_j + x_{n+m+j}\right).$$

Step 3: Write down the necessary conditions for the maximum points of $f(x_1, x_2, \ldots, x_n)$ according to the Lagrange's method, viz.,

$$\frac{\partial L}{\partial x_j} = 0;$$

$$j = 1, 2, \ldots, n, n+1, n+2, \ldots, n+m, \tag{1}$$

and

$$\frac{\partial L}{\partial \lambda_j} = 0;$$

$$i = 1, 2, \ldots, m, m+1, m+2, \ldots, m+n, \tag{2}$$

Step 4: Split the set of equations (1) into two sets of equations

$$\frac{\partial L}{\partial x_j} = 0,$$

for

$$j = 1, 2, \ldots, n \tag{3}$$

and

$$\frac{\partial L}{\partial x_j} = 0,$$

for

$$j = n+1, n+2, \ldots, (n+m) \tag{4}$$

Split the set of equations (2) into two sets of equations

$$\frac{\partial L}{\partial \lambda_i} = 0,$$

for $\quad i = 1, 2, \ldots, m \qquad (5)$

and $$\frac{\partial L}{\partial \lambda_i} = 0,$$

for $\quad i = m+1, m+2, \ldots, (m+n) \qquad (6)$

Step 5: Solve the system of linear equations in (1) and (2) by the two-phase Simplex method, and by treating them as constraints for optimising a dummy objective function

$$z = 0 \cdot x_1 + 0 \cdot x_2 + \cdots + 0 \cdot x_{n+m} + 0 \cdot \lambda_1 + 0 \cdot \lambda_2 + \cdots + 0 \cdot \lambda_{m+n}$$

Step 6: Modify the dummy objective function by adding $y_1, y_2, \ldots, y_n$ (all ≥ 0) to z (call it z^*) and by adding them to the n constraints given by (3)

Step 7: Maximise $z^* = y_1 + y_2 + \cdots + y_n$, by assuming the cost coefficients as -1 each or minimise $z^* = y_1 + y_2 + \cdots + y_n$, by assuming the cost coefficients as $+1$ each.

Step 8: The optimum solution obtained in Step 7 is the maximum solution of the given QPP.

Note: The set of equations (4) results in $\lambda_i = 0$,

for $\quad i = 1, 2, \ldots, m$

and hence, $\quad \lambda_i x_{n+i} = 0,$

for $\quad i = 1, 2, \ldots, m \qquad (7)$

The set of equations (6) results in

$$x_{n+m+j} = x_j,$$

for $\quad j = 1, 2, \ldots, n$

and hence $\quad \lambda_{m+j} x_j = 0,$

for $\quad j = 1, 2, \ldots, n \qquad (8)$

Each pair λ_i and $x_{n+i} (i = 1, 2, \ldots, m)$ and similarly, each pair λ_{m+j} and $x_j (j = 1, 2, \ldots, n)$ are called complementary variables. Conditions (7) and (8) are equivalent to (4) and (6) and are called *complementary slackness conditions*.

While solving the QPP by Wolfe's method (viz., by the two-phase Simplex method), care should be taken to see that the basis at any stage of iteration contains only one complementary variable.

Beale's method for solving a quadratic programming problem

Beale's technique of solving a QPP uses neither Lagrange's multiplier method nor the Kunh–Tucker conditions. It consists of partitioning variables into basic and non-basic ones. The step-by-step procedure involved in Beale's algorithms can be easily derived after understanding the worked examples given here.

WORKED EXAMPLES 7(B)

Example 1

Use Wolfe's method to maximise $z = 4x_1 + 6x_2 - x_1^2 + 3x_2^2$, subject to the constraint $x_1 + 2x_2 \le 4;\ x_1, x_2 \ge 0$.

SOLUTION

Rewriting the constraints, we have $x_1 + 2x_2 + x_3 = 4$, $-x_1 + x_4 = 0$ and $-x_2 + x_5 = 0$

Lagrange's function for this problem is

$$L = 4x_1 + 6x_2 - x_1^2 + 3x_2^2 - \lambda_1(x_1 + 2x_2 + x_3 - 4) - \lambda_2(-x_1 + x_4) - \lambda_3(-x_2 + x_5)$$

The necessary conditions for the maximum of z are given by $\frac{\partial L}{\partial x_1} = 0, \frac{\partial L}{\partial x_2} = 0, \frac{\partial L}{\partial \lambda_1} = 0$ and $\frac{\partial L}{\partial x_3} = 0, \frac{\partial L}{\partial x_4} = 0, \frac{\partial L}{\partial x_5} = 0, \frac{\partial L}{\partial \lambda_2} = 0, \frac{\partial L}{\partial \lambda_3} = 0$

viz.,
$$4 - 2x_1 - \lambda_1 + \lambda_2 = 0$$

or
$$2x_1 + \lambda_1 - \lambda_2 = 4 \qquad (1)$$

$$6 - 6x_2 - 2\lambda_1 + \lambda_3 = 0$$

or
$$6x_2 + 2\lambda_1 - \lambda_3 = 6 \qquad (2)$$

$$x_1 + 2x_2 + x_3 = 4 \qquad (3) \qquad (A)$$

and
$$\lambda_1 = 0, \lambda_2 = 0, \lambda_3 = 0, \quad -x_1 + x_4 = 0$$

and
$$-x_2 + x_5 = 0 \qquad (B)$$

The constraints in (B) are equivalent to the complementary slackness conditions $\lambda_i x_{n+i} = 0$, for $i = 1, 2, \ldots, m$ and $\lambda_{m+j} x_j = 0$, for $j = 1, 2, \ldots, n$.

viz.,
$$\lambda_1 x_3 = 0,$$

$$\lambda_2 x_1 = 0$$

and
$$\lambda_3 x_2 = 0 \qquad (C)$$

To solve the equations in (A) and (B) simultaneously, let us maximise $z^* = 0 \cdot x_1 + 0 \cdot x_2 + 0 \cdot x_3 + 0 \cdot \lambda_1 + 0 \cdot \lambda_2 + 0 \cdot \lambda_3 + y_1 + y_2$, subject to the constraints

$$2x_1 + \lambda_1 - \lambda_2 + y_1 = 4, \qquad (1)',$$

$$6x_2 + 2\lambda_1 - \lambda_3 + y_2 = 6, \qquad (2)',$$

equation (3) and equations in (B) and also the complementary slackness conditions in (C), by using the two-phase Simplex method, where $x_r \ge 0, \lambda_r \ge 0,\ y_r \ge 0$ are

	c_j		0	0	0	0	0	0	−1	−1	
B	C_B	X_B	x_1	x_2	x_3	λ_1	λ_2	λ_3	y_1	y_2	RR
y_1	−1	4	2	0	0	1	−1	0	1	0	-
y_2	−1	6	0	[6]	0	2	0	−1	0	1	1 →
x_3	0	4	1	2	1	0	0	0	0	0	2
$z^* = -10$			−2	−6 ↑	0	−3	1	1	0	0	-

B	C_B	X_B	x_1	x_2	x_3	λ_1	λ_2	λ_3	y_1	y_2	RR
y_1	−1	4	[2]	0	0	1	−1	0	1	0	2 →
x_2	0	1	0	1	0	$\frac{1}{3}$	0	$-\frac{1}{6}$	0	$\frac{1}{6}$	–
x_3	0	2	1	0	1	$-\frac{2}{3}$	0	$\frac{1}{3}$	0	$-\frac{1}{3}$	2
$z^* = -4$			−2 ↑	0	0	−1	1	0	0	1	–

B	C_B	X_B	x_1	x_2	x_3	λ_1	λ_2	λ_3	y_1	y_2
x_1	0	2	1	0	0	$\frac{1}{2}$	$-\frac{1}{2}$	0	$\frac{1}{2}$	0
x_2	0	1	0	1	0	$\frac{1}{3}$	0	$-\frac{1}{6}$	0	$\frac{1}{6}$
x_3	0	0	0	0	1	$-\frac{7}{6}$	$\frac{1}{2}$	$\frac{1}{3}$	$-\frac{1}{2}$	$-\frac{1}{3}$
$z^* = 0$			0	0	0	0	0	0	1	1

Optimality has been attained at $x_1 = 2$ and $x_2 = 1$

Maximum value of $z = 8 + 6 - 4 - 3 = 7$

***Note*:** We can also minimise $z^* = y_1 + y_2$, with the cost coefficients for y_1 and y_2 as 1 each.

Example 2

Use Wolfe's method to minimise $z' = x_1^2 - x_1x_2 + 2x_2^2 - x_1 - x_2$, subject to the constraint $2x_1 + x_2 \leq 1$; $x_1, x_2 \geq 0$.

SOLUTION

We convert the problem into one of maximisation as given here:

To maximise $z = x_1 + x_2 - x_1^2 + x_1x_2 - 2x_2^2$, subject to $2x_1 + x_2 \leq 1; x_1, x_2 \geq 0$ so that $\min(z') = -(\max z)$.

Rewriting the constraints, we have $2x_1 + x_2 + x_3 = 1$, $-x_1 + x_4 = 0$ and $-x_2 + x_5 = 0$.

Lagrange's function for the problem is

$$L = x_1 + x_2 - x_1^2 + x_1x_2 - 2x_2^2 - \lambda_1(2x_1 + x_2 + x_3 - 1) - \lambda_2(-x_1 + x_4) - \lambda_3(-x_2 + x_5)$$

The necessary conditions for the maximum of z are given by

$$\frac{\partial L}{\partial x_1} = 0,$$

viz., $$2x_1 - x_2 + 2\lambda_1 - \lambda_2 = 1 \quad (1)$$

$$\frac{\partial L}{\partial x_2} = 0,$$

i. e., $$-x_1 + 4x_2 + \lambda_1 - \lambda_3 = 1 \quad (2)$$

$$\frac{\partial L}{\partial \lambda_1} = 0,$$

i. e., $$2x_1 + x_2 + x_3 = 1 \quad (3)$$

$$\frac{\partial L}{\partial x_3} = 0,$$

i. e., $$\lambda_1 = 0 \quad (4);$$

$$\frac{\partial L}{\partial x_4} = 0,$$

i. e., $$\lambda_2 = 0 \quad (5);$$

$$\frac{\partial L}{\partial x_5} = 0,$$

i. e., $$\lambda_3 = 0 \quad (6)$$

$$\frac{\partial L}{\partial \lambda_2} = 0,$$

i. e., $$-x_1 + x_4 = 0 \quad (7)$$

$$\frac{\partial L}{\partial \lambda_3} = 0,$$

i. e., $$-x_2 + x_5 = 0 \quad (8)$$

Conditions (4) to (7) are equivalent to $\lambda_i x_{n+i} = 0$, i. e., $\lambda_1 x_3 = 0$ and $\lambda_{m+j} x_j = 0$, i.e., $\lambda_2 x_1 = 0$ and $\lambda_3 x_2 = 0$

Using two-phase Simplex method, we minimise $Z^* = \sum_{r=1}^{3} 0, x_r + \sum_{r=1}^{3} 0, \lambda_r + y_1 + y_2$, subject to

$2x_1 - x_2 + 2\lambda_1 - \lambda_2 + y_1 = 1, -x_1 + 4x_2 + \lambda_1 - \lambda_3 + y_2 = 1, 2x_1 + x_2 + x_3 = 1, x_r \geq 0, \lambda_r \geq 0$ and $y_r \geq 0$.

		C_j:	0	0	0	0	0	0	1	1		
B	C_B	X_B	x_1	x_2	x_3	λ_1	λ_2	λ_3	y_1	y_2	RR	
y_1	1	1	2	−1	0	2	−1	0	1	0	–	
y_2	1	1	−1	[4]	0	1	0	−1	0	1	$\frac{1}{4}$	→
x_3	0	1	2	1	1	0	0	0	0	0	1	
	$z^* = 2$		1	3	0	3	−1	−1	0	0	–	
				↑								

Note: The index row entries corresponding to the variables x_2 and λ_1 are most positive, each equal to 3. If we decide to bring λ_1 into the basis, then y_1 leaves the basis. This results in a situation where the basis will contain the complementary variables pair λ_1 and x_3 in the second stage of iteration. As this is to be avoided, we make x_2 to enter the basis and proceed further.

		C_j	0	0	0	0	0	0	1	1		
B	C_B	X_B	x_1	x_2	x_3	λ_1	λ_2	λ_3	y_1	y_2	RR	
y_1	1	$\frac{5}{4}$	$\frac{7}{4}$	0	0	$\frac{9}{4}$	−1	$-\frac{1}{4}$	1	$\frac{1}{4}$	$\frac{5}{7}$	
x_2	0	$\frac{1}{4}$	$-\frac{1}{4}$	1	0	$\frac{1}{4}$	0	$-\frac{1}{4}$	0	$\frac{1}{4}$	–	
x_3	0	$\frac{3}{4}$	[$\frac{9}{4}$]	0	1	$-\frac{1}{4}$	0	$\frac{1}{4}$	0	$-\frac{1}{4}$	$\frac{1}{3}$	→
	$z^* = \frac{5}{4}$		$\frac{7}{4}$	0	0	$\frac{9}{4}$	−1	$-\frac{1}{4}$	0	$-\frac{3}{4}$	–	
			↑									

B	C_B	X_B	x_1	x_2	x_3	λ_1	λ_2	λ_3	y_1	y_2	R. R	
y_1	1	$\frac{2}{3}$	0	0	$-\frac{7}{9}$	[$\frac{22}{9}$]	−1	$-\frac{4}{9}$	1	$\frac{4}{9}$	$\frac{3}{11}$	→

(Contd)

Table (*Contd*)

B	C_B	X_B	x_1	x_2	x_3	λ_1	λ_2	λ_3	y_1	y_2	R. R
x_2	0	$\frac{1}{3}$	0	1	$\frac{1}{9}$	$\frac{2}{9}$	0	$-\frac{2}{9}$	0	$\frac{2}{9}$	$\frac{3}{2}$
x_1	0	$\frac{1}{3}$	1	0	$\frac{4}{9}$	$-\frac{1}{9}$	0	$\frac{1}{9}$	0	$-\frac{1}{9}$	–
	$z^* = \frac{2}{3}$		0	0	$-\frac{7}{9}$	$\frac{22}{9}$	−1	$-\frac{4}{9}$	0	$-\frac{5}{9}$	–
						↑					

B	C_B	X_B	x_1	x_2	x_3	λ_1	λ_2	λ_3	y_1	y_2
λ_1	0	$\frac{3}{11}$	0	0	$-\frac{7}{22}$	1	$-\frac{9}{22}$	$-\frac{2}{11}$	$\frac{9}{22}$	$\frac{2}{11}$
x_2	0	$\frac{3}{11}$	0	1	$\frac{2}{11}$	0	$\frac{1}{11}$	$-\frac{2}{11}$	$-\frac{1}{11}$	$\frac{2}{11}$
x_1	0	$\frac{4}{11}$	1	0	$\frac{9}{22}$	0	$-\frac{1}{22}$	$\frac{1}{11}$	$\frac{1}{22}$	$-\frac{1}{11}$
	$z^* = 0$		0	0	0	0	0	0	−1	−1

Optimality has been attained corresponding to $x_1 = \frac{4}{11}$ and $x_2 = \frac{3}{11}$

Maximum of $Z = \frac{4}{11} + \frac{3}{11} - \frac{16}{121} + \frac{12}{121} - \frac{18}{121} = \frac{5}{11}$.

Minimum of $Z' = -\frac{5}{11}$.

Example 3

Use Wolfe's method to maximise $z = x_1 + 2x_2 - x_2^2$, subject to the constraints $3x_1 + 2x_2 \le 6$, $x_1 + 2x_2 \le 4$; $x_1, x_2 \ge 0$.

Solution

Rewriting constraints, we have $3x_1 + 2x_2 + x_3 = 6$,

$$x_1 + 2x_2 + x_4 = 4,\ -x_1 + x_5 = 0,\ -x_2 + x_6 = 0.$$

Lagrange's function for the problem is

$$L = x_1 + 2x_2 - x_2^2 - \lambda_1(3x_1 + 2x_2 + x_3 - 6) - \lambda_2(x_1 + 2x_2 + x_4 - 4) - \lambda_3(-x_1 + x_5) - \lambda_4(-x_2 + x_6)$$

The necessary conditions for the maximum of z are given by

$$\frac{\partial L}{\partial x_1} = 0,$$

viz., $$3\lambda_1 + \lambda_2 - \lambda_3 = 1 \qquad (1)$$

$$\frac{\partial L}{\partial x_2} = 0,$$

viz., $$2x_2 + 2\lambda_1 + 2\lambda_2 - \lambda_4 = 2 \qquad (2)$$

$$\frac{\partial L}{\partial \lambda_1} = 0,$$

viz., $$3x_1 + 2x_2 + x_3 = 6 \qquad (3)$$

$$\frac{\partial L}{\partial \lambda_2} = 0,$$

viz., $$x_1 + 2x_2 + x_4 = 4 \qquad (4)$$

and the complementary slackness conditions

$$\lambda_1 x_3 = 0, \lambda_2 x_4 = 0, \lambda_3 x_1 = 0$$

and $$\lambda_4 x_2 = 0 \qquad (5)$$

Using two-phase Simplex method, let us minimise $z^* = y_1 + y_2$, subject to

$$3\lambda_1 + \lambda_2 - \lambda_3 + y_1 = 1 \qquad (1'),$$

$$-2x_2 + 2\lambda_1 + 2\lambda_2 - \lambda_4 = 2 \qquad (2'),$$

and (3), (4) and (5), taking the cost coefficients for y_1 and y_2 as 1 each

		C_j	0	0	0	0	0	0	0	0	1	1		
B	C_B	X_B	x_1	x_2	x_3	x_4	λ_1	λ_2	λ_3	λ_4	y_1	y_2	RR	
y_1	1	1	0	0	0	0	3	1	−1	0	1	0	–	
y_2	1	2	0	2	0	0	[2]	2	0	−1	0	1	1	→
x_3	0	6	3	2	1	0	0	0	0	0	0	0	3	
x_4	0	4	1	2	0	1	0	0	0	0	0	0	2	
	$z^* = 3$		0	2	0	0	5	3	−1	−1	0	0	–	
				↑										

Note: As λ_1 and λ_2 cannot enter the basis, x_2 enters.

B	C_B	X_B	x_1	x_2	x_3	x_4	λ_1	λ_2	λ_3	λ_4	y_1	
y_1	1	1	0	0	0	0	3	1	−1	0	1	
x_2	0	1	0	1	0	0	1	1	0	$-\frac{1}{2}$	0	
x_3	0	4	[3]	0	1	0	−2	−2	0	1	0	→

(Contd)

Table (*Contd*)

B	C_B	X_B	x_1	x_2	x_3	x_4	λ_1	λ_2	λ_3	λ_4	y_1
x_4	0	2	1	0	0	1	−2	−2	0	1	0
	$z^*=1$		0	0	0	0	3	1	−1	0	0
			↑								

Note 1 As λ_1 and λ_2 cannot enter the basis, x_1 enters.

Note 2 As y_2 leaves the basis, the column corresponding to it has been deleted from the tables in the first and subsequent iterations.

B	C_B	X_B	x_1	x_2	x_4	λ_1	λ_2	λ_3	λ_4	y_1	
y_1	1	1	0	0	0	3	1	−1	0	1	→
x_2	0	1	0	1	0	1	1	0	$-\frac{1}{2}$	0	
x_1	0	$\frac{4}{3}$	1	0	0	$-\frac{2}{3}$	$-\frac{2}{3}$	0	$\frac{1}{3}$	0	
x_4	0	$\frac{2}{3}$	0	0	1	$-\frac{4}{3}$	$-\frac{4}{3}$	0	$\frac{2}{3}$	0	
	$z^*=1$		0	0	0	3	1	−1		0	
						↑					

B	C_B	X_B	x_1	x_2	x_4	λ_1	λ_2	λ_3	λ_4
λ_1	0	$\frac{1}{3}$	0	0	0	1	$\frac{1}{3}$	$-\frac{1}{3}$	0
x_2	0	$\frac{2}{3}$	0	1	0	0	$\frac{2}{3}$	$\frac{1}{3}$	$-\frac{1}{2}$
x_1	0	$\frac{14}{9}$	1	0	0	0	$-\frac{4}{9}$	$-\frac{2}{9}$	$\frac{1}{3}$
x_4	0	$\frac{10}{9}$	0	0	1	0	$-\frac{8}{9}$	$-\frac{4}{9}$	$\frac{2}{3}$
$z^*=0$			0	0	0	0	0	0	0

Optimality has been attained corresponding to

$$x_1 = \frac{14}{9},$$

$$x_2 = \frac{2}{3},$$

$$x_3 = 0,$$

$$x_4 = \frac{10}{9},$$

$$\lambda_1 = \frac{1}{3},$$

$$\lambda_2 = \lambda_3 = \lambda_4 = 0$$

Maximum value of $$Z = \frac{14}{9} + \frac{4}{3} - \frac{4}{9} = \frac{22}{9}.$$

Example 4

Use Wolfe's method to maximise $z = 6x_1 + 3x_2 - x_1^2 + 4x_1x_2 - 4x_2^2$, subject to the constraints, $x_1 + x_2 \leq 3, 4x_1 + x_2 \leq 9; x_1, x_2 \geq 0$

Solution

Rewriting the constraints, we have $x_1 + x_2 + x_3 = 3$, $4x_1 + x_2 + x_4 = 9$, $-x_1 + x_5 = 0$, $-x_2 + x_6 = 0$.

Lagrange's function for the problem is

$$L = 6x_1 + 3x_2 - x_1^2 + 4x_1x_2 - 4x_2^2 - \lambda_1(x_1 + x_2 + x_3 - 3) - \lambda_2(4x_1 + x_2 + x_4 - 9) - \lambda_3(-x_1 + x_5) - \lambda_4(-x_2 + x_6)$$

The necessary conditions for the maximum of z are

$$\frac{\partial L}{\partial x_1} = 0,$$

viz., $$2x_1 - 4x_2 + \lambda_1 + 4\lambda_2 - \lambda_3 = 6 \quad (1)$$

$$\frac{\partial L}{\partial x_2} = 0,$$

viz., $$-4x_1 + 8x_2 + \lambda_1 + \lambda_2 - \lambda_4 = 3 \quad (2)$$

$$\frac{\partial L}{\partial \lambda_1} = 0,$$

viz., $$x_1 + x_2 + x_3 = 3 \quad (3)$$

$$\frac{\partial L}{\partial \lambda_2} = 0,$$

viz., $$4x_1 + x_2 + x_4 = 9 \quad (4)$$

and the complementary slackness conditions

$$\lambda_1 x_3 = 0,$$

$$\lambda_2 x_4 = 0,$$

$$\lambda_3 x_1 = 0$$

and $$\lambda_4 x_2 = 0 \tag{5}$$

Using two-phase Simplex method, let us maximise

$$z^* = y_1 + y_2,$$

subject to $$2x_1 - 4x_2 + \lambda_1 + 4\lambda_2 - \lambda_3 + y_1 = 6 \qquad (1)',$$

$$-4x_1 + 8x_2 + \lambda_1 + \lambda_2 - \lambda_4 + y_2 = 3 \qquad (2')$$

and (3),(4) and (5), taking the cost coefficients for y_1 and y_2 as -1 each

		C_j:	0	0	0	0	0	0	0	0	−1	−1	
B	C_B	X_B	x_1	x_2	x_3	x_4	λ_1	λ_2	λ_3	λ_4	y_1	y_2	
y_1	−1	6	2	−4	0	0	1	4	−1	0	1	0	
y_2	−1	3	−4	[8]	0	0	1	1	0	−1	0	1	→
x_3	0	3	1	1	1	0	0	0	0	0	0	0	
x_4	0	9	4	1	0	1	0	0	0	0	0	0	
$z^*=-9$			2	−4	0	0	−2	−5	1	1	0	0	
				↑									

B	C_B	X_B	x_1	x_2	x_3	x_4	λ_1	λ_2	λ_3	λ_4	y_1	
y_1	−1	$\frac{15}{2}$	0	0	0	0	$\frac{3}{2}$	$\frac{9}{2}$	−1	$-\frac{1}{2}$	1	
x_2	0	$\frac{3}{8}$	$-\frac{1}{2}$	1	0	0	$\frac{1}{8}$	$\frac{1}{8}$	0	$-\frac{1}{8}$	0	
x_3	0	$\frac{21}{8}$	[$\frac{3}{2}$]	0	1	0	$-\frac{1}{8}$	$-\frac{1}{8}$	0	$\frac{1}{8}$	0	→
x_4	0	$\frac{69}{8}$	$\frac{9}{2}$	0	0	1	$-\frac{1}{8}$	$-\frac{1}{8}$	0	$\frac{1}{8}$	0	
$z^* = -\frac{15}{2}$			0	0	0	0	$-\frac{3}{2}$	$-\frac{9}{2}$	1	$\frac{1}{2}$	0	
			↑									

B	C_B	X_B	x_1	x_2	x_4	λ_1	λ_2	λ_3	λ_4	y_1	
y_1	−1	$\frac{15}{2}$	0	0	0	$\frac{3}{2}$	$\frac{9}{2}$	−1	$-\frac{1}{2}$	1	
x_2	0	$\frac{5}{4}$	0	1	0	$\frac{1}{12}$	$\frac{1}{12}$	0	$-\frac{1}{12}$	0	
x_1	0	$\frac{7}{4}$	1	0	0	$-\frac{1}{12}$	$-\frac{1}{12}$	0	$\frac{1}{12}$	0	
x_4	0	$\frac{3}{4}$	0	0	1	[$\frac{1}{4}$]	$\frac{1}{4}$	0	$-\frac{1}{4}$	0	→
$z^* = -\frac{15}{2}$			0	0	0	$-\frac{3}{2}$	$-\frac{9}{2}$	1	$\frac{1}{2}$	0	
						↑					

B	C_B	X_B	x_1	x_2	λ_1	λ_2	λ_3	λ_4	y_1	
y_1	−1	3	0	0	0	3	−1	1	1	→
x_2	0	1	0	1	0	0	0	0	0	
x_1	0	2	1	0	0	0	0	0	0	
λ_1	0	3	0	0	1	1	0	−1	0	
	$z^* = -3$		0	0	0	−3	1	−1	0	
						↑				

B	C_B	X_B	x_1	x_2	λ_1	λ_2	λ_3	λ_4
λ_2	0	1	0	0	0	1	$-\frac{1}{3}$	$\frac{1}{3}$
x_2	0	1	0	1	0	0	0	0
x_1	0	2	1	0	0	0	0	0
λ_1	0	2	0	0	1	0	$\frac{1}{3}$	$-\frac{4}{3}$
		$z^* = 0$	0	0	0	0	0	0

Optimality has been attained with $x_1 = 2, x_2 = 1(x_3 = x_4 = 0, \lambda_1 = 2, \lambda_2 = 1, \lambda_3 = \lambda_4 = 0)$
Maximum value of $z = 12 + 3 - 4 + 8 - 4 + 15$.

Example 5

Use Wolfe's method to minimise $z' = 2x_1^2 + 4x_1x_2 + 3x_2^2 - 6x_1 - 3x_2$, subject to the constraints $x_1 + x_2 \leq 1, 2x_1 + 3x_2 \leq 4; x_1, x_2 \geq 0$.

Solution

Converting the problem into one of maximisation, it becomes:

Maximise $z = 6x_1 + 3x_2 - 2x_1^2 - 4x_1x_2 - 3x_2^2$, subject to the given constraints, so that min $(z') = -\max(z)$.

Proceeding as usual, the necessary conditions for the maximum of z are given by

$$4x_1 + 4x_2 + \lambda_1 + 2\lambda_2 - \lambda_3 = 6 \quad (1)$$

$$4x_1 + 6x_2 + \lambda_1 + 3\lambda_2 - \lambda_4 = 3 \quad (2)$$

$$x_1 + x_2 + x_3 - 1 = 0 \quad (3)$$

$$2x_1 + 3x_2 + x_4 - 4 = 0 \quad (4)$$

and the complementary slackness conditions

$$\lambda_1 x_3 = 0,$$

$$\lambda_2 x_4 = 0,$$

$$\lambda_3 x_1 = 0$$

and

$$\lambda_4 x_2 = 0 \quad (5)$$

Using the two-phase Simplex method, let us minimise $z^* = y_1 + y_2$, where y_1 and y_2 (the artificial variables with cost coefficient 1 for each) are introduced in the constraints (1) and (2) respectively.

		C_j	0	0	0	0	0	0	0	0	1	1	
B	C_B	X_B	x_1	x_2	x_3	x_4	λ_1	λ_2	λ_3	λ_4	y_1	y_2	
y_1	1	6	4	4	0	0	1	2	−1	0	1	0	
y_2	1	3	4	[6]	0	0	1	3	0	−1	0	1	→
x_3	0	1	1	1	1	0	0	0	0	0	0	0	
x_4	0	4	2	3	0	1	0	0	0	0	0	0	
$z^* = 9$			8	10	0	0	2	5	−1	−1	0	0	
				↑									

B	C_B	X_B	x_1	x_2	x_3	x_4	λ_1	λ_2	λ_3	λ_4	y_1	
y_1	1	4	$\frac{4}{3}$	0	0	0	$\frac{1}{3}$	0	−1	$\frac{2}{3}$	1	
x_2	0	$\frac{1}{2}$	[$\frac{2}{3}$]	1	0	0	$\frac{1}{6}$	$\frac{1}{2}$	0	$-\frac{1}{6}$	0	→
x_3	0	$\frac{1}{2}$	$\frac{1}{3}$	0	1	0	$-\frac{1}{6}$	$-\frac{1}{2}$	0	$\frac{1}{6}$	0	
x_4	0	$\frac{5}{2}$	0	0	0	1	$-\frac{1}{2}$	$-\frac{3}{2}$	0	$\frac{1}{2}$	0	
	$z^* = 4$		$\frac{4}{3}$	0	0	0	$\frac{1}{3}$	0	−1	$\frac{2}{3}$	0	
			↑									

B	C_B	X_B	x_1	x_3	x_4	λ_1	λ_2	λ_3	λ_4	y_1	
y_1	1	3	0	0	0	0	−1	−1	1	1	
x_1	0	$\frac{3}{4}$	1	0	0	$\frac{1}{4}$	$\frac{3}{4}$	0	$-\frac{1}{4}$	0	
x_3	0	$\frac{1}{4}$	0	1	0	$-\frac{1}{4}$	$-\frac{3}{4}$	0	[$\frac{1}{4}$]	0	→
x_4	0	$\frac{5}{2}$	0	0	1	$-\frac{1}{2}$	$-\frac{3}{2}$	0	$\frac{1}{2}$	0	
	$z^* = 3$		0	0	0	0	−1	−1	1	0	
									↑		

B	C_B	X_B	x_1	x_4	λ_1	λ_2	λ_3	λ_4	y_1	
y_1	1	2	0	0	[1]	2	−1	0	1	→
x_1	0	1	1	0	0	0	0	0	0	
λ_4	0	1	0	0	−1	−3	0	1	0	

(Contd)

Table (*Contd*)

B	C_B	X_B	x_1	x_4	λ_1	λ_2	λ_3	λ_4	y_1
x_4	0	2	0	1	0	0	0	0	0
$z^* = 2$			0	0	1	2	−1	0	0
					↑				

B	C_B	X_B	x_1	x_4	λ_1	λ_2	λ_3	λ_4
λ_1	0	2	0	0	1	2	−1	0
x_1	0	1	1	0	0	0	0	0
λ_4	0	3	0	0	0	−1	−1	1
x_4	0	2	0	1	0	0	0	0
$z^* = 0$			0	0	0	0	0	0

Optimality has been attained with $x_1 = 1$ and $x_2 = 0$ $(x_3 = x_4 = 0)$

Maximum value of $z = 6 - 2 = 4$

$\therefore$ Minimum value of $z' = -4$

Example 6

Use Beale's method to maximise $z = 3x_1 + 6x_2 - 3x_1^2 - 4x_1x_2 - 2x_2^2$, subject to the constraint $x_1 + x_2 \leq 1$; $x_1 \geq 0, x_2 \geq 0$.

SOLUTION

Introducing the slack variable $x_{3,}$ the constraint is made an equality viz., the constraint becomes

$$x_1 + x_2 + x_3 = 1 \tag{1}$$

Interation 1

Step1: Choose one variable (as there is only constraint), say x_3, as a basic variable and the remaining variables x_1 and x_2 as non-basic variables (NBV). The partitioning of the variables is arbitrary.

Step 2 Express z in terms of the NBVs x_1, x_2

Thus, $$z = f(x_1, x_2) = 3x_1 + 6x_2 - 3x_1^2 - 4x_1x_2 - 2x_2^2$$

Step 3: Find $\frac{\partial f}{\partial x_1}$ and $\frac{\partial f}{\partial x_2}$ and evaluate $\left(\frac{\partial f}{\partial x_1}\right)_{(0,\,0)}$ and $\left(\frac{\partial f}{\partial x_2}\right)_{(0,\,0)}$, as we try to determine whether (0, 0) could be the optimal point of f. If both $\left(\frac{\partial f}{\partial x_1}\right)_{(0,\,0)}$ and $\left(\frac{\partial f}{\partial x_2}\right)_{(0,\,0)}$ are ≤ 0, the optimum of f has been attained at (0, 0).

In this problem, $$\left(\frac{\partial f}{\partial x_1}\right)_{(0,\,0)} = (3 - 6x_1 - 4x_2)_{(0,\,0)} = 3$$

$$\left(\frac{\partial f}{\partial x_2}\right)_{(0,\,0)} = (6 - 4x_1 - 4x_2)_{(0,0)} = 6.$$

$\left(\frac{\partial f}{\partial x_1}\right)_{(0,0)}$ and $\left(\frac{\partial f}{\partial x_2}\right)_{(0,0)}$ are both > 0, the variable x_2 corresponding to the greater value, namely 6, enters the basis and x_1 remains as a non-basic variable with value 0.

Step 4: To determine by what quantity x_2 may increase in order to maximise z, we proceed as follows:

$$[x_3]_{x_1=0} = [1 - x_1 - x_2]_{x_1=0} = 1 - x_2 \qquad \text{using (1)}$$

As $$(x_3)_{x_1=0} \geq 0,$$

$$1 - x_2 \geq 0$$

or $$x_2 \leq 1,$$

viz,.the upper bound (UB) of $x_2 = 1$ (2)

With $x_1 = 0$, f can be considered as a function of x_2 only.

For f to be maximum, $$\frac{df}{dx_2} = 0$$

or $$\left(\frac{\partial f}{\partial x_2}\right)_{x_1=0} = 0$$

viz., $$(6 - 4x_1 - 4x_2)_{x_1=0} = 0$$

$\therefore$ $$x_2 = \frac{3}{2} \qquad (3)$$

Since (2) and (3) are to be satisfied, the improved value of $x_2 = \min\left(1, \frac{3}{2}\right) = 1$

As this improved value of x_2 was derived by considering the nature of x_3, x_3 is replaced by x_2.

Step 5: Steps 2, 3 and 4 are repeated with the new basic and non-basic variables, until the optimum basic feasible solution is obtained.

Iteration 2

Basic variable is x_2 and the non-basic variables are (x_1, x_3)

$$z = f(x_1, x_3) = 3x_1 + 6(1 - x_1 - x_3) - 3x_1^2 - 4x_1(1 - x_1 - x_3) - 2(1 - x_1 - x_3)^2$$

$$\frac{\partial f}{\partial x_1} = 3 - 6 - 6x_1 - 4 + 8x_1 + 4x_3 + 4(1 - x_1 - x_3); \left(\frac{\partial f}{\partial x_1}\right)_{(0,0)} = -3$$

$$\frac{\partial f}{\partial x_3} = -6 + 4x_1 + 4(1 - x_1 - x_3); \left(\frac{\partial f}{\partial x_3}\right)_{(0,0)} = -2$$

Since both $\left(\frac{\partial f}{\partial x_1}\right)_{(0,0)}$ and $\left(\frac{\partial f}{\partial x_3}\right)_{(0,0)}$ are ≤ 0, optimality has been attained with $x_1 = 0$ and $x_3 = 0$ or $1 - x_1 - x_2 = 0$ i.e., $x_2 = 1$

$\therefore$ Maximum of z occurs at $x_1 = 0$ and $x_2 = 1$ and maximum value of $z = 6 - 2 = 4$.

Example 7

Use Beale's method to maximise $z = 2x_1 + x_2 - x_1^2$, subject to the constraint $2x_1 + 3x_2 \leq 6$; $x_1 \geq 0, x_2 \geq 0$

SOLUTION

The constraint is converted as

$$2x_1 + 3x_2 + x_3 = 6 \tag{1}$$

where $x_3 \geq 0$ is a slack variable.

Iteration 1

Basic variable is x_3 and non-basic variables are x_1 and x_2

$$z = f(x_1, x_2) = 2x_1 + x_2 - x_1^2$$

$$\frac{\partial f}{\partial x_1} = 2 - 2x_1;$$

$$\frac{\partial f}{\partial x_2} = 1$$

and

$$\left(\frac{\partial f}{\partial x_1}\right)_{(0,0)} = 2;$$

$$\left(\frac{\partial f}{\partial x_2}\right)_{(0,0)} = 1$$

$\therefore$ x_1 enters the basis and x_2 remains as an NBV.

We express the BV x_3 in terms of NBVs x_1 and x_2

Using the constraint (1); i.e.; $x_3 = 6 - 2x_1 - 3x_2$,

$$\therefore \quad [x_3]_{x_2=0} = 6 - 2x_1 \geq 0,$$

$$\therefore \quad x_1 \leq 3 \tag{2}$$

$$\left(\frac{\partial f}{\partial x_1}\right)_{x_2=0} = 2 - 2x_1 = 0$$

$$\Rightarrow \quad x_1 = 1 \tag{3}$$

$\therefore$ Improved value of $x_1 = \min(3, 1) = 1$.

As this improved value of x_1 was obtained from (3), viz. by consideration of $\left(\frac{\partial f}{\partial x_1}\right)_{x_2=0}$, we introduce a free variable $x_4 = \frac{\partial f}{\partial x_1}$

i.e.,

$$x_4 = 2 - 2x_1$$

or

$$2x_1 + x_4 = 2 \tag{4},$$

which is taken as an additional constraint.

Iteration 2

BV is x_1, NBVs are x_2 and x_4; x_3 is kept in reserve.

$$z = f(x_2, x_4) = (2 - x_4) + x_2 - \frac{1}{4}(2 - x_4)^2, \text{ using (4)}$$

$$\left(\frac{\partial f}{\partial x_2}\right)_{(0,0)} = 1;$$

$$\left(\frac{\partial f}{\partial x_4}\right)_{(0,0)} = \left[-1 + \frac{1}{2}(2 - x_4)\right]_{(0,0)} = 0.$$

$\therefore$ x_2 enters the basis and x_4 remains as NBV.

Expressing the BV x_1 in terms of NBVs x_2 and x_4, we have from (4)

$$x_1 = \frac{1}{2}(2 - x_4)$$

$\therefore$ $$[x_1]_{x_4=0} = 1,$$

$\therefore$ the upper bound for x_2 cannot be found.

Also $\left(\frac{\partial f}{\partial x_2}\right)_{x_4=0} = 1$, which does not provide any value for x_2.

$\therefore$ Improved value of x_2 cannot be found out.

Iteration 3

BV is x_2, NBVs are x_3 and x_4; x_1 leaves the basis and is kept in reserve.

$$z = f(x_3, x_4) = (2 - x_4) + \frac{1}{3}(4 - x_3 + x_4) - \frac{1}{4}(2 - x_4)^2,$$

since $$\begin{aligned} 3x_2 &= 6 - 2x_1 - x_3, \text{ by (1)} \\ &= 6 - (2 - x_4) - x_3, \text{ by (4)} \\ &= 4 - x_3 + x_4. \end{aligned}$$

$$\frac{\partial f}{\partial x_3} = -\frac{1}{3};$$

$$\frac{\partial f}{\partial x_4} = -1 + \frac{1}{3} + \frac{1}{2}(2 - x_4)$$

$\therefore$ $$\left(\frac{\partial f}{\partial x_3}\right)_{(0,0)} = -\frac{1}{3};$$

$$\left(\frac{\partial f}{\partial x_4}\right)_{(0,0)} = \frac{1}{3}$$

$\therefore$ x_4 enters the basis and x_3 remains as NBV.

Expressing the BV x_2 in terms of the NBVs x_3 and x_4, we have $x_2 = \frac{1}{3}(4 - x_3 + x_4)$

$\therefore$ $[x_2]_{x_3=0} = \frac{1}{3}(4 + x_4)$;

$\therefore$ $$\text{UB}(x_4) = \infty.$$

$$\frac{\partial f}{\partial x_4} = \left(\frac{\partial f}{\partial x_4}\right)_{x_3=0}$$

$$= \frac{1}{3} - \frac{1}{2}x_4 = 0$$

gives $$x_4 = \frac{2}{3}$$

Improved value of $x_4 = \min\left(\infty, \frac{2}{3}\right) = \frac{2}{3}$, which corresponds to $\left(\frac{\partial f}{\partial x_4}\right)_{x_3=0}$.

$\therefore$ We introduce another free variable $x_5 = \frac{\partial f}{\partial x_4}$ i.e., $x_5 = \frac{1}{3} - \frac{1}{2}x_4$ or $3x_4 + 6x_5 = 2$ (5) which is taken as an additional constraint.

Iteration 4

B.V. is x_4, NBVs are x_3 and x_5; x_2 leaves the basis and is kept in reserve.

$$z = f(x_3, x_5) = \left(\frac{4}{3} + 2x_5\right) + \frac{1}{9}(14 - 3x_3 - 6x_5) - \left(\frac{2}{3} + x_5\right)^2, \text{ by using (4) and (5)}$$

$\therefore$
$$\left(\frac{\partial f}{\partial x_3}\right) = -\frac{1}{3}; \frac{\partial f}{\partial x_5} = 2 - \frac{2}{3} - 2\left(\frac{2}{3} + x_5\right)$$

and
$$\left(\frac{\partial f}{\partial x_3}\right)_{(0,0)} = -\frac{1}{3} \text{ and } \left(\frac{\partial f}{\partial x_5}\right)_{(0,0)} = 0$$

$\therefore$ Optimality has been attained with $x_3 = 0$ and $x_5 = 0$

viz., $$2x_1 + 3x_2 = 6, \qquad \text{by (1)}$$

and $$3x_4 + 6x_5 = 2 \qquad \text{by (5)}$$

viz., $$2x_1 + 3x_2 = 6,$$

and $$x_4 = \frac{2}{3}$$

viz., $$x_1^* = \frac{1}{2}(2 - x_4) = \frac{2}{3}$$

and $$x_2^* = \frac{1}{3}(6 - 2x_1) = \frac{14}{9}$$

The maximum value of $z = z^* = \frac{22}{9}$.

Example 8

Use Beale's method to maximise $z = x_1 + x_2 - x_1^2$, subject to the constraint $x_1 + x_2 \leq 1$; $x_1 \geq 0, x_2 \geq 0$.

SOLUTION

The constraint is converted as

$$x_1 + x_2 + x_3 = 1 \quad (1),$$

where $x_3 \geq 0$ is a slack variable.

Iteration 1

Let x_3 be the BV and (x_1, x_2) be NBVs

$$z = f(x_1, x_2) = x_1 + x_2 - x_1^2$$

$$\frac{\partial f}{\partial x_1} = 1 - 2x_1;$$

$$\frac{\partial f}{\partial x_2} = 1$$

$\therefore$
$$\left(\frac{\partial f}{\partial x_1}\right)_{(0,0)} = 1;$$

and
$$\left(\frac{\partial f}{\partial x_2}\right)_{(0,0)} = 1$$

Once there is a tie for the variable to enter the basis, we choose x_1 as the BV and (x_2, x_3) as the NBVs

$$z = f(x_2, x_3) = (1 - x_2 - x_3) + x_2 - (1 - x_2 - x_3)^2$$

$$\frac{\partial f}{\partial x_2} = -1 + 1 + 2(1 - x_2 - x_3);$$

$$\frac{\partial f}{\partial x_3} = -1 + 2(1 - x_2 - x_3)$$

$$\left(\frac{\partial f}{\partial x_2}\right)_{(0,0)} = 2;$$

$$\left(\frac{\partial f}{\partial x_3}\right)_{(0,0)} = 1;$$

$\therefore$ x_2 enters the basis and x_3 remains as NBV.

We express the current BV x_3 in terms of NBVs x_1 and x_2

i.e.,
$$x_3 = 1 - x_1 - x_2$$

$\therefore$
$$[x_3]_{x_1=0} = 1 - x_2 \geq 0$$

$\therefore$
$$x_2 \leq 1 \quad (2)$$

$$\left(\frac{\partial f}{\partial x_2}\right)_{x_1=0} = 2(1 - x_2) = 0$$

$\Rightarrow$
$$x_2 = 1 \quad (3)$$

$\therefore$ Improved value of $x_2 = \min(1, 1) = 1$.

If we assume that the improved value of x_2 has been derived, from (3), we have to introduce a free variable $x_4 = \frac{\partial f}{\partial x_2}$ i.e., $x_4 = 2(1 - x_2 - x_3)$ or

$$2x_2 + 2x_3 + x_4 = 2 \quad (4)$$

as an additional constraint and proceed further by taking x_2 as BV and (x_3, x_4) as NBVs.

Instead, if we assume that the improved value of x_2 has been derived from (1), x_3 is replaced by x_2

Iteration 2

BV is x_2 and NBVs are x_1 and x_3.

$$z = f(x_1, x_3) = x_1 + (1 - x_1 - x_3) - x_1^2$$

$$\left(\frac{\partial f}{\partial x_1}\right) = -2x_1,$$

$$\frac{\partial f}{\partial x_3} = -1;$$

$$\therefore \quad \left(\frac{\partial f}{\partial x_1}\right)_{(0,0)} = 0;$$

$$\left(\frac{\partial f}{\partial x_3}\right)_{(0,0)} = -1$$

$\therefore$ Optimality has been attained with $x_1 = 0$ and $x_3 = 0$

viz., $x_1 = 0$ and $(1 - x_1 - x_2) = 0$

viz., $x_1^* = 0$ and $x_2^* = 1$ and the maximum value of $z = z^* = 1$.

Alternatively,

Iteration 2

BV is x_2 and NBVs are x_3 and x_4

$$z = f(x_3, x_4) = \frac{1}{2}x_4 + 1 - x_3 - \frac{1}{2}x_4 - \frac{1}{4}x_4^2, \quad \text{from (1) and (4)}$$

$$\left(\frac{\partial f}{\partial x_3}\right) = -1,$$

$$\frac{\partial f}{\partial x_4} = -\frac{1}{2}x_4;$$

$$\left(\frac{\partial f}{\partial x_3}\right)_{(0,0)} = -1,$$

$$\left(\frac{\partial f}{\partial x_4}\right)_{(0,0)} = 0$$

$\therefore$ Optimality has been attained with $x_3 = 0$ and $x_4 = 0$

viz., $\quad 1 - x_1 - x_2 = 0$

and $$1 - x_2 - x_3 = 0$$

viz., $x_1^* = 0$ and $x_2^* = 1$ and the maximum value of $z = z^* = 1$.

Example 9

Use Beale's method to maximise $z = x_1 + 3x_2 - x_1^2$, subject to the constraints $x_1 + x_2 \le 1$, $x_1 + 4x_2 \le 3$; $x_1 \ge 0, x_2 \ge 0$.

SOLUTION

Introducing the slack variables $x_3 \ge 0$ and $x_4 \ge 0$, the constraints become

$$x_1 + x_2 + x_3 = 1 \tag{1}$$

and $$x_1 + 4x_2 + x_4 = 3 \tag{2}$$

Iteration 1

Let (x_3, x_4) be taken as BVs and (x_1, x_2) as NBVs.

$$z = f(x_1, x_2) = x_1 + 3x_2 - x_1^2$$

$$\left(\frac{\partial f}{\partial x_1}\right) = 1 - 2x_1;$$

$$\frac{\partial f}{\partial x_2} = 3;$$

$\therefore$ $$\left(\frac{\partial f}{\partial x_1}\right)_{(0,0)} = 1,$$

$$\left(\frac{\partial f}{\partial x_2}\right)_{(0,0)} = 3$$

$\therefore$ x_2 enters the basis and x_1 remains as NBV.

$$[x_3]_{x_1=0} = [1 - x_1 - x_2]_{x_1=0} = 1 - x_2 \ge 0$$

$\therefore$ $$x_2 \le 1$$

or $$\text{UB }(x_2) = 1$$

$$[x_4]_{x_1=0} = [3 - x_1 - 4x_2]_{x_1=0} = 3 - 4x_2$$

$\therefore$ $$x_2 \le \frac{3}{4}$$

or $$\text{UB }(x_2) = \frac{3}{4}$$

$\left(\frac{\partial f}{\partial x_2}\right)_{x_1=0} = 3$, which does not give any information about x_2.

$\therefore$ Improved value of $x_2 = \min\left(1, \frac{3}{4}\right) = \frac{3}{4}$

As this improved value of x_2 is derived from the nature of x_4, x_4 leaves the basis and become N.B.V.

Iteration 2

BVs are x_2 and x_3; NBVs are x_1 and x_4.

$$z = f(x_1, x_4) = x_1 + \frac{3}{4}(3 - x_1 - x_4) - x_1^2$$

$$\left(\frac{\partial f}{\partial x_1}\right) = 1 - \frac{3}{4} - 2x_1;$$

$$\frac{\partial f}{\partial x_4} = -\frac{3}{4}.$$

$\therefore$ $$\left(\frac{\partial f}{\partial x_1}\right)_{(0,0)} = \frac{1}{4},$$

$$\left(\frac{\partial f}{\partial x_4}\right)_{(0,0)} = -\frac{3}{4}$$

$\therefore$ x_1 enters the basis and x_4 remains as NBV.
Expressing the BVs x_2 and x_3 in terms of the NBVs x_1 and x_4,

we have $$x_2 = \frac{1}{4}(3 - x_1 - x_4),$$ from (2)

and $$x_3 = 1 - x_1 - x_2 = 1 - x_1 - \frac{1}{4}(3 - x_1 - x_4),$$

from (1) and (2)

$$= \frac{1}{4} - \frac{3}{4}x_1 + \frac{1}{4}x_4$$

$\therefore$ $$[x_2]_{x_4=0} = \frac{1}{4}(3 - x_1) \geq 0$$

$\therefore$ $$x_1 \leq 3$$

or UB for $$x_1 = 3$$

$$[x_3]_{x_4=0} = \frac{1}{4}(1 - 3x_1) \geq 0$$

$\therefore$ $$x_1 \leq \frac{1}{3}$$

or UB for $$x_1 = \frac{1}{3}$$

$\therefore$ the lesser of the two upper bounds for $x_1 = \frac{1}{3}$, which is derived from $[x_3]_{x_4=0} \geq 0$. $\therefore$ x_3 leaves the basis.

$$\left(\frac{\partial f}{\partial x_1}\right)_{x_4=0} = \frac{1}{4} - 2x_1 = 0$$

$$\Rightarrow \qquad x_1 = \frac{1}{8}$$

Now $$\min\left(\frac{1}{3},\frac{1}{8}\right) = \frac{1}{8}.$$

$\therefore$ Improved value of $x_1 = \frac{1}{8}$, which is derived from $\left(\frac{\partial f}{\partial x_1}\right)_{x_4=0} = 0.$

$\therefore$ the leaving variable x_3 is kept in reserve and a free variable $x_5 = \frac{\partial f}{\partial x_1} = \frac{1}{4} - 2x_1$ is taken as NBV with x_4. The new constraint is $8x_1 + 4x_5 = 1$ (3)

Iteration 3

BVs are x_1 and x_2 and the NBVs are x_4 and x_5.

$$z = f(x_4, x_5) = \frac{1}{8}(1-4x_5) + \frac{3}{32}(23 - 8x_4 + 4x_5) - \frac{1}{64}(1-4x_5)^2,$$

since $$4x_2 = 3 - x_1 - x_4, \qquad \text{by (2)}$$

$$= 3 - \frac{1}{8}(1-4x_5) - x_4, \qquad \text{by (3)}$$

$$= \frac{1}{8}(23 - 8x_4 + 4x_5)$$

or $$x_2 = \frac{1}{32}(23 - 8x_4 + 4x_5)$$

$$\frac{\partial f}{\partial x_4} = -\frac{3}{4};$$

$$\frac{\partial f}{\partial x_5} = -\frac{1}{2} + \frac{3}{8} + \frac{1}{8}(1-4x_5)$$

$$\left(\frac{\partial f}{\partial x_4}\right)_{(0,0)} = -\frac{3}{4}$$

and $$\left(\frac{\partial f}{\partial x_5}\right)_{(0,0)} = 0.$$

$\therefore$ Optimality has been attained with $x_4 = 0$ and $x_5 = 0$

viz., $$x_1 + 4x_2 = 3$$

and $$8x_1 = 1$$

viz., $$x_1^* = \frac{1}{8}$$

and $$x_2^* = \frac{1}{4}(3 - x_1^*) = \frac{23}{32}.$$

The maximum value of $z = z^* = \frac{145}{64}.$

Example 10

Use Beale's method to maximise $z = 6x_1 + 3x_2 - x_1^2 + 4x_1x_2 - 4x^2$, subject to the constraints $x_1 + x_2 \leq 3$, $4x_1 + x_2 \leq 9$; $x_1 \geq 0$, $x_2 \geq 0$.

SOLUTION

Introducing the slack variables $x_3 \geq 0$ and $x_4 \geq 0$, the constraints become

$$x_1 + x_2 + x_3 = 3 \quad (1)$$

and

$$4x_1 + x_2 + x_4 = 9 \quad (2)$$

Iteration 1

Let (x_3, x_4) be taken as BVs and (x_1, x_2) as NBVs.

$$z = f(x_1, x_2) = 6x_1 + 3x_2 - x_1^2 + 4x_1x_2 - 4x^2$$

$$\left(\frac{\partial f}{\partial x_1}\right) = 6 - 2x_1 + 4x_2;$$

$$\left(\frac{\partial f}{\partial x_2}\right) = 3 + 4x_1 - 8x_2$$

$$\left(\frac{\partial f}{\partial x_1}\right)_{(0,0)} = 6$$

and

$$\left(\frac{\partial f}{\partial x_2}\right)_{(0,0)} = 3.$$

$\therefore$ x_1 enters the basis and x_2 remains as NBV.

$$[x_3]_{x_2=0} = [3 - x_1 - x_2]_{x_2=0} = 3 - x_1 \geq 0$$

$\Rightarrow$

$$x_1 \leq 3$$

or UB for

$$x_1 = 3$$

$$[x_4]_{x_2=0} = [9 - 4x_1 - x_2]_{x_2=0} = 9 - 4x_1 \geq 0$$

$\Rightarrow$

$$x_1 \leq \frac{9}{4}$$

or UB for

$$x_1 = \frac{9}{4}$$

The lesser of the two upper bounds of $x_1 = \dfrac{9}{4}$

$$\left(\frac{\partial f}{\partial x_1}\right)_{x_2=0} = (6 - 2x_1 + 4x_2)_{x_2=0} = 0$$

$\Rightarrow$

$$x_1 = 3$$

$\therefore$ Improved value of $x_1 = \min\left(\frac{9}{4}, 3\right) = \frac{9}{4}$, which corresponds to $[x_4]_{x_2=0} \geq 0$.

$\therefore$ x_4 leaves the basis giving place to x_1 and becomes a NBV.

Iteration 2

Basic variables are (x_1, x_3) and NBVs are (x_2, x_4)

$$z = f(x_2, x_4) = \frac{6}{4}(9 - x_2 - x_4) + 3x_2 - \frac{1}{16}(9 - x_2 - x_4)^2 + x_2(9 - x_2 - x_4) - 4x_2^2$$

$$\frac{\partial f}{\partial x_2} = -\frac{3}{2} + 3 + \frac{1}{8}(9 - x_2 - x_4) - x_2 + (9 - x_2 - x_4) - 8x_2$$

$$\frac{\partial f}{\partial x_4} = -\frac{3}{2} + \frac{1}{8}(9 - x_2 - x_4) - x_2$$

$\therefore$
$$\left(\frac{\partial f}{\partial x_2}\right)_{(0,0)} = -\frac{3}{2} + 3 + \frac{9}{8} + 9 = \frac{93}{8};$$

$$\left(\frac{\partial f}{\partial x_4}\right)_{(0,0)} = -\frac{3}{2} + \frac{9}{8} = -\frac{3}{8}$$

$\therefore$ x_2 enters the basis and x_4 remains as NBV.

$$[x_1]_{x_4=0} = \left[\frac{1}{4}(9 - x_2 - x_4)\right]_{x_4=0}$$

$$= \frac{1}{4}(9 - x_2) \geq 0$$

$\therefore$
$$x_2 \leq 9$$

or UB for
$$x_2 = 9.$$

$$[x_3]_{x_4=0} = \left[\frac{1}{4}(3 - 3x_2 + x_4)\right]_{x_4=0}$$

$$= \frac{1}{4}(3 - 3x_2) \geq 0$$

$\therefore$
$$x_2 \leq 1$$

or UB for
$$x_2 = 1.$$

Note:
$$4x_1 + 4x_2 + 4x_3 = 12 \quad (1)'$$

and
$$4x_1 + x_2 + x_4 = 9 \quad (2)$$

Eliminating x_1 from (1)′ and (2), we have $3x_2 + 4x_3 - x_4 = 3$ and so $x_3 = \frac{1}{4}(3 - 3x_2 + x_4)]$

The lesser of the two upper bounds of $x_2 = 1$, which is derived from $[x_3]_{x_4=0} \geq 0$

$$\left(\frac{\partial f}{\partial x_2}\right)_{x_4=0} = \frac{93}{8} - \frac{81}{8}x_2 = 0$$

$$\Rightarrow \qquad x_2 = \frac{31}{27}$$

Improved value of $x_2 = \min\left(1, \frac{31}{27}\right) = 1.$

As this improved value has been obtained by consideration of $[x_3]_{x_4=0} \geq 0$, x_3 leaves the basis, giving place to x_2 and becomes NBV.

Iteration 3

Basic variables are (x_1, x_2) and NBVs are (x_3, x_4)

$$z = f(x_3, x_4) = 2(6 + x_3 - x_4) + (3 - 4x_3 + x_4) - \frac{1}{9}(6 + x_3 - x_4)^2$$
$$+ \frac{4}{9}(6 + x_3 - x_4)(3 - 4x_3 + x_4) - \frac{4}{9}(3 - 4x_3 + x_4)^2,$$

got by solving (1) and (2).

$$\left(\frac{\partial f}{\partial x_3}\right)_{(0,0)} = -2; \left(\frac{\partial f}{\partial x_4}\right)_{(0,0)} = -1;$$

$\therefore$ Optimality has been attained with $x_3 = 0$ and $x_4 = 0$
viz., $x_1 + x_2 = 3$ and $4x_1 + x_2 = 9$
viz., $x_1^* = 2$ and $x_2^* = 1$
Maximum value of $z = z^* = 12 + 3 - 4 + 8 - 4 = 15$

EXERCISE 7(B)

Part A (Short answer questions)

1. Define a quadratic programming problem and name two standard methods used to solve it.
2. Give the step-by-step procedure of Wolfe's method of solving a quadratic programming problem.
3 Explain the complementary slackness conditions used in Wolfe's method.
4. When we maximise $z = f(x_1, x_2)$ by Beale's method of solving a QPP, how will you find the non-basic variable (x_1 or x_2) to enter the basis?
5. When solving a QPP, in which (x_1, x_2) are NBVs and x_3 is the basic variable and x_1 has to enter the basis in the next iteration, how will you find the improved value of x_1?
6. When solving a QPP by Beale's method, when is a free variable introduced?

Part B

Solve the following quadratic programming problems by Wolfe's method:

7. Maximise $z = 2x_1 + 3x_2 - x_1^2$, subject to the constraint $x_1 + 2x_2 \leq 4$; $x_1, x_2 \geq 0$.
8. Maximise $z = 4x_1 + 2x_2 - x_1^2 - x_2^2 - 5$, subject to the constraint $x_1 + x_2 \leq 4$; $x_1, x_2 \geq 0$.
9. Minimise $z = 6 - 6x_1 + 2x_1^2 - 2x_1x_2 + 2x_2^2$, subject to the constraint $x_1 + x_2 \leq 2$; $x_1, x_2 \geq 0$.
10. Maximise $z = 4x_1 + 6x_2 - 2x_1^2 - 2x_1x_2 - 2x_2^2$, subject to the constraint $x_1 + 2x_2 \leq 2$; $x_1, x_2 \geq 0$.
11. Maximise $z = 2x_1 + 3x_2 - 2x_1^2$, subject to the constraints $x_1 + 4x_2 \leq 4$, $x_1 + x_2 \leq 2$, $x_1, x_2 \geq 0$.
12. Maximise $z = 2x_1 + 3x_2 - 2x_2^2$, subject to the constraints $x_1 + 4x_2 \leq 4$, $x_1 + x_2 \leq 2$, $x_1, x_2 \geq 0$.
13. Minimise $z = 2x_1 + x_2 - x_1^2$, subject to the constraints $2x_1 + 3x_2 \leq 6$, $2x_1 + x_2 \leq 4$, $x_1, x_2 \geq 0$

Solve the following quadratic programming problems by Beale's method:

14. Maximise $z = 6x_1 + 4x_2 - 3x_1^2 - x_2^2$, subject to the constraint $2x_1 + x_2 \le 4$; $x_1, x_2 \ge 0$.
15. Maximise $z = x_1 + x_2 - x_1^2 - x_2^2$, subject to the constraint $x_1 + x_2 \le 10$; $x_1, x_2 \ge 0$.
16. Minimise $z = 183 - 44x_1 - 42x_2 + 8x_1^2 - 12x_1x_2 + 17x_2^2$, subject to the constraint $2x_1 + x_2 \le 10$; $x_1, x_2 \ge 0$.
17. Maximise $z = x_1 + 2x_2 - x_2^2$, subject to the constraints $3x_1 + 2x_2 \le 6$, $x_1 + 2x_2 \le 4$, $x_1, x_2 \ge 0$.
18. Maximise $z = 3x_1 + 6x_2 - 3x_1^2 - 4x_1x_2 - 2x_2^2$, subject to the constraints $x_1 + x_2 \le 1$, $3x_1 + 2x_2 \le 4$, $x_1, x_2 \ge 0$.
19. Maximise $z = 10x_1 + 25x_2 - 10x_1^2 - x_2^2 - 4x_1x_2$, subject to the constraints $x_1 + 2x_2 \le 10$, $x_1 + x_2 \le 9$, $x_1, x_2 \ge 0$.
20. Minimise $z = x_1^2 - x_1 - 3x_2$, subject to the constraints $x_1 + x_2 \le 1$, $x_1 + 4x_2 \le 2$, $x_1, x_2 \ge 0$.

ANSWERS

Exercise 7(A)

Part B

11. $\lambda = 0.24;\ x_1 = 94,\ x_2 = 13;\ z^* = 208.9$
12. $\lambda = \frac{36}{11};\ x_1 = \frac{3}{31},\ x_2 = \frac{18}{31};\ z* = \frac{54}{31}$
13. $\lambda = 30;\ x_1 = 5,\ x_2 = 11;\ x_3 = 4;\ z^* = 381.$
14. $\lambda = \frac{120}{17};\ x_1 = \frac{120}{17},\ x_2 = \frac{30}{17};$ $x_3 = \frac{120}{17}\ z* = 105.88$
15. $\lambda = 1;\ (x_1, x_2, x_3) = (2, 2, 1);\ \min z = 9$ and $\lambda = 1.4;\ (x_1, x_2, x_3) = (2.8, 0, 1.4);\ \max z = 9.8$
16. $x_1 = \frac{12}{11},\ x_2 = \frac{35}{11};\ z^* = \frac{1997}{121} \simeq 16.5$
17. $\lambda_1 = 10,\ \lambda_2 = 0;\ (x_1, x_2, x_3) = (5, 5, 5);$ $z^* = 150.$
18. $\lambda_1 = \frac{40}{9},\ \lambda_2 = \frac{52}{9};\ (x_1, x_2, x_3) = \left(\frac{11}{3}, \frac{10}{3}, 8\right);$ $z^* = 91.111$
19. $\lambda = \frac{146}{3};\ x_1 = 22,\ x_2 = 1;\ z^* = 1752$
20. $\lambda = 6;\ x_1 = 6,\ x_2 = 0;\ z^* = 36.$
21. $\lambda_1 = 0,\ \lambda_2 = 1,\ x_1 = 0,\ x_2 = 4;\ z^* = 4.$
22. $\lambda_1 = \frac{2081}{25},\ \lambda_2 = \frac{1429}{25};\ x_1 = \frac{48}{5},\ x_2 = \frac{1}{5};$ $z^* = 703.72.$
23. $\lambda_1 = \frac{8}{17};\ \lambda_2 = 0;\ x_1 = \frac{13}{17},\ x_2 = \frac{18}{17};\ z^* = \frac{1173}{289}$
24. $\lambda_1 = 0,\ \lambda_2 = 0;\ x_1 = 1,\ x_2 = 1$, and $z^* = 1.$
25. $\lambda_1 = \lambda_2 = \frac{2}{3};\ x_1 = 1,\ x_2 = 1;\ z^* = 1$

Exercise 7(B)

Part B

7. $x_1 = \frac{1}{4},\ x_2 = \frac{15}{8};\ z^* = \frac{97}{16}$
8. $x_1 = 2,\ x_2 = 1;\ z^* = 0.$
9. $x_1 = \frac{3}{2},\ x_2 = \frac{1}{2};\ z^* = -\frac{1}{2}.$
10. $x_1 = \frac{1}{3},\ x_2 = \frac{5}{6};\ z^* = \frac{25}{6}.$
11. $x_1 = 0,\ x_2 = 1,\ z^* = 3.$
12. $x_1 = \frac{7}{4},\ x_2 = \frac{1}{4};\ z^* = \frac{33}{8}.$
13. $x_1 = \frac{2}{3},\ x_2 = \frac{14}{9};\ z^* = \frac{22}{9}.$
14. $x_1 = 1,\ x_2 = 2,\ z^* = 7.$
15. $x_1 = \frac{1}{2},\ x_2 = \frac{1}{2};\ z^* = \frac{1}{2}..$
16. $x_1 = 3.8,\ x_2 = 2.4;\ z^* = 19.$
17. $x_1 = \frac{14}{9},\ x_2 = \frac{2}{3};\ z^* = \frac{22}{9}.$
18. $x_1 = 0,\ x_2 = 1;\ z^* = 4.$
19. $x_1 = 0,\ x_2 = 5;\ z^* = 100.$
20. $x_1 = \frac{1}{8},\ x_2 = \frac{15}{32};\ z^* = -\frac{97}{64}.$

Chapter 8 TRANSPORTATION AND ASSIGNMENT PROBLEMS

8.0 INTRODUCTION

The transportation problem is a special kind of LP problem in which the objective is to transport various quantities of the same type of product available at various origins or sources (say, factories) to different destinations or sinks (say warehouses) such that the total transportation cost may be minimum.

The transportation problem can be described mathematically as follows. Let there be m origins O_i $(i = 1, 2, \ldots, m)$ from where products are to be transported to n destinations D_j $(j = 1, 2, \ldots, n)$.

Let the number of units of the product available at the origin O_i be a_i $(i = 1, 2, \ldots, m)$ and let the number of units of the product required at the destination D_j be b_j $(j = 1, 2, \ldots, n)$, so that the total number of units of the product available is equal to that required viz., $\sum_{i=1}^{m} a_i = \sum_{j=1}^{n} b_j$.

Let the cost of transporting one unit of the product from O_i to D_j be C_{ij}.

The objective is to determine x_{ij}, viz., the number of units of the product to be transported from O_i to D_j such that the total transportation cost $z = \sum_{i=1}^{m}\sum_{j=1}^{n} c_{ij}x_{ij}$ is minimum.

The linear programming model representing the transportation problem is given as follows:

$$\text{Minimise } z = \sum_{i=1}^{m}\sum_{j=1}^{n} c_{ij}x_{ij},$$

subject to the constraints

$$\sum_{j=1}^{n} x_{ij} = a_i,\ i = 1, 2, \ldots, m$$

$$\sum_{i=1}^{m} x_{ij} = b_j,\ j = 1, 2, \ldots, n$$

$$\sum_{i=1}^{m} a_i = \sum_{j=1}^{n} b_j. \text{ and } x_{ij} \geq 0, \text{ for all } i, j.$$

Usually the unit costs of transportation are given in the form of a table as given here.

Destination

Origin	D_1	D_2	......	D_j	......	D_n	Availability
O_1	C_{11}	C_{12}	......	C_{1j}	......	C_{1n}	a_1
O_2	C_{21}	C_{22}	......	C_{2j}	......	C_{2n}	a_2
.		....	....	....	....	....	.
.		....	....	....	....	....	.
.		....	....	....	....	....	.
.		....	....	....	....	....	.
O_i	C_{i1}	C_{i2}	......	C_{ij}	......	C_{in}	a_i
.		....	....	....	....	....	.
.		....	....	....	....	....	.
.		....	....	....	....	....	.
O_m	C_{m1}	C_{m2}	......	C_{mj}	....	C_{mn}	a_m
Requirement	b_1	b_2	....	b_j	......	b_n	

Note: The unit cost matrix will be assumed to be divided into mn cells and the unit cost corresponding to the $(i-j)$th cell is C_{ij}. The value of x_{ij} corresponding to C_{ij} is called the allocation to the $(i-j)$th cell.

8.1 DEFINITIONS

1. A *feasible solution* of a transportation problem is a set of non-negative values (cell allocations) x_{ij} $(i = 1, 2, \ldots, m; j = 1, 2, \ldots, n)$ that satisfies the row and column sum restrictions
2. A *basic feasible solution* of an $m \times n$ transportation problem is a feasible solution containing no more than $(m + n - 1)$ non-negative allocations.
3. An *optimal solution* of a transportation problem is a feasible solution (not necessarily basic) that minimises the total transportation cost.
4. A *non-degenerate basic feasible solution* is a feasible solution containing exactly $(m + n - 1)$ non-negative allocations in independent cells (explained later)
5. A *degenerate basic feasible solution* is a feasible solution containing less than $(m + n - 1)$ non-negative allocations.

A *loop in a transportation table* is a closed circuit consisting of horizontal and vertical line segments, in which the right-angled turns occur only at occupied cells, viz., the cells in which non-negative allocations have been made (Refer to the diagrams given here). The line segments may skip occupied cells.

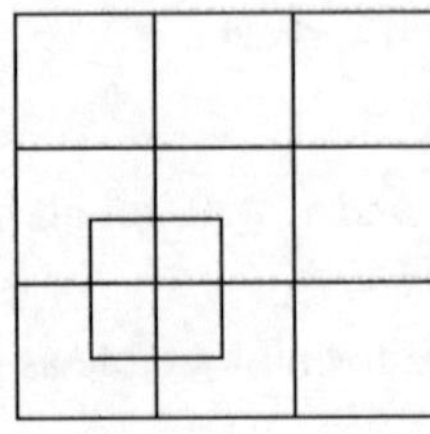

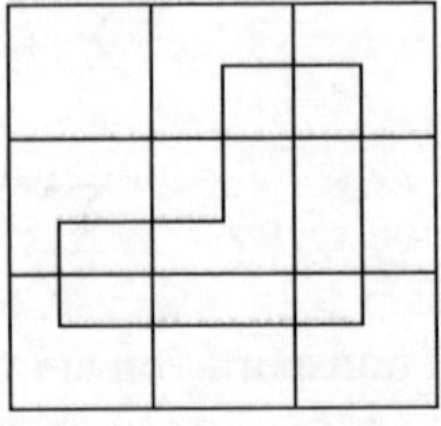

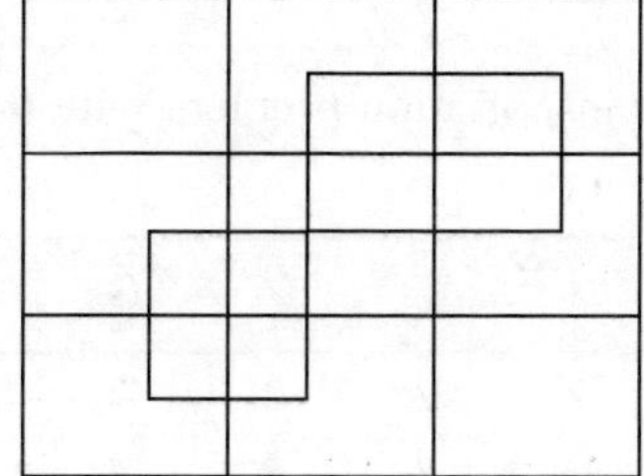

Independent cells or allocations

If the occupied cells in which non-negative allocations have been made do not form a closed loop, they are said to be independent cells or allocations.

Balanced and unbalanced transportation problems

If the total availability in all origins equals the total requirement in all the destinations, viz., if $\sum_{i=1}^{m} a_i = \sum_{j=1}^{n} b_j$, then the transportation problem is said to be a **balanced transportation problem**.

Otherwise, viz., if $\sum_{i=1}^{m} a_i \neq \sum_{j=1}^{n} b_j$, then the problem is said to be an unbalanced one.

8.2 STAGES IN THE SOLUTION OF A TRANSPORTATION PROBLEM

Even though a transportation problem can be regarded as a special kind of LP problem, to solve it by using the Simplex method will be tedious and time consuming. Hence, a more powerful algorithm called the transportation algorithm has been developed to solve transportation problems.

The three stages in the transportation algorithm are as follows:

Step I: Find an initial basic feasible solution to the transportation problem by any of the three available methods (the methods will be discussed in this chapter).

Step II: Apply the optimality test, viz., test if the initial BFS is an optimal solution.

Step III: If the initial BFS is not optimal, make successive improvements to it by iterative procedures until the optimal solution is obtained.

Step I: Initial BFS by north-west corner rule

Step 1: The first allocation is made to the upper most left-hand corner (North-west corner) cell (1, 1) of the transportation table. This allocation x_{11} is made such that the availability in O_1, i.e., a_1 is exhausted or the requirement of D_1, i.e., b_1 is satisfied or both.

Step 2(a): If a_1 is exhausted first, we move down vertically to the cell (2, 1) and the second allocation x_{21} is made to this cell such that a_2 is exhausted or the remaining requirement of D_2, i.e., $b_1 - x_{11}$ is satisfied.

Step 2(b): If b_1 is satisfied, we first move right horizontally to the cell (1, 2) and the second allocation x_{12} is made to this cell such that $a_1 - x_{11}$ is exhausted or b_2 is satisfied.

Step 2(c): If a_1 and b_1 are satisfied simultaneously (this will be the case when $a_1 = b_1$), we move down diagonally to the cell (2, 2) and the second allocation x_{22} is made to this cell as per the procedure in Step 1. In this case, the initial BFS will be degenerate.

Step 3: Steps 1 and 2 are repeated until we make the last allocation to the lowest right-hand corner cell, satisfying all the rim requirements.

Example

Let us obtain the initial BFS to the transportation problem with the following unit cost matrix by the north-west corner rule.

	D_1	D_2	D_3	D_4	a_i
O_1	2	2	2	1	3
O_2	10	8	5	4	7
O_3	7	6	6	8	5
b_j	4	3	4	4	-

SOLUTION

As per the north-west corner rule, the first allocation is equal to min (a_1, b_1) = min (3, 4) = 3 made to the cell (1, 1). Now a_1 has been exhausted. So we move down vertically to the cell (2, 1).

The second allocation equal to (a_2, balance of b_1) = min (7, 1) = 1 is made to this cell. Since b_1 has been satisfied, we move right horizontally to the cell (2, 2)

The third allocation equal to min (balance of a_2, b_2) = min (6, 3) = 3 is made to this cell. Since b_2 has been satisfied, we move further right horizontally to the cell (2, 3).

The fourth allocation equal to min (balance of a_2, b_3) = min (3, 4) = 3 is made to this cell. Since a_2 has been exhausted, we move down vertically to the cell (3, 3).

The fifth allocation equal to min (a_3, balance of b_3) = min (5, 1) = 1 is made to this cell. Since b_3 has been satisfied, we move right horizontally to the cell (3, 4).

The last allocation equal to min (balance of a_3, b_4) = min (4, 4) = 4 is made to this cell.

The initial BFS. obtained is displayed in the following table. The unit costs of the cells are written in the left-hand top corners and enclosed by arcs of circles or rectangles. The allocations in the cells are written in the centre and circled.

	D_1	D_2	D_3	D_4
O_1	2 (3)	2	2	1
O_2	10 (1)	8 (3)	5 (3)	4
O_3	7	6	6 (1)	8 (4)

The total transportation cost according to this initial BFS is given by $z = 3 \times 2 + 1 \times 10 + 3 \times 8 + 3 \times 5 + 1 \times 6 + 4 \times 8 = 93$.

Note:

1. The initial BFS obtained in this example is a non-degenerate BFS. since the number of occupied cells $= m + n - 1 = 3 + 4 - 1 = 6$.
2. The solution obtained here need not be the optimal solution.
3. The cells to which non-zero allocations are made are called ***occupied cells*** or ***basic cells***.

Initial BFS. by the lowest cost entry method or matrix minima method

Step 1: The cell with the smallest unit transportation cost is chosen from the unit cost matrix. If there is a tie, any cell with the least cost is chosen arbitrarily.

Step 2: The maximum possible allocation is made to this cell, consistent with the row availability and the column requirement.

Step 3: The row or column for which the availability or requirement is satisfied is eliminated from the transportation table.

Step 4: The balance availability or balance requirement is noted for future allocations.

Step 5: Steps 1–4 are repeated for the resulting reduced transportation table until all the rim requirements are satisfied.

Example

Let us obtain the initial BFS to the same transportation problem taken in the previous example by the lowest cost entry method.

	D_1	D_2	D_3	D_4	a_i
O_1	2	2	2	1 (3)	3/0
O_2	10 (2)	8	5 (4)	4 (1)	7/6/2/0
O_3	7 (2)	6 (3)	6	8	5/2/0
b_j	4/2/0	3/0	4/0	4/1/0	

SOLUTION

The least cost cell is (1, 4) cell. So we make the maximum possible allocation = min (3, 4) = 3 to this cell.

Since the first row availability, namely 3, has been exhausted, we eliminate this row from the table. This is indicated by putting a 'O' on the right of the original availability 3 after a vertical bar as shown. The balance requirement of the fourth column, namely '1' is written below the original requirement 4 after a horizontal bar as shown.

The next least cost cell is (2, 4) cell, since all the cells in the first row are ignored. The maximum possible allocation = min (7, 1) = 1 is made in this cell. Then adjustments in the second row and the fourth column totals are made as shown.

The third allocation of 4 units is made to the (2, 3) cell and adjustments are made in the totals of the row and column concerned.

The fourth allocation of 3 units is made to the (3, 2) cell and adjustments are made in the totals of the row and column concerned.

The fifth allocation of 2 units is made to the (3, 1) cell and adjustments are made in the totals of the row and column concerned.

The sixth and the last allocation of 2 units is made to the (2, 1) cell. This exhausts both row and column totals.

As before, the allocations are written in the centre of the cells and circled. The initial BFS obtained by this method is also a non-degenerate BFS., as there are $m + n - 1 = 6$ occupied cells.

The total transportation cost according to this initial BFS is given by $z = 3 \times 1 + 2 \times 10 + 4 \times 5 + 1 \times 4 + 2 \times 7 + 3 \times 6 = 79$.

Note: The total cost got by this method is less than that got by the north-west corner cell method. A still more powerful method for getting the initial BFS is Vogel's approximation method which is explained here.

Initial BFS by Vogel's approximation method (VAM) or unit cost penalty method

Step 1: From the unit cost matrix of the transportation problem, we compute the absolute difference between the lowest and the second lowest costs (called penalty) in each column and write it below the corresponding column.

Step 2: Similarly, the penalty for each row is computed and written to the right of the corresponding row.

Step 3: The row or column for which the penalty is the largest is selected. If there is a tie, any tied row or column is selected arbitrarily.

Step 4: In the row or column selected, the cell which has the least cost is chosen and the maximum possible allocation is made to the cell.

Step 5: The row or column for which the availability or requirement is satisfied is eliminated from the transportation table. If both availability and column are simultaneously satisfied, both the row and column are eliminated.

Step 6: The row and column penalties for the reduced cost matrix are re-computed and Steps 3-5 are repeated until all the rows and columns are eliminated.

Note: For clarity, the availabilities will be written on the left and the requirements will be written on the top of the unit cost matrix, when we get the initial BFS by VAM

Example

Let us take the same transportation problem considered in the earlier examples and apply VAM to get the initial BFS.

a_i \ b_j	1/4	3/0	0/1/4	0/4	Rowwise penalties			
0 / 3	2 (3)	2	2	1	1	–	–	–
0 / 3 / 7	10	8	5 (3)	4 (4)	1	1	3	–
0 / 5	7 (1)	6 (3)	6 (1)	8	1	1	1	1
Columnwise penalties	5	4	3	3				
	3	2	1	4				
	3	2	1	–				

Solution

The first round of row-wise penalties and column-wise penalties are computed and written in the first column on the right of the transportation table and in the first row below the transportation table respectively.

The largest penalty, viz., 5 corresponds to the first column. The least cost in this column occurs in the (1, 1) cell. We allocate the maximum possible units (3) to this cell. Once this is done, the first row availability is exhausted. Hence, it is struck off by a dotted line or indicated by putting a 'O' on the left of the original availability 3 after a vertical bar as shown. The balance requirement of the first column (1) is noted.

Omitting the first row that has been struck off, we compute the second round of penalties and write in the second column or row as before. The second allocation is made as before. The procedure is continued until all the rows and columns of the transportation table are struck off.

The total transportation cost according to the initial BFS obtained by this method is given by $z = 3 \times 2 + 3 \times 5 + 4 \times 4 + 1 \times 7 + 3 \times 6 + 1 \times 6 = 68$

Note: Of the three methods discussed here, the VAM gives the least initial total transportation cost and so the initial solution obtained by this method is very near the optimal solution viz., if we start with the initial solution got by VAM, we can get the optimal solution in lesser number of iterations. Hence if the choice is left to us, we have to prefer only VAM for finding the initial BFS

Step II: Optimality test – Computational procedure

The optimality test described here can be applied to a basic feasible solution of an $(m \times n)$ transportation problem, iff the solution contains exactly $(m + n - 1)$ number of non-negative allocations in independent cells.

The step by step procedure to test whether an initial BFS which consists of $(m + n - 1)$ non-negative allocations in independent cells is optimal or not, is given here.

Step 1: We find a set of $m + n$ arbitrary numbers $u_i (i = 1, 2, \ldots, m)$ $v_j (j = 1, 2, \ldots, n)$ such that for each occupied cell

$$u_i + v_j = c_{ij} \tag{1}$$

Since there are $(m + n - 1)$ occupied cells, there will be $(m + n - 1)$ equations of the type (1), while there are $(m + n)$ unknown variables u_i and v_j. Hence, one of the unknowns u_i and v_j can be assigned any value arbitrarily. It will be advantageous to assign the value zero to u_i or v_j which corresponds to the maximum row total or maximum column total.

Step 2: Then, we calculate the net calculations (also called opportunity costs) for each unoccupied cell given by

$$d_{ij} = u_i + v_j - c_{ij}$$

and enter them in the center of the unoccupied cells uncircled.

Step 3: The signs of d_{ij} are examined.

(i) If all $d_{ij} \leq 0$ for the unoccupied cells, then the current BFS is an optimal solution.

(ii) If all $d_{ij} \leq 0$ with atleast one $d_{ij} = 0$, then the current B.F.S. is an optimal solution and an alternative optimal solution exists.

(iii) If at least one $d_{ij} > 0$, then the current solution is not optimal and it can be improved iteratively as explained in Stage 3.

Example

Let us consider the same transportation problem considered earlier and test the optimality of the solutions obtained by the matrix minima method and by Vogel's approximation method.

SOLUTION

First, we take the initial BFS obtained by the matrix minima method. It is given here.

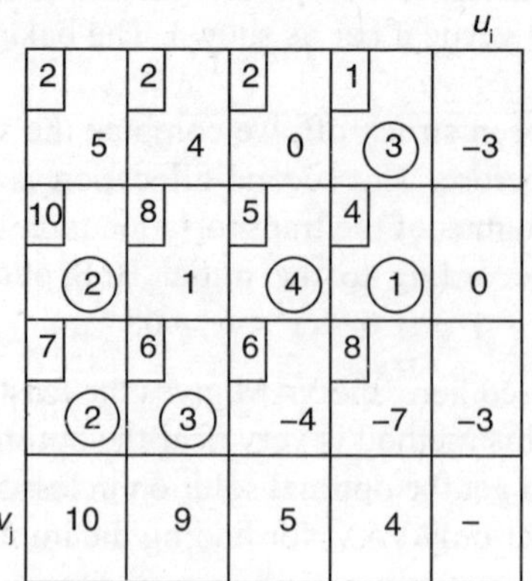

					u_i
	2 / 5	2 / 4	2 / 0	1 / (3)	−3
	10 / (2)	8 / 1	5 / (4)	4 / (1)	0
	7 / (2)	6 / (3)	6 / −4	8 / −7	−3
v_j	10	9	5	4	–

Since the total of the second row is maximum (= 7), we assign $u_2 = 0$.

Since (2, 4) cell with unit cost 4 is occupied, we put $u_2 + v_4 = 4$ and so $v_4 = 4$.

Since (2, 3) cell with unit cost 5 is occupied, we put $u_2 + v_3 = 5$ and so $v_3 = 5$.

Since (2, 1) cell with unit cost 10 is occupied, we put $u_2 + v_1 = 10$ and so $v_1 = 10$,

Since (3, 1) cell with unit cost 7 is occupied, we put $u_3 + v_1 = 7$ and so $u_3 = -3$,

Since (3, 2) cell with unit cost 6 is occupied, we put $u_3 + v_2 = 6$ and so $v = 9$,

Finally, since (1, 4) cell with unit cost 1 is occupied, we put $u_1 + v_4 = 1$ and so $u_1 = -3$

Thus, the values of u_i and v_j are obtained without any ambiguity for the initially assumed value, $u_2 = 0$.

For the unoccupied cells, the net evaluations are computed as follows. They are entered at the centres of the cells concerned and not circled.

For the (1, 1) cell, $d_{11} = u_1 + v_1 - C_{11} = -3 + 10 - 20 = 5$

For the (1, 2) cell, $d_{12} = u_1 + v_2 - C_{12} = -3 + 9 - 2 = 4$ and so on.

Since d_{ij} values for the (1, 1), (1, 2) and (2, 2) cells are positive, the solution in consideration is not optimal.

Now let us consider the initial BFS obtained by VAM that is given here and apply the optimality test.

					u_i
	2 / (3)	2 / −1	2 / −1	1 / −1	−4
	10 / −4	8 / 3	5 / (3)	4 / (4)	0
	7 / (1)	6 / (3)	6 / (1)	8 / −3	1
v_j	6	5	5	4	–

The u_i, v_j and d_{ij} values are found out as before.

Since d_{ij} values for all the unoccupied cells are negative, the solution in consideration is optimal.

Let us consider the following transportation problem for which the initial BFS was obtained by the north-west corner rule.

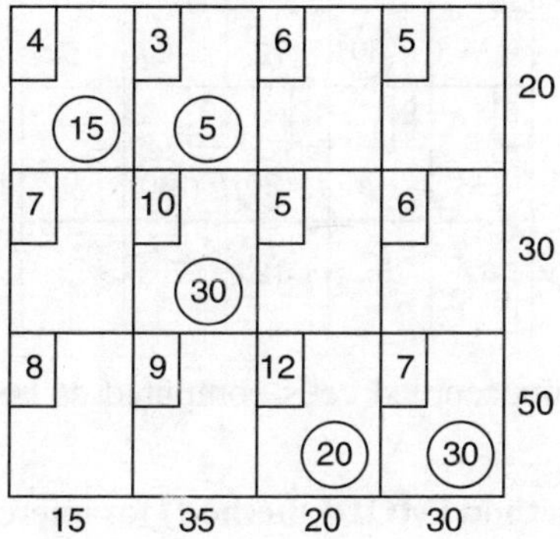

4 (15)	3 (5)	6	5	20
7	10 (30)	5	6	30
8	9	12 (20)	7 (30)	50
15	35	20	30	

The optimality test cannot be applied to this solution, as it is a degenerate BFS, containing only 5 ($< m + n - 1$) occupied cells. Before we apply the optimality test, we convert the degenerate BFS into a non-degenerate BFS by making one of the empty or unoccupied cells occupied by allocating ε ($= 0$) units.

Recalling that the optimality test can be applied to a non-degenerate BFS, provided that the allocations are made in independent cells, we allocate ε units to any cell which does not form a loop with the cells which have been already occupied. Accordingly, we should not allocate ε units to the (2, 1) cell, since it forms a loop with the (1, 1), (1, 2) and (2, 2) cells which are occupied.

If we allocate ε units to the (2, 1) cell carelessly, we cannot find the u_i and v_j values unambiguously and hence, cannot apply the optimality test as explained here.

					u_i
	4 (15)	3 (5)	6	5	2
	7 (ϵ)	10 (30)	5	6	9
	8	9	12 (20)	7 (30)	0
v_j	−2	1	12	7	–

We note that the condition $u_i + v_j = c_{ij}$ is satisfied for all the occupied cells except the (1, 1) cell. This will be the case whatever be the choice of u_i and v_j.

On the other hand, if we allocate ε units to any of the other unoccupied cells, u_i and v_j values can be found out unambiguously and hence, the optimality test can be applied as explained here.

				u_i
[4] (15)	[3] (5)	[6] 4	[5] (ε)	−2
[7] 4	[10] (30)	[5] 12	[6] 6	5
[8] −2	[9] −4	[12] (20)	[7] (30)	0
v_j 6	5	12	7	–

Since all the d_{ij} values for the unoccupied cells computed as before are not negative, the current solution is not optimal.

Step III: Modified distribution method (MODI method) for improving the non-optimal solution and getting the optimal solution by iteration.

Step 1: If the current solution is not optimal, at least one net evaluation d_{ij} is positive. We select the empty cell for which d_{ij} is the largest. If there is a tie, we may select the empty cell with lesser unit cost.

Step 2: We put a tick (√) mark in this selected cell, indicating that this cell is to be made occupied.

Step 3: We then construct a loop with one of the corners at this √ marked cell and other corners at the already occupied cells as usual.

Step 4: At the corners of the loop, we put plus (+) sign and minus (−) sign alternately beginning with (and of course ending with) + sign for the cell selected at Step 1.

Step 5: We observe the allocations in the occupied cells at the corners of the loop with (−) sign and select the smallest allocation. This smallest allocation is added to the allocation in the cells with (+) sign and subtracted from the allocations in the cells with (−) sign. Thus, we have got a revised BFS. If there is a tie for the smallest allocation, the revised BFS will be degenerate. Before we proceed to Step 6, we make the BFS into a non-degenerate BFS by allocating ε units to one of the recently vacated (emptied) cells, preferably to that cell with lesser unit cost.

Step 6: We apply the optimality test to this revised non-degenerate BFS. If the revised solution is still not optimal, we proceed to Step 1 and repeat the procedure, until the solution becomes optimal.

Example

Let us consider the transportation problem for which the initial BFS obtained by the matrix minima method was not optimal.

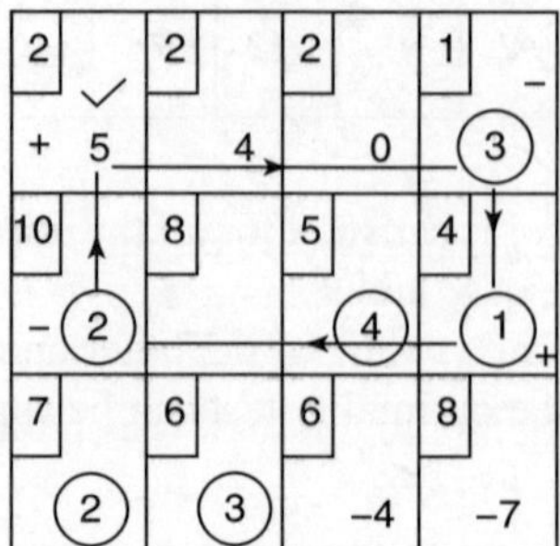

SOLUTION

First we identify the (1, 1) cell in which d_{ij} is positive and maximum. It is this cell which is to be made occupied and so we put a √ mark in that cell. We construct a loop having the occupied cells (1, 4), (2, 4), (2, 1) and of course, the cell (1, 1) to be made occupied as corners.

Note: The sides of the loop are taken as curved lines so as not to cut the entries in the cells

We put (+), (−), (+) and (−) sign alternately in the (1, 1), (1, 4), (2, 4) and (2, 1) cells respectively. The arrows may be marked clockwise or anti-clockwise.

We note down the allocations in the cells with (−) sign. In this problem, they are 3 and 2, the smaller of which is 2. It is this 2 which is to be added to the already existing allocations in the cells with (+) sign and subtracted from the already existing allocations in the cells with (−) sign. Consequently, the revised allocations in the cells (1, 1), (1, 4) (2, 4) and (2, 1) will be respectively 2, 1, 3 and 0. It is to be noted that the revised allocations will satisfy the rim requirements

Let us now apply the optimality test to the revised solution as before.

					u_i
	[2] +(2)	[2] −1	[2] 0	[1] (1) −	−3
	[10] −5	[8] −4	[5] − (4)	[4] + (3)	0
	[7] −(2)	[6] (3)	[6] ✓ 1 +	[8] −2	2
v_j	5	4	5	4	–

Even this revised solution is not optimal, as there is one d_{ij} value that is positive. Hence, there is still scope for improving the solution. We revise this solution by making (3, 3) cell occupied as before. Now the revised solution is subjected to the optimality test.

					u_i
	[2] (3)	[2] −1	[2] −1	[1] −1	−4
	[10] −4	[8] −3	[5] (3)	[4] (4)	0
	[7] (1)	[6] (3)	[6] (1)	[8] −3	1
v_j	6	5	5	4	

Since all the d_{ij} values are negative, the optimal solution has been obtained. This optimal solution is the same as that got by VAM and the minimum total transportation cost $= 3 \times 2 + 3 \times 5 + 4 \times 4 + 1 \times 7 + 3 \times 6 + 1 \times 6 = 68$.

Example

Let us consider another example in which the revised solution got by the MODI method will result in degeneracy.

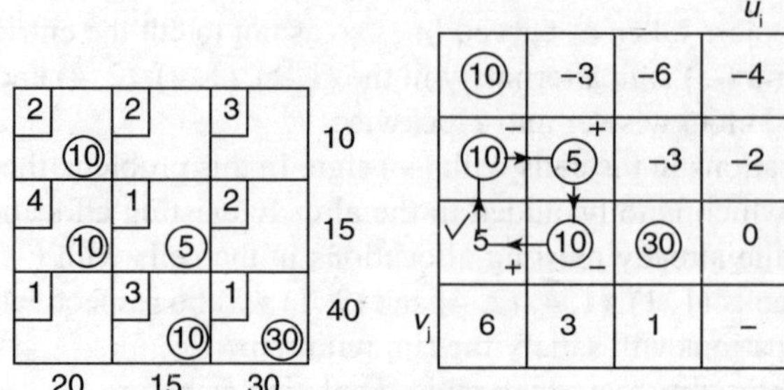

2 (10)	2	3	10
4 (10)	1 (5)	2	15
1	3 (10)	1 (30)	40
20	15	30	

			u_i
(10)	−3	−6	−4
(10) −	(5) +	−3	−2
5 + ✓	(10) −	(30)	0
v_j: 6	3	1	–

Solution

The following revised solution is a degenerate solution.

(10)		
	(15)	
(10)	(10)	(30)

(10)		
	(15)	
(10)	(ε)	(30)

To make it non-degenerate, we allocate ε units to any of the recently vacated cells, namely (2, 1) and (3, 2) cells, preferably to the cell (3, 2) with a smaller unit cost. Even this solution can be seen to be not optimal. Revising this solution once again by the MODI method, we get the optimal solution that is given here.

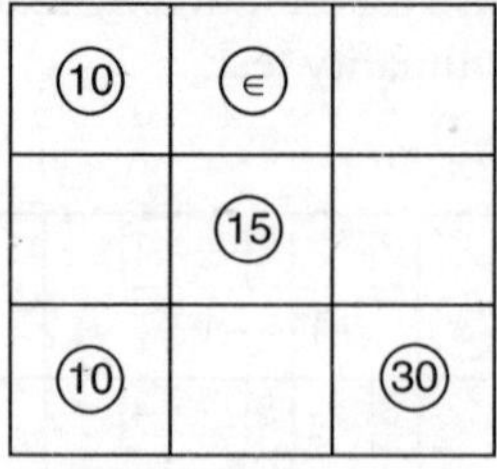

(10)	(ε)	
	(15)	
(10)		(30)

Unbalanced Transportation Problem

A transportation problem is said to be an unbalanced transportation problem, if the total availability in all the origins is not equal to the total requirement at all the destinations, viz., if $\sum_{i=1}^{m} a_i \neq \sum_{j=1}^{n} b_j$.

An unbalanced transportation problem is first converted into a balanced transformation problem as explained here and then solved by the method discussed earlier.

When the total availability in all the origins is greater than the total requirement, viz., if $\sum a_i > \sum b_j$ a dummy destination (or column) is introduced in the transportation matrix. The unit cost of transporting from each origin to this dummy destination is assumed to be zero. The excess availability is extended as the rim requirement for the dummy destination.

On the other hand, if the total requirement in all the destinations is greater than the total availability, viz, if $\sum b_j > \sum a_i$, a dummy origin (or row) is introduced in the transportation matrix with zero unit transportation costs. The excess requirement is entered as the rim total for the dummy origin.

Alternative Optimal Solution of a Transportation Problem

When we discussed the optimality test, it was pointed out that the transportation problem has a unique optimal solution, if the net evaluation d_{ij} for all the unoccupied cells are strictly negative. If one or more d_{ij} values are equal to zero, the optimal solution has been obtained. However, it also indicates that the problem has an alternative optimal solution.

To get the alternative optimal solution, we treat the current optimal solution as non-optimal and improve it by the MODI method so as to convert the zero entry unoccupied cell into an occupied cell. In both the optimal solutions, the minimum total transportation cost will be the same.

Maximisation Case in Transportation Problem

The algorithms discussed so far hold good only when the objective is to minimise the total transportation cost. There are situations where the objective will be maximise the total profit due to transportation, when the profits obtained by transporting one unit of the product from various origins to different destinations are given.

To solve the maximisation type of transportation problems, a different algorithm need not be developed, as a maximisation problem can be converted into a minimisation problem as explained here.

The cell with maximum unit profit is first identified and the unit profits in all other cells are subtracted from this maximum value. The resulting transportation problem is treated as a minimisation problem and solved by the algorithm developed earlier. The optimal solution of the minimisation problem is also the optimal solution of the original maximisation problem. However, the total maximum profit is got by multiplying the allocated units and the unit profits of occupied cells.

WORKED EXAMPLES 8 (A)

Example 1

Solve the transportation problem, the unit cost matrix of which is given below so as to minimise the total transportation cost. Obtain the initial BFS by the north-west corner rule.

		Warehouse				
		W_1	W_2	W_3	W_4	Factory capacity
Factory	F_1	2	3	4	5	40
	F_2	3	2	3	4	50
	F_3	4	3	3	4	60
	F_4	6	4	4	5	70
Warehouse demand		65	55	55	45	

Solution

The initial BFS got by the north-west cell (NWC) rule is given in the following table.

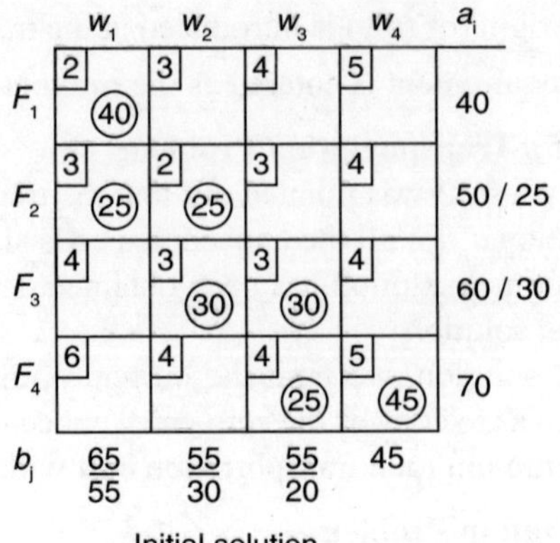

	W_1	W_2	W_3	W_4	a_i
F_1	2 (40)	3	4	5	40
F_2	3 (25)	2 (25)	3	4	50 / 25
F_3	4	3 (30)	3 (30)	4	60 / 30
F_4	6	4	4 (25)	5 (45)	70
b_j	65 / 55	55 / 30	55 / 20	45	

Initial solution

Let us now apply the optimality test by finding the opportunity costs of the unoccupied cells.

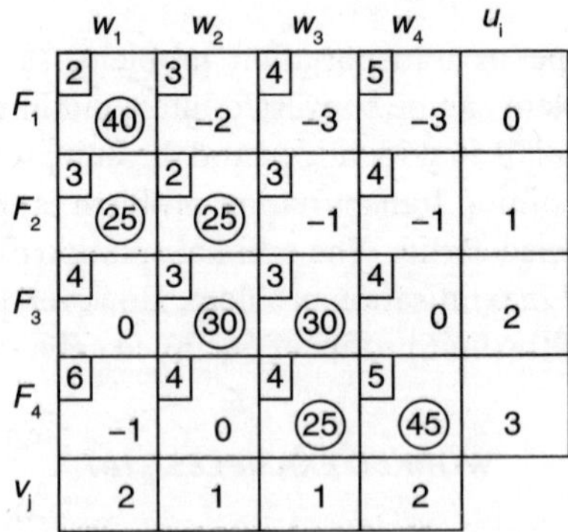

	W_1	W_2	W_3	W_4	u_i
F_1	2 (40)	3 −2	4 −3	5 −3	0
F_2	3 (25)	2 (25)	3 −1	4 −1	1
F_3	4 0	3 (30)	3 (30)	4 0	2
F_4	6 −1	4 0	4 (25)	5 (45)	3
v_j	2	1	1	2	

Starting (optimal) table

Since the opportunity costs (d_{ij} values) of all the unoccupied cells are ≤ 0, optimality has been attained.

Hence, the initial solution itself is the optimal solution which is given as
40 units from F_1 to W_1, 25 units from F_2 to W_1,
25 units from F_2 to W_2, 30 units from F_3 to W_2,
30 units from F_3 to W_3, 25 units from F_4 to W_3 and
45 units from F_4 to W_4 must be transported to make the total transportation cost minimum.

The minimum total transportation cost $= 40 \times 2 + 25 \times 3 + 25 \times 2 + 30 \times 3 + 30 \times 3 + 25 \times 4 + 45 \times 5 = 710$.

Example 2

A company has 4 warehouses and 3 retail stores. The warehouses altogether have a surplus of 34 units of a given commodity divided among them as follows:

Warehouses:	A	B	C	D
Surplus:	5	8	7	14

The retail stores altogether require 34 units of this commodity divided among them as follows:

Store:	P	Q	R
Requirement:	7	9	18

The costs of shipping one unit of the commodity from the warehouses to the stores are given in the following unit cost matrix.

	P	Q	R
A	2	7	4
B	3	3	1
C	5	4	7
D	1	6	2

Obtain the optimal solution of this transportation problem and find the minimum total cost of shipping. Use the NWC rule to get the initial solution.

Solution

The initial BFS. obtained by the NWC rule is given in the following matrix.

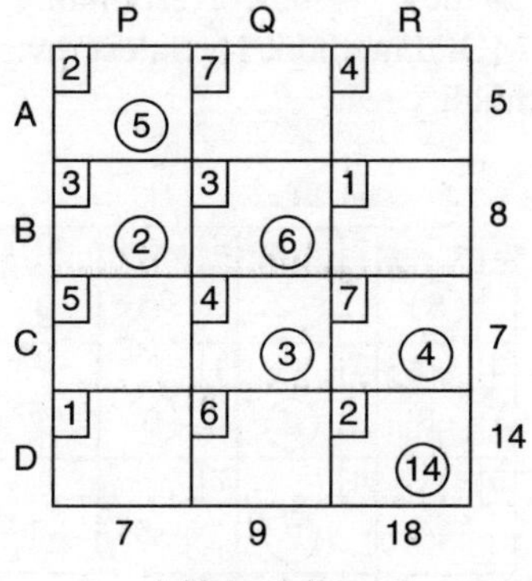

	P	Q	R	
A	[2] (5)	[7]	[4]	5
B	[3] (2)	[3] (6)	[1]	8
C	[5]	[4] (3)	[7] (4)	7
D	[1]	[6]	[2] (14)	14
	7	9	18	

Initial solution

	P	Q	R	u_i
A	[2] (5)	[7] −5	[4] 1	0
B	[3] (2)	[3] (6) →	[1] ✓ 5	1
C	[5] −1	[4] (3) ↑	[7] ← (4)	2
D	[1] −2	[6] −7	[2] (14)	−3
v_j	2	2	5	–

Starting table

Since the initial solution contains $m + n - 1 = 4 + 3 - 1 = 6$ occupied cells, it is a non-degenerate BFS and hence, the optimality test is applied by finding d_{ij} values for the unoccupied cells.

Since the d_{ij} values for the A – R and B – R cells are positive, optimality has not been obtained.

Hence, we revise the current solution by the MODI method so as to make the B – R cell (for which d_{ij} value is maximum) occupied.

The revised solution is given in the following matrix and it is then subjected to the optimality test.

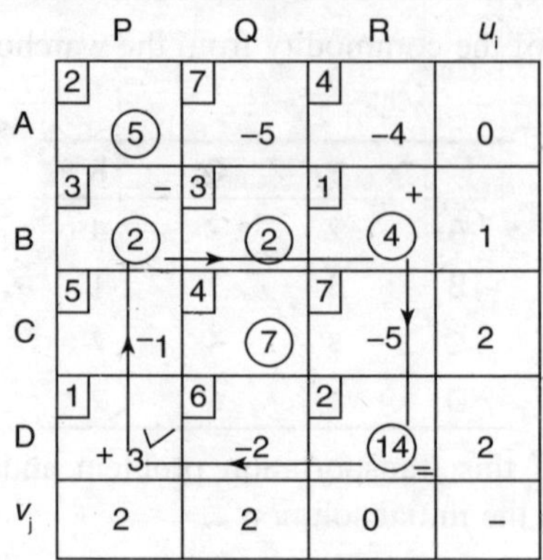

	P	Q	R	u_i
A	[2] (5)	[7] −5	[4] −4	0
B	[3] (2) –	[3] (2)	[1] (4) +	1
C	[5] −1	[4] (7)	[7] −5	2
D	[1] + 3	[6] −2	[2] (14) –	2
v_j	2	2	0	–

First iteration table

Even this solution is not optimal, as the d_{ij} value corresponding to the D – P cell is positive. Hence, this solution is further revised by the MODI method. The latest revised solution is given in the following matrix. It is then subjected to the optimality test.

	P	Q	R	u_i
A	[2] (5)	[7] −2	[4] −1	0
B	[3] −3	[3] (2)	[1] (6)	−2
C	[5] −4	[4] (7)	[7] −5	−1
D	[1] (2)	[6] −2	[2] (12)	−1
v_j	2	5	3	–

Optimal table

Since all the d_{ij} values for the empty cells are negative, the optimal solution has been obtained, which is given as follows:

A to P: 5 units; B to Q: 2 units; B to R: 6 units;

C to Q: 7 units; D to P: 2 units; D to R: 12 units;

The minimum total cost of transportation cost $= 5 \times 2 + 2 \times 3 + 6 \times 1 + 7 \times 4 + 2 \times 1 + 12 \times 2 = 76$.

Example 3

Solve the transportation problem with the following unit cost matrix by obtaining the initial solution using matrix minima method.

	D_1	D_2	D_3	D_4	D_5	a_i
O_1	3	4	6	8	9	20
O_2	2	10	1	5	8	30
O_3	7	11	20	40	3	15
O_4	2	1	9	14	16	13
	40	6	8	18	6	–

SOLUTION

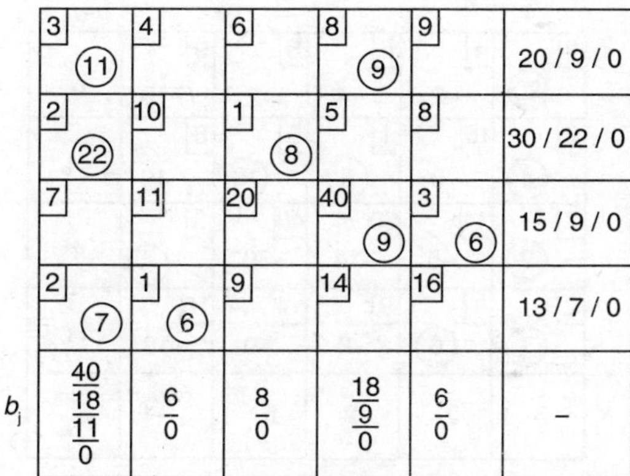

	[3] (11)	[4]	[6]	[8] (9)	[9]	20 / 9 / 0
	[2] (22)	[10]	[1] (8)	[5]	[8]	30 / 22 / 0
	[7]	[11]	[20]	[40] (9)	[3] (6)	15 / 9 / 0
	[2] (7)	[1] (6)	[9]	[14]	[16]	13 / 7 / 0
b_j	40 / 18 / 11 / 0	6 / 0	8 / 0	18 / 9 / 0	6 / 0	–

Initial table

	[3] − (11)	[4] −2	[6] 4	[8] + (9)	[9] −38	0
	[2] (22)	[10] −9	[1] (8)	[5] 2	[8] −38	−1
	[7] ✓ 28 +	[11] 23	[20] 14	[40] − (9)	[3] (6)	a_i 32
	[2] (7)	[1] (6)	[9] −8	[14] −7	[16] −46	−1
v_j	3	2	2	8	−29	–

Starting table

Optimal solution is given as follows as per the optimal table given on the next page.

$O_1 - D_1$: 20 units; $O_2 - D_1$: 4 units; $O_2 - D_3$: 8 units; $O_2 - D_4$: 18 units; $O_3 - D_1$: 9 units; $O_3 - D_5$: 6 units; $O_4 - D_1$: 7 units; $O_4 - D_2$: 6 units.

					u_i
	[3] + (2)	[4] 2	[6] −4	[8] − (18)	[9] −10	0
	[2] − (22)	[10] −9	[1] (8)	[5] + ✓ 2	[8] −10	−1
	[7] (9)	[11] −5	[20] −14	[40] −28	[3] (6)	4
	[2] (7)	[1] (6)	[9] −8	[14] −7	[16] −18	−1
v_j	3	2	2	8	−1	–

Table after first iteration

						u_i
	[3] (20)	[4] −2	[6] −4	[8] −2	[9] −10	0
	[2] (4)	[10] −9	[1] (8)	[5] (18)	[8] −10	−1
	[7] (9)	[11] −5	[20] −14	[40] −30	[3] (6)	4
	[2] (7)	[1] (6)	[9] −8	[14] −9	[16] −18	−1
v_j	3	2	2	6	−1	–

Optimal table

Minimum total transportation cost $= 20 \times 3 + 4 \times 2 + 8 \times 1 + 18 \times 5 + 9 \times 7 + 6 \times 3 + 7 \times 2 + 6 \times 1 = 267$.

Example 4

Solve the transportation problem with the following unit cost matrix by obtaining the initial solution using VAM.

	D_1	D_2	D_3	D_4	a_i
O_1	2	3	11	7	6
O_2	1	0	6	1	1
O_3	5	8	15	9	10
b_j	7	5	3	2	

Solution

a_i \ b_j	0 / 6 / 7	0 / 5	0 / 3	0 / 1 / 0	Row-wise penalties			
0 / 1 / 6	[2] (1)	[3] (5)	[11]	[7]	1	1	5 ✓	–
0 / 1	[1]	[0]	[6]	[1] (1)	1	–	–	–
0 / 1 / 4 / 10	[5] (6)	[8]	[15] (3)	[9] (1)	3	3	4	4 ✓
Column-wise penalties	1	3	5	6 ✓				
	3	5 ✓	4	2				
	3	–	4	2				

Initial table got by VAM

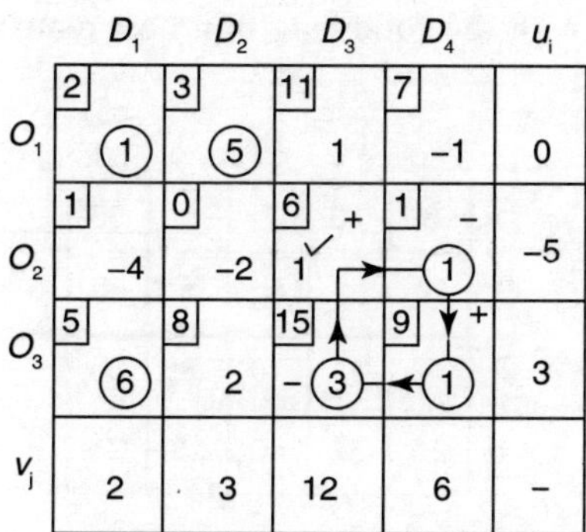

	D_1	D_2	D_3	D_4	u_i
O_1	[2] (1)	[3] (5)	[11] 1	[7] −1	0
O_2	[1] −4	[0] −2	[6] + 1 ✓	[1] − (1)	−5
O_3	[5] (6)	[8] 2	[15] − (3)	[9] + (1)	3
v_j	2	3	12	6	–

Starting solution

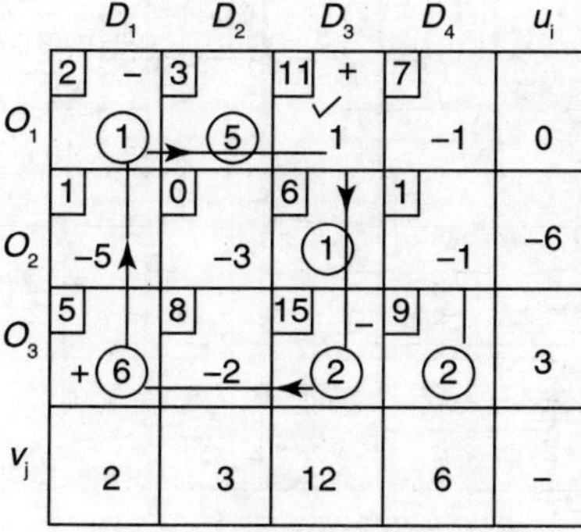

	D_1	D_2	D_3	D_4	u_i
O_1	[2] − (1)	[3] (5)	[11] + ✓ 1	[7] −1	0
O_2	[1] −5	[0] −3	[6] (1)	[1] −1	−6
O_3	[5] + (6)	[8] −2	[15] − (2)	[9] (2)	3
v_j	2	3	12	6	–

Table after first iteration

	D_1	D_2	D_3	D_4	u_i
O_1	[2] –1	[3] (5)	[11] (1)	[7] –2	0
O_2	[1] –5	[0] –2	[6] (1)	[1] –1	–5
O_3	[5] (7)	[8] –1	[15] (1)	[9] (2)	4
v_j	1	3	11	5	–

Optimal table

Optimal solution is: $O_1 - D_2$: 5 units; $O_1 - D_3$: 1 unit; $O_2 - D_3$: 1 unit; $O_3 - D_1$: 7 units; $O_3 - D_3$: 1 unit; $O_3 - D_4$: 2 units.

Minimum total cost $= 15 + 11 + 6 + 35 + 15 + 18 = 100$.

Example 5

Solve the transportation problem with the following unit cost matrix by obtaining the initial solution using matrix minima method:

	D_1	D_2	D_3	D_4	a_i
O_1	8	10	7	6	50
O_2	12	9	4	7	40
O_3	9	11	10	8	30
b_j	25	32	40	23	120

SOLUTION

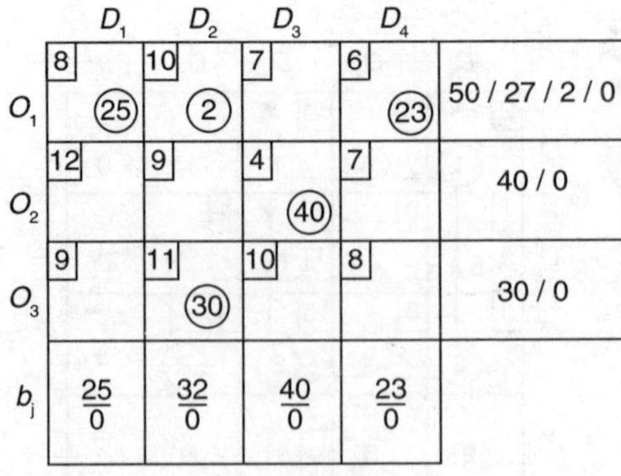

	D_1	D_2	D_3	D_4	
O_1	[8] (25)	[10] (2)	[7]	[6] (23)	50 / 27 / 2 / 0
O_2	[12]	[9]	[4] (40)	[7]	40 / 0
O_3	[9]	[11] (30)	[10]	[8]	30 / 0
b_j	25 / 0	32 / 0	40 / 0	23 / 0	

Initial table

It is noted that the number of occupied cells $= 5 < (= m + n - 1)$. Hence, the initial solution is degenerate. So the optimality test cannot be applied without converting it into a non-degenerate solution by introducing $\varepsilon = (0)$ allocation to any of the unoccupied cells. The cell chosen for ε allocation should be independent, viz., the ε allocated cell should not form a loop with the already existing occupied cells. Among the possible unoccupied cells, we prefer O_1D_3 cell for ε allocation, as the unit cost corresponding to this cell is the least. We then go ahead with the optimality test.

	D_1	D_2	D_3	D_4	u_i
O_1	[8] (25)	[10] (2)	[7] (ε)	[6] (23)	0
O_2	[12] −7	[9] −2	[4] (40)	[7] −4	−3
O_3	[9] 0	[11] (30)	[10] −2	[8] −1	1
v_j	8	10	7	6	

Starting table / Optimal table

Optimal solution is:

$O_1 - D_1$: 25 units; $O_1 - D_2$: 2 units; $O_1 - D_3$: 23 units; $O_2 - D_3$: 40 units; $O_3 - D_2$: 30 units; Minimum total cost $= 25 \times 8 + 2 \times 10 + 23 \times 6 + 40 \times 4 + 30 \times 11 = 848$.

Example 6

Solve the transportation problem with the following unit cost matrix by obtaining the initial solution using the north-west corner rule.

	D_1	D_2	D_3	D_4	a_i
O_1	3	4	6	6	30
O_2	6	5	4	2	40
O_3	2	4	4	3	50
b_j	20	30	40	30	

SOLUTION

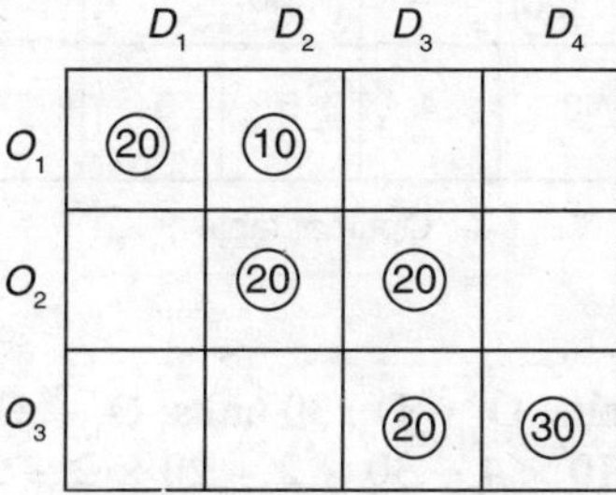

	D_1	D_2	D_3	D_4
O_1	(20)	(10)		
O_2		(20)	(20)	
O_3			(20)	(30)

Initial solution got by NWC rule

	D_1	D_2	D_3	D_4	u_i
O_1	[3] – (20)	[4] + (10)	[6] –3	[6] –4	0
O_2	[6] –2	[5] – (20)	[4] + (20)	[2] 1	1
O_3	[2] + ✓ 2	[4] 1	[4] – (20)	[3] (30)	1
v_j	3	4	3	2	–

Starting solution

	D_1	D_2	D_3	D_4	u_i
O_1	[3] (∈)	[4] (30)	[6] –1	[6] –2	0
O_2	[6] –4	[5] –2	[4] – (40)	[2] ✓ + 1	–1
O_3	[2] + (20)	[4] –1	[4] + (∈)	[3] – (30)	–1
v_j	3	4	5	4	–

Table after first iteration

We note that the number of occupied cells now is only $4 < m + n - 1$ $(= 6)$. Thus, this solution is degenerate. This is a case of degeneracy in the intermediate stage. To resolve this intermediate degeneracy, we allocate ε units to two unoccupied cells, preferably to two unoccupied with less unit costs. Thus, we allocate ε units to each of O_1D_1 and O_3D_3 cells.

	D_1	D_2	D_3	D_4	u_i
O_1	[3] (∈)	[4] (30)	[6] –1	[6] –3	0
O_2	[6] –4	[5] –2	[4] (10)	[2] (30)	–1
O_3	[2] + (20)	[4] –1	[4] (30)	[3] –1	–1
v_j	3	4	5	3	–

Optimal table

Optimal solution is
$O_1 - D_2$: 30 units; $O_2 - D_3$: 10 units; $O_2 - D_4$: 30 units; $O_3 - D_1$: 20 units; $O_3 - D_3$: 30 units.
Minimum total cost $= 30 \times 4 + 10 \times 4 + 30 \times 2 + 20 \times 2 + 30 \times 4 = 380$.

Example 7

Find the optimal solution of the transportation problem with the following unit cost matrix by obtaining the initial solution using VAM. Is there an alternative optimal solution for this problem? If so, find it also.

SOLUTION

	D_1	D_2	D_3	D_4	D_5	D_6	a_i
O_1	1	2	1	4	5	2	30
O_2	3	3	2	1	4	3	50
O_3	4	2	5	9	6	2	75
O_4	3	1	7	3	4	6	20
b_j	20	40	30	10	50	25	–

	0 / 20	0 / 40	0 / 20 / 30	0 / 10	0 / 50	0 / 25	Penalties					
0 / 10 / 30	1 (20)	2	1 (10)	4	5	2	1	1	1	3	–	–
0 / 20 / 30 / 50	3	3	2 (20)	1 (10)	4 (20)	3	1	1	1	1	1	3
0 / 10 / 35 / 75	4	2 (40)	5	9	6 (10)	2 (25)	2	3	3	1	1	3
0 / 20	3	1	7	3	4 (20)	6	2	2	1	1	1	1
Penalties	2	1	1	2	1	1						
	–	1	1	2	1	1						
	–	–	1	2	1	1						
	–	–	1	2	2	–						
	–	–	3	2	2	–						
	–	–	–	2	2	–						

Initial solution got by VAM

	D_1	D_2	D_3	D_4	D_5	D_6	u_i
O_1	[1] – (20)	[2] –3	[1] + (10)	[4] –4	[5] –2	[2] –3	0
O_2	[3] –1	[3] –3	[2] – (20)	[1] (10)	[4] + (20)	[3] –3	1
O_3	[4] ✓ + 0	[2] (40)	[5] –1	[9] –6	[6] – (10)	[2] (25)	3
O_4	[3] –1	[1] –1	[7] –5	[3] –2	[4] (20)	[6] –6	1
v_j	1	–1	1	0	3	–1	–

Starting table / Optimal table

Since the d_{ij} value corresponding to O_3D_1 cell is zero, we conclude that the problem has an alternate optimal solution. To obtain the alternative optimal solution, we treat that current (optimal) solution as non-optimal and revise it by the MODI method so as to convert the zero entry unoccupied cell O_3D_1 to an occupied cell.

	D_1	D_2	D_3	D_4	D_5	D_6	u_i
O_1	[1] (10)	[2] –3	[1] (20)	[4] –4	[5] –2	[2] –3	0
O_2	[3] –1	[3] –3	[2] (10)	[1] (10)	[4] (30)	[3] –3	1
O_3	[4] (10)	[2] (40)	[5] –1	[9] –6	[6] 0	[2] (25)	3
O_4	[3] –1	[1] –1	[7] –5	[3] –2	[4] (20)	[6] –6	1
v_j	1	–1	1	0	3	–1	–

Alternative optimal solution

The two optimal solutions are:

1. $O_1 - D_1$: 20 units; $O_1 - D_3$: 10 units; $O_2 - D_3$: 20 units; $O_2 - D_4$: 10 units; $O_2 - D_5$: 20 units. $O_3 - D_2$: 40 units. $O_3 - D_5$: 10 units. $O_3 - D_6$: 25 units. $O_4 - D_5$: 20 units.
2. $O_1 - D_1$: 10 units; $O_1 - D_3$: 20 units; $O_2 - D_3$: 10 units; $O_2 - D_4$: 10 units; $O_2 - D_5$: 30 units. $O_3 - D_1$: 10 units. $O_3 - D_2$: 40 units. $O_3 - D_6$: 25 units. $O_4 - D_5$: 20 units.

Corresponding to both the solutions, the minimum total transportation cost = 430

Example 8

Three warehouses A, B and C supply four markets P, Q, R and S. The following table gives information about the supply available in each warehouse, the requirement of each market and the unit transportation cost from each warehouse to each market.

		Market				
		P	Q	R	S	Supply
Warehouse	A	5	2	4	3	22
	B	4	8	1	6	15
	C	4	6	7	5	8
Requirement		7	12	17	9	

The shipping clerk has worked out the following schedule from experience: 12 units from A to Q; 1 unit from A to R; 9 units from A to S; 15 units from B to R; 7 units from C to P and 1 unit from C to R.

(a) Check if the clerk has the optimal schedule.

(b) If this schedule is not optimal, find the optimal solution and minimum total shipping cost.

(c) If the clerk is approached by a carrier of route C to Q who offers to reduce his rate in the hope of getting some business, by how much must the rate be reduced before the clerk should consider giving him an order?

SOLUTION

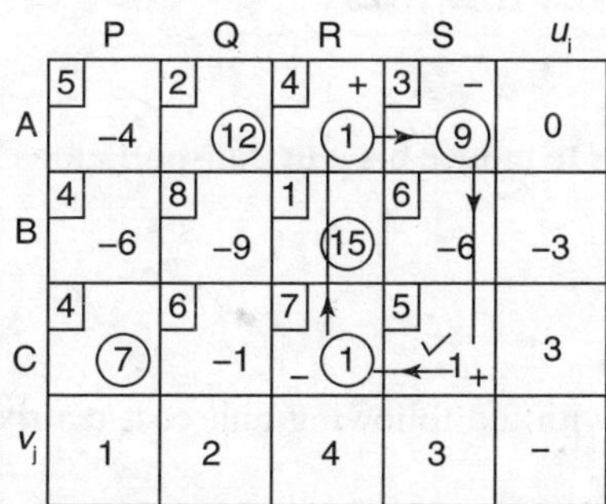

	P	Q	R	S	u_i
A	[5] –4	[2] (12)	[4] (1) +	[3] (9) –	0
B	[4] –6	[8] –9	[1] (15)	[6] –6	–3
C	[4] (7)	[6] –1	[7] – (1)	[5] ✓ 1 +	3
v_j	1	2	4	3	–

Starting solution / non-optimal solution

Since the d_{ij} value of the C – S cell is positive, there is still scope for improving the solution. Hence, the solution suggested by the clerk is not optimal. Let us now revise the solution so as to make the C – S cell occupied.

	P	Q	R	S	u_i
A	[5] –3	[2] (12) –	[4] (2)	[3] (8) +	0
B	[4] –5	[8] –9	[1] (15)	[6] –6	–3
C	[4] (7)	[6] + ✓ –2	[7] –1	[5] (1)	2
v_j	2	2	4	3	–

First iterated solution / Optimal solution

The total transportation cost according to the clerk's schedule $= 12 \times 2 + 1 \times 4 + 9 \times 3 + 15 \times 1 + 7 \times 4 + 1 \times 7 = 105$, whereas the minimum total cost is $= 12 \times 2 + 2 \times 4 + 8 \times 3 + 15 \times 1 + 7 \times 4 + 1 \times 5 = 104$

If the clerk wishes to give some order to the C–Q carrier without increasing the minimum total cost, let the reduced unit cost for the C–Q route be assumed as x. Let us find an alternative optimal solution so as to make C–Q cell occupied.

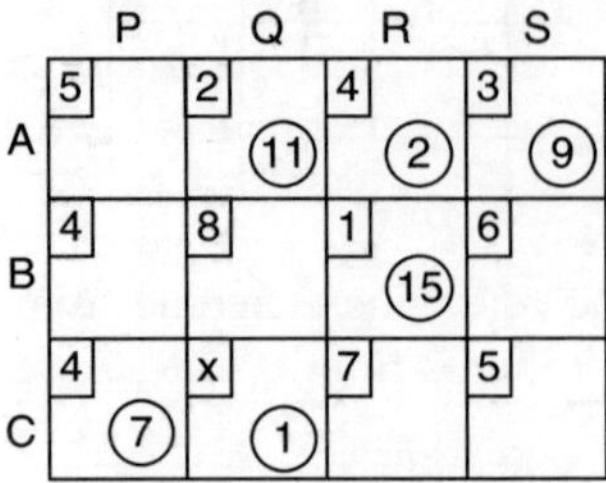

Revised optimal solution

The total transportation cost according to the revised solution $= 11 \times 2 + 2 \times 4 + 9 \times 3 + 15 \times 1 + 7 \times 4 + 1 \times x = 100 + x$

As this total cost must be equal to the minimum total cost, we have

$$100 + x = 104$$

$$\therefore \quad x = 4$$

Hence, if the C – Q carrier offers to reduce his unit transportation cost to 4, he can be given the order for transporting 1 unit.

Example 9

Solve the transportation problem with the following unit cost matrix by obtaining the initial solution using VAM.

	D_1	D_2	D_3	D_4	a_i
O_1	2.75	3.50	4.25	2.25	350
O_2	3.00	3.25	4.50	1.75	250
O_3	2.50	3.50	4.75	2.00	150
b_j	100	100	75	250	

SOLUTION

As $\sum a_i = 750 > 525 = \sum b_j$, the given problem is an unbalanced transportation problem. We will first convert it into a balanced transportation problem by introducing a dummy destination D_5 for which the requirement is $750 - 525 = 225$ units and the unit cost of transportation from each origin is zero, and then proceed to solve it.

	0 / 100	0 / 50 / 100	0 / 75	0 / 250	0 / 225	Penalties			
0 / 125 / 350	2.75	3.50 (50)	4.25 (75)	2.25	0 (225)	2.25	0.50	0.75	0.75
0 / 250	3.00	3.25	4.50	1.75	0 (250)	1.75	1.25	–	–
0 / 50 / 150	2.50 (100)	3.50 (50)	4.75	2.00	0	2.00	0.50	1.00	1.25
	0.25	0.25	0.25	0.25	0				
	0.25	0.25	0.25	0.25	–				
Penalties	0.25	0.00	0.50	–	–				
	–	0.00	0.50	–	–				

Initial solution got by VAM

	D_1	D_2	D_3	D_4	D_5	a_i
O_1	2.75 −0.25	3.50 (50) –	4.25 (75)	2.25 0.50	0 (225) +	0
O_2	3.00 −0.50	3.25 ✓ 0.25 +	4.50 −0.25	1.75 (250) –	0 (∈)	0
O_3	2.50 (100)	3.50 (50)	4.75 −0.50	2.00 −0.25	0 0	0
v_j	2.50	3.50	4.25	1.75	0	–

Starting table

The optimal solution given on the next page is

$O_1 - D_2$: 50 units; $O_1 - D_3$: 75 units; $O_1 - D_5$: 225 units; $O_2 - D_4$: 250 units; $O_3 - D_1$: 100 units; $O_3 - D_2$: 50 units;

The minimum total transportation cost $= 50 \times 3.50 + 75 \times 4.25 + 225 \times 0 + 250 \times 1.75 + 100 \times 2.50 + 50 \times 3.50 = 1356.25$

	D_1	D_2	D_3	D_4	D_5	a_i
O_1	2.75 −0.25	3.50 (50)	4.25 (75)	2.25 −0.25	0 (225)	0
O_2	3.00 −0.75	3.25 (∈)	4.50 −0.50	1.75 (250)	0 −0.25	−0.25
O_3	2.50 (100)	3.50 (50)	4.75 −0.50	2.00 0	0 0	0
v_j	2.50	3.50	4.25	2.00	0	–

First iterated solution / Optimal solution

Note: $O_1 - D_5$: 225 units means 225 units are retained in O_1 and not transported to any destination.

Example 10

Solve the transportation problem with the following unit cost matrix by obtaining the initial solution using the lowest cost entry method.

	D_1	D_2	D_3	D_4	D_5	a_i
O_1	20	19	14	21	16	40
O_2	15	20	13	9	16	60
O_3	18	15	18	20	30	90
b_j	30	40	70	40	60	–

Solution

Since $\sum a_i = 190 < 240 = \sum b_j$, the given problem is an unbalanced transportation problem.

It is first converted into a balanced transportation problem by introducing a dummy origin O_4 in which the availability is 240 – 190 = 50 and the unit cost of transportation to each destination is taken as zero, and then solve as shown on the next page.

The optimal solution is $O_1 - D_3$: 30 units; $O_1 - D_5$: 10 units; $O_2 - D_3$: 20 units; $O_2 - D_4$: 40 units; $O_3 - D_1$: 30 units; $O_3 - D_2$: 40 units; $O_3 - D_3$: 20 units; $O_4 - D_5$: 50 units

The minimum total transportation cost $= 30 \times 14 + 10 \times 16 + 20 \times 13 + 40 \times 9 + 30 \times 18 + 40 \times 15 + 20 \times 18 + 50 \times 0 = 2700$.

	D_1	D_2	D_3	D_4	D_5	a_i
O_1	[20]	[19]	[14]	[21]	[16] (40)	40 / 0
O_2	[15]	[20]	[13] (20)	[9] (40)	[16]	60 / 20 / 0
O_3	[18] (30)	[15] (40)	[18]	[20]	[30] (20)	90 / 50 / 20
O_4	[0]	[0]	[0] (50)	[0]	[0]	50 / 0
b_j	30 / 0	40 / 0	70 / 20 / 0	40 / 0	60 / 20	

Initial solution

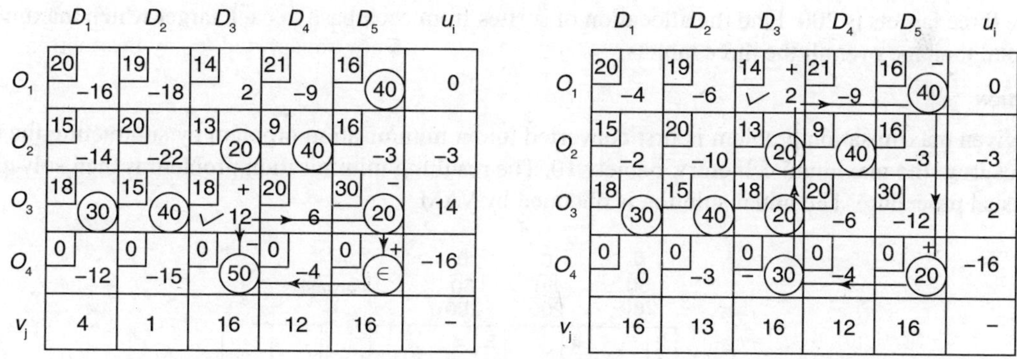

	D_1	D_2	D_3	D_4	D_5	u_i
O_1	[20] −16	[19] −18	[14] 2	[21] −9	[16] (40)	0
O_2	[15] −14	[20] −22	[13] (20)	[9] (40)	[16] −3	−3
O_3	[18] (30)	[15] (40)	[18] + ✓ 12	[20] 6	[30] − (20)	14
O_4	[0] −12	[0] −15	[0] − (50)	[0] −4	[0] + (∈)	−16
v_j	4	1	16	12	16	–

Starting table

	D_1	D_2	D_3	D_4	D_5	u_i
O_1	[20] −4	[19] −6	[14] + ✓ 2	[21] −9	[16] − (40)	0
O_2	[15] −2	[20] −10	[13] (20)	[9] (40)	[16] −3	−3
O_3	[18] (30)	[15] (40)	[18] (20)	[20] −6	[30] −12	2
O_4	[0] 0	[0] −3	[0] − (30)	[0] −4	[0] + (20)	−16
v_j	16	13	16	12	16	–

First iterated solution

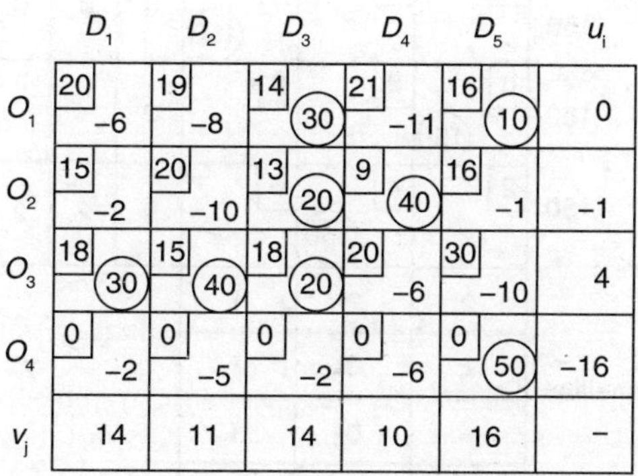

	D_1	D_2	D_3	D_4	D_5	u_i
O_1	[20] −6	[19] −8	[14] (30)	[21] −11	[16] (10)	0
O_2	[15] −2	[20] −10	[13] (20)	[9] (40)	[16] −1	−1
O_3	[18] (30)	[15] (40)	[18] (20)	[20] −6	[30] −10	4
O_4	[0] −2	[0] −5	[0] −2	[0] −6	[0] (50)	−16
v_j	14	11	14	10	16	–

Second iterated solution / Optimal solution

Note: $O_4 - D_5$: 50 units means that the destination D_5 cannot be allocated 50 units

Example 11

The tonnes of bombs that can be delivered by an aircraft from any of the four bases B_1, B_2, B_3, B_4 to any of the three targets T_1, T_2, T_3 are given in the following table:

	T_1	T_2	T_3
B_1	8	6	5
B_2	6	6	6
B_3	10	8	4
B_4	8	6	4

The daily sortie capability of each of the four bases is 150. The daily requirement in sorties over each of the three targets is 200. Find the allocation of sorties from each base to each target which maximises the total tonnage over all the three targets.

Solution

The given maximisation problem is first converted into a minimisation problem by subtracting the cell entries from the maximum cell entry, namely 10. The resulting minimisation problem is then solved by the usual procedure. The initial solution is obtained by VAM.

	0 / 50 / 200	0 / 50 / 200	0 / 50 / 200	Penalties			
0 / 100 / 150	[2] (50)	[4] (50)	[5] (50)	2	2	1	1
150	[4]	[4]	[4] (150)	0	0	0	0
150	[0] (150)	[2]	[6]	2	–	–	–
150	[2]	[4] (150)	[6]	2	2	2	–
Penalties	2	2	1				
	2	0	1				
	–	0	1				
	–	0	1				

Initial solution

The optimal solution as per optimal table given on the next page is

$B_1 - T_1$: 50 sorties; $B_1 - T_2$: 50 sorties; $B_1 - T_3$: 50 sorties; $B_2 - T_3$: 150 sorties; $B_3 - T_1$: 150 sorties; $B_4 - T_2$: 150 sorties;

	T_1	T_2	T_3	u_i
B_1	[2] (50)	[4] (50)	[5] (50)	0
B_2	[4] −3	[4] −1	[4] (150)	−1
B_3	[0] (150)	[2] 0	[6] −3	−2
B_4	[2] 0	[4] (150)	[6] −1	0
v_j	2	4	5	

Starting table / Optimal solution

The total maximum total tonnage = 50 × 8 + 50 × 6 + 50 × 5 + 150 × 6 + 150 × 10 + 150 × 6 = 4250.

(***Note:*** The presence of two zero d_{ij} values in the optimal table indicates that there are two alternative optimal solutions. they are

1. $B_1 - T_1 = 100$; $B_1 - T_3 = 50$; $B_2 - T_3 = 150$; $B_3 - T_1 = 100$; $B_3 - T_2 = 50$ and $B_4 - T_2 = 150$.
2. $B_1 - T_2 = 100$; $B_1 - T_3 = 50$; $B_2 - T_3 = 150$; $B_3 - T_1 = 150$; $B_4 - T_1 = 50$; $B_4 - T_2 = 100$)

Example 12

Solve the following transportation problem to maximise the profit: The entries in the body of the matrix represent profit (in Rs) per unit.

	D_1	D_2	D_3	D_4	a_i
O_1	20	5	2	13	100
O_2	24	15	10	10	30
O_3	18	18	8	10	70
b_j	40	20	60	30	

SOLUTION

Since $\sum a_i = 200 > 150 = \sum b_j$, the given problem is an unbalanced transportation problem. To make it balanced, a dummy destination D_5 with unit profit for each cell equal to zero and with the requirement equal to 200 − 150 = 50 units is introduced.

The maximisation problem is then converted into a minimisation problem by subtracting all the cell entries (unit profits) from the maximum cell entry, namely 24. The resulting transportation matrix is given here:

	D_1	D_2	D_3	D_4	D_5	a_i
O_1	4	19	22	11	24	100
O_2	0	9	14	14	24	30
O_3	6	6	16	14	24	70
b_j	40	20	60	30	50	

The initial solution is obtained by VAM and the usual procedure is followed as given here.

	0 / 10 / 40	0 / 20	0 / 20 / 60	0 / 30	0 / 50	Penalties			
0 /70 / 100	4	19 (20)	22 (30)	11	24 (50)	7	7	7	11 ✓
0 / 30	0 (30)	9	14	14	24	9 ✓	–	–	–
0 / 40 / 50 / 70	6 (10)	6 (20)	16 (40)	14	24	8	8	8 ✓	2
Penalties	4	3	2	3	0				
	2	13 ✓	6	3	0				
	2	–	6	3	0				
	–	–	6	3	0				

Initial solution got by VAM

	D_1	D_2	D_3	D_4	D_5	u_i
O_1	4 + 8 ✓	19 7	22 – (20)	11 (30)	24 (50)	0
O_2	0 (30)	9 –9	14 –4	14 –15	24 –12	–12
O_3	6 – (10)	6 (20)	16 + (40)	14 –9	24 –6	–6
v_j	12	12	22	11	24	–

Starting table

	D_1	D_2	D_3	D_4	D_5	u_i
O_1	4 + (10)	19 −7	22 − (10)	11 (30)	24 (50)	0
O_2	0 − (30)	9 −1	14 + 4	14 −7	24 −4	−4
O_3	6 −8	6 (20)	16 (50)	14 −9	24 −6	−6
v_j	12	12	22	11	24	–

First iterated solution

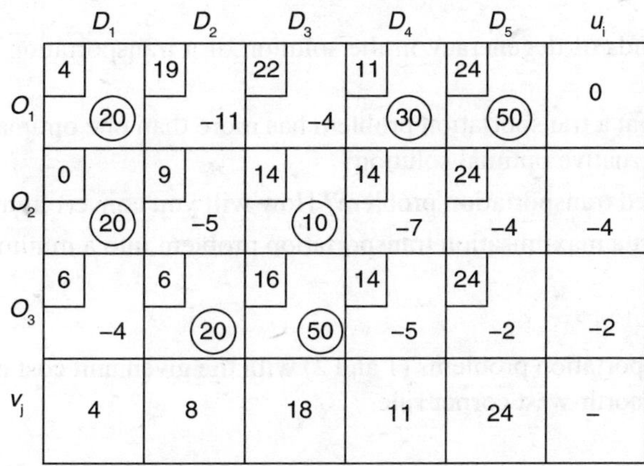

	D_1	D_2	D_3	D_4	D_5	u_i
O_1	4 (20)	19 −11	22 −4	11 (30)	24 (50)	0
O_2	0 (20)	9 −5	14 (10)	14 −7	24 −4	−4
O_3	6 −4	6 (20)	16 (50)	14 −5	24 −2	−2
v_j	4	8	18	11	24	–

Second iterated solution / Optimal solution

Optimal solution is given by:

$O_1 - D_1$: 20 units; $O_1 - D_4$: 30 units; $O_1 - D_5$: 50 units; $O_2 - D_1$: 20 units; $O_2 - D_3$: 10 units; $O_3 - D_2$: 20 units; $O_3 - D_3$: 50 units;

To get the total maximum profit, we have to multiply the units allocated with the corresponding original unit profits of the cells and add. Thus, the total maximum profit is equal to $20 \times 20 + 30 \times 13 + 50 \times 0 + 20 \times 24 + 10 \times 10 + 20 \times 18 + 50 \times 8 = 2130$ Rs.

Note:

1. $O_1 - D_1$: 50 units means 50 units are retained in O_1 and not transported to any destination.
2. Instead of transforming the maximisation problem into a balanced problem and then converting it into a minimisation problem, we could have also converted the given problem into a minimisation problem and then made it a balanced one.

EXERCISE 8(A)

Part A (Short answer questions)

1. What is a transportation problem?
2. Express a transportation problem as a linear programming model.

3. Define feasible and basic feasible solutions of a transportation problem.
4. Define non-degenerate and degenerate basic feasible solutions of a transportation problem.
5. What do you mean by independent allocations in a transportation problem?
6. What are the three major stages in the solution of a transportation problem?
7. Explain briefly the north-west corner rule of finding the initial BFS of a transportation problem with a sample example.
8. Express briefly the matrix minima method of finding the initial BFS of a transportation problem with a sample example.
9. Explain briefly Vogel's approximation method of finding the initial BFS. of a transportation problem with a sample example.
10. Explain how you will test whether a given BFS of a transportation problem is optimal or not with a simple example.
11. Explain the MODI method for improving a non-optimal solution of a transportation problem with a simple example.
12. What are the two kinds of degeneracy in the solution of a transportation problem? How will you resolve them?
13. How will you find that a transportation problem has more than one optimal solution? Explain how you will find the alternative optimal solution?
14. What is an unbalanced transportation problem? How will you convert it into a balanced problem?
15. How will you convert a maximisation transportation problem into a minimisation problem?

Part B

Solve the following transportation problems (1 and 2) with the given unit cost matrices by obtaining the initial solution using the north-west corner rule.

16.

	D_1	D_2	D_3	D_4	a_i
O_1	1	2	1	4	30
O_2	3	3	2	1	50
O_3	4	2	5	9	20
b_j	20	40	30	10	

17.

	1	2	3	4	5	Availability
P	10	15	20	20	40	50
Q	20	40	15	30	30	100
R	30	35	40	55	25	150
Requirement	25	115	60	30	70	

Solve the following transportation problems (3 and 4) with the given unit cost matrices by obtaining the initial solutions using the matrix minima method:

18.

	W_1	W_2	W_3	W_4	Factory capacity
Factory F_1	11	30	50	10	7
Factory F_2	70	30	40	60	9
Factory F_3	40	8	70	20	18
Warehouse requirement	5	8	7	14	

19.

	D_1	D_2	D_3	D_4	a_i
O_1	10	20	5	7	10
O_2	13	9	12	8	20
O_3	4	15	7	9	30
O_4	14	7	1	0	40
O_5	3	12	5	19	50
b_j	60	60	20	10	–

Solve the following transportation problems (5 and 6) with the following unit cost matrices by obtaining the initial solution using VAM.

20.

	D_1	D_2	D_3	D_4	a_i
O_1	13	11	15	20	2
O_2	17	14	12	13	6
O_3	18	18	15	12	7
b_j	3	3	4	5	

21.

	D_1	D_2	D_3	D_4	D_5	a_i
O_1	73	40	9	79	20	8
O_2	62	93	96	8	13	7
O_3	96	65	80	50	65	9
O_4	57	58	29	12	87	5
O_5	56	23	87	18	12	3
b_j	6	8	10	5	3	

22. Solve the following transportation problem where cell entries are unit costs by finding the initial solution using the matrix minima method.

	D_1	D_2	D_3	D_4	D_5	a_i
O_1	4	3	1	2	6	40
O_2	5	2	3	4	5	30
O_3	3	5	6	3	2	20
O_4	2	4	4	5	3	10
b_j	30	30	15	20	5	–

23. Solve the following transportation problem where cell entries are unit costs by finding the initial solution using VAM.

	D_1	D_2	D_3	D_4	D_5	D_6	a_i
O_1	9	12	9	6	9	10	5
O_2	7	3	7	7	5	5	6
O_3	6	5	9	11	3	11	2
O_4	6	8	11	2	2	10	9
b_j	4	4	6	2	4	2	–

24. Solve the following transportation problem where the cell entries are unit costs by finding the initial solution using the matrix minima method.

	D_1	D_2	D_3	D_4	D_5	a_i
O_1	4	7	3	8	2	4
O_2	1	4	7	3	8	7
O_3	7	2	4	7	7	9
O_4	4	8	2	4	7	2
b_j	8	3	7	2	2	–

25. Solve the following transportation problem where the cell entries are unit costs by finding the initial solution using the north-west corner rule.

	D_1	D_2	D_3	D_4	D_5	a_i
O_1	2	1	6	4	5	90
O_2	7	5	4	3	2	50
O_3	4	2	3	1	7	80
O_4	6	5	2	6	4	30
b_j	60	20	50	40	80	–

26. Solve the following transportation problem where the cell entries are unit costs by finding the initial solution using Vogel's approximation method. Find an alternative optimal solution, if there is one.

	D_1	D_2	D_3	D_4	D_5	a_i
O_1	5	3	4	3	7	8
O_2	7	6	5	8	4	14
O_3	6	5	6	3	2	10
b_j	4	5	7	8	8	–

27. Solve the following transportation problem where the cell entries are unit costs, by finding the initial solution using the lowest cost entry method. Find an alternative optimal solution, if it exists.

	D_1	D_2	D_3	D_4	D_5	D_6	a_i
O_1	8	0	0	9	6	4	90
O_2	9	1	9	7	3	4	20
O_3	3	3	5	5	1	5	60
O_4	8	7	4	7	10	7	50
b_j	20	40	20	60	40	40	–

28. Solve the following transportation problem where the cell entries are unit costs by finding the initial solution using VAM. Also find two alternative optimal solutions.

	D_1	D_2	D_3	D_4	a_i
O_1	8	9	5	11	19
O_2	2	6	7	13	21
O_3	3	2	14	8	5
O_4	15	13	9	6	17
O_5	9	10	11	4	8
b_j	28	17	15	10	–

29. Find the minimal total cost of transportation for the following transportation problem:

Warehouse

	W_1	W_2	W_3	W_4	Production
F_1	19	30	50	10	7
Factory F_2	70	30	40	60	9
F_3	40	8	70	20	18
Capacity	5	8	7	11	

30. Cities Allahabad, Bombay, Calcutta and Delhi are used as central warehouses for north Indian distribution of a scooter company in Tamil Nadu. At a given time, the inventories at these places are 500, 2000, 800 and 600 units respectively. A demand occurs at places E, F, G, H and J for 300, 400, 200, 200, and 400 units respectively. The shipping costs per scooter between the various places are given in the following matrix. How can shipments be made so as to minimise the total cost of transportation? Find also the total optimum cost.

	A	B	C	D
E	27.5	30.0	25.0	32.5
F	35.0	32.5	35.0	27.5
G	42.5	45.0	47.5	40.0
H	22.5	17.5	20.0	25.0
J	15.0	10.0	12.5	17.5

31. Consider the following unbalanced transportation problem:

To

	1	2	3	Supply
1	5	1	7	10
From 2	6	4	6	80
3	3	2	5	15
Demand	75	20	50	

Since there is not enough supply, some of the demands at these destinations may not be satisfied. Suppose there are penalty costs for every unit of unsatisfied demand which are given by 5, 3 and 2 for destination 1, 2 and 3 respectively, find the optimal solution.

(Hint: Introduce a dummy source 4, from which the unit transportation cost to the destinations 1, 2 and 3 are 5, 3 and 2 respectively and from which the supply is $145 - 105 = 40$ units)

32. Solve the following transportation problem to maximise the total profit:

	D_1	D_2	D_3	D_4	a_i
O_1	50	56	30	80	33
O_2	71	16	32	70	44
O_3	23	32	41	5	23
b_j	30	16	31	23	

33. Solve the following transportation problem for maximum profit:

		Market				
		A	B	C	D	Availability
	P	12	18	6	25	200
Warehouse	Q	8	7	10	18	500
	R	14	3	11	20	300
Demand		180	320	100	400	

34. Solve the following transportation problem for maximum profit:

	D_1	D_2	D_3	D_4	D_5	a_i
O_1	14	12	9	6	8	10
O_2	15	10	8	9	10	125
O_3	15	15	10	7	5	50
b_j	40	50	10	30	30	–

35. Solve the following transportation problem for maximum profit. Find as many optimal solutions as possible.

	D_1	D_2	D_3	D_4	D_5	a_i
O_1	175	250	400	275	325	200
O_2	125	200	475	350	250	150
O_3	100	175	450	325	300	250
O_4	150	225	425	350	275	300
b_j	200	250	75	100	150	–

8.3 ASSIGNMENT PROBLEM

The assignment problem is a special kind of transportation problem in which the objective is to assign a number of origins (jobs) to an equal number of destinations (machines) at a minimum total cost (or maximum profit), using the given cost (profit) matrix$[C_{ij}]$, where C_{ij} is the cost of assigning the ith job to the jth machine.

Mathematical model of AP

To find $x_{i,j}$ $i, j = 1, 2, \ldots, n$ so as to minimise $z = \sum_{i=1}^{n}\sum_{j=1}^{n} x_{ij}\, c_{ij}$,

subject to the constraints $\sum_{i=1}^{n} x_{ij} = 1 : j = 1, 2, \ldots, n,$

$$\sum_{j=1}^{n} x_{ij} = 1 : i = 1, 2, \ldots, n$$

and $x_{ij} = 0$ or 1.

Assignment Algorithm (Hungarian method or Flood's technique)

1. Subtract the minimum entry of each row of the cost matrix from all the entries of the respective row. Then repeat, this operation w.r.t. each column. We get the reduced matrix.
2. Draw the minimum number N of horizontal and vertical lines to cover all the zeros of the reduced matrix.
   (Row and column scanning: If there is only one zero in a row, draw a vertical line through it. If all the zeros are covered, go to the next step. If not, do column scanning. If there is exactly one zero in a column, draw a *horizontal* line through it. If all the zeros are covered, go to the next step, otherwise repeat this procedure)
3. (i) If $N = n$ (the order of the cost matrix), optimal assignment could be made.
   (ii) If $N < n$, go to the next step.
4. Find the smallest uncovered entry. Subtract it from all the uncovered entries. Add it to all entries at the intersection of covering lines. Leave the other entries on the covering lines as they are.
5. Repeat Steps 1, 2, 3 and 4 until $N = n$.
6. When $N = n$, we make optimal assignment as follows. Find a row with exactly one 0 entry. Encircle it. Cancel the row and column through the encircled 0. Repeat it for columns also.
7. Continue Step 6 until
   (i) no unmarked zero is left
   (ii) there lies more than one unmarked zero in any column or row.
8. In case (i), the procedure terminates. In case (ii), an alternative optimum exists. To get it, encircle any one unmarked zero arbitrarily and cancel the remaining zeros in that row and column. Repeat the process until no unmarked zero is left in the matrix.

WORKED EXAMPLES 8(B)

Example 1

A marketing manager has five salesmen and five sales districts. The sales per month (in 100s of rupees) for salesmen in each district are given in the following cost matrix. Find the assignment of the salesmen to districts that will result in minimum sales.

		District				
		A	B	C	D	E
	1	32	38	40	28	40
Salesman	2	40	24	28	21	36
	3	41	27	33	30	37
	4	22	38	41	36	36
	5	29	33	40	35	39

SOLUTION

We subtract the minimum entry of each row of the given cost matrix from all the entries of the concerned row and get the following row reduced matrix:

	A	B	C	D	E
1	4	10	12	0	12
2	19	3	7	0	15
3	14	0	6	3	10
4	0	16	19	14	14
5	0	4	11	6	10

We repeat this operation column-wise and get the following fully reduced matrix:

	A	B	C	D	E
1	4	10	6	0	2
2	19	3	1	0	5
3	14	0	0	3	0
4	0	16	13	14	4
5	0	4	5	6	0

Row scanning is now done. There is only one zero in the first row, viz, in the (1–D) position. We draw a vertical line through this zero. This vertical line also covers the zero in the (2–D) position, though it is a single zero in the second row. Since the (4–A) position zero is a single zero in the fourth row, a vertical line is drawn through the first column. This line also covers the zero in the (5–A) position. Now as there is a single zero in the (5–E) position, a vertical line is drawn through this zero. After the row scanning has been completed, we proceed to perform the column scanning. Corresponding to a single zero in the second column, a horizontal line is drawn through this zero. All the zeros have been covered. The fully scanned matrix will become as shown.

	A	B	C	D	E	
1	4	10	6	0	2	
2	19	3	1	0	5	
3	-14-	-0-	-0-	-3-	-0-	C_2
4	0	16	13	14	4	
5	0	4	5	6	0	
	R_4			R_1	R_5	

We note that the number of covering lines, viz., 4 is less than the order of the cost matrix, viz., 5. Hence, optimal assignment cannot be made at this stage. Therefore, we revise the last matrix as follows: The smallest uncovered element in the last matrix is 1. We subtract 1 from each of the uncovered elements, add 1 to each of the elements at the intersection of the covering lines and leave the other elements on the covering lines as they are. We get the revised cost matrix as given here:

	A	B	C	D	E
1	4	9	5	0	2
2	19	2	0	0	5
3	15	0	0	4	1
4	0	15	12	14	4
5	0	3	4	6	0
	R_4	R_3	R_2	R_1	R_5

As the number of covering lines for this revised matrix equals 5, optimal assignment can be made now as explained here:

As there is only one zero in the first row, we put a cross on this zero or encircle it and cancel the row and column passing through it. Similarly, we find a single zero in the second row; we cross it and cancel the row and column passing through it. The procedure is continued until no uncrossed zero is left. In this problem, the optimal assignment is made to the positions in which the zeros are encircled.

	A	B	C	D	E
1	4	9	5	(0)	2
2	19	2	(0)	0	5
3	15	(0)	0	4	1
4	(0)	15	12	14	4
5	0	3	4	6	(0)

The optimal assignment is given by:

1 → D; 2 → C; 3 → B; 4 → A; 5 → E.

∴ The minimum total sales = 28 + 28 + 27 + 22 + 39 = 144.

Example 2

A car hire company has one car at each of 6 depots A, B, C, D, E, F. A customer in each of 6 towns 1, 2, 3, 4, 5, 6 requires a car. The distance (in kms) between the depots and the towns is given in the following distance matrix. How should the cars be assigned to the towns so as to minimise the distance travelled?

	1	2	3	4	5	6
A	20	41	50	30	30	72
B	23	20	16	29	30	19
C	33	27	42	38	40	35
D	22	22	41	50	29	17
E	40	71	55	39	19	12
F	41	15	42	29	62	31

Row reduced matrix

$$\begin{pmatrix} 0 & 21 & 30 & 10 & 10 & 52 \\ 7 & 4 & 0 & 13 & 14 & 3 \\ 6 & 0 & 15 & 11 & 13 & 8 \\ 5 & 5 & 24 & 33 & 12 & 0 \\ 28 & 59 & 43 & 27 & 7 & 0 \\ 26 & 0 & 27 & 14 & 47 & 16 \end{pmatrix}$$

Fully reduced matrix

$$\begin{pmatrix} 0 & 21 & 30 & 0 & 3 & 52 \\ 7 & 4 & 0 & 3 & 7 & 3 \\ 6 & 0 & 15 & 1 & 6 & 8 \\ 5 & 5 & 24 & 23 & 5 & 0 \\ 28 & 59 & 43 & 17 & 0 & 0 \\ 26 & 0 & 27 & 4 & 40 & 16 \end{pmatrix}$$

Covering lines after row scanning

$$\begin{pmatrix} 0 & 21 & 30 & 0 & 3 & 52 \\ 7 & 4 & 0 & 3 & 7 & 3 \\ 6 & 0 & 15 & 1 & 6 & 8 \\ 5 & 5 & 24 & 23 & 5 & 0 \\ 28 & 59 & 43 & 17 & 0 & 0 \\ 26 & 0 & 27 & 4 & 40 & 16 \end{pmatrix}$$

$\qquad R_2 \quad R_1 \qquad\qquad R_4 \quad R_3$

Covering lines after full scanning

$$C_1 \begin{pmatrix} 0 & 21 & 30 & 0 & 3 & 52 \\ 7 & 4 & 0 & 3 & 7 & 3 \\ 6 & 0 & 15 & 1 & 6 & 8 \\ 5 & 5 & 24 & 23 & 5 & 0 \\ 28 & 29 & 43 & 17 & 0 & 0 \\ 26 & 0 & 27 & 4 & 40 & 16 \end{pmatrix}$$

$\qquad R_2 \quad R_1 \qquad\qquad R_4 \quad R_3$

$$n(=5) < N(=6)$$

$\therefore$ Distance matrix is revised

Revised distance matrix

$$\begin{pmatrix} 0 & 22 & 31 & 0 & 4 & 53 \\ 6 & 4 & 0 & 2 & 7 & 3 \\ 5 & 0 & 15 & 0 & 6 & 8 \\ 4 & 5 & 24 & 22 & 5 & 0 \\ 27 & 29 & 43 & 16 & 0 & 0 \\ 25 & 0 & 27 & 3 & 40 & 16 \end{pmatrix}$$

Covering lines after full scanning of revised matrix.

$$\begin{matrix} C_1 \\ \\ \\ C_4 \\ \\ \\ \end{matrix} \begin{pmatrix} 0 & 22 & 31 & 0 & 4 & 53 \\ 6 & 4 & 0 & 2 & 7 & 3 \\ 5 & 0 & 15 & 0 & 6 & 8 \\ 4 & 5 & 24 & 22 & 5 & 0 \\ 27 & 29 & 43 & 16 & 0 & 0 \\ 25 & 0 & 27 & 3 & 40 & 16 \end{pmatrix}$$

$\qquad R_6 \quad R_2 \qquad\qquad R_5 \quad R_4$

$$n(=6) = N(=6)$$

Hence, optimal assignment could be made. It is shown here:

$$\begin{pmatrix} \textcircled{0} & 22 & 31 & 0 & 4 & 53 \\ 6 & 4 & \textcircled{0} & 2 & 7 & 3 \\ 5 & 0 & 15 & \textcircled{0} & 6 & 8 \\ 4 & 5 & 24 & 22 & 5 & \textcircled{0} \\ 27 & 29 & 43 & 16 & \textcircled{0} & 0 \\ 25 & \textcircled{0} & 27 & 3 & 40 & 16 \end{pmatrix}$$

The optimal assignment is as follows:

A $\rightarrow$ 1: B $\rightarrow$ 3: C $\rightarrow$ 4: D $\rightarrow$ 6: E $\rightarrow$ 5: F $\rightarrow$ 2

The minimum total distance $= 26 + 16 + 38 + 17 + 19 + 12 = 125$ km.

Example 3

Solve the following minimal assignment problem. Point out the special feature, if any in the optimal solution.

	I	II	III	IV	V
A	6	5	8	11	16
B	1	3	16	1	10
C	16	11	8	8	8
D	9	14	12	10	16
E	10	13	11	8	16

Row reduced and also fully reduced matrix

$$\begin{pmatrix} 1 & 0 & 3 & 6 & 11 \\ 0 & 2 & 15 & 0 & 9 \\ 8 & 3 & 0 & 0 & 0 \\ 0 & 5 & 3 & 1 & 7 \\ 2 & 5 & 3 & 0 & 8 \end{pmatrix}$$

Covering lines after row scanning

$$\begin{pmatrix} 1 & 0 & 3 & 6 & 11 \\ 0 & 12 & 15 & 0 & 9 \\ 8 & 3 & 0 & 0 & 0 \\ 0 & 5 & 3 & 1 & 7 \\ 2 & 5 & 3 & 0 & 8 \end{pmatrix}$$

$R_4 \quad R_1 \qquad R_5$

Covering lines after full scanning

$$C_3 \begin{pmatrix} 1 & 0 & 3 & 6 & 11 \\ 0 & 12 & 15 & 0 & 9 \\ 8 & 3 & 0 & 0 & 0 \\ 0 & 5 & 3 & 1 & 7 \\ 2 & 5 & 3 & 0 & 8 \end{pmatrix}$$

$R_4 \quad R_1 \qquad R_5$

$n(=4) < N(=5)$

$\therefore$ Cost matrix is revised.

Revised cost matrix

$$\begin{pmatrix} 1 & 0 & 0 & 6 & 8 \\ 0 & 2 & 12 & 0 & 6 \\ 11 & 6 & 0 & 3 & 0 \\ 0 & 5 & 0 & 1 & 4 \\ 2 & 5 & 0 & 0 & 5 \end{pmatrix}$$

Covering lines after full scanning of the revised matrix

$$\begin{matrix} C_2 \\ \\ C_5 \\ \\ \\ \end{matrix} \begin{pmatrix} 1 & 0 & 0 & 6 & 8 \\ 0 & 2 & 12 & 0 & 6 \\ 11 & 6 & 0 & 3 & 0 \\ 0 & 5 & 0 & 1 & 4 \\ 2 & 5 & 0 & 0 & 5 \end{pmatrix}$$

$R_2 \qquad\qquad R_3 \; R_2$

Note: First, we have done column scanning, as there is no row containing a single zero. After column scanning, we find 3 rows containing two zeros. So we resort to row scanning with the rows containing two zeros, starting with the second row. Corresponding to this row scanning, the first and fourth columns are covered. Then, the fourth row is scanned, as there is only one zero in the (D–III) position. Corresponding to this row scanning, the third column is covered and hence all the zeros have been covered.) Since $n\,(=5) = N\,(=5)$, the optimal assignment could be made at this stage.

$$\begin{pmatrix} 1 & 0 & 0 & 6 & 8 \\ 0 & 2 & 12 & 0 & 6 \\ 11 & 6 & 0 & 3 & 0 \\ 0 & 5 & 0 & 1 & 4 \\ 2 & 5 & 0 & 0 & 5 \end{pmatrix}$$

Note: There is no row containing only one zero, so we resort to column-wise assignment, starting with the second column (A–II) and (C–V) positions.

At this stage, we find neither a row nor a column containing a single zero. We choose the second row for the next assignment. There are zeros in the (B–I) and (B–IV) positions. This is an indication that the problem has an alternative optimal solution.

We choose the (B–I) position for the next assignment. The remaining two assignments are made in the (D–III) and (E–IV) positions.

Had we chosen (B–IV) position for the third assignment, the remaining two assignments would have been made in the (E–III) and (D–I) positions.

The two alternative optimal assignment matrices are given here:

$$\begin{pmatrix} 1 & \textcircled{0} & 0 & 6 & 8 \\ \textcircled{0} & 2 & 12 & 0 & 6 \\ 11 & 6 & 0 & 3 & \textcircled{0} \\ 0 & 5 & \textcircled{0} & 1 & 4 \\ 2 & 5 & 0 & \textcircled{0} & 5 \end{pmatrix} \qquad \begin{pmatrix} 1 & \textcircled{0} & 0 & 6 & 8 \\ 0 & 2 & 12 & \textcircled{0} & 6 \\ 11 & 6 & 0 & 3 & \textcircled{0} \\ \textcircled{0} & 5 & 0 & 1 & 4 \\ 2 & 5 & \textcircled{0} & 0 & 5 \end{pmatrix}$$

Optimal assignment is

A → II: B → I; C → V: D → III: and E → IV

The minimum total costThe minimum total cost
$= 5 + 1 + 8 + 12 + 8 = 34.$

Alternative optimal assignment is

A → II: B → IV; C → V: D → I and E → III

$= 5 + 1 + 8 + 9 + 11 = 34.$

Example 4

Five men are available to do five different jobs. The time (in hours) that each man takes to do each job is known and is given in the following time matrix. Find the assignment that will minimise the total time taken. Give also the alternative assignments, if any.

	J_1	J_2	J_3	J_4	J_5
M_1	2	9	2	7	1
M_2	6	8	7	6	1
M_3	4	6	5	3	1
M_4	4	2	7	3	1
M_5	5	3	9	5	1

SOLUTION

Row reduced matrix

$$\begin{pmatrix} 1 & 8 & 1 & 6 & 0 \\ 5 & 7 & 6 & 5 & 0 \\ 3 & 5 & 4 & 2 & 0 \\ 3 & 1 & 6 & 2 & 0 \\ 4 & 2 & 8 & 4 & 0 \end{pmatrix}$$

Fully reduced matrix

$$\begin{pmatrix} 0 & 7 & 0 & 4 & 0 \\ 4 & 6 & 5 & 3 & 0 \\ 2 & 4 & 3 & 0 & 0 \\ 2 & 0 & 5 & 0 & 0 \\ 3 & 1 & 7 & 2 & 0 \end{pmatrix}$$

Covering lines after row scanning

$$\begin{pmatrix} 0 & 7 & 0 & 4 & 0 \\ 4 & 6 & 5 & 3 & 0 \\ 2 & 4 & 3 & 0 & 0 \\ 2 & 0 & 5 & 0 & 0 \\ 3 & 1 & 7 & 2 & 0 \end{pmatrix}$$
$$\quad R_4 \qquad R_3 \; R_2$$

Covering lines after full scanning

$$C_1 \begin{pmatrix} 0 & 7 & 0 & 4 & 0 \\ 4 & 6 & 5 & 3 & 0 \\ 2 & 4 & 3 & 0 & 0 \\ 2 & 0 & 5 & 0 & 0 \\ 3 & 1 & 7 & 2 & 0 \end{pmatrix}$$
$$\quad R_4 \qquad R_3 \; R_2$$

$$n(=4) < N(=5)$$

∴ The time matrix is revised.

Revised time matrix

$$\begin{matrix} C_3 \\ \\ C_4 \\ C_2 \\ \\ \end{matrix} \begin{pmatrix} 0 & 9 & 0 & 6 & 2 \\ 2 & 6 & 3 & 3 & 0 \\ 0 & 4 & 1 & 0 & 0 \\ 0 & 0 & 3 & 0 & 0 \\ 1 & 1 & 5 & 2 & 0 \end{pmatrix}$$
$$\qquad\qquad R_1$$

New revised time matrix

$$C_3 \begin{pmatrix} 0 & 9 & 0 & 6 & 3 \\ 1 & 5 & 2 & 2 & 0 \\ 0 & 4 & 1 & 0 & 1 \\ 0 & 0 & 3 & 0 & 1 \\ 0 & 0 & 4 & 1 & 0 \end{pmatrix}$$
$$R_3 \; R_5 \qquad R_4 \; R_2$$

Note: This revised time matrix remains unchanged after row and column reductions. So we have subjected it to scanning. Even now we see that n (= 4) < N (= 5).
The latest time matrix is further revised and scanned. Since at this stage, n (= 5) = N (= 5), we make optimal assignment.

Three optimal assignments are possible, which are given here.

(i) $M_1 \to J_3$; $M_2 \to J_5$; $M_3 \to J_1$; $M_4 \to J_4$; $M_5 \to J_2$.
(ii) $M_1 \to J_3$; $M_2 \to J_5$; $M_3 \to J_4$; $M_4 \to J_1$; $M_5 \to J_2$.
(iii) $M_1 \to J_3$; $M_2 \to J_5$; $M_3 \to J_4$; $M_4 \to J_2$; $M_5 \to J_1$.

Minimum total time in all the three cases = 13 hours.

$$\begin{pmatrix} 0 & 9 & \textcircled{0} & 6 & 3 \\ 1 & 5 & 2 & 2 & \textcircled{0} \\ \textcircled{0} & 4 & 1 & 0 & 1 \\ 0 & 0 & 3 & \textcircled{0} & 1 \\ 0 & \textcircled{0} & 4 & 1 & 0 \end{pmatrix} \quad \begin{pmatrix} 0 & 9 & \textcircled{0} & 6 & 3 \\ 1 & 5 & 2 & 2 & \textcircled{0} \\ \textcircled{0} & 4 & 1 & \textcircled{0} & 1 \\ 0 & \textcircled{0} & 3 & 0 & 1 \\ 0 & 0 & 4 & 1 & 0 \end{pmatrix} \quad \begin{pmatrix} 0 & 9 & \textcircled{0} & 6 & 3 \\ 1 & 5 & 2 & 2 & \textcircled{0} \\ 0 & 4 & 1 & \textcircled{0} & 1 \\ 0 & \textcircled{0} & 3 & 0 & 1 \\ \textcircled{0} & 0 & 4 & 1 & 0 \end{pmatrix}$$

Example 5

Solve the following unbalanced minimal assignment problem:

	A	B	C	D	E
1	4	3	6	2	7
2	10	12	11	14	16
3	4	3	2	1	5
4	8	7	6	9	6

SOLUTION

Note: The Hungarian method of solving an assignment problem holds good only when the cost matrix is a square matrix. In the current problem, the number of rows is less than the number of columns. So we convert the cost matrix into a square matrix by introducing a dummy row (job) (row 5) with zero cost for each column and then apply the Hungarian method.

Converted (balanced) cost matrix

	A	B	C	D	E
1	4	3	6	2	7
2	10	12	11	14	16
3	4	3	2	1	5
4	8	7	6	9	6
5	0	0	0	0	0

Fully reduced matrix

$$\begin{pmatrix} 2 & 1 & 4 & 0 & 5 \\ 0 & 2 & 1 & 4 & 6 \\ 3 & 2 & 1 & 0 & 4 \\ 2 & 1 & 0 & 3 & 0 \\ 0 & 0 & 0 & 0 & 0 \end{pmatrix}$$

Covering lines after full scanning

$$\begin{matrix} \\ \\ \\ C_4 \\ C_2 \end{matrix}\begin{pmatrix} 2 & 1 & 4 & 0 & 5 \\ 0 & 2 & 1 & 4 & 6 \\ 3 & 2 & 1 & 0 & 4 \\ 2 & 1 & 0 & 3 & 0 \\ 0 & 0 & 0 & 0 & 0 \end{pmatrix}$$
$$\quad R_2 \qquad\qquad R_1$$

Revised cost matrix

$$\begin{pmatrix} 2 & 0 & 3 & 0 & 4 \\ 0 & 1 & 0 & 4 & 5 \\ 3 & 1 & 0 & 0 & 3 \\ 3 & 1 & 0 & 4 & 0 \\ 1 & 0 & 0 & 1 & 0 \end{pmatrix}$$

Note: The fully reduced matrix corresponding to the aforementioned revised cost matrix is the same.)

Covering lines after full scanning

$$\begin{matrix} \\ C_1 \\ \\ \\ \\ \end{matrix}\begin{pmatrix} 2 & 0 & 3 & 0 & 4 \\ 0 & 1 & 0 & 4 & 5 \\ 3 & 1 & 0 & 0 & 3 \\ 3 & 1 & 0 & 4 & 0 \\ 1 & 0 & 0 & 1 & 0 \end{pmatrix}$$
$$\qquad R_1 \; R_5 \; R_3 \; R_4$$

Since $n\ (= 5) = N\ (= 5)$, optimal assignment is made.

Two optimal assignments are possible namely 1 → B; 2 → A; 3 → D; 4 → C (or E) and 5 → E (or C)
The corresponding solutions are given here:
No job is assigned to C or E if assignments are made to E or C. Minimum total cost = 20.

$$\begin{pmatrix} 2 & \textcircled{0} & 3 & 0 & 4 \\ \textcircled{0} & 1 & 0 & 4 & 5 \\ 3 & 1 & 0 & \textcircled{0} & 3 \\ 3 & 1 & \textcircled{0} & 4 & 0 \\ 1 & 0 & 0 & 1 & \textcircled{0} \end{pmatrix} \quad \begin{pmatrix} 2 & \textcircled{0} & 3 & 0 & 4 \\ \textcircled{0} & 1 & 0 & 4 & 5 \\ 3 & 1 & 0 & \textcircled{0} & 3 \\ 3 & 1 & 0 & 4 & \textcircled{0} \\ 1 & 0 & \textcircled{0} & 1 & 0 \end{pmatrix}$$

Example 6

Solve the following unbalanced minimal assignment problem given here:

		Machine		
		A	B	C
Job	1	9	26	15
	2	13	27	6
	3	35	20	15
	4	18	30	20

SOLUTION

We introduce a dummy machine (column) D with zero cost for each job (row) to convert the cost matrix into a square matrix and proceed further.

Balanced cost matrix

	A	B	C	D
1	9	26	15	0
2	13	27	6	0
3	35	20	15	0
4	18	30	20	0

Fully reduced matrix

$$\begin{pmatrix} 0 & 6 & 9 & 0 \\ 4 & 7 & 0 & 0 \\ 26 & 0 & 9 & 0 \\ 9 & 10 & 14 & 0 \end{pmatrix}$$

Covering lines after full scanning

$$\begin{matrix} C_1 \\ C_2 \\ C_3 \\ \\ \end{matrix} \begin{pmatrix} 0 & 6 & 9 & 0 \\ 4 & 7 & 0 & 0 \\ 26 & 0 & 9 & 0 \\ 9 & 10 & 14 & 0 \end{pmatrix}$$
$$R_4$$
$$n(=4) = N = (4)$$

Optimal assignment matrix

$$\begin{pmatrix} \textcircled{0} & 6 & 9 & 0 \\ 4 & 7 & \textcircled{0} & 0 \\ 26 & \textcircled{0} & 9 & 0 \\ 9 & 10 & 14 & \textcircled{0} \end{pmatrix}$$

Optimal assignments are: 1 → A; 2 → C; 3 → B; 4 → D
∴ Job 4 is not assigned to any machine. Minimum total cost = 9 + 6 + 20 = 35.

Example 7

A company has 5 jobs to be done. The following matrix shows the return (profit) in Rs. of assigning the *i*th machine to the *j*th job. Assign the 5 jobs to the 5 machines so as to maximise the total profit.

		Job				
		1	2	3	4	5
	1	5	11	10	12	4
Machine	2	2	4	6	3	5
	3	3	1	5	14	6
	4	6	14	4	11	7
	5	7	9	8	12	8

Solution

Note: The Hungarian method of solving an assignment problem holds good only for a minimisation problem, the entries in the cost matrix of which are non-negative. So to convert the problem into a minimisation problem, we subtract all entries of the profit matrix from the maximum entry. In this problem, it is 14. After solving the corresponding minimisation problem and identifying the locations for the minimum assignment, we add the original profit entries of these locations to get the maximum total profit.

Corresponding cost matrix

		Job				
		1	2	3	4	5
	1	9	3	4	2	10
Machine	2	12	10	8	11	9
	3	11	13	9	0	8
	4	8	0	10	3	7
	5	7	5	6	2	6

Fully reduced matrix, that also gives covering lines after full scanning

$$\begin{array}{c} \\ C_1 \\ \\ \\ \\ \end{array}\begin{pmatrix} 3 & 1 & 2 & 0 & 7 \\ 0 & 2 & 0 & 3 & 0 \\ 7 & 13 & 9 & 0 & 7 \\ 4 & 0 & 10 & 3 & 6 \\ 1 & 3 & 4 & 0 & 3 \end{pmatrix}$$
$$\qquad\quad R_4 \qquad R_1$$

Revised cost matrix with covering lines after full scanning

$$C_2\begin{pmatrix} 2 & 1 & 1 & 0 & 6 \\ 0 & 3 & 0 & 4 & 0 \\ 6 & 13 & 8 & 0 & 6 \\ 3 & 0 & 9 & 3 & 5 \\ 0 & 3 & 3 & 0 & 2 \end{pmatrix}$$

$R_5 \quad R_4 \quad R_1$

New revised 'cost' matrix with covering lines after full scanning

$$C_5\begin{pmatrix} 2 & 1 & 0 & 0 & 5 \\ 1 & 4 & 0 & 5 & 0 \\ 6 & 13 & 7 & 0 & 5 \\ 3 & 0 & 8 & 3 & 4 \\ 0 & 3 & 2 & 0 & 1 \end{pmatrix}$$

$R_5 \quad R_4 \quad R_1 \quad R_3 \qquad n(=5) = N(=5)$

Optimal assignment matrix

$$\begin{pmatrix} 2 & 1 & \textcircled{0} & 0 & 5 \\ 1 & 4 & 0 & 5 & \textcircled{0} \\ 6 & 13 & 7 & \textcircled{0} & 5 \\ 3 & \textcircled{0} & 8 & 3 & 4 \\ \textcircled{0} & 3 & 2 & 0 & 1 \end{pmatrix}$$

The optimal assignments are given by

$M_1 \to J_3$; $M_2 \to J_5$; $M_3 \to J_4$; $M_4 \to J_2$; $M_5 \to J_1$.

Maximum total profit $= 10 + 5 + 14 + 14 + 7 = 50$.

Example 8

Five machines are available for processing four jobs. Any machine can process any job with the resulting profit matrix (in rupees) as follows: what is the maximum profit that may be expected if an optimal assignment is made. Which machine has to remain idle?

	M_1	M_2	M_3	M_4	M_5
J_1	12	18	20	8	20
J_2	20	4	8	1	16
J_3	21	7	13	10	17
J_4	2	18	21	16	16

Solution

Corresponding cost matrix

	M_1	M_2	M_3	M_4	M_5
J_1	9	3	1	13	1
J_2	1	17	13	20	5
J_3	0	14	8	11	4
J_4	19	3	0	5	5

Balanced cost matrix

	M_1	M_2	M_3	M_4	M_5
J_1	9	3	1	13	1
J_2	1	17	13	20	5
J_3	0	14	8	11	4
J_4	19	3	0	5	5
J_5	0	0	0	0	0

Fully reduced cost matrix

$$\begin{pmatrix} 8 & 2 & 0 & 12 & 0 \\ 0 & 16 & 12 & 19 & 4 \\ 0 & 14 & 8 & 11 & 4 \\ 19 & 3 & 0 & 5 & 5 \\ 0 & 0 & 0 & 0 & 0 \end{pmatrix}$$

Covering lines after full scanning

$$\begin{array}{c} C_5 \\ \\ \\ \\ C_2 \end{array} \begin{pmatrix} 8 & 2 & 0 & 12 & 0 \\ 0 & 16 & 12 & 19 & 4 \\ 0 & 14 & 8 & 11 & 4 \\ 19 & 3 & 0 & 5 & 5 \\ 0 & 0 & 0 & 0 & 0 \end{pmatrix}$$
$$\quad R_2 \qquad R_4$$

Revised 'cost' matrix with covering lines

$$\begin{array}{c} \\ \\ \\ C_3 \\ C_4 \end{array} \begin{pmatrix} 11 & 2 & 3 & 12 & 0 \\ 0 & 13 & 12 & 16 & 1 \\ 0 & 11 & 8 & 8 & 1 \\ 19 & 0 & 0 & 2 & 2 \\ 3 & 0 & 3 & 0 & 0 \end{pmatrix}$$
$$R_2 \qquad\qquad R_1$$

New revised 'cost' matrix with covering lines

$$\begin{array}{c} C_5 \\ \\ \\ C_3 \\ C_4 \end{array} \begin{pmatrix} 11 & 0 & 1 & 10 & 0 \\ 0 & 11 & 10 & 14 & 1 \\ 0 & 9 & 6 & 6 & 1 \\ 21 & 0 & 0 & 2 & 4 \\ 5 & 0 & 3 & 0 & 2 \end{pmatrix}$$
$$R_2$$

Latest revised cost matrix with covering lines.

$$\begin{array}{c} \\ \\ \\ C_3 \\ C_4 \end{array} \begin{pmatrix} 12 & 0 & 1 & 10 & 0 \\ 0 & 10 & 9 & 13 & 0 \\ 0 & 8 & 5 & 5 & 0 \\ 22 & 0 & 0 & 2 & 4 \\ 6 & 0 & 3 & 0 & 2 \end{pmatrix}$$
$$R_3 \quad R_1 \qquad\quad R_2$$

Optimal assignment matrix

$$\begin{pmatrix} 12 & \textcircled{0} & 1 & 10 & 0 \\ \textcircled{0} & 10 & 9 & 13 & 0 \\ 0 & 8 & 5 & 5 & \textcircled{0} \\ 22 & 0 & \textcircled{0} & 2 & 4 \\ 6 & 0 & 3 & \textcircled{0} & 2 \end{pmatrix} \text{ OR } \begin{pmatrix} 12 & \textcircled{0} & 1 & 10 & 0 \\ 0 & 10 & 9 & 13 & \textcircled{0} \\ \textcircled{0} & 8 & 5 & 5 & 0 \\ 22 & 0 & \textcircled{0} & 2 & 4 \\ 6 & 0 & 3 & \textcircled{0} & 2 \end{pmatrix}$$

The optimal assignments are given by

$J_1 \to M_2$; $J_2 \to M_1$; $J_3 \to M_5$; $J_4 \to M_3$; $J_5 \to M_4$. (or) $J_1 \to M_2$; $J_2 \to M_5$; $J_3 \to M_1$; $J_4 \to M_3$; $J_5 \to M_4$

Maximum total profit $= 18 + 20 + 17 + 21 = 76$. Machine 4 has to remain idle.

Example 9

Solve the restricted minimal assignment problem, the cost matrix of which is given here: The dashes in the matrix mean that J_2 cannot be assigned to M_1, J_3 cannot be assigned to M_3 and so on.

	J_1	J_2	J_3	J_4	J_5
M_1	3	-	8	-	8
M_2	4	7	15	18	8
M_3	8	12	-	-	12
M_4	5	5	8	3	6
M_5	10	12	15	10	-

Solution

Since the dashes indicate that a job cannot be assigned to the corresponding machine, we replace the dashes by infinity each so that the job cannot be assigned to the corresponding machine as the cost is prohibitive. The rest of the procedure follows the Hungarian method.

Fully reduced matrix   Covering lines after full scanning

$$\begin{pmatrix} 0 & \infty & 0 & \infty & 2 \\ 0 & 1 & 6 & 14 & 1 \\ 0 & 2 & \infty & \infty & 1 \\ 2 & 0 & 0 & 0 & 0 \\ 0 & 0 & 0 & 0 & \infty \end{pmatrix} \qquad \begin{matrix} C_3 \\ \\ \\ C_5 \\ C_2 \end{matrix}\begin{pmatrix} 0 & \infty & 0 & \infty & 2 \\ 0 & 1 & 6 & 14 & 1 \\ 0 & 2 & \infty & \infty & 1 \\ 2 & 0 & 0 & 0 & 0 \\ 0 & 0 & 0 & 0 & \infty \end{pmatrix}$$

R_2

$n(=4) < N(5)$

Revised matrix with covering lines

$$\begin{pmatrix} 1 & \infty & 0 & \infty & 2 \\ 0 & 0 & 5 & 13 & 0 \\ 0 & 1 & \infty & \infty & 0 \\ 3 & 0 & 0 & 0 & 0 \\ 1 & 0 & 0 & 0 & \infty \end{pmatrix}$$

$R_3 \quad R_4 \quad R_1 \quad R_4 \quad R_3$

One optimal assignment matrix

$$\begin{pmatrix} 1 & \infty & \textcircled{0} & \infty & 2 \\ 0 & 0 & 5 & 13 & \textcircled{0} \\ \textcircled{0} & 1 & \infty & \infty & 0 \\ 3 & 0 & 0 & \textcircled{0} & 0 \\ 1 & \textcircled{0} & 0 & 0 & \infty \end{pmatrix}$$

One optimal assignment is $M_1 \to J_3$; $M_2 \to J_5$; $M_3 \to J_1$; $M_4 \to J_4$; $M_5 \to J_2$.
Maximum total profit $= 8 + 8 + 8 + 3 + 12 = 39$.

Example 10

A machine operator processes four types of works on his machine each week and must choose a sequence for them. The set up cost per change depends on the work presently processed on the machine and the work to be processed next, according to the following table. If he produces each type of work only once each week, how should he sequence the works on his machine in order to minimise the total set up cost?

		Process next			
		A	B	C	D
	A	∞	45	75	35
Process first	B	45	∞	65	35
	C	75	65	∞	75
	D	35	35	75	∞

SOLUTION

We first solve the problem as an assignment problem.

Fully reduced matrix with covering lines One optimal assignment matrix

$$\begin{pmatrix} \infty & 10 & 10 & 0 \\ 10 & \infty & 0 & 0 \\ 10 & 0 & \infty & 10 \\ 0 & 0 & 10 & \infty \end{pmatrix} \quad R_4 \; R_3 \; R_2 \; R_1 \qquad \begin{pmatrix} \infty & 10 & 10 & \textcircled{0} \\ 10 & \infty & \textcircled{0} & 0 \\ 10 & \textcircled{0} & \infty & 10 \\ \textcircled{0} & 0 & 10 & \infty \end{pmatrix}$$

The optimal assignments is given by A → D; B → C; C → B; D → A.

The optimal assignment means that the job D is to be processed after A and the job A is to be processed after D. Viz., the jobs B and C are not at all processed.

∴ The optimal assignment does not provide the solution to the sequencing problem according to which all jobs should be processed one after the other at minimum total set up cost.

Now we try to find the solution to the sequencing problem as follows: Instead of making the assignment to (A → D) in the first row, we make the assignment to the next minimum location (A → B) or (A → C) with entry 10. Then in the second row, we may make the assignment to the location (B → C) or (B → D) and so on.

In order to get the solution to the sequencing problem, there are two possibilities as shown in the following matrices:

$$\begin{array}{c} \\ A \\ B \\ C \\ D \end{array}\begin{array}{c} \begin{matrix} A & B & C & D \end{matrix} \\ \begin{pmatrix} \infty & \textcircled{10} & 10 & 0 \\ 10 & \infty & \textcircled{0} & 0 \\ 10 & 0 & \infty & \textcircled{10} \\ \textcircled{0} & 0 & 10 & \infty \end{pmatrix} \end{array} \qquad \begin{array}{c} \\ A \\ B \\ C \\ D \end{array}\begin{array}{c} \begin{matrix} A & B & C & D \end{matrix} \\ \begin{pmatrix} \infty & 10 & \textcircled{10} & 0 \\ 10 & \infty & 0 & \textcircled{0} \\ 10 & \textcircled{0} & \infty & 10 \\ \textcircled{0} & 0 & 10 & \infty \end{pmatrix} \end{array}$$

Corresponding to the first matrix, the sequencing is A → B → C → D → A. The total set-up cost corresponding to this sequencing = 45 + 65 + 75 + 35 = 220.

Corresponding to the second matrix, the sequencing is A → C → B → D → A. The total set up cost corresponding to this sequencing = 75 + 65 + 35 + 35 = 210.

∴ The minimum set up cost corresponding to the sequencing A → C → B → D → A

EXERCISE 8 (B)

Part A: Short answer questions

1. What is an assignment problem?
2. Give the mathematical model of an assignment problem.
3. How will you get the reduced cost matrix when the Hungarian method is used to solve an assignment problem?
4. When the Hungarian method is applied to solve an assignment problem, how will you find out the minimum number of lines that cover all the zeros in the reduced matrix?
5. Explain the procedure to get the revised cost matrix which permits optimal assignment.
6. When optimality has been attained, how will you make optimal assignment?
7. Explain unbalanced assignment problem. How can it be made a balanced one?
8. When solving an assignment problem by the Hungarian method, how will you identify whether the problem has an alternative optimal solution?
9. Explain how you will convert a maximal assignment problem into minimal assignment problem.
10. What is meant by a restricted assignment problem? How is the cost matrix of such a problem assumed?

Part B

Solve the minimal assignment problems represented by the following cost matrices: Also find the minimal total cost in each case.

11.

		Machines				
		A	B	C	D	E
	1	13	8	16	18	19
Jobs	2	9	15	24	9	12
	3	12	9	4	4	4
	4	6	12	10	8	13
	5	15	17	18	12	20

12.

		Machines					
		M_1	M_2	M_3	M_4	M_5	M_6
	O_1	10	12	8	10	8	12
	O_2	9	10	8	7	8	9
Operators	O_3	8	7	8	8	8	6
	O_4	12	13	14	14	15	14
	O_5	9	9	9	8	8	10
	O_6	7	8	9	9	9	8

13.

	1	2	3	4	5	6
A	42	72	39	52	25	51
B	22	29	49	65	81	50
C	27	39	60	51	32	32
D	45	50	48	52	37	43
E	29	40	39	26	30	33
F	82	40	40	60	51	30

14.

	J_1	J_2	J_3	J_4	J_5	J_6
M_1	20	23	18	10	16	20
M_2	50	20	17	16	15	11
M_3	60	30	40	55	8	7
M_4	6	7	10	20	100	9
M_5	18	19	28	17	60	70
M_6	9	10	20	30	40	55

15. Solve the following minimal assignment problems represented by the cost matrices: Also find the alternative optimal solutions, if they exist.

	M_1	M_2	M_3	M_4	M_5
J_1	20	30	25	15	35
J_2	25	10	40	12	28
J_3	15	18	22	32	24
J_4	29	8	34	10	40
J_5	35	23	17	26	45

16.

	P	Q	R	S	T
A	6	10	7	10	14
B	10	6	6	14	15
C	12	14	12	17	7
D	6	6	6	9	8
E	9	7	9	7	12

17.

	1	2	3	4	5	6
A	7	20	56	9	17	25
B	41	76	70	48	61	46
C	39	26	89	35	43	31
D	72	40	25	47	37	30
E	34	9	55	20	23	16
F	1	54	51	29	15	26

18. Solve the unbalanced minimal assignment problems represented by the following cost matrices. Indicate which machine/job is not assigned.

	J_1	J_2	J_3	J_4	J_5
M_1	30	39	31	38	40
M_2	43	37	32	35	38
M_3	34	41	33	41	34
M_4	39	36	43	32	36
M_5	32	49	35	40	37
M_6	36	42	35	44	42

19.

	M_1	M_2	M_3	M_4	M_5	M_6
J_1	27	28	28	27	25	26
J_2	23	27	26	25	20	21
J_3	22	21	24	21	23	21
J_4	24	26	25	24	26	24
J_5	27	24	28	24	26	27

20.

	M_1	M_2	M_3	M_4	M_5
J_1	2.5	5	1	6	1
J_2	2	5	1.5	7	3
J_3	3	6.5	2	8	3
J_4	3.5	7	2	9	4.5
J_5	4	7	3	9	6
J_6	6	9	5	10	6

21. Solve the maximal assignment problem represented by the following profit matrices. Also find the maximum total profit in each case.

	A	B	C	D	E
1	30	37	40	28	40
2	40	24	27	21	36
3	40	32	33	30	35
4	25	38	40	36	36
5	29	62	41	34	39

22.

	A	B	C	D	E
1	29	33	40	35	39
2	22	38	41	36	36
3	41	27	33	30	37
4	40	24	28	21	6
5	32	38	40	28	40

23.

		Sales region			
		R_1	R_2	R_3	R_4
	M_1	10	22	12	14
Salesman	M_2	16	18	22	10
	M_3	24	20	12	18
	M_4	16	14	24	20

24. Solve the maximal assignment problems represented by the following profit matrices. Indicate alternative optimal solutions if they are available.

	1	2	3	4	5
A	9	2	9	4	10
B	5	3	4	5	10
C	7	5	6	8	10
D	7	9	4	8	10
E	6	8	2	6	10

25.

	1	2	3	4	5
A	21	23	27	25	20
B	19	28	24	24	23
C	22	27	28	21	25
D	23	22	20	19	22
E	28	24	27	28	24

26. Solve the unbalanced maximal assignment problems represented by the following profit matrices. Indicate which jobs cannot be assigned to any machine.

		Job				
		A	B	C	D	E
	1	62	78	50	101	82
Machine	2	71	84	61	73	59
	3	87	92	111	71	81
	4	48	64	87	77	80

27.

		Machine			
		A	B	C	D
	1	15	17	16	14
	2	15	12	14	13
Job	3	16	14	13	17
	4	13	18	15	18
	5	17	11	14	14
	6	13	16	12	16

28. Solve the following restricted minimal assignment problems

	A	B	C	D	E
1	-	4	7	3	4
2	4	-	6	3	4
3	7	6	-	7	5
4	3	3	7	-	7
5	4	4	5	7	-

29.

	J_1	J_2	J_3	J_4	J_5
M_1	4	6	10	5	6
M_2	7	4	-	5	4
M_3	-	6	9	6	2
M_4	9	3	7	2	3

30. Find the sequence of production so as to minimise the total set up cost per cycle, using the following set up cost matrix;

	A	B	C	D	E
A	∞	2	5	7	1
B	6	∞	3	8	2
C	8	7	∞	4	7
D	12	4	6	∞	5
E	1	3	2	8	∞

ANSWERS

Exercise 8(A)

Part B

16. $O_1 - D_1$: 20; $O_1 - D_2$: 10; $O_2 - D_2$: 10; $O_2 - D_3$: 30; $O_2 - D_4$: 10. $O_3 - D_2$: 20. Min. total cost (MTC) = 180.
17. P-2:50, Q-1:10, Q-3:60, Q-4:30, R-1:15, R-2:65, R-5:70, MTC = 7325.
18. F_1W_1: 5, F_1W_4: 2, F_2W_2: 2, F_2W_3: 7, F_3W_2: 6, F_3W_4: 12, MTC = 703.
19. O_1D_3: 10, O_2D_2: 20, O_3D_1: 30, O_4D_2: 30, O_4D_4: 10, O_5D_1: 30, O_5D_2: 10, O_5D_3: 10, MTC = 820.
20. O_1D_1: 1, O_1D_2: 1, O_2D_2: 2, O_2D_3: 4, O_3D_1: 2, O_3D_4: 5, MTC = 196.
21. O_1D_3: 8, O_2D_4: 5, O_2D_5: 2, O_3D_1: 3, O_3D_2: 6, O_4D_1: 3, O_4D_3: 2, O_5D_2: 2, O_5D_5: 1, MTC = 1103.
22. Initial degeneracy O_1D_1: 5, O_1D_3: 15, O_1D_4: 20, O_2D_2: 30, O_3D_1: 15, O_3D_5: 5, O_4D_1: 10, MTC = 210.
23. Initial degeneracy O_1D_3: 5, O_2D_2: 4, O_2D_6: 2, O_3D_1: 1, O_3D_3: 1, O_4D_1: 3, O_4D_4: 2, O_4D_5: 4, MTC = 112.
24. Degeneracy during iteration O_1D_1: 1, O_1D_3: 1, O_1D_5: 2, O_2D_1: 7, O_3D_2: 3, O_3D_3: 6, O_4D_4: 2, MTC = 56.
25. Degeneracy during iteration O_1D_1: 60, O_1D_2: 20, O_1D_5: 10, O_2D_5: 50, O_3D_3: 40, O_3D_4: 40, O_4D_3: 10, O_4D_5: 20, MTC = 550.
26. O_1D_1: 3, O_1D_2: 5, O_2D_1: 1, O_2D_3: 7, O_2D_5: 6, O_3D_4: 8, O_3D_5: 2, Alternative optimal solution: O_1D_2: 5, O_1D_4: 3, O_2D_1: 4, O_2D_3: 7, O_2D_5: 3, O_3D_4: 5, O_3D_5: 5, MTC = 124.
27. O_1D_2: 40, O_1D_3: 20, O_1D_6: 30, O_2D_5: 10, O_2D_6: 10, O_3D_1: 20, O_3D_4: 10, O_3D_5: 30, O_4D_4: 50, Alternative optimal solution:O_1D_2: 40, O_1D_3: 20, O_1D_6: 30, O_2D_4: 10, O_2D_6: 10, O_3D_1: 20, O_3D_5: 40, O_4D_4: 50, MTC = 680.
28. (i) O_1D_2: 4, O_1D_3: 15, O_2D_1: 21, O_3D_2: 5, O_4D_2: 7, O_4D_4: 10, O_5D_1: 7, O_5D_2: 1
    (ii) O_1D_1: 4, O_1D_3: 15, O_2D_1: 21, O_3D_2: 5, O_4D_2: 7, O_4D_4: 10, O_5D_1: 3, O_5D_2: 5
    (iii) O_1D_2: 11, O_1D_3: 8, O_2D_1: 21, O_3D_2: 5, O_4D_2: 7, O_4D_4: 10, O_5D_1: 7, O_5D_2: 1, MTC = 387
29. F_1W_1: 5, F_1W_4: 2, F_2W_3: 7, F_3W_2: 8, F_3W_4: 9, MTC = 639.
30. BH: 200, BJ: 400, CE: 300, DF: 400, DG: 200, MTC = 34,000.
31. (1, 2): 10, (2, 1): 60, (2, 2): 10, (2, 3): 10, (3, 1): 15, (4, 3): 40, Minimum total (transportation + penalty) cost = 595.
32. O_1D_2: 16, O_1D_4: 17, O_2D_1: 30, O_2D_3: 8, O_2D_4: 6, O_3D_3: 23, Max. total profit (MTP) = 6005.

33. PB: 200, QB: 120, QC: 100, QD: 280, RA: 180, RD: 120, MTP = 15,400.
34. O_1D_3: 10, O_2D_1: 40, O_2D_4: 30, O_2D_5: 30, O_3D_2: 50, (MTC) = 2010.
35. O_1D_2: 200, O_2D_3: 75, O_2D_4: 75, O_3D_5: 150, O_4D_1: 200, O_4D_2: 50, O_4D_4: 25. (MTP) = 2,06,875.
Alternative solutions are
(i) O_1D_1: 200, O_2D_3: 75, O_2D_4: 75, O_3D_5: 150, O_4D_2: 250, O_4D_4: 25
(ii) O_1D_2: 175, O_1D_5: 25, O_2D_3: 75, O_2D_4: 75, O_3D_5: 125, O_4D_1: 200, O_4D_2: 75, O_4D_4: 25
(iii) O_1D_2: 200, O_2D_3: 75, O_2D_4: 50, O_3D_5: 150, O_4D_1: 200, O_4D_2: 50, O_4D_4: 50

Exercise 8(B)

Part B

11. 1 → B : 2 → E : 3 → C : 4 → A : 5 → D; MTC: 42.
12. $O_1 \to M_3 : O_2 \to M_4 : O_3 \to M_6 : O_4 \to M_1 : O_5 \to M_5 : O_6 \to M_2$: MTC = 49
13. A → 5 : B → 2 : C → 1 : D → 3 : E → 4 : F → 6 : MTC: 185.
14. $M_1 \to J_4 : M_2 \to J_6 : M_3 \to J_5 : M_4 \to J_3 : M_5 \to J_1 : M_6 \to J_2$: MTC = 67
15. $M_1 \to J_1 : M_2 \to J_2 : M_3 \to J_5 : M_4 \to J_4 : M_5 \to J_3$ (or) $M_1 \to J_1 : M_2 \to J_4 : M_3 \to J_5 : M_4 \to J_2 : M_5 \to J_3$: MTC = 81
16. A → P : B → Q : C → T : D → R : E → S (or) A → P : B → R : C → T : D → Q : E → S ; MTC = 32
17. A → 4 : B → 1 : C → 6 : D → 3 : E → 2 F → 5 (or) A → 4 : B → 6 : C → 2 : D → 3 : E → 5 : F → 1 : MTC = 130
18. $M_1 \to J_3 : M_2 \to J_2 : M_3 \to J_5 : M_4 \to J_4 : M_5 \to J_1$ $M_6 \to J_6$: M_6 not assigned any job; MTC = 166
19. $M_1 \to J_4 : M_2 \to J_3 : M_3 \to J_6 : M_4 \to J_5 : M_5 \to J_1$ $M_6 \to J_2$ (or) $M_1 \to J_4 : M_2 \to$ J5 : $M_3 \to J_6 : M_4 \to J_3 : M_5 \to J_2$ $M_6 \to J_1$: M_3 not assigned any job; MTC = 115
20. $J_1 \to M_4 : J_2 \to M_1 : J_3 \to M_5 : J_4 \to M_3 : J_5 \to M_2$ (or) $J_1 \to M_4 : J_2 \to M_2 : J_3 \to M_5 : J_4 \to M_3 : J_5 \to M_1$: J_6 is not assigned to any machine; MTP = 20
21. 1 → C : 2 → E : 3 → A : 4 → D : 5 → B : MTP: 214.
22. 1 → D : 2 → C : 3 → E : 4 → A : 5 → B : MTP: 191.
23. $M_1 \to R_2 : M_2 \to R_3 : M_3 \to R_1 : M_4 \to R_4$: MTP = 88
24. A → 3 : B → 5 : C → 1 : D → 4 : E → 2 (or) A → 3 : B → 5 : C → 4 : D → 1 : E → 2 (OR) A → 3 : B → 5 : C → 4 : D → 2 : E → 1 : MTP = 42
25. A → 3 : B → 2 : C → 5 : D → 1 : E → 4 (or) A → 4 : B → 2 : C → 3 : D → 5 : E → 1 : MTP = 131
26. 1 → D : 2 → B : 3 → C : 4 → E Job A cannot be assigned to any machine. MTP: 376.
27. 1 → C : 2 → F : 3 → D : 4 → B : 5 → A : 6 → E : MTP: 68; jobs 2 and 6 cannot be assigned to any machine
28. 1 → D : 2 → E : 3 → A : 4 → B : 5 → C : MTC = 22
29. $J_1 \to M_1 : J_2 \to M_2 : J_4 \to M_4 : J_5 \to M_3$: J_3 is not assigned to any machine; MTP = 12
30. A → B → C → D → E → A : Minimum total set up cost = 15.

Chapter 9

SEQUENCING

9.0 INTRODUCTION

By sequencing, we mean selecting an appropriate order (sequence) in which a finite number of jobs are to be performed or processed on a finite number of machines (service facilities) so as to minimise the total involved time (cost).

A general sequencing problem may be defined as follows:

Let there be n jobs $(1, 2, \ldots, n)$ which are to be processed one at a time on each of the m machines $A, B, C, \ldots$. The order of the machines in which each job is to be processed and the time required for the jobs on each machine are given. Then the problem of sequencing is to find for each machine a sequence of jobs which minimises the total time elapsed from the start of the first job to the completion of the last job.

9.1 ASSUMPTIONS FOR THE SOLUTION OF SEQUENCING PROBLEMS

1. All machines are of different types.
2. No machine can process more than one job at a time.
3. The processing times of jobs on machines are independent of their processing order.
4. Each job, once started on a machine, must be continued till the completion on that machine.
5. The time taken by the jobs in going from one machine to another is negligible.
6. All jobs are completely known and are ready for start of the processing.

9.2 OPTIMAL SEQUENCE ALGORITHMS

We shall discuss four kinds of sequencing problems which are listed here and the iterative procedures adopted to determine the optimal sequence of jobs.

1. Processing n jobs on two machines A and B, with all the jobs processed in the order $A \to B$.
2. Processing n jobs on three machines A, B and C, with all the jobs processed in the order $A \to B \to C$.
3. Processing n jobs on m machines $A, B, C, \ldots, K$ and L, with all the jobs processed in the order $A \to B \to C \to \ldots \to K \to L$.
4. Processing two jobs on m machines, each job to be processed on the machines in a prescribed order which is not necessarily the same for the two jobs.

9.3 I OPTIMAL SEQUENCE OF *n* JOBS TO BE PROCESSED ON TWO MACHINES

Let the n jobs to be processed be numbered 1, 2, 3, …, n.

Let A_i and B_i be the processing times of the ith job on the machines A and B respectively (i = 1, 2, …, n)

The problem is to find the sequence of jobs to be processed on the two machines so that the total time T elapsed from the start of the first job to the completion of the last job is minimised.

Iterative procedure

Step 1: Find the smallest value in the lists ($A_1, A_2, \ldots, A_n$) and ($B_1, B_2, \ldots, B_n$). If there is a tie, select either of these values.

Step 2: If the smallest value is A_r, process rth job first on the machines A and B in the order $A \to B$. If it is B_s, process sth job last in the order $A \to B$.

Step 3: Delete the job already assigned and the corresponding processing time.

Step 4: Repeat Steps 1 to 3 for the remaining jobs until all the jobs are assigned in the sequence.

Step 5: Finally, the minimum time T elapsed can be computed by forming a table.

9.4 II OPTIMAL SEQUENCE OF *n* JOBS TO BE PROCESSED ON THREE JOBS

Let the n jobs to be processed be numbered 1, 2, 3, …, n. Let A_i, B_i, and C_i be the processing times of the ith job on the machines A, B and C respectively ($i = 1, 2, \ldots, n$).

The problem is to find the sequence of jobs to be processed on the three machines so that the total time T elapsed from the start of the first job to the completion of the last job is minimised.

Note: As no method is available for the solution of the general problem of this kind, the solution procedure for processing n jobs on two machines is extended as explained here, provided either or both of the following conditions are satisfied.

(i) The minimum of $A_i \geq$ maximum of B_i; $i = 1, 2, \ldots, n$

(ii) The minimum of $C_i \geq$ maximum of B_i; $i = 1, 2, \ldots, n$.

We proceed as in the case of n jobs on two machines by considering two equivalent (fictitious) machines G and H with the processing times G_i and H_i for the ith job ($i = 1, 2, \ldots, n$) given by $G_i = A_i + B_i$ and $H_i = B_i + C_i$.

We find the optimal sequence of the n jobs to be processed on the two machines G and H in the order $G \to H$ as explained earlier. The optimal sequence on G and H will also be the optimal sequence on the three machines A, B and C.

9.5 III OPTIMAL SEQUENCE OF *n* JOBS TO BE PROCESSED ON *m* MACHINES

Let the n jobs to be processed be numbered 1, 2, 3, …, n and the m machines be named $A, B, C, \ldots, K, L$

Note: The name of the last machine is assumed as L, which is taken as the mth machine and that of the last but one machine viz., k as the $(m - 1)$th machine.

Let $A_i, B_i, \ldots, K_i, L_i$ be the processing times of the ith job on the machines $A, B, \ldots, K, L$ respectively ($i = 1, 2, \ldots, n$). The problem is to find the sequence of jobs to be processed on the m machines so that the total time T elapsed from the start of the first job to the completion of the last job is minimised.

The method used in II is extended to m machines. Here we consider two equivalent (fictitious) machines X and Y with the processing times X_i and Y_i for the ith job ($i = 1, 2, \ldots, n$) given by $X_i = A_i + B_i + \ldots + K_i$ = sum of the processing times of the ith job on the first machine to the $(m - 1)$th machine and $Y_i = B_i + C_i + \ldots + L_i$ = sum of the processing times of the ith job on the second machine to the mth machine.

We find the optimal sequence of the n jobs to be processed on the two machines X and Y in the order $X \rightarrow Y$. The method holds good provided either or both of the following conditions are satisfied.

(i) The minimum of $A_i \geq$ maximum of $B_i, C_i, \ldots, K_i$ $(i = 1, 2, \ldots, n)$
(ii) The minimum of $L_i \geq$ maximum of $B_i, C_i, \ldots, K_i$ $(i = 1, 2, \ldots, n)$

The optimal sequence on X and Y will also be the optimal sequence on the m machines $A, B, \ldots, L$.

Note: Sometimes, it may happen that $(B_i + C_i + \ldots + K_i)$ = a fixed positive constant for all $i = 1, 2, \ldots, n$. In this case, it is enough we find the optimal sequence of n jobs on the two machines A and L in the order $A \rightarrow L$ as in method I.

9.6 IV OPTIMAL SEQUENCE OF TWO JOBS TO BE PROCESSED ON *m* MACHINES

The graphical method is best suited for this kind of sequencing problems.

We draw two perpendicular lines – the horizontal line (x-axis) representing the processing time for job 1 and the vertical line (y-axis) representing that for job 2. We mark the processing times on different machines on the corresponding axes in the given order. We construct shaded rectangular blocks representing different machines in the relevant positions on the graph sheet.

Then, we draw a line consisting of horizontal, vertical and 45° lines (lines making an angle of 45° with the horizontal) that starts from the origin, avoids the shaded rectangular blocks and ends with the final point.

Movement to the right along a horizontal line means that job 1 is processed while job 2 is idle, movement upwards means that job 2 is processed while job 1 is idle and movement along the 45° line means that both jobs are processed simultaneously. The optimal path is that which minimises the idle times for job 1 and job 2, viz., which minimises the horizontal and vertical segments in the path. Hence, we should try to move as much as we can along the inclined line.

The total elapsed time (T) = Processing time for job i + Idle time for job i (i = 1 or 2)

WORKED EXAMPLES 9

Example 1

Six jobs are to be processed first on machine A and then on machine B. The order of completion of jobs has no significance. The following table gives the processing times in hours for the six jobs on the two machines.

Job no. (i):	1	2	3	4	5	6
Processing time on A (A_i):	4	5	2	3	2	1
Processing time on B (B_i):	1	8	6	2	3	5

Find the sequence of jobs that minimises the total elapsed time to complete the jobs. Find also the idle times for the two machines.

Solution

The minimum processing time on the machines is 1, which corresponds to the job 6 on machine A and to the job 1 on machine B. Hence, job 6 will be processed first and job 1 will be processed last.

The sequence of jobs is represented by 6 cells in a row, each cell representing the order of a particular job to be processed as given here.

6					1

Deleting jobs 1 and 6 which have been assigned, results in the following reduced data for the remaining four jobs.

Job:	2	3	4	5
Machine A:	5	2*	3	2*
Machine B:	8	6	2*	3

Now the processing time 2 is minimum. It corresponds to the jobs 3 and 5 on machine A and the job 4 on machine B. Hence, the jobs 3 and 5 are placed in the second and third sequence cells and the job 4 is placed in the first cell (viz., the last but one cell). The only remaining job 2 is placed in the remaining fourth sequence cell. Thus, the optimal sequence of jobs is as given in the following sequence of cells.

6	3	5	2	4	1

Note: The alternative optimal sequence is

6	5	3	2	4	1

The minimum elapsed time from the start of the first job to the completion of the last job for the first optimal sequence is computed as follows:

Job	Machine A		Machine B		Idle time for B
	Time in	Time out	Time in	Time out	
6	0	1	1	6	1
3	1	3	6	12	–
5	3	5	12	15	–
2	5	10	15	23	–
4	10	13	23	25	–
1	13	17	25	26	–

∴ The minimum elapsed time is 26 hours and the idle time of machine B is 1 hour.

Example 2

Determine the optimal sequence of five jobs that optimises the total elapsed time based on the information given in the following table.

Job (i)	1	2	3	4	5
Processing time in hrs. on machine A (A_i):	3	8	7	5	2
(B_i):	3	4	2	1	5
(C_i):	5	8	10	7	6

SOLUTION

From the data, we find that the minimum of C_i (= 5) ≥ the maximum of B_i (= 5). Hence, the given problem can be converted into a 5 jobs and 2 machines (X and Y) problem, such that $X_i = A_i + B_i$ and $Y_i = B_i + C_i$; $i = 1,2 ,.., 5$.

Then the given problem can be re-written as follows.

Job (*i*):	1	2	3	4	5
X_i:	3 + 3 = 6	8 + 4 = 12	7 + 2 = 9	5 + 1 = 6	2 + 5 = 7
Y_i:	3 + 5 = 8	4 + 8 = 12	2 + 10 = 12	1 + 7 = 8	5 + 6 = 11

Proceeding as in example 1(using the optimal sequence algorithm), the optimal sequence of 5 jobs is given by

1	4	5	3	2

or

4	1	5	3	2

The computation of the minimum elapsed time for the first optimal sequence is given in the following table:

Note: The optimal sequence on X and Y is also that on the machines A, B and C

Job	Machine A		Machine B		Machine C		Idle time for M/c. B	Idle time for M/c. C
	Time in	Time out	Time in	Time out	Time in	Time out		
1	0	3	3	6	6	11	3	6
4	3	8	8	9	11	18	2	–
5	8	10	10	15	18	24	1	–
3	10	17	17	19	24	34	2	–
2	17	25	25	29	34	42	6 + 13	–

The total elapsed time = 42 hours
Idle time for machine A = 17 hours
Idle time for machine B = 27 hours
Idle time for machine C = 6 hours

Example 3

Find the optimal sequence of seven jobs that minimises the total elapsed time based on the following information. Processing time on the machines is given in hours and passing is not allowed.

Job:	J_1	J_2	J_3	J_4	J_5	J_6	J_7
M/c. A:	4	7	9	8	8	3	7

(Contd)

Table (*Contd*)

M/c. B:	5	2	1	3	4	4	3
M/c. C:	11	5	5	7	6	6	12

The condition 'min. $C_i \geq$ max. B_i' is satisfied so that the problem can be converted into a two-machine problem. Introducing two fictitious machines G and H such that $G_i = A_i + B_i$ and $H_i = B_i + C_i$; $i = 1, 2, \ldots, 7$, the problem is rewritten as a 7 jobs and 2 machines problem as given here:

Job (J_i):	J_1	J_2	J_3	J_4	J_5	J_6	J_7
G_i:	9	9	10	11	12	7	10
H_i:	16	7	6	10	10	10	15

Using the optimal sequence algorithm, we get the following optimal sequence

J_6	J_1	J_7	J_4	J_5	J_2	J_3

or

J_6	J_1	J_7	J_5	J_4	J_2	J_3

The computation of the minimum elapsed time for the first optimal sequence is given in the following table:

Job	Machine A		Machine B		Machine C		Idle time for M/c. B	Idle time for M/c. C
	Time in	Time out	Time in	Time out	Time in	Time out		
J_6	0	3	3	7	7	13	3	7
J_1	3	7	7	12	13	24	–	–
J_7	7	14	14	17	24	36	2	–
J_4	14	22	22	25	36	43	5	–
J_5	22	30	30	34	43	49	5	–
J_2	30	37	37	39	49	54	3	–
J_3	37	46	46	47	54	59	7 + 12	–

Total elapsed time = 59 hours
Idle time for machine A = 59 – 46 = 13 hours
Idle time for machine B = 37 hours
Idle time for machine C = 7 hours

Example 4

Find an optimal sequence for the following sequencing problem of 4 jobs and 5 machines when passing is not allowed for which the processing time (in hours) is given here:

Job:	1	2	3	4
M/c. A:	11	12	10	8
M/c. B:	5	7	8	4
M/c. C:	4	6	7	5
M/c. D:	7	8	6	5
M/c. E:	9	8	7	10

Since the minimum of A_i (= 8) ≥ the maximum of B_is, C_is and D_is, the problem can be converted into a 4 jobs and 2 machines sequencing problem by introducing 2 fictitious machines X and Y such that

$$X_i = A_i + B_i + C_i + D_i;$$
$$i = 1, 2, 3, 4$$

and

$$Y_i = B_i + C_i + D_i + E_i;$$
$$i = 1, 2, 3, 4$$

Job (i):	1	2	3	4
X_i:	27	33	31	22
Y_i:	25	29	28	24

Using the optimal sequence algorithm, the following optimal sequence is obtained

4	2	3	1

The computation of the minimum elapsed time is given in the following table:

Job	Machine A		Machine B		Machine C		Machine D		Machine E	
	in	out	in	out	in	out	in	out	in	out
4	0	8	8	12	12	17	17	22	22	32
2	8	20	20	27	27	33	33	41	41	49
3	20	30	30	38	38	45	45	51	51	58
1	30	41	41	46	46	50	51	58	58	67

Total elapsed time = 67 hours
Idle time for machine A = 26 hours
Idle time for machine B = 8 + 8 + 3 + 3 + 21 = 43 hours
Idle time for machine C = 12 + 10 + 5 + 1 + 17 = 45 hours
Idle time for machine D = 17 + 11 + 4 + 9 = 41 hours
Idle time for machine E = 22 + 9 + 2 = 33 hours

Example 5

Find the optimal sequence for a sequencing problem of 5 jobs and 4 machines for which the processing times (in hours) are given here:

Job	Machine			
	A	B	C	D
J_1	9	6	5	10
J_2	7	7	4	11
J_3	10	7	6	8
J_4	11	3	7	14
J_5	5	4	3	9

Since the minimum of D_i (= 8) ≥ the maximum of B_is and C_is, the problem can be converted to a 5 jobs and 2 machines sequencing problem, by introducing two fictitious machines X and Y, such that

$$X_i = A_i + B_i + C_i$$

and
$$Y_i = B_i + C_i + D_i;$$

$$i = 1, 2, 3, 4, 5.$$

Job (J_i):	J_1	J_2	J_3	J_4	J_5
X_i:	20	18	23	21	12
Y_i:	21	22	21	24	16

Using the optimal sequence algorithm, the following optimal sequence is obtained:

J_5	J_2	J_1	J_4	J_3

The computation of the minimum total elapsed time is given in the following table.

Job	Machine			
	A	B	C	D
J_5	0–5	5–9	9–12	12–21
J_2	5–12	12–19	19–23	23–34
J_1	12–21	21–27	27–32	34–44
J_4	21–32	32–35	35–42	44–58
J_3	32–42	42–49	49–55	58–66

Total elapsed time = 66 hours
Idle time for machine A = 24 hours
Idle time for machine B = 5 + 3 + 2 + 5 +7 + 17 =39 hours
Idle time for machine C = 9 + 7 + 4 + 3 + 7 + 11 = 41 hours
Idle time for machine D = 12 + 2 = 14 hours.

Example 6

Solve the following sequencing problem giving an optimal solution when passing is not allowed:

Job No. (i):	1	2	3	4	5
Processing Time on A (A_i):	18	11	16	12	14
(B_i):	9	10	8	11	7
(C_i):	7	6	8	5	9
(D_i):	14	16	12	19	20

SOLUTION

Minimum of A_i (= 11) $\geq$ maximum of B_is and C_is. Also minimum of D_i (= 12) $\geq$ maximum of B_is and C_is ($i = 1, 2, \ldots, 5$)

Hence, the problem can be converted into a 5 jobs and 2 machines problem

Moreover, we note that $B_i + C_i = 16$, for $i = 1, 2, \ldots, 5$

Hence, we can treat the problem as a 5 jobs and 2 machines (A and D) problem, with reference to the note given under the discussion of III.

Using the optimal sequence algorithm, the following optimal sequence is obtained.

2	4	5	1	3

Example 7

Determine an optimal sequence for the following sequencing problem:

Job	Machines (Time in hours)			
	A	B	C	D
1	21	5	1	21
2	17	6	4	17
3	13	11	3	13
4	16	5	6	16

Minimum of A_i (= 13) $\geq$ maximum of B_is and C_is and

Minimum of D_i (= 13) $\geq$ minimum of B_is and C_is

Hence, the problem can be converted into a 4 jobs and 2 machines problem by introducing two fictitious machines X and Y, such that

$$X_i = A_i + B_i + C_i$$

and

$$Y_i = B_i + C_i + D_i;$$

$$i = 1, 2, 3, 4.$$

Job (i):	1	2	3	4
X_i:	27	27	27	27
Y_i:	27	27	27	27

Since, $X_i = Y_i$; $i = 1, 2, 3, 4$ we will get $\lfloor 4 = 24$ (the number of permutations of A, B, C, D) optimal sequences (when we use the optimal sequence algorithm) which are listed here:

ABCD, ABDC, ACBD, ACDB, ADBC, ADCB,
BACD, BADC, BCAD, BCDA, BDAC, BDCA,

CABD, CADB, CBAD, CBDA, CDAB, CDBA,
DABC, DACB, DBAC, DBCA, DCAB, DCBA.
Each of the aforementioned optimal sequences will give the same total elapsed time 189.

Example 8

Use the graphical method to minimise the time required to process the two jobs on four machines in the sequence given in the table. Find the total elapsed time to complete both the jobs.

Job 1	Sequence of M/c's	A	B	C	D
	Time in hours	2	5	6	3
Job 2	Sequence of M/c's	C	A	D	B
	Time in hours	5	4	2	2

SOLUTION

Draw the rectangle representing machine A whose boundaries are $x = 0, x = 2, y = 5$ and $y = 9$

Draw the rectangle representing B whose boundaries are $x = 2, x = 7, y = 11$ and $y = 13$

Draw the rectangle representing C whose boundaries are $x = 7, x = 13, y = 0$ and $y = 5$

Draw the rectangle representing D whose boundaries are $x = 13, x = 16, y = 9$ and $y = 11$

The optimal path connecting the origin and the finish point is OEF, which consists of inclined segment OE and the horizontal segment EF.

During the movement along OE, both the jobs are processed and during the movement along EF, job 1 is processed, while job 2 is idle.

Total elapsed time = Processing time of job 1 + Idle time of job + Vertical distance covered
= Horizontal distance covered by the optimal path
= (16 + 0) hours = 16 hours

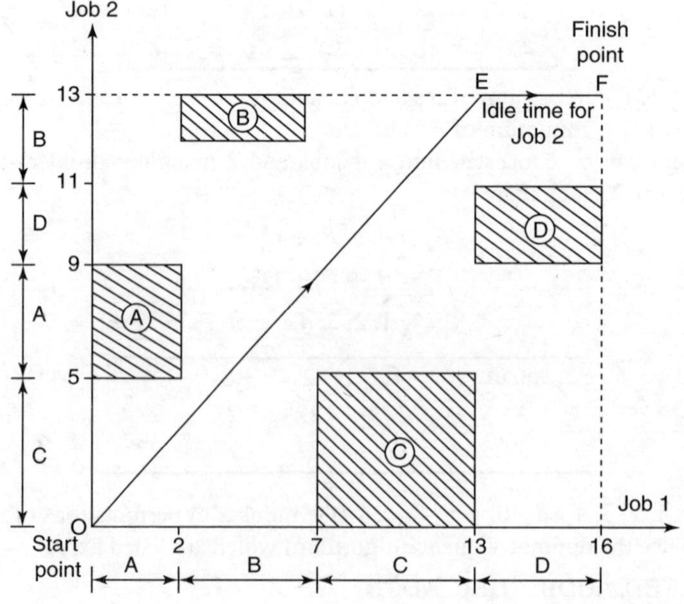

or

Total elapsed time = Processing time of job 2 + Idle time of job 2

= Vertical distance covered by the optimal path + Pure horizontal distance covered

= 13 + 3 = 16 hours

Example 9

Use the graphical method to find the minimum elapsed time sequence of 2 jobs and 5 machines using the information given here:

Job 1	Sequence of M/c's	A	B	C	D	E
	Time in hours	2	6	4	3	2
Job 2	Sequence of M/c's	D	A	B	E	C
	Time in hours	6	2	3	5	4

Solution

If we consider job 1,

Total elapsed time = Processing time + Idle time

= Horizontal distance covered + Pure vetical distances covered

= 17 + 1 + 2 = 20 hours.

If we consider job 2,

Total elapsed time = Processing time + Idle time

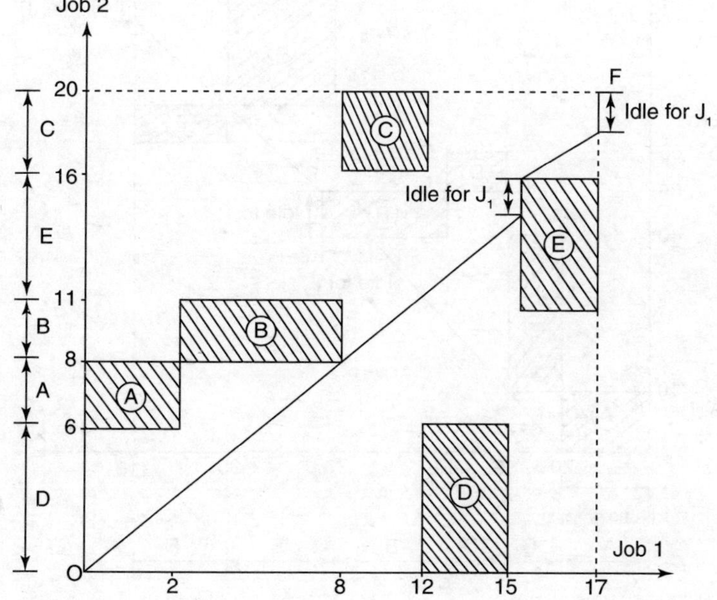

= Total vertical distance covered by the optimal path from O to F + Pure horizontal distance covered by it

= 20 + 0 = 20 hours.

Example 10

The time (in hours) spent in processing two jobs on 6 machines A, B, C, D, E, F and the necessary technological orderings of the machines are as follows

Job 1:	A-20	C-10	D-10	B-30	E-25	F-15
Job 2:	A-10	C-30	B-15	D-10	F-15	E-20

Use graphical method to determine an optimal sequence of jobs, which minimises the total elapsed time.

SOLUTION

If we consider job 1,

Total elapsed time = Processing time + Idle time

= Total horizontal distance covered by the optimal path from O to L + Pure vertical distances covered by it.

= 110 + (15 + 5) = 130 hours.

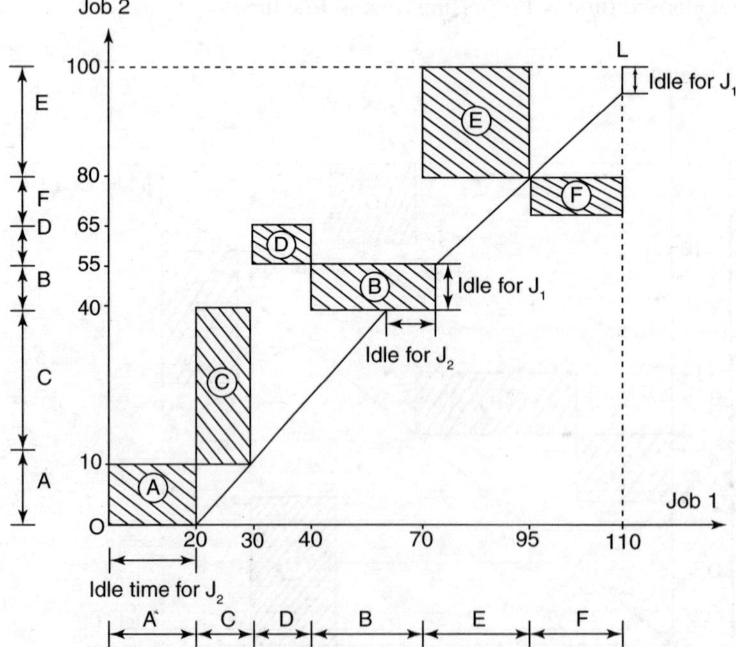

If we consider job 2,

Total elapsed time = Total vertical distance covered by the optimal path from O to L + Pure horizontal distance covered by it

= 100 + (20 + 10) = 130 hours.

EXERCISE 9

Part A (Short answer questions)

1. Define a general sequencing problem.
2. State the assumptions used while solving sequencing problems.
3. Give the iterative procedure for determining an optimal sequence for processing n jobs on two machines.
4. Give the iterative procedure for determining an optimal sequence for processing n jobs on three machines.
5. What are the conditions to be satisfied by the processing times on the three machines, if the procedure in Question 4 is to be used?
6. Give the iterative procedure for determining an optimal sequence for processing n jobs on m machines.
7. What are the conditions to be satisfied by the processing times on the m machines if the procedure in Question 6 is to be used?
8. When can the sequencing problem with n jobs and m machines be reduced to a sequencing problem with n jobs and the first and the last machines only?
9. State the conditions under which all the optimal sequences in respect to a sequencing problem with n jobs and m machines will yield the same total elapsed time.
10. How will you identify the optimal path connecting the start and finish points when solving a sequencing problem with two jobs and m machines graphically?
11. We have 4 jobs, each of which has to go through 2 machines M_1 and M_2. Processing times (in hours) are given as:

Job:	1	2	3	4
Machine 1:	4	3	2	1
Machine 2:	1	2	3	5

Find the optimal sequence of jobs that minimises the total elapsed time for the 4 jobs

12. Use the data for a sequencing problem with 3 jobs and 2 machines given below to find the total elapsed time (in hours) for the jobs and the idle times for the two machines:

	J_1	J_2	J_3
M_1	8	6	5
M_2	8	3	4

Part B

Find the sequence that minimises the total elapsed time (in hours) required to complete the following jobs on two machines A and B in the order AB. Find also the idle time for each machine:

13.

Job:	1	2	3	4	5
M/c. A:	5	1	9	3	10
M/c. B:	2	6	7	8	4

14.

Job:	J_1	J_2	J_3	J_4	J_5	J_6
A:	5	9	4	7	8	6
B:	7	4	8	3	9	5

15.

Job:	1	2	3	4	5	6
A:	30	120	50	20	90	110
B:	80	100	90	60	30	10

16.

Job:	J_1	J_2	J_3	J_4	J_5	J_6	J_7
A:	3	12	15	6	10	11	9
B:	8	10	10	6	12	1	3

Determine the optimal sequence of jobs that minimises the total elapsed time based on the following information. Processing time on machines is given in hours and passing is not allowed. Find also the idle time for each machine:

17.

Job:	1	2	3	4	5	6
Machine A:	8	3	7	2	5	1
Machine B:	3	4	5	2	1	6
Machine C:	8	7	6	9	10	9

18.

Job	1	2	3	4	5	6	7
Machine A:	3	8	7	4	9	8	7
Machine B:	4	3	2	5	1	4	3
Machine C:	6	7	5	11	5	6	12

19.

Job:	1	2	3	4	5	6	7	8
Machine A:	5	6	2	3	4	9	15	11
Machine B:	4	6	7	4	5	3	6	2
Machine C:	8	10	7	8	11	8	9	13

Determine the optimal sequence of jobs that minimises the total elapsed time (in hours) when passing is not allowed.

20.

Job	M_1	M_2	M_3	M_4
A	24	7	7	29
B	16	9	5	15
C	22	8	6	14
D	21	6	8	32

21.

Job	M/c. A	M/c. B	M/c. C	M/c. D	M/c. E
1	6	4	1	2	8
2	5	5	3	4	9
3	4	3	4	5	7
4	7	2	2	1	5

22.

Job	Machines					
	A	B	C	D	E	F
1	18	8	7	2	10	25
2	17	6	9	6	8	19
3	11	5	8	5	7	15
4	20	4	3	4	8	12

Two jobs are to be processed on the given machines in the technological sequence indicated. Use graphical method to minimise the time required to process them and hence, find the total elapsed time:

23.

Job 1	Sequence	A	B	C	D
	Time (hrs)	2	5	6	3

(Contd)

Table (*Contd*)

Job 2	Sequence	C	A	D	B
	Time (hrs)	5	4	2	2

24.

Job 1	Sequence	A	B	C	D	E
	Time (hrs)	2	3	4	6	2
Job 2	Sequence	C	A	D	E	B
	Time (hrs)	4	5	3	2	6

25.

Job 1	Sequence	A	B	C	D	E
	Time (hrs)	3	4	2	6	2
Job 2	Sequence	B	C	A	D	E
	Time (hrs)	5	4	3	2	6

ANSWERS

Exercise 9

Part A

11. $4 \to 3 \to 2 \to 1$.

12. $J_1 \to J_3 \to J_2$; Total elapsed time = 23 hours; Idle time for M_1 = 4 hours; Idle time for M_2 = 8 hours.

Part B

13. $2 \to 4 \to 3 \to 5 \to 1$; TET = 30 hours; Idl. T(A) = 2 hours; Idl. T(B) = 3 hours.

14. $3 \to 1 \to 5 \to 6 \to 2 \to 4$; TET = 42 hours; Idl. T(A) = 3 hours; Idl. T(B) = 6 hours.

15. $4 \to 1 \to 3 \to 2 \to 5 \to 6$; TET = 430 hours; Idl. T(A) = 10 hours; Idl. T(B) = 60 hours.

16. $1 \to 4 \to 5 \to 3 \to 2 \to 7 \to 6$; TET = 67 hours; Idl. T(A) = 1 hour; Idl. T(B) = 17 hours.

17. $4 \to 5 \to 2 \to 6 \to 1 \to 3$; TET = 53 hours; Idl. T(A) = 27 hours; Idl. T(B) = 32 hours; Idl. T(C) = 4 hours.

18. $1 \to 4 \to 7 \to 6 \to 2 \to 3 \to 5$; TET = 59 hours; Idl. T(A) = 13 hours; Idl. T(B) = 37 hours; Idl. T(C) = 7 hours.

19. $4 \to 1 \to 3 \to 5 \to 2 \to 8 \to 7 \to 6$; TET = 82 hours; Idl. T(A) = 27 hours; Idl. T(B) = 44 hours; Idl. T(C) = 8 hours.

20. $D \to A \to B \to C$; TET = 125 hours.

21. $1 \to 3 \to 2 \to 4$; TET = 43 hours.

22. $3 \to 1 \to 2 \to 4$; TET = 112 hours.

23. TET = $PT_1 + IT_1 = 16 + 0 = 16$ hours or $PT_2 + IT_2 = 13 + 3 = 16$ hours.

24. TET = $PT_1 + IT_1 = 17 + 3 = 20$ hours or $PT_2 + IT_2 = 20 + 0 = 20$ hours.

25. TET = $PT_1 + IT_1 = 17 + 5 = 22$ hours or $PT_2 + IT_2 = 20 + 2 = 22$ hours.

Chapter 10

GAME THEORY

10.0 INTRODUCTION

Game theory is a mathematical theory that deals with taking decisions in a competitive situation (such as political campaigns during elections, marketing advertisements by competing business firms, etc.) where there are two or more opposing parties with conflicting interests and the action of one depends upon the action taken by the opponents.

A competitive situation with a finite number of competitors (hereafter called *players*) is called a *competitive game*. In competitive games, such as chess and card games, each player adopts a number of choices, finite and infinite, called *strategies*. The different strategies adopted by the players determine the respective outcomes or pay-offs (or payments) which may be positive (gain), negative (loss) or zero (draw) due to them.

Note: Games of gambling are not games of strategy, as a player playing such games is simply betting against the odds.

We shall only deal with the theory of games of strategy in this chapter.

10.1 TWO-PERSON ZERO-SUM GAME

A game in which the algebraic sum of the gains and losses of all the players is zero is called a *zero-sum game*. A game with two players in which the gain of one player equals the loss of the other is known as a two-person zero-sum game or *rectangular game*.

If A and B are the two players for whom the numbers of strategies available are m and n respectively, the game is described by the following ($m \times n$) matrix, called the *pay-off matrix* of the game for the player A, whose aim is to maximise his gain:

$$\begin{array}{cc} & \text{Player B} \\ & \begin{array}{ccccccc} B_1 & B_2 & \dots & B_j & \dots & B_n \end{array} \\ \text{Player A} \begin{array}{c} A_1 \\ A_2 \\ \vdots \\ A_i \\ \vdots \\ A_m \end{array} & \begin{bmatrix} a_{11} & a_{12} & \dots & a_{1j} & \dots & a_{1n} \\ a_{21} & a_{22} & \dots & a_{2j} & \dots & a_{2n} \\ \dots & \dots & \dots & \dots & \dots & \dots \\ a_{i1} & a_{i1} & \dots & a_{ij} & \dots & a_{in} \\ \dots & \dots & \dots & \dots & \dots & \dots \\ a_{m1} & a_{m2} & \dots & a_{mj} & \dots & a_{mn} \end{bmatrix} \end{array}$$

The row headings $A_1, A_2, \dots, A_m$ are the strategies available to the player A, while the column headings $B_1, B_2, \dots, B_n$ are those available to the player B. The cell entries a_{ij} ($i = 1, 2, \dots, m; j = 1, 2, \dots, n$) represent the pay-offs to the player A when he adopts the ith strategy and B adopts the jth strategy. Clearly, the pay-off matrix for the player B whose aim is to minimise his loss is the matrix $\{-a_{ij}\}$, so that the sum of the pay-off matrices for the players A and B is a null matrix. Usually in any game theory problem, the pay-off matrix of the player whose strategies are displayed on the left side of the matrix will be given.

10.2 SOLUTION OF A GAME (PROBLEM) AND THE VALUE OF THE GAME

By solving a game, we mean finding the optimal strategies that maximise the gain of one player (A) and minimise the loss of the other (B).

The value of the game is defined as the maximum gain guaranteed to the player A or the minimum possible loss to the player B when both the players use their best (optimal) strategies.

10.3 PURE AND MIXED STRATEGIES

As pointed out earlier, the strategy of a player is a predetermined procedure selected from his or her list of courses of action during the game. There are two types of strategies, namely, pure strategy and mixed strategy.

Pure strategy is a strategy chosen from the strategies available to the player before the start of the plays (rounds) of the game.

Mixed strategy is a strategy chosen by using a known probability distribution of the strategies available to the player before the start of each play (round) of the game.

10.4 MAXIMIN–MINIMAX PRINCIPLE OF OPTIMALITY

This principle states that an optimal solution of a two-person, zero-sum game can be obtained, if the player A selects the strategy (pure or mixed) which maximises his minimum gain, the minimum being taken over all the strategies of B and the player B selects his strategy which minimises his maximum loss.

For example, let us consider the game for which the pay-off matrix is given here:

$$\begin{array}{c} \\ A_1 \\ A_2 \\ A_3 \end{array} \begin{array}{c} \begin{array}{cccc} B_1 & B_2 & B_3 & B_4 \end{array} \\ \begin{bmatrix} \textcircled{-5} & 3 & 1 & \boxed{20} \\ \boxed{5} & \boxed{5} & \boxed{\textcircled{4}} & 6 \\ -4 & -2 & 0 & \textcircled{-5} \end{bmatrix} \end{array}$$

The minimum elements of the A_1, A_2, A_3 rows are -5, 4 and -5 that are enclosed by small circles as shown.

As A likes to maximise his minimum gain, he has to use the strategy A_2 and the value of the game for A is 4. The maximum elements of the B_1, B_2, B_3, B_4 columns are 5, 5, 4 and 20 that are enclosed by small squares as shown. As B is interested in minimising his loss, he has to use the strategy B_3 and the value of the game for B is 4. Hence, the value of the game is 4.

10.5 SADDLE POINT

The position of the element in the pay-off matrix at which the maximum value coincides with the minimax value is called a saddle point of the pay-off matrix. In other words, the point that is enclosed by both the circle ○ and the square □ is a saddle point. In this example, the (2, 3)th position is the saddle point. The value of the element (pay-off) at the saddle point is the value of the game.

The saddle point of a pay-off matrix and hence, the value of the game need not be unique.

Solution of (2 × 2) games [two-person, zero-sum game] without saddle point:
In this case, the optimal strategies are mixed strategies. Let the pay-off matrix of the game be

$$\text{Player A}\quad \begin{array}{c} \\ A_1 \\ A_2 \end{array}\!\!\begin{array}{c} \text{Player B} \\ \begin{array}{cc} B_1 & B_2 \end{array} \\ \begin{bmatrix} a_{11} & a_{12} \\ a_{21} & a_{22} \end{bmatrix} \end{array}$$

Let the probability distributions of the strategies of A and B be

A_1	p_1
A_2	p_2

and

B_1	B_2
q_1	q_2

,

where $$p_1 + p_2 = 1 \text{ and } q_1 + q_2 = 1$$

When B uses his strategy B_1, the expected gain of A is given by

$$E(G_A/B \to B_1) = a_{11}p_1 + a_{21}p_2 \tag{1}$$

Similarly, $$E(G_A/B \to B_2) = a_{12}p_1 + a_{22}p_2 \tag{2}$$

When A uses his strategy A_1, the expected loss of B is given by

$$E(L_B/A \to A_1) = a_{11}q_1 + a_{12}q_2 \tag{3}$$

Similarly, $$E(L_B/A \to A_2) = a_{21}q_1 + a_{22}q_2 \tag{4}$$

It can be proved that a two-person, zero-sum game always possesses a value when mixed strategies are used. Assuming this result, let the value of the game be taken as V.

Since A is interested in maximising his gain, from (1) and (2), we get

$$a_{11}p_1 + a_{21}p_2 \geq V \tag{5}$$

and $$a_{12}p_1 + a_{22}p_2 \geq V \tag{6}$$

Since B is interested in minimising his loss, from (3) and (4), we get

$$a_{11}q_1 + a_{12}q_2 \leq V \tag{7}$$

and $$a_{21}q_1 + a_{22}q_2 \leq V \tag{8}$$

As A does not know the strategy used by B beforehand and he has to be satisfied with the minimum of his maximum gain,

$$a_{11}p_1 + a_{21}p_2 = V \tag{9}$$

and $$a_{12}p_1 + a_{21}p_2 = V \tag{10}$$

Similarly, $$a_{11}q_1 + a_{12}q_2 = V \tag{11}$$

and $$a_{21}q_1 + a_{22}q_2 = V \tag{12}$$

Solving (9), (10), (11) and (12), we get

$$p_1 = \frac{a_{22} - a_{21}}{Dr},\ p_2 = \frac{a_{11} - a_{12}}{Dr},$$

$$q_1 = \frac{a_{22} - a_{12}}{Dr},\ q_2 = \frac{a_{11} - a_{21}}{Dr}$$

and $$V = \frac{a_{11}a_{22} - a_{12}a_{21}}{Dr},$$

where $$Dr = (a_{11} + a_{22}) - (a_{12} + a_{21})$$

Note: The formulas derived for p_1, p_2, q_1, q_2 and V are valid only for (2×2) games without saddle points.

Dominance property, in the case of non-existence of a saddle point.

Definitions: If all the elements in a row, say the rth row, of a pay-off matrix are less than or equal to the corresponding elements of any other row, say the sth row, then the rth row is said to be dominated by the sth row (or the sth row dominates the rth row) and the player A will never choose the rth strategy, considered as an inferior strategy.

Similarly, if all the elements in a column, say the ith column of a pay-off matrix are greater than or equal to the corresponding elements of any other column, say the jth column, then the ith column is said to be dominated by the jth column (or the jth column dominates the ith column) and the player B will never choose the ith strategy, considered as an inferior strategy.

By using the dominance rules (given as definitions previously), the size of the pay-off matrix is reduced to a (2×2) matrix by deleting the inferior strategies and then the solution of the game is obtained by the procedures discussed earlier.

Note 1: The value of the game and the non-zero choice of probabilities remain unchanged even after the deletion of inferior strategies.

Note 2: In case the pay-off matrix reduces to a $(m \times 2)$ or $(2 \times n)$ matrix, the graphical method (discussed later) has to be used to find the solution of the game.

Modified dominance property

The dominance property need not be based on the superiority of pure strategies only. A given strategy can also be said to be dominated if it is inferior to a convex linear combination of two or more pure strategies or to an average of two or more pure strategies. In such situations, one of the strategies involved in the linear combination / the average is deleted to reduce the size of the pay-off matrix.

WORKED EXAMPLES 10(A)

Example 1

Find the range of values of λ and μ that will make the (2, 2) position of the following pay-off matrix a saddle point for the game:

$$\begin{array}{cc} & \textit{Player B} \\ & \begin{array}{ccc} B_1 & B_2 & B_3 \end{array} \\ \textit{Player A} \begin{array}{c} A_1 \\ A_2 \\ A_3 \end{array} & \begin{bmatrix} 2 & 4 & 5 \\ 10 & 7 & \lambda \\ 4 & \mu & 6 \end{bmatrix} \end{array}$$

Ignoring the values of λ and μ, we note the row minima (enclose by a small circle) and the column maxima (enclose by a small rectangle) and obtain the following pay-off matrix:

$$\begin{array}{c} \\ A_1 \\ A_2 \\ A_3 \end{array} \begin{array}{c} \begin{array}{ccc} B_1 & B_2 & B_3 \end{array} \\ \begin{bmatrix} (2) & 4 & 5 \\ \boxed{(10)} & \boxed{(7)} & \lambda \\ (4) & \mu & \boxed{(6)} \end{bmatrix} \end{array}$$

The element 7 that occurs in the (2, 2) position is enclosed by both ○ and □.

If this position corresponds to a saddle point of the game, we get $7 \leq \lambda$ and $7 \geq \mu$ or $\mu \leq 7 \leq \lambda$.

Example 2

Find the pure strategies for the players A and B, the saddle point and the value of the game represented by the following pay-off matrix:

$$\begin{array}{c} \\ A_1 \\ A_2 \\ A_3 \\ A_4 \end{array} \begin{array}{c} \begin{array}{ccccc} B_1 & B_2 & B_3 & B_4 & B_5 \end{array} \\ \begin{bmatrix} 3 & 5 & 4 & 9 & 6 \\ 5 & 6 & 3 & 7 & 8 \\ 8 & 7 & 9 & 8 & 7 \\ 4 & 2 & 8 & 5 & 3 \end{bmatrix} \end{array}$$

SOLUTION

After enclosing the row minima by small circles and the column maxima by small squares, the pay-off matrix becomes as shown in the following matrix:

$$\begin{array}{c} \\ A_1 \\ A_2 \\ A_3 \\ A_4 \end{array} \begin{array}{c} \begin{array}{ccccc} B_1 & B_2 & B_3 & B_4 & B_5 \end{array} \\ \begin{bmatrix} (3) & 5 & 4 & \boxed{9} & 6 \\ 5 & 6 & (3) & 7 & \boxed{8} \\ \boxed{8} & \boxed{(7)} & \boxed{9} & 8 & (7) \\ 4 & (2) & 8 & 5 & 3 \end{bmatrix} \end{array}$$

As the entry in the (3, 2) position is enclosed by both ○ and □, the (3, 2) position is a saddle point of the game.

The optimum pure strategies of the players A and B are A_3 and B_2 respectively. The value of the game is 7 (for A) and −7 (for B).

Example 3

Find the optimal strategies of the players for the game with the following pay-off matrix and the value of the game:

$$\begin{array}{c} \\ A1 \\ A2 \end{array} \begin{array}{c} \begin{array}{cc} B1 & B2 \end{array} \\ \begin{bmatrix} 3 & 1 \\ 2 & 4 \end{bmatrix} \end{array}$$

SOLUTION

$$\begin{array}{c} \\ A_1 \\ A_2 \end{array} \begin{array}{c} \begin{array}{cc} B_1 & B_2 \end{array} \\ \begin{bmatrix} \boxed{3} & (1) \\ (2) & \boxed{4} \end{bmatrix} \end{array}$$

As the maximum value 2 is not equal to the minimax value 3, there is no saddle point.

Let the probability distributions of the strategies of A and B be

A_1	p_1
A_2	p_2

and

B_1	B_2
q_1	q_2

,

where $p_1 + p_2 = 1$ and $q_1 + q_2 = 1$

If A selects the strategy A_1 and B selects the strategies B_1, the expected pay-off $= 3p_1q_1$ and so on.

The expected pay-off function P of the game is the sum of the expected pay-offs corresponding to various combinations of the strategies.

$$\therefore \quad E(p) = 3p_1q_1 + 2p_2q_1 + p_1q_2 + 4p_2q_2$$

$$= 3p_1q_1 + 2(1-p_1)q_1 + p_1(1-q_1) + 4(1-p_1)(1-q_1)$$

$$= 4p_1q_1 - 3p_1 - 2q_1 + 4$$

$$= 4\left(p_1q_1 - \frac{3}{4}p_1 - \frac{2}{4}q_1\right) + 4$$

$$= 4\left(p_1 - \frac{1}{2}\right)\left(q_1 - \frac{3}{4}\right) + \frac{5}{2}$$

The minimum of the maximum gain of A is got when $p_1 = \frac{1}{2}$ and the maximum of the minimum loss of B is got when $q_1 = \frac{3}{4}$.

$\therefore$ The optimal strategies of A and B are given by

A_1	$\frac{1}{2}$
A_2	$\frac{1}{2}$

and

B_1	B_2
$\frac{3}{4}$	$\frac{1}{4}$

and the value of the game $= \frac{5}{2}$.

Example 4

In a game of matching coins with two players A and B, A wins one unit of value when there are 2 heads, wins nothing when there are 2 tails and loses $\frac{1}{2}$ unit of value when there are one head and one tail. Write down the pay-off matrix, the best strategies for each player and the value of the game to A.

SOLUTION

The pay-off matrix of the game is clearly

$$\begin{array}{cc} & \textit{Player B} \\ & \begin{array}{cc} H & T \end{array} \\ \textit{Player A} \begin{array}{c} H \\ T \end{array} & \begin{bmatrix} 1 & -\frac{1}{2} \\ -\frac{1}{2} & 0 \end{bmatrix} = \begin{bmatrix} a_{11} & a_{12} \\ a_{21} & a_{22} \end{bmatrix} \end{array}$$

The pay-off matrix does not have a saddle point. Hence, the players have to use mixed strategies, which are given as follows

$$\text{Player } A \begin{array}{|c|c|}\hline H & p_1 \\ \hline T & p_2 \\ \hline \end{array} \text{ and } \overset{\text{Player } B}{\begin{array}{|c|c|}\hline H & T \\ \hline q_1 & q_2 \\ \hline \end{array}},$$

where $$p_1 + p_2 = 1 \text{ and } q_1 + q_2 = 1$$

$$p_1 = \frac{a_{22} - a_{21}}{Dr},$$

where $$Dr = (a_{11} + a_{22}) - (a_{12} + a_{21})$$

$$= (1+0) - \left(-\frac{1}{2} - \frac{1}{2}\right) = 2$$

$$\therefore p_1 = \frac{0 - \left(-\frac{1}{2}\right)}{2} = \frac{1}{4}$$

$$p_2 = \frac{a_{11} - a_{12}}{Dr} = \frac{1 - \left(-\frac{1}{2}\right)}{2} = \frac{3}{4}$$

$$q_1 = \frac{a_{22} - a_{12}}{Dr} = \frac{0 - \left(-\frac{1}{2}\right)}{2} = \frac{1}{4}$$

and $$q_2 = 1 - q_1 = \frac{3}{4}$$

Value of the game is given by

$$V = \frac{a_{11}a_{22} - a_{12}a_{21}}{Dr} = \frac{(1)(0) - \left(-\frac{1}{2}\right)\left(-\frac{1}{2}\right)}{2}$$

i.e., the value of the game to $A = -\frac{1}{8}$.

Example 5

A and B each take out one or two tokens and guess how many tokens the opponent has taken. If one of the players guesses correctly, then the opponent has to pay as many rupees as the sum of the number of tokens taken by both the players. Otherwise, the pay-out is zero. Write down the pay-off matrix and the optimal strategies of both players by two different methods.

Solution

The pay-off matrix of the game is clearly as shown:

$$\begin{array}{c} \text{Player } B \\ \begin{array}{cc} & \begin{array}{cc} 1 & 2 \end{array} \\ \text{Player } A & \begin{array}{c} 1 \\ 2 \end{array}\begin{bmatrix} 2 & 0 \\ 0 & 4 \end{bmatrix} = \begin{bmatrix} a_{11} & a_{12} \\ a_{21} & a_{22} \end{bmatrix} \end{array} \end{array}$$

The pay-off matrix does not have a saddle point. Hence, the players have to use the mixed strategies as given here.

Player A

1	p_1
2	p_2

and Player B

1	2
q_1	q_2

,

where $p_1 + p_2 = 1$

and $q_1 + q_2 = 1.$

Method 1: The expected pay-off function P of the game that is given by the sum of the expected pay-offs of the four cells is given by

$$\begin{aligned} E(p) &= 2p_1q_1 + 4p_2q_2 \\ &= 2p_1q_1 + 4(1-p_1)(1-q_1) \\ &= 6p_1q_1 - 4p_1 - 4q_1 + 4 \\ &= 6\left(p_1q_1 - \frac{2}{3}p_1 - \frac{2}{3}q_1\right) + 4 \\ &= 6\left(p_1 - \frac{2}{3}\right)\left(q_1 - \frac{2}{3}\right) + \left(4 - \frac{24}{9}\right) \\ &= 6\left(p_1 - \frac{2}{3}\right)\left(q_1 - \frac{2}{3}\right) + \frac{4}{3} \end{aligned}$$

$\therefore$ The optimal strategies of A are given by $p_1 = \frac{2}{3}$ and $p_2 = \frac{1}{3}$ and the optimal strategies of B are given by $q_1 = \frac{2}{3}$ and $q_2 = \frac{1}{3}$.

The value of the game to $A = \frac{4}{3}$.

Method 2:

$$p_1 = \frac{a_{22} - a_{21}}{Dr} = \frac{4-0}{Dr},$$

where

$$\begin{aligned} Dr &= (a_{11} + a_{22}) - (a_{12} + a_{21}) \\ &= (2+4) - (0+0) = 6 \end{aligned}$$

$$\text{i.e., } p_1 = \frac{2}{3}$$

and hence $$p_2 = \frac{1}{3}$$

$$q_1 = \frac{a_{22} - a_{12}}{Dr}$$

$$= \frac{4-0}{6} = \frac{2}{3}$$

and hence $$q_2 = \frac{1}{3}.$$

Value of the game to $$A = \frac{a_{11}a_{22} - a_{12}a_{21}}{Dr}$$

$$= \frac{(2\times4)-(0\times0)}{6} = \frac{4}{3}$$

Example 6

Use dominance property to reduce the following game to a (2 × 2) game and hence, find the optimum strategies and the value of the game

$$\begin{array}{c} \\ A_1 \\ A_2 \\ A3 \end{array}\begin{array}{c} \begin{array}{cccc} B_1 & B_2 & B_3 & B_4 \end{array} \\ \begin{bmatrix} -5 & 3 & 1 & 20 \\ 5 & 5 & 4 & 6 \\ -4 & -2 & 0 & -5 \end{bmatrix} \end{array}$$

SOLUTION

The elements of the A_2 row are $\geq$ the corresponding elements of the A_3 row. Hence, A_3 is an inferior strategy for A and is hence deleted from the pay-off matrix, which gets reduced as given here:

$$\begin{array}{c} \\ A_1 \\ A_2 \end{array}\begin{array}{c} \begin{array}{cccc} B_1 & B_2 & B_3 & B_4 \end{array} \\ \begin{bmatrix} -5 & 3 & 1 & 20 \\ 5 & 5 & 4 & 6 \end{bmatrix} \end{array}$$

The elements of B_1 column are $\leq$ the corresponding elements of the B_2 and B_4 columns. Hence, B_2 and B_4 strategies are inferior strategies for B and they are deleted from the reduced pay-off matrix, which gets reduced further as given here:

$$\begin{array}{c} \\ A_1 \\ A_2 \end{array}\begin{array}{c} \begin{array}{cc} B_1 & B_3 \end{array} \\ \begin{bmatrix} \textcircled{-5} & 1 \\ \boxed{5} & \boxed{\textcircled{4}} \end{bmatrix} \end{array}$$

There is a saddle point at the (2, 3) position.

Hence, the A_2 and B_3 are the best strategies for A and B respectively and the value of the game is 4 for A.

Example 7

Solve the following game by dominance rules:

$$\begin{array}{c} \\ A_1 \\ A_2 \\ A_3 \end{array}\begin{array}{c} \begin{array}{ccc} B_1 & B_2 & B_3 \end{array} \\ \begin{bmatrix} 1 & 7 & 2 \\ 6 & 2 & 7 \\ 5 & 1 & 6 \end{bmatrix} \end{array}$$

SOLUTION

The element of A_2 row are $\geq$ the corresponding elements of the A_3 row. Hence, the inferior strategy A_3 is deleted from the pay-off matrix, which becomes

$$\begin{array}{c} \\ A_1 \\ A_2 \end{array}\begin{array}{c} \begin{array}{ccc} B_1 & B_2 & B_3 \end{array} \\ \begin{bmatrix} 1 & 7 & 2 \\ 6 & 2 & 7 \end{bmatrix} \end{array}$$

Now the elements of B_1column are $\leq$ the corresponding elements of the B_3 column. Hence, the inferior strategy B_3 is deleted from the reduced pay-off matrix, which becomes

$$\begin{array}{c} \\ A_1 \\ A_2 \end{array}\begin{array}{c} \begin{array}{cc} B_1 & B_2 \end{array} \\ \begin{bmatrix} 1 & 7 \\ 6 & 2 \end{bmatrix} \end{array} = \begin{bmatrix} a_{11} & a_{12} \\ a_{21} & a_{22} \end{bmatrix}$$

As there is no saddle point for this reduced matrix, the players have to use mixed strategies as given here:

A_1	p_1
A_2	p_2

and

B_1	B_2
q_1	q_2

,

where $$p_1 + p_2 = 1$$

and $$q_1 + q_2 = 1.$$

$$p_1 = \frac{a_{22} - a_{21}}{Dr} = \frac{2-6}{Dr},$$

where $$Dr = (a_{11} + a_{22}) - (a_{12} + a_{21}) = -10$$

$\therefore$ $$p_1 = -\frac{4}{(-10)}$$

or $$\frac{2}{5}$$

and hence $$p_2 = \frac{3}{5}$$

$$q_1 = \frac{a_{11} - a_{12}}{Dr} = \frac{2-7}{-10} = \frac{1}{2}$$

and hence $$q_2 = \frac{1}{2}$$

The value of the game is given by

$$V = \frac{a_{11}a_{22} - a_{12}a_{21}}{Dr} = \frac{2-42}{-10} = 4$$

$\therefore$ The optimal strategies for A are $\left(\frac{2}{5}, \frac{3}{5}, 0\right)$ and those for B are $\left(\frac{1}{2}, \frac{1}{2}, 0\right)$.

Example 8

A and B play a game in which each has a 1 Re coin, 2 Rs coin and 5 Rs coin. Each selects a coin without the knowledge of the other's choice. If the sum of the values of the coins is an odd amount, A wins an amount equal to B's coin value. If the sum is an even amount, B wins an amount equal to A's coin value. Find the best strategies for each player and the value of the game.

Solution

Let us find the elements of the I row of the pay-off matrix of A, which corresponds to the choice of 1Re by A. If B also selects the 1Re coin, the sum of the values = 2 (even)

$\therefore$ $$a_{11} = -1,$$

since B wins 1Re

Similarly, a_{12} (corresponding to A choosing 1Re and B choosing 2Rs) = 2 and a_{13} (corresponding to A choosing 1 Re and B choosing 5 Rs) = −1

Similarly, $$a_{21} = 1,$$

$$a_{22} = -2$$

and $$a_{23} = 5$$

$$a_{31} = -5,$$

$$a_{32} = 2$$

and $$a_{33} = -5$$

$\therefore$ Pay-off matrix of player A is given by

		Player B		
		1	2	5
Player A	1	−1	2	−1
	2	1	−2	5
	5	−5	2	−5

On noting the minimax and maxmin values of the pay-off matrix as given in the following matrix, we note that the game has no saddle-point and hence, A and B have to use mixed strategies.

	1	2	5
1	(−1)	[2]	(−1)
2	[1]	(−2)	[5]
5	(−5)	[2]	(−5)

Comparing the I and III rows, we note that the III row gives an inferior strategy, which is deleted.

In the revised matrix, the III column gives an inferior strategy in comparison with the I column and hence, is deleted.

Therefore, the reduced (2 × 2) matrix, without a saddle point is found as

$$\begin{array}{cc} & \textit{Player B} \\ \textit{Player A} & \begin{bmatrix} -1 & 2 \\ 1 & -2 \end{bmatrix} = \begin{bmatrix} a_{11} & a_{12} \\ a_{21} & a_{22} \end{bmatrix} \end{array}$$

Let the strategies of A and B be taken as

$$S_A = \begin{array}{|c|c|} \hline A_1 & p_1 \\ A_2 & p_2 \\ \hline \end{array} \text{ and } S_B = \begin{array}{|cc|} \hline B_1 & B_2 \\ \hline q_1 & q_2 \\ \hline \end{array},$$

where $p_1 + p_2 = 1$

and $q_1 + q_2 = 1.$

$$p_1 = \frac{a_{22} - a_{21}}{Dr},$$

where $Dr = (a_{11} + a_{22}) - (a_{12} + a_{21})$

$$= (-1 - 2) - (2 + 1) = -6$$

$$\text{i.e., } p_1 = -\frac{-2-1}{-6} = \frac{1}{2}$$

and hence, $p_2 = \dfrac{1}{2}$

$$q_1 = \frac{a_{22} - a_{12}}{Dr}$$

$$= \frac{-2-2}{-6} = \frac{2}{3}$$

and hence, $q_2 = \dfrac{1}{3}$

Value of the game given by $V = \dfrac{a_{11}a_{22} - a_{12}a_{21}}{Dr} = 0$

The optimal strategies of the two players are

$$S_A = \begin{array}{|c|c|} \hline 1 & \frac{1}{2} \\ 2 & \frac{1}{2} \\ 5 & 0 \\ \hline \end{array} \text{ and } S_B = \begin{array}{|ccc|} \hline 1 & 2 & 5 \\ \hline \frac{2}{3} & \frac{1}{3} & 0 \\ \hline \end{array}$$

Example 9

Use the notion of dominance to simplify the rectangular game with the following pay-off matrix as a (2 × 2) game and then solve it.

$$\begin{array}{c} \\ A_1 \\ A_2 \\ A3 \end{array}\begin{array}{c} \begin{array}{cccc} B_1 & B_2 & B_3 & B_4 \end{array} \\ \begin{bmatrix} 19 & 15 & 17 & 16 \\ 0 & 20 & 15 & 5 \\ 10 & 12 & 15 & 8 \end{bmatrix} \end{array}$$

SOLUTION

The elements of A_1 row are ≥ the corresponding elements of A_3 row. Hence, after deleting the inferior strategy row A_3, the pay-off matrix becomes

$$\begin{array}{c} \\ A_1 \\ A_2 \end{array}\begin{array}{c} \begin{array}{cccc} B_1 & B_2 & B_3 & B_4 \end{array} \\ \begin{bmatrix} 19 & 15 & 17 & 16 \\ 0 & 20 & 15 & 5 \end{bmatrix} \end{array}$$

Now the elements of the B_4 column are ≤ the corresponding elements of the B_3 column. Hence, we delete the inferior strategy column B_3 and get the pay-off matrix as

$$\begin{array}{c} \\ A_1 \\ A_2 \end{array}\begin{array}{c} \begin{array}{ccc} B_1 & B_2 & B_4 \end{array} \\ \begin{bmatrix} 19 & 15 & 16 \\ 0 & 20 & 5 \end{bmatrix} \end{array}$$

Now the elements of the B_4 column are ≤ the averages of the corresponding elements of the B_1 and B_2 columns (viz., $16 \le \frac{19+15}{2}$ and $5 \le \frac{0+20}{2}$). Hence, either B_1 column or B_2 column is deleted. The resulting pay-off matrix, after deleting B_1 column, becomes

$$\begin{array}{c} \\ A_1 \\ A_2 \end{array}\begin{array}{c} \begin{array}{cc} B_2 & B_4 \end{array} \\ \begin{bmatrix} 15 & 16 \\ 20 & 5 \end{bmatrix} \end{array} = \begin{bmatrix} a_{11} & a_{12} \\ a_{21} & a_{22} \end{bmatrix}$$

Let the strategies of A and B be taken as

$$S_A = \begin{array}{|c|c|} \hline A_1 & p_1 \\ \hline A_2 & p_2 \\ \hline \end{array} \text{ and } S_B = \begin{array}{|cc|} \hline B_2 & B_4 \\ \hline q_1 & q_2 \\ \hline \end{array},$$

where $p_1 + p_2 = 1$

and $q_1 + q_2 = 1$

$$p_1 = \frac{a_{22} - a_{21}}{Dr},$$

where $Dr = (a_{11} + a_{22}) - (a_{12} + a_{21})$

$$= 20 - 36 = -16$$

$$i.e., p_1 = \frac{5 - 20}{-16} = \frac{15}{16}$$

and hence, $$p_2 = \frac{1}{16}$$

$$q_1 = \frac{a_{22} - a_{12}}{Dr}$$

$$= \frac{5-16}{-16} = \frac{11}{16}$$

and hence, $$q_2 = \frac{5}{16}$$

The value of the game is given by $$V = \frac{a_{11}a_{22} - a_{12}a_{21}}{Dr}$$

$$= \frac{75-320}{-16} = \frac{245}{16}$$

The optimal strategies of the two players are given by

$$S_A = \left(\frac{15}{16}, \frac{1}{16}, 0\right) \text{ and } S_B = \left(0, \frac{11}{16}, 0, \frac{5}{16}\right)$$

Example 10

Solve the following game, by using dominance rule:

$$\begin{array}{c} \\ A_1 \\ A_2 \\ A_3 \\ A4 \end{array} \begin{array}{c} \begin{matrix} B_1 & B_2 & B_3 & B_4 \end{matrix} \\ \begin{bmatrix} 3 & 2 & 4 & 0 \\ 3 & 4 & 2 & 4 \\ 4 & 2 & 4 & 0 \\ 0 & 4 & 0 & 8 \end{bmatrix} \end{array}$$

SOLUTION

Comparing A_1 and A_3, we note that A_1 is an inferior strategy. After deleting the A_1 row, we compare B_1 and B_3, and note that B_1 is an inferior strategy. After deleting the B_1 column, the resulting pay-off is obtained as:

$$\begin{array}{c} \\ A_2 \\ A_3 \\ A_4 \end{array} \begin{array}{c} \begin{matrix} B_2 & B_3 & B_4 \end{matrix} \\ \begin{bmatrix} 4 & 2 & 4 \\ 2 & 4 & 0 \\ 4 & 0 & 8 \end{bmatrix} \end{array}$$

Now elements of $A_2 \geq \frac{1}{2}$ (corresponding elements of A_3) + $\frac{1}{2}$ (corresponding elements of A_4).

Hence, after deleting the A_4 row, the pay-off matrix becomes

$$\begin{array}{c} \\ A_2 \\ A_3 \end{array} \begin{array}{c} \begin{matrix} B_2 & B_3 & B_4 \end{matrix} \\ \begin{bmatrix} 4 & 2 & 4 \\ 2 & 4 & 0 \end{bmatrix} \end{array}$$

Elements of the B_4 column $\leq$ the corresponding elements of the B_2 column. Hence, the B_2 column is deleted and the resulting pay-off matrix is

$$\begin{array}{c} \\ A_2 \\ A_3 \end{array}\begin{array}{c} \begin{matrix} B_3 & B_4 \end{matrix} \\ \begin{bmatrix} 2 & 4 \\ 4 & 0 \end{bmatrix} \end{array} = \begin{bmatrix} a_{11} & a_{12} \\ a_{21} & a_{22} \end{bmatrix}$$

Let the strategies of A and B be assumed as

A_2	p_1
A_3	p_2

and

B_3	B_4
q_1	q_2

,

where $$p_1 + p_2 = 1$$

and $$q_1 + q_2 = 1$$

$$p_1 = \frac{a_{22} - a_{21}}{Dr},$$

where $$Dr = (a_{11} + a_{22}) - (a_{12} + a_{21})$$

$$= 2 - 8 = -6$$

$$\therefore p_1 = -\frac{0-4}{-6} = \frac{2}{3}$$

and $$p_2 = \frac{1}{3}$$

$$q_1 = \frac{a_{22} - a_{12}}{Dr}$$

$$= \frac{0-4}{-6} = \frac{2}{3}$$

and $$q = \frac{1}{3}$$

Value of the game $$V = \frac{a_{11}a_{22} - a_{12}a_{21}}{Dr}$$

$$= \frac{0-16}{-6} = \frac{8}{3}$$

The optimal strategies of the two players are given by $S_A = \left(0, \frac{2}{3}, \frac{1}{3}, 0\right)$ and $S_B = \left(0, 0, \frac{2}{3}, \frac{1}{3}\right)$

EXERCISE 10(A)

Part A (Short answer questions)

1. Define game and strategy in the context of game theory.
2. Define a rectangular game.
3. Distinguish between pure and mixed strategies.
4. Define pay-off matrix and saddle point.

5. What do you mean by solution of a game and value of a game?
6. State the dominance rules in game theory.
7. Find the optimum strategies of the players A and B and the value of the game, if the pay-off matrix is

$$\begin{array}{c} \\ A_1 \\ A_2 \end{array}\begin{array}{c} \begin{array}{cc} B_1 & B_2 \end{array} \\ \begin{bmatrix} a & -b \\ -c & d \end{bmatrix} \end{array}$$

8. Find the optimal strategies of the players A and B and the value of the game given by

$$\begin{array}{c} \\ A_1 \\ A_2 \\ A_3 \end{array}\begin{array}{c} \begin{array}{ccc} B_1 & B_2 & B_3 \end{array} \\ \begin{bmatrix} -3 & -2 & 6 \\ 2 & 0 & 2 \\ 5 & -2 & -4 \end{bmatrix} \end{array}$$

9. Find the optimal strategies of the players A and B and the value of the game given by

$$\begin{array}{c} \\ A_1 \\ A_2 \\ A_3 \end{array}\begin{array}{c} \begin{array}{ccc} B_1 & B_2 & B_3 \end{array} \\ \begin{bmatrix} 1 & 3 & 1 \\ 0 & -4 & -3 \\ 1 & 5 & -1 \end{bmatrix} \end{array}$$

10. Use dominance property to solve the game given by

$$\begin{array}{c} \\ A_1 \\ A_2 \end{array}\begin{array}{c} \begin{array}{ccc} B_1 & B_2 & B_3 \end{array} \\ \begin{bmatrix} 6 & 8 & 6 \\ 4 & 12 & 2 \end{bmatrix} \end{array}$$

Part B

11. Solve the game with the following pay-off matrix:

$$\begin{array}{c} \\ A_1 \\ A_2 \\ A_3 \\ A_4 \\ A_5 \end{array}\begin{array}{c} \begin{array}{ccccc} B_1 & B_2 & B_3 & B_4 & B_5 \end{array} \\ \begin{bmatrix} 4 & 0 & 1 & 7 & -1 \\ 0 & -3 & -5 & -7 & 5 \\ 3 & 2 & 3 & 4 & 3 \\ -6 & 1 & -1 & 0 & 5 \\ 0 & 0 & 6 & 0 & 0 \end{bmatrix} \end{array}$$

12. For what value of λ is the game with the following pay-off matrix strictly determinable?

$$\begin{array}{c} \\ A_1 \\ A_2 \\ A_3 \end{array}\begin{array}{c} \begin{array}{ccc} B_1 & B_2 & B_3 \end{array} \\ \begin{bmatrix} \lambda & 6 & 2 \\ -1 & \lambda & -7 \\ -2 & 4 & \lambda \end{bmatrix} \end{array}$$

Hint: The game is determinable, if the maximin and minimax values are equal.

13. Solve the game with the following pay-off matrix in two ways:

$$\begin{array}{c} \\ A_1 \\ A_2 \end{array}\begin{array}{c} \begin{array}{cc} B_1 & B_2 \end{array} \\ \begin{bmatrix} 2 & 5 \\ 7 & 3 \end{bmatrix} \end{array}$$

14. Solve the game with the following pay-off matrix in two ways:

$$\begin{array}{c} \\ A_1 \\ A_2 \end{array}\begin{array}{c} \begin{array}{cc} B_1 & B_2 \end{array} \\ \begin{bmatrix} -4 & 6 \\ 2 & -3 \end{bmatrix} \end{array}$$

15. Two players A and B toss a fair coin each. If the outcomes are the same, A wins 1Re; otherwise B wins 1Re. Obtain the optimum strategies for the players and the value of the game.
16. Solve the following game by using dominance rule:

$$\begin{array}{c} \\ A_1 \\ A_2 \\ A_3 \end{array}\begin{array}{c} \begin{array}{ccc} B_1 & B_2 & B_3 \end{array} \\ \begin{bmatrix} 1 & 7 & 2 \\ 6 & 2 & 7 \\ 0 & 1 & 6 \end{bmatrix} \end{array}$$

17. Solve the following game after reducing it to a (2×2) game:

$$\begin{array}{c} \\ A_1 \\ A_2 \\ A_3 \end{array}\begin{array}{c} \begin{array}{ccc} B_1 & B_2 & B_3 \end{array} \\ \begin{bmatrix} 1 & 7 & 2 \\ 0 & 2 & 7 \\ 5 & 1 & 6 \end{bmatrix} \end{array}$$

18. Use dominance to reduce the following game to a (2×2) game and hence, solve it:

$$\begin{array}{c} \\ A_1 \\ A_2 \\ A_3 \end{array}\begin{array}{c} \begin{array}{ccc} B_1 & B_2 & B_3 \end{array} \\ \begin{bmatrix} 3 & -2 & 4 \\ -1 & 4 & 2 \\ 2 & 2 & 6 \end{bmatrix} \end{array}$$

19. Use dominance to solve the following game:

$$\begin{array}{c} \\ A_1 \\ A_2 \\ A_3 \\ A_4 \end{array}\begin{array}{c} \begin{array}{cccc} B_1 & B_2 & B_3 & B_4 \end{array} \\ \begin{bmatrix} 6 & 8 & 3 & 13 \\ 4 & 1 & 5 & 3 \\ 8 & 10 & 4 & 12 \\ 3 & 2 & 7 & 2 \end{bmatrix} \end{array}$$

20. Use dominance to solve the following game:

$$\begin{array}{c} \\ A_1 \\ A_2 \\ A_3 \\ A_4 \\ A_5 \end{array}\begin{array}{c} \begin{array}{cccccc} B_1 & B_2 & B_3 & B_4 & B_5 & B_6 \end{array} \\ \begin{bmatrix} 4 & 2 & 0 & 2 & 1 & 1 \\ 4 & 3 & 1 & 3 & 2 & 2 \\ 4 & 3 & 7 & -5 & 1 & 2 \\ 4 & 3 & 4 & -1 & 2 & 2 \\ 4 & 3 & 3 & -2 & 2 & 2 \end{bmatrix} \end{array}$$

10.5.1 Graphical Solution of (2 × *n*) and (*m* × 2) Games

The graphical method of solution helps us to reduce a $(2 \times n)$ or $(m \times 2)$ game to a Simpler (2×2) game. This (2×2) game is then solved in the usual manner.

Let us consider the $(2 \times n)$ game whose pay-off matrix is given by

$$\begin{array}{c} \\ A_1 \\ A_2 \end{array}\begin{array}{c} \begin{array}{cccc} B_1 & B_2 & \cdots & B_n \end{array} \\ \begin{bmatrix} a_{11} & a_{12} & \cdots & a_{1n} \\ a_{21} & a_{22} & \cdots & a_{2n} \end{bmatrix} \end{array}$$

We shall assume that the game does not have a saddle point.

Let two vertical lines representing the strategies A_1 and A_2 be marked with (a_{ij}) values on some convenient scale. Then, we join the points a_{1j} (lying on the A_1 line) and A_{2j} (lying on the A_2 line) representing the strategy B_j $(j = 1, 2, \ldots, n)$.

The lower envelope of these lines represent the region of minimum expected pay-offs of the player A. The highest point on this lower envelope represents the maximin expected pay-off of the player A. If this maximin point happens to be the point of intersection of the strategy line B_r and B_s, then the reduced (2×2) game is given by

$$\begin{array}{c} \\ A_1 \\ A_2 \end{array}\begin{array}{c} \begin{array}{cc} B_r & B_s \end{array} \\ \begin{bmatrix} a_{1r} & a_{1s} \\ a_{2r} & a_{2s} \end{bmatrix} \end{array}$$

In a similar manner, we proceed for a $(m \times 2)$ game. The lowest point on the upper envelope represents the minimax of the expected pay-off of the player B.

10.5.2 General Solution of (*m* × *n*) Rectangular Games

In a rectangular game with an $(m \times n)$ pay-off matrix, if there does not exist any saddle point and it is not possible to reduce the size of the game to a (2×2) game by using the concept of dominance, the following methods are generally used to solve the game.

(a) Algebraic method
(b) Linear programming method
(c) Iterative method for approximate solution

Algebraic method

Let $[a_{ij}]_{m\times n}$ be the pay-off matrix of a rectangular game between two players A and B.

Let $S_A = \begin{bmatrix} A_1 & A_2 & \dots & A_m \\ p_1 & p_2 & \dots & p_m \end{bmatrix}$, $p_1 + p_2 + \dots + p_m = 1$ and $S_B = \begin{bmatrix} B_1 & B_2 & \dots & B_n \\ q_1 & q_2 & \dots & q_n \end{bmatrix}$, $q_1 + q_2 + \dots + q_n = 1$ be the optimal mixed strategies of A and B respectively.

When B uses his pure strategies $B_1, B_2, \dots, B_n$, the expected gains of A are $\sum_{i=1}^{m} a_{i1} p_i, \sum_{i=1}^{m} a_{i2} p_i, \cdots \sum_{i=1}^{m} a_{in} p_i$.

If the value of the game is V, then the minimum expected gains of A satisfy

$$\sum_{i=1}^{m} a_{i1} p_i \geq V, \sum_{i=1}^{m} a_{i2} p_i \geq V, \cdots, \sum_{i=1}^{m} a_{in} p_i \geq V, \tag{1}$$

Similarly, if we consider B's expected losses and note that his maximum loss is V, we get

$$\sum_{j=1}^{n} a_{1j} q_j \leq V, \sum_{j=1}^{n} a_{2j} q_j \leq V, \cdots, \sum_{j=1}^{n} a_{mj} q_j \leq V, \tag{2}$$

Also $$p_1 + p_2 + \dots + p_m = 1; q_1 + q_2 + \dots + q_n = 1; p_i \geq 0; q_j \geq 0 \tag{3}$$

To get the values of p_i ($i = 1, 2, \dots, m$) and q_j ($j = 1, 2, \dots, n$), we treat the inequalities in (1) and (2) as equalities and solve them subject to (3).

Note: If the system of equations (1), (2) (considered as equalities) and (3) and (4) are inconsistent (viz., does not possess a solution), it means that at least one of the inequalities is a strict inequality. In such cases, we try to solve the system by taking one or more values of p_is and q_is and zeros until we get a solution. Of course, we have to rely on a trial and error approach.

Linear programming method

Using the same assumption as in the algebraic method, the expected gains g_j ($j = 1, 2, \dots, n$) of player A against B's moves are given by

$$g_j = a_{1j} p_1 + a_{2j} p_2 + \dots + a_{mj} p_m; (j = 1, 2, 3, \dots, n)$$

The expected losses l_i ($i = 1, 2, \dots, m$) of player B against A's moves are given by

$$l_i = a_{i1} q_1 + a_{i2} q_2 + \dots + a_{in} q_n; (i = 1, 2, \dots, m)$$

If we put $u = \min \{g_1, g_2, \dots, g_n\}$ and $v = \max \{l_1, l_2, \dots, l_m\}$, then the problem for player A is to maximise u subject to the constraints $g_j = a_{1j} p_1 + a_{2j} p_2 + \dots + a_{mj} p_j \geq u$

$$\sum_{i=1}^{m} p_i = 1 \text{ and } p_i \geq 0 \text{ for all } i$$

and the problem for player B is to minimise v subject to the constraints $l_i = a_{i1} q_1 + a_{i2} q_2 + \dots + a_{in} q_n \leq v$

$$\sum_{j=1}^{n} q_j = 1 \text{ and } q_j \geq 0 \text{ for all } j.$$

We can make every element a_{ij} strictly positive by adding some constant to all the original values of a_{ij}.

After the optimum solution is obtained, the true value of the game is obtained by subtracting that constant.

Thus, we can assume that $u > 0$ and $v > 0$ and introduce new variables $p_i' = \frac{p_i}{u}$, $i = 1, 2, 3, \ldots, m$ and $q_j' = \frac{q_j}{v}$, $j = 1, 2, \ldots, n$.

Then, the problem for player A becomes:

$$\text{Maximise } u \text{ or minimise } \frac{1}{u} = \sum_{i=1}^{m} \frac{p_i}{u} \text{ or } \sum_{i=1}^{m} p_i'$$

viz., minimise $$p_0 = p_1' + p_2' + \ldots + p_m',$$

subject to constraints

$$a_{1j}p_1' + a_{2j}p_2' + \ldots + a_{mj}p_j' \geq 1; \; j = 1, 2, \ldots, n$$

$$p_i' \geq 0, \; i = 1, 2, \ldots, m$$

The problem for player B becomes:

$$\text{Minimise } v \text{ or maximise } \frac{1}{v} = \sum_{j=1}^{n} \frac{q_j}{v} \text{ or } \sum_{j=1}^{n} q_j'$$

viz., maximise $$q_0 = q_1' + q_2' + \ldots + q_n',$$

subject to constraints

$$a_{i1}q_1' + a_{i2}q_2' + \ldots + a_{in}q_n' \leq 1; \; i = 1, 2, \ldots, m$$

$$q_j' \geq 0, \; j = 1, 2, \ldots, n$$

Using the Simplex method, the optimum solution is obtained and then, the original optimum values can be obtained. We note that *B*'s problem is actually the dual of *A*'s problem. Thus, if one problem is solved either by Simplex or dual Simplex method, the solution to the other can be easily obtained.

Iterative method for approximate solution

Let A and B be the maximising and minimising players of a game. Let A start the game by selecting his strategy A_i arbitrarily and write it below the pay-off matrix. Player B then will examine the strategy A_i selected by the player A and will choose the strategy (column) B_j which corresponds to the smallest entry in A_i. The column corresponding to B_j is then placed to the right of the matrix.

Player A will then examine this column and will choose the row corresponding to the largest entry in this column. This row will be added to the row last chosen and the sum of the two rows will be placed below the row last chosen. Player B will then choose the column corresponding to the smallest entry in the new row and add this column to the column last chosen.

(Note: If there is a tie, the player will select the row or column which is different from his last choice.)

The procedure is repeated a finite number of times (say, 10 times) and the iterations are presented with the smallest entries in the rows and with the largest entries in the columns being encircled. The approximate strategies after 10 iterations are found by dividing the number of encircled elements in each row (for A) and in each column (for B) by the total number (10) of iterations.

The upper bound for the value of the game is taken as the largest element in the last column by the total number of iterations. The lower bound for V is taken as the smallest element in the last row by the total number of iterations.

WORKED EXAMPLES 10(B)

Example 1

Reduce the following game to a (2 × 2) game graphically and hence solve it

$$\begin{array}{c} \textit{Player B} \\ \textit{Player A} \begin{array}{c} \\ A_1 \\ A_2 \end{array} \begin{array}{c} \begin{array}{ccc} B_1 & B_2 & B_3 \end{array} \\ \begin{bmatrix} 0 & -2 & 2 \\ 5 & 4 & -3 \end{bmatrix} \end{array} \end{array}$$

SOLUTION

The lower envelope is the region that is striped.

The highest point on this lower envelope, representing the maximum expected gain of player A, is the point of intersection of the strategy lines B_2 and B_3.

Hence, the reduced (2 × 2) game is given by

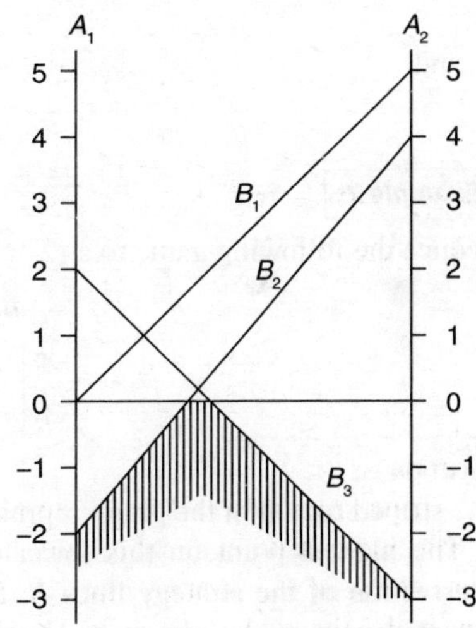

$$\begin{array}{c} \begin{array}{cc} B_2 & B_3 \end{array} \\ \begin{array}{c} A_1 \\ A_2 \end{array} \begin{bmatrix} -2 & 2 \\ 4 & -3 \end{bmatrix} = \begin{bmatrix} a_{11} & a_{12} \\ a_{21} & a_{22} \end{bmatrix} \end{array}$$

Note: We could have got this reduced (2 × 2) game by using the dominance rule.

Let the strategies of A and B be taken as

A_1	p_1
A_2	p_2

and

B_2	B_3
q_1	q_2

,

where $$p_1 + p_2 = 1$$

and $$q_1 + q_2 = 1.$$

$$p_1 = \frac{a_{22} - a_{21}}{Dr},$$

where $$Dr = (a_{11} + a_{22}) - (a_{12} + a_{21})$$

$$= -5 - 6 = -11$$

$$i.e., p_1 = -\frac{-3-(4)}{-11} = \frac{7}{11}$$

and $$p_2 = \frac{4}{11}$$

$$q_1 = \frac{a_{22} - a_{12}}{Dr}$$

$$= \frac{-3-2}{-11} = \frac{5}{11}$$

and $$q_2 = \frac{6}{11}$$

Value of the game given by $$V = \frac{a_{11}a_{22} - a_{12}a_{21}}{Dr}$$

$$= \frac{6-8}{-11} = \frac{2}{11}$$

∴ The optimal strategies of the two players are given by

$$S_A = \left(\frac{7}{11}, \frac{4}{11}\right)$$

and $$S_B = \left(0, \frac{5}{11}, \frac{6}{11}\right)$$

Example 2

Reduce the following game to a (2 × 2) game graphically and hence solve it.

$$\begin{array}{c} \\ A_1 \\ A_2 \end{array}\begin{array}{c} \begin{matrix} B_1 & B_2 & B_3 & B_4 & B_5 \end{matrix} \\ \begin{bmatrix} 2 & -1 & 5 & -2 & 6 \\ -2 & 4 & -3 & 1 & 0 \end{bmatrix} \end{array}$$

Solution

The striped region in the graph represents the lower envelope.

The highest point on this lower envelope, which is the point of intersection of the strategy lines B_1 and B_4, represents the maximum expected gain of the player A. Hence, the reduced (2 × 2) game is given by

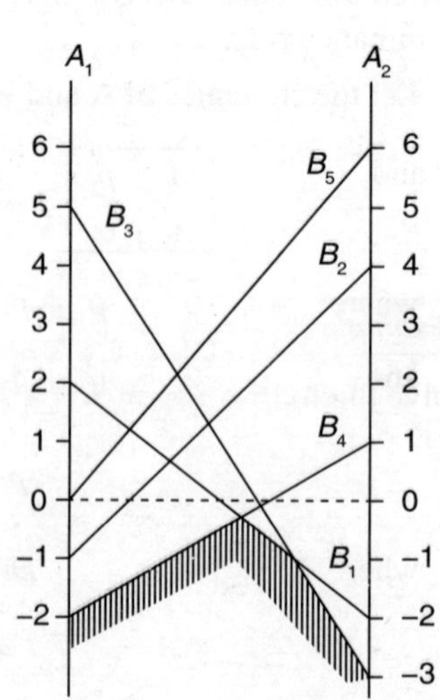

$$\begin{array}{c} \\ A_1 \\ A_2 \end{array}\begin{array}{c} \begin{matrix} B_1 & B_4 \end{matrix} \\ \begin{bmatrix} 2 & -2 \\ -2 & 1 \end{bmatrix} \end{array} = \begin{bmatrix} a_{11} & a_{12} \\ a_{21} & a_{22} \end{bmatrix}$$

Let the strategies of A and B be

A_1	p_1
A_2	p_2

and

B_1	B_4
q_1	q_2

,

where $$p_1 + p_2 = 1$$

and $$q_1 + q_2 = 1.$$

$$p_1 = \frac{a_{22} - a_{21}}{Dr},$$

where $$Dr = (a_{11} + a_{22}) - (a_{12} + a_{21})$$

$$= 3 - (-4) = 7$$

$$\therefore p_1 = \frac{1-(-2)}{7} = \frac{3}{7}$$

and $$p_2 = \frac{4}{7}$$

$$q_1 = \frac{a_{22} - a_{12}}{Dr}$$

$$= \frac{1-(-2)}{7} = \frac{3}{7}$$

and $$q_2 = \frac{4}{7}$$

Value of the game $$V = \frac{a_{11}a_{22} - a_{12}a_{21}}{Dr}$$

$$= \frac{2-4}{7} = -\frac{2}{7}$$

$\therefore$ The optimal strategies of the two players are given by

$$S_A = \left(\frac{3}{7}, \frac{4}{7}\right)$$

and $$S_B = \left(\frac{3}{7}, 0, 0, \frac{4}{7}, 0\right)$$

Example 3

Solve the following game graphically.

$$\begin{array}{c} \\ A_1 \\ A_2 \\ A_3 \\ A_4 \end{array}\begin{array}{c} \begin{array}{cc} B_1 & B_2 \end{array} \\ \begin{bmatrix} 2 & 4 \\ 2 & 3 \\ 3 & 2 \\ -2 & 6 \end{bmatrix} \end{array}$$

SOLUTION

The striped region is the upper envelope bounded by the strategy lines of A.

The lowest point on this upper envelope represents the minimax expected loss of player B. It is the point of intersection of the strategy lines A_1 and A_3. Hence, the reduced game is given by

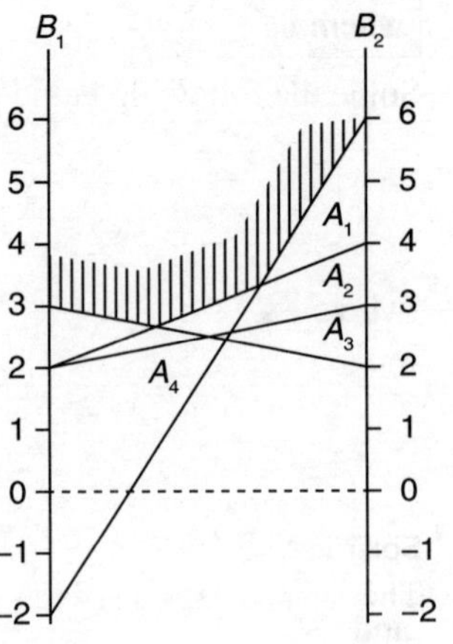

$$\begin{array}{c} \quad B_1 \; B_2 \\ \begin{array}{c} A_1 \\ A_3 \end{array}\begin{bmatrix} 2 & 4 \\ 3 & 2 \end{bmatrix} = \begin{bmatrix} a_{11} & a_{12} \\ a_{21} & a_{22} \end{bmatrix} \end{array}$$

Let the strategies of A and B be taken as

A_1	p_1
A_3	p_2

and

B_1	B_2
q_1	q_2

,

where $\quad p_1 + p_2 = 1$

and $\quad q_1 + q_2 = 1.$

$$p_1 = \frac{a_{22} - a_{21}}{Dr},$$

where
$$Dr = (a_{11} + a_{22}) - (a_{12} + a_{21})$$
$$= 4 - 7 = -3$$

$$i.e., p_1 = \frac{2-3}{-3} = \frac{1}{3}$$

and
$$p_2 = \frac{2}{3}$$

$$q_1 = \frac{a_{22} - a_{12}}{Dr}$$
$$= \frac{2-4}{-3} = \frac{2}{3}$$

and
$$q_2 = \frac{1}{3}$$

$$V = \frac{a_{11}a_{22} - a_{12}a_{21}}{Dr}$$
$$= \frac{4-12}{-3} = \frac{8}{3}$$

$$S_A = \left(\frac{1}{3}, 0, \frac{2}{3}, 0\right)$$

and
$$S_B = \left(\frac{2}{3}, \frac{1}{3}\right)$$

Example 4

Solve the following game graphically

$$\begin{array}{c} \\ A_1 \\ A_2 \\ A_3 \\ A_4 \\ A_5 \end{array}\begin{array}{c} \begin{array}{cc} B_1 & B_2 \end{array} \\ \begin{bmatrix} -6 & 7 \\ 4 & -5 \\ -1 & -2 \\ -2 & 5 \\ 7 & -6 \end{bmatrix} \end{array}$$

SOLUTION

The striped region is the upper envelope bounded by the strategy lines of A.

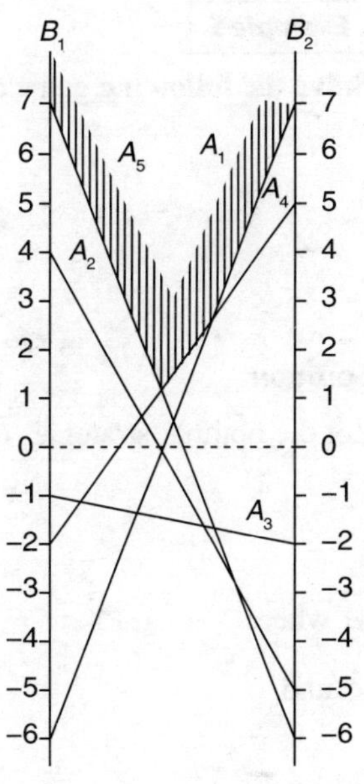

The lowest point on this upper envelope represents the minimax loss of player B. It is the point of intersection of the strategy lines A_4 and A_5. Hence, the reduced (2 × 2) game is given by

$$\begin{array}{c} \\ A_4 \\ A_5 \end{array}\begin{array}{c} \begin{array}{cc} B_1 & B_2 \end{array} \\ \begin{bmatrix} -2 & 5 \\ 7 & -6 \end{bmatrix} \end{array} = \begin{bmatrix} a_{11} & a_{12} \\ a_{21} & a_{22} \end{bmatrix}$$

Let the strategies of A and B be taken as

A_4	p_1
A_5	p_2

and

B_1	B_2
q_1	q_2

,

where $p_1 + p_2 = 1$

and $q_1 + q_2 = 1.$

$$p_1 = \frac{a_{22} - a_{21}}{Dr},$$

where

$$Dr = (a_{11} + a_{22}) - (a_{12} + a_{21})$$
$$= -8 - 12 = -20$$

$$\therefore p_1 = \frac{-6-7}{-20} = \frac{13}{20}$$

and

$$p_2 = \frac{7}{20}$$

$$q_1 = \frac{a_{22} - a_{12}}{Dr}$$

$$= \frac{-6-5}{-20} = \frac{11}{20}$$

and

$$q_2 = \frac{9}{20}$$

$$V = \frac{a_{11}a_{22} - a_{12}a_{21}}{Dr}$$

$$= \frac{12-35}{-20} = \frac{23}{20}$$

$$S_A = \left(0, 0, 0, \frac{13}{20}, \frac{7}{20}\right)$$

and $$S_B = \left(\frac{11}{20}, \frac{9}{20}\right)$$

Example 5

Solve the following game algebraically:

$$\begin{array}{c} \\ A_1 \\ A_2 \\ A_3 \end{array} \begin{array}{c} \begin{array}{ccc} B_1 & B_2 & B_3 \end{array} \\ \begin{bmatrix} 9 & 1 & 4 \\ 0 & 6 & 3 \\ 5 & 2 & 8 \end{bmatrix} \end{array}$$

SOLUTION

Let the optimal strategies of players A and B be assumed as

$$S_A = \begin{bmatrix} A_1 & A_2 & A_3 \\ p_1 & p_2 & p_3 \end{bmatrix} \text{ and } S_B = \begin{bmatrix} B_1 & B_2 & B_3 \\ q_1 & q_2 & q_3 \end{bmatrix}$$

where $$p_1 + p_2 + p_3 = 1$$

and $$q_1 + q_2 + q_3 = 1;$$

$$p_i \geq 0, q_i \geq 0 \quad (1)$$

The values of p_is and q_is are given by the equations

$$9p_1 + 0p_2 + 5p_3 = V \quad (2);$$

$$p_1 + 6p_2 + 2p_3 = V \quad (3);$$

$$4p_1 + 3p_2 + 8p_3 = V \quad (4);$$

$$9q_1 + q_2 + 4q_3 = V \quad (5)$$

$$0q_1 + 6q_2 + 3q_3 = V \quad (6)$$

$$5q_1 + 2q_2 + 8q3 = V \quad (7)$$

where V is the value of the game for player A. Solving (2), (3) and (4), we get

$$p_1 = \frac{9}{91}V, p_2 = \frac{13}{91}V, p_3 = \frac{2}{91}V$$

Using these values in (1),

$$\left(\frac{9}{91} + \frac{13}{91} + \frac{2}{91}\right)V = 1$$

viz.,
$$\frac{24}{91}V = 1$$

$\therefore$
$$V = \frac{91}{24}$$

$\therefore$
$$p_1 = \frac{3}{8}, p_2 = \frac{13}{24}, p_3 = \frac{1}{12}$$

Solving (5), (6) and (7), we get

$$q_1 = \frac{21}{273}\text{V}$$
$$= \frac{21}{273} \times \frac{91}{24} = \frac{7}{24};$$
$$q_3 = \frac{11}{273}\text{V}$$
$$= \frac{11}{273} \times \frac{91}{24} = \frac{11}{72};$$
$$q_2 = \frac{40}{273}\text{V}$$
$$= \frac{40}{273} \times \frac{91}{24} = \frac{5}{9};$$

$\therefore$
$$S_A = \left(\frac{3}{8}, \frac{13}{24}, \frac{1}{12}\right)$$

and
$$S_B = \left(\frac{7}{24}, \frac{5}{9}, \frac{11}{72}\right)$$

Example 6

Solve the following game algebraically:

$$\begin{array}{c} \\ A_1 \\ A_2 \\ A_3 \end{array} \begin{array}{c} \begin{array}{ccc} B_1 & B_2 & B_3 \end{array} \\ \begin{bmatrix} 6 & 0 & 3 \\ 8 & -2 & 3 \\ 4 & 6 & 5 \end{bmatrix} \end{array}$$

Solution

Let the optimal strategies of the players A and B be assumed as $S_A = \begin{bmatrix} A_1 & A_2 & A_3 \\ p_1 & p_2 & p_3 \end{bmatrix}$ and $S_B = \begin{bmatrix} B_1 & B_2 & B_3 \\ q_1 & q_2 & q_3 \end{bmatrix}$,

where $\quad p_1 + p_2 + p_3 = 1$

and $\quad q_1 + q_2 + q_3 = 1;$

$$p_i \geq 0, q_i \geq 0 \tag{1}$$

The values of p_is and q_is are given by the equations

$$6p_1 + 8p_2 + 4p_3 = V \qquad (2);$$

$$-2p_2 + 3p_3 = V \qquad (3);$$

$$3p_1 + 3p_2 + 5p_3 = V \qquad (4);$$

$$6q_1 + 3q_3 = V \qquad (5)$$

$$8q_1 - 2q_2 + 3q_3 = V \qquad (6)$$

$$4q_1 + 6q_2 + 5q3 = V \qquad (7)$$

We note that the determinant of the coefficients of the system (2), (3) and (4) and also that of the system (5), (6) and (7) is equal to zero each.

This is an indication that one or two values of p_is are zero and one or two values of q_is are zero.

Let us try to solve the equations (2), (3) and (4), by assuming $p_1 = 0$.

Then the equations become

$$8p_2 + 4p_3 = V \qquad (2)',$$

$$-2p_2 + 6p_3 = V \qquad (3)'$$

and $$3p_2 + 5p_3 = V \qquad (4)'$$

Solving (2)′ and (3)′,

$$\begin{array}{cccc} & p_2 & p_3 & V \\ 4 & -1 & 8 & 4 \\ 6 & -1 & -2 & 6 \end{array}$$

we get

viz., $$\frac{p_2}{2} = \frac{p_3}{10} = \frac{V}{56}$$

viz., $$p_2 = \frac{V}{28} \text{ and } p_3 = \frac{5V}{28}$$

Using these values in (1),

$$0 + \frac{V}{28} + \frac{5V}{28} = 1$$

$$\therefore \quad V = \frac{14}{3} \qquad (8)$$

$$\therefore \quad p_1 = 0,\ p_2 = \frac{1}{6},\ p_3 = \frac{5}{6},$$

(***Note:*** Using these values of p_2 and p_3 in (4)′, we may get $3p_2 + 5p_3 \geq V$, which is acceptable. If $3p_2 + 5p_3 < V$, the assumption that $p_1 = 0$ is not correct; in which case, we have to try $p_2 = 0$ or $p_3 = 0$)

$$\therefore \quad S_A = \left(0, \frac{1}{6}, \frac{5}{6}\right)$$

$$V = \frac{14}{3}$$

Now let us solve the equation (5), (6) and (7) by assuming $q_1 = 0$.

Then, the equations become

$$3q_3 = V; \tag{5)'$$

$$-2q_2 = 0 \tag{6)'$$

Using these values in (1),

$$0+0+\frac{V}{3}=1$$

$$V = 3, \text{ which violates (8).}$$

Let us now assume that $q_2 = 0$ and use it in equations (6) and (7).

Note: Equation (5) is not used, since it does not contain the q_2 term.

Then the equations become $\quad 8q_1 + 3q_3 = V \qquad (6)''$

and $\quad 4q_1 + 5q_3 = V \qquad (7)''$

Solving them, by the rule of cross-multiplication, we get

$$\begin{array}{cccc} & q_1 & q_3 & V \\ 3 & -1 & 8 & 3 \\ 5 & -1 & 4 & 5 \end{array}$$

viz.,
$$\frac{q_1}{2}=\frac{q_3}{4}=\frac{V}{28}$$

viz.,
$$q_1=\frac{2V}{28};\ q_3=\frac{4V}{28}$$

Using these values in (1),

we have
$$\frac{2}{28}V+0+\frac{4V}{28}=1.$$

$\therefore$
$$V=\frac{14}{3}$$

$\therefore$
$$S_B=\left(\frac{1}{3}, 0, \frac{2}{3}\right)$$

and
$$V=\frac{14}{3}$$

Now let us assume that $q_3 = 0$ and use it in (6) and (7) to find an alternative solution.

Equations (6) and (7) becomes

$$8q_1 - 2q_2 = V \tag{6)''$$

and

$$4q_1 + 6q_2 = V \tag{7)''.$$

Solving them, by the rule of cross-multiplication, we get

$$\begin{array}{cccc} & q_1 & q_2 & V \\ -2 & -1 & 8 & -2 \\ 6 & -1 & 4 & 6 \end{array}$$

viz., $$\frac{q_1}{8} = \frac{q_2}{4} = \frac{V}{56}$$

viz., $$q_1 = \frac{V}{7}$$

and $$q_2 = \frac{V}{14}$$

Using these values in (1),

we have $$\frac{V}{7} + \frac{V}{14} + 0 = 1$$

$\therefore$ $$V = \frac{14}{3}$$

$\therefore$ An alternative strategy of B is given by

$$S_B = \left(\frac{2}{3}, \frac{1}{3}, 0\right)$$

and $$V = \frac{14}{3}$$

Example 7

Solve the following game by using the Simplex method:

$$\begin{array}{c} \\ A_1 \\ A_2 \\ A_3 \end{array}\begin{array}{c} \begin{array}{ccc} B_1 & B_2 & B_3 \end{array} \\ \begin{bmatrix} 1 & 7 & 2 \\ 6 & 2 & 7 \\ 5 & 1 & 6 \end{bmatrix} \end{array}$$

SOLUTION

Let the strategies of the players A and B be assumed to be

$$S_A = \begin{bmatrix} A_1 & A_2 & A_3 \\ p_1 & p_2 & p_3 \end{bmatrix} \text{ and } S_B = \begin{bmatrix} B_1 & B_2 & B_3 \\ q_1 & q_2 & q_3 \end{bmatrix}$$

If u is the minimum expected gain for player A, then we have to minimise $p_0 = \frac{1}{u} = p_1' + p_2' + p_3'$, subject to the constraints

$$p_1' + 6p_2' + 5p_3' \geq 1$$
$$7p_1' + 2p_2' + p_3' \geq 1$$
$$2p_1' + 7p_2' + 6p_3' \geq 1,$$

where $$p_i' = \frac{p_i}{u} \geq 0; \; i = 1, 2, 3$$

If v is the maximum expected loss for player B, then we have to maximise $q_0 = \frac{1}{v} = q_1' + q_2' + q_3'$, subject to the constraints

$$q_1' + 7q_2' + 2q_3' \leq 1$$
$$6q_1' + 2q_2' + 7q_3' \leq 1$$
$$5q_1' + q_2' + 6p_3' \leq 1,$$

where $$q_i' = \frac{q_i}{v} \geq 0; i = 1,2,3$$

Let us solve the problem of the player B by introducing the slack variables s_1, s_2, s_3 and using Simplex method. The iterative tables are given as follows:

		C_j	1	1	1	0	0	0	
B	C_B	X_B	q_1'	q_2'	q_3'	s_1	s_2	s_3	
s_1	0	1	1	7	2	1	0	0	
s_2	0	1	(6)	2	7	0	1	0	→
s_3	0	1	5	1	6	0	0	1	
$\frac{1}{v} = q_0 = 0$			−1	−1	−1	0	0	0	
			↑						

B	C_B	X_B	q_1'	q_2'	q_3'	s_1	s_2	s_3	
s_1	0	$5/6$	0	($40/6$)	$5/6$	1	$-1/6$	0	→
q_1'	1	$1/6$	1	$2/6$	$7/6$	0	$1/6$	0	
s_3	0	$1/6$	0	$-4/6$	$1/6$	0	$-5/6$	1	
$\frac{1}{v} = q_0 = \frac{1}{6}$			0	$-4/6$	$1/6$	0	$1/6$	0	
				↑					

B	C_B	X_B	q_1'	q_2'	q_3'	s_1	s_2	s_3
q_2'	1	$\frac{1}{8}$	0	1	$\frac{1}{8}$	$\frac{3}{20}$	$-\frac{1}{40}$	0
q_1'	1	$\frac{1}{8}$	1	0	$\frac{9}{8}$	$-\frac{1}{20}$	$\frac{7}{40}$	0
s_3	0	$\frac{1}{4}$	0	0	$1/4$	$1/10$	$\frac{17}{20}$	1
$\frac{1}{v} = q_0 = \frac{1}{4}$			0	0	$\frac{1}{4}$	$\frac{1}{10}$	$\frac{3}{20}$	0

Optimality has been reached with $S_B = \begin{pmatrix} B_1 & B_2 & B_3 \\ vq_1' & vq_2' & 0 \end{pmatrix} = \begin{pmatrix} B_1 & B_2 & B_3 \\ 1/2 & 1/2 & 0 \end{pmatrix}$ and the value of the game for B $= \dfrac{1}{q_0} = 4$ viz., loss for player B = 4

By duality, $S_A = \begin{pmatrix} A_1 & A_2 & A_3 \\ up_1' & up_2' & 0 \end{pmatrix} = \begin{pmatrix} A_1 & A_2 & A_3 \\ 2/5 & 3/5 & 0 \end{pmatrix}$ and the value of the game for A $= \dfrac{1}{p_0} = 4$

viz., gain for player A = 4.

Example 8

A and B play a game as follows: They write one of the three numbers 1, 2, 3 simultaneously and independently, without being noticed by the opponent. If the sum of the numbers written is even, B pays to A this sum in rupees. If it is odd, A pays the sum to B in rupees. Form the pay-off matrix of the game for A. Solve the game by the Simplex method.

SOLUTION

It can be easily seen that the pay-off matrix of the game is given by

$$\begin{array}{cc} & B \\ & \begin{array}{ccc} 1 & 2 & 3 \end{array} \\ A \begin{array}{c} 1 \\ 2 \\ 3 \end{array} & \begin{bmatrix} 2 & -3 & 4 \\ -3 & 4 & -5 \\ 4 & -5 & 6 \end{bmatrix} \end{array}$$

Adding a constant 6 to all the elements of the matrix, so that all the elements of the matrix are strictly positive, the transformed matrix becomes as given here:

$$\begin{array}{cc} & B \\ & \begin{array}{ccc} 1 & 2 & 3 \end{array} \\ A \begin{array}{c} 1 \\ 2 \\ 3 \end{array} & \begin{bmatrix} 8 & 3 & 10 \\ 3 & 10 & 1 \\ 10 & 1 & 12 \end{bmatrix} \end{array}$$

Let
$$S_A = \begin{bmatrix} A_1 & A_2 & A_3 \\ p_1 & p_2 & p_3 \end{bmatrix}$$

and
$$S_B = \begin{bmatrix} B_1 & B_2 & B_3 \\ q_1 & q_2 & q_3 \end{bmatrix}$$

If u is the minimum expected gain for player A, then the problem is to maximise u or minimise $\dfrac{1}{u} = p_0 = p_1' + p_2' + p_3'$, subject to the constraints

$$8p_1' + 3p_2' + 10p_3' \geq 1,$$

$$3p_1' + 10p_2' + p_3' \geq 1,$$
$$10p_1' + p_2' + 12p_3' \geq 1;$$
$$p_1', p_2', p_3' \geq 0,$$

where $$p_i' = \frac{p_i}{u}$$

Similarly, if v is the maximum expected loss for player B, then the problem is to minimise v or maximise $\frac{1}{v} = q_0 = q_1' + q_2' + q_3'$, subject to the constraints

$$8q_1' + 3q_2' + 10q_3' \leq 1$$
$$3q_1' + 10q_2' + q_3' \leq 1$$
$$10q_1' + q_2' + 12q_3' \leq 1;$$
$$q_i' \geq 0,$$

where $$q_i' = \frac{q_i}{v}$$

Let us solve the problem of the player B by introducing the slack variables s_1, s_2, s_3 and using Simplex method. The iterative tables are given as follows:

		C_j:	1	1	1	0	0	0	
B	C_B	X_B	q_1'	q_2'	q_3'	s_1	s_2	s_3	
s_1	0	1	8	3	10	1	0	0	
s_2	0	1	3	10	1	0	1	0	
s_3	0	1	(10)	1	12	0	0	1	→
$\frac{1}{v} = q_0 = 0$			−1	−1	−1	0	0	0	
			↑						

B	C_B	X_B	q_1'	q_2'	q_3'	s_1	s_2	s_3	
s_1	0	$2/10$	0	$22/10$	$4/10$	1	0	$-8/10$	
s_2	0	$7/10$	0	($97/10$)	$-26/10$	0	1	$-3/10$	→
q_1'	1	$1/10$	1	$1/10$	$12/10$	0	0	$1/10$	
$\frac{1}{v} = q_0 = \frac{1}{10}$			0	$-9/10$	$2/10$	0	0	$1/10$	
				↑					

B	C_B	X_B	q_1'	q_2'	q_3'	s_1	s_2	s_3
s_1	0	$4/97$	0	0	$(96/97)$	1	$-\frac{22}{97}$	$-\frac{71}{97}$ →
q_2'	1	$7/97$	0	1	$-26/97$	0	$10/97$	$-3/97$
q_1'	1	$9/97$	1	0	$119/97$	0	$-1/97$	$10/97$
$\frac{1}{v}=q_0=1$			0	0	$-4/97$ ↑	0	$9/97$	$7/97$

B	C_B	X_B	q_1'	q_2'	q_3'	s_1	s_2	s_3
q_3'	1	$1/24$	0	0	1	$97/96$	$-22/96$	$-71/96$
q_2'	1	$2/24$	0	1	0	$13/48$	$1/24$	$-11/48$
q_1'	1	$1/24$	1	0	0	$-119/96$	$26/96$	$97/96$
$\frac{1}{v}=q_0=\frac{1}{6}$			0	0	0	$1/24$	$1/12$	$1/24$

Value of the game for the player B is given by $v^* = \frac{1}{q_0} = 6$ and

$$S_B = \begin{pmatrix} B_1 & B_2 & B_3 \\ q_1 & q_2 & q_3 \end{pmatrix} = \begin{pmatrix} B_1 & B_2 & B_3 \\ v^*q_1' & v^*q_2' & v^*q_3' \end{pmatrix} = \begin{pmatrix} B_1 & B_2 & B_3 \\ 1/4 & 1/2 & 1/4 \end{pmatrix}$$

The entries in the index row under the columns s_1, s_2, s_3 give the values of p_1', p_2', p_3', since the problem of A is the dual of the problem of B.

$\therefore$ $$u^* = 6$$

and $$S_A = \begin{pmatrix} A_1 & A_2 & A_3 \\ p_1 & p_2 & p_3 \end{pmatrix}$$

$$= \begin{pmatrix} A_1 & A_2 & A_3 \\ u^*p_1' & u^*p_2' & u^*p_3' \end{pmatrix} = \begin{pmatrix} A_1 & A_2 & A_3 \\ 1/4 & 1/2 & 1/4 \end{pmatrix}$$

Since we have added 6 to each element of the pay-off matrix, the true value of the game V is got by subtracting 6 from v^* for B and from u^* for A. Hence the value of the game $V = 0$.

Example 9

Solve the following game approximately by iteration:

$$\text{Player } A \begin{array}{c} \text{Player } B \\ \begin{bmatrix} -1 & 2 & 1 \\ 1 & -2 & 2 \\ 3 & 4 & -3 \end{bmatrix} \end{array}$$

SOLUTION

	B_1	B_2	B_3	I_1	I_2	I_3	I_4	I_5	I_6	I_7	I_8	I_9	I_{10}	
A_1	−1	2	1	−1	0	1	(3)	2	4	(5)	6	7	(9)	3/10
A_2	1	−2	2	1	(3)	(5)	3	(4)	2	4	(6)	(8)	6	5/10
A_3	3	4	−3	(3)	0	−3	1	4	(8)	5	2	−1	3	2/10
I_1:	(−1)	2	1											
I_2:	2	6	(−2)											
I_3:	3	4	(0)											
I_4:	4	(2)	2											
I_5:	(3)	4	3											
I_6:	4	(2)	5											
I_7:	7	6	(2)											
I_8:	6	8	(3)											
I_9:	7	6	(5)											
I_{10}:	8	(4)	7											
	2/10	3/10	5/10											

The numbers of encircled elements in the A_1, A_2 and A_3 rows are 3, 5 and 2 respectively

∴ The approximate strategies of A are given by

$$S_A = \left(\frac{3}{10}, \frac{5}{10}, \frac{2}{10}\right)$$

The numbers of encircled elements in the B_1, B_2 and B_3 columns are 2, 3 and 5 respectively.

∴ The approximate strategies of B are given by

$$S_B = \left(\frac{2}{10}, \frac{3}{10}, \frac{5}{10}\right)$$

The lower bound of the value V of the game corresponds to the minimum of elements in the I_{10} row, viz., $4/10$

The upper bound of the value V of the game corresponds to the maximum of the elements in the I_{10} column, viz., $9/10$

∴ $$\frac{4}{10} \le V \le \frac{9}{10}$$

Example 10

Find the bounds for the value of the following game approximately by iteration:

$$\begin{array}{c} \text{Player } B \\ \begin{array}{cc} & \begin{array}{cccc} B_1 & B_2 & B_3 & B_4 \end{array} \\ \text{Player } A \begin{array}{c} A_1 \\ A_2 \\ A_3 \end{array} & \begin{bmatrix} 3 & 7 & -1 & 3 \\ 4 & 8 & 0 & -6 \\ 6 & -9 & -2 & 4 \end{bmatrix} \end{array} \end{array}$$

Solution

(Encircled elements are shown in bold.)

	B_1	B_2	B_3	B_4	I_1	I_2	I_3	I_4	I_5	I_6	I_7	I_8	I_9	I_{10}	
A_1	3	7	−1	3	−1	**2**	5	**4**	**3**	**2**	**1**	**0**	**−1**	**−2**	8/10
A_2	4	8	0	−6	**0**	−6	−12	−12	−12	−12	−12	−12	−12	−12	1/10
A_3	6	−9	−2	4	−2	2	**6**	4	2	0	−2	−4	−6	−8	1/10
I_1:	3	7	**−1**	3											
I_2:	7	15	−1	**−3**											
I_3:	10	22	**−2**	0											
I_4:	16	13	**−4**	4											
I_5:	19	20	**−5**	7											
I_6:	22	27	**−6**	10											
I_7:	25	34	**−7**	13											
I_8:	28	41	**−8**	16											
I_9:	31	48	**−9**	19											
I_{10}:	34	55	**−10**	22											
	0	0	9/10	1/10											

The numbers of encircled elements in the A_1, A_2 and A_3 rows are 8, 1 and 1 respectively.

∴ The approximate strategies of A are given by

$$S_A = \left(\frac{8}{10}, \frac{1}{10}, \frac{1}{10}\right)$$

The numbers of encircled elements in the B_1, B_2, B_3 and B_4 columns are 0, 0, 9 and 1 respectively.

∴ The approximate strategies of B are given by

$$S_B = \left(0, 0, \frac{9}{10}, \frac{1}{10}\right)$$

The lower bound of the value V of the game corresponds to the minimum of elements in the I_{10} row, viz., $-\frac{10}{10} = -1$.

The upper bound of the value V of the game corresponds to the maximum of elements in the I_{10} column, viz., $-\frac{2}{10}$

$$\therefore \qquad -1 \le V \le -\frac{1}{5}$$

EXERCISE 10(B)

Part A (Short answer questions)

1. Explain briefly how a $(2 \times n)$ game is reduced to a (2×2) game graphically.
2. Explain briefly how a $(m \times 2)$ game is reduced to a (2×2) game graphically.
3. Explain briefly the algebraic method of solving a $(m \times n)$ rectangular game.
4. Explain briefly the linear programming method of solving a $(m \times n)$ rectangular game.
5. Explain briefly the iterative method of solving a $(m \times n)$ rectangular game approximately.

Part B

Reduce the following games to (2×2) games graphically and solve them.

6

$$\begin{array}{cc} & \textit{Player } B \\ & \begin{array}{ccc} B_1 & B_2 & B_3 \end{array} \\ \textit{Player } A: \begin{array}{c} A_1 \\ A_2 \end{array} & \begin{bmatrix} 1 & 3 & 11 \\ 8 & 5 & 2 \end{bmatrix} \end{array}$$

7.

$$\begin{array}{cc} & \textit{Player } B \\ & \begin{array}{cccc} B_1 & B_2 & B_3 & B_4 \end{array} \\ \textit{Player } A: \begin{array}{c} A_1 \\ A_2 \end{array} & \begin{bmatrix} 2 & 1 & 0 & -2 \\ 1 & 0 & 3 & 2 \end{bmatrix} \end{array}$$

8.

$$\begin{array}{cc} & \textit{Player } B \\ & \begin{array}{cccc} B_1 & B_2 & B_3 & B_4 \end{array} \\ \textit{Player } A: \begin{array}{c} A_1 \\ A_2 \end{array} & \begin{bmatrix} 1 & 3 & -3 & 7 \\ 2 & 5 & 4 & -6 \end{bmatrix} \end{array}$$

9.

$$\begin{array}{cc} & \textit{Player } B \\ & \begin{array}{cccc} B_1 & B_2 & B_3 & B_4 \end{array} \\ \textit{Player } A: \begin{array}{c} A_1 \\ A_2 \end{array} & \begin{bmatrix} 2 & 2 & 3 & -1 \\ 4 & 3 & 2 & 6 \end{bmatrix} \end{array}$$

10.

$$\begin{array}{cc} & \textit{Player } B \\ & \begin{array}{cccc} B_1 & B_2 & B_3 & B_4 \end{array} \\ \textit{Player } A: \begin{array}{c} A_1 \\ A_2 \end{array} & \begin{bmatrix} -6 & -1 & 4 & 3 \\ 7 & -2 & -5 & 7 \end{bmatrix} \end{array}$$

11.
$$\text{Player } A: \begin{array}{c} \\ A_1 \\ A_2 \end{array} \overset{\textit{Player } B}{\begin{array}{c} \begin{array}{ccccc} B_1 & B_2 & B_3 & B_4 & B_5 \end{array} \\ \begin{bmatrix} -2 & -5 & 0 & -3 & 1 \\ 5 & 3 & -2 & 0 & -4 \end{bmatrix} \end{array}}$$

12.
$$\begin{array}{c} \\ A_1 \\ A_2 \\ A_3 \end{array} \begin{array}{c} \begin{array}{cc} B_1 & B_2 \end{array} \\ \begin{bmatrix} 2 & 7 \\ 3 & 5 \\ 11 & 2 \end{bmatrix} \end{array}$$

13.
$$\begin{array}{c} \\ A_1 \\ A_2 \\ A_3 \\ A_4 \end{array} \begin{array}{c} \begin{array}{cc} B_1 & B_2 \end{array} \\ \begin{bmatrix} 3 & -2 \\ -1 & 3 \\ 1 & 2 \\ 2 & 3 \end{bmatrix} \end{array}$$

14.
$$\begin{array}{c} \\ A_1 \\ A_2 \\ A_3 \\ A_4 \\ A_5 \end{array} \begin{array}{c} \begin{array}{cc} B_1 & B_2 \end{array} \\ \begin{bmatrix} 1 & -3 \\ 3 & 5 \\ -1 & 6 \\ 4 & 1 \\ 2 & 2 \end{bmatrix} \end{array}$$

Solve the following games algebraically:

15.
$$\begin{array}{c} \\ A_1 \\ A_2 \\ A_3 \end{array} \begin{array}{c} \begin{array}{ccc} B_1 & B_2 & B_3 \end{array} \\ \begin{bmatrix} -1 & 2 & 1 \\ 1 & -2 & 2 \\ 3 & 4 & -3 \end{bmatrix} \end{array}$$

16.
$$\begin{array}{c} \\ A_1 \\ A_2 \\ A_3 \end{array} \begin{array}{c} \begin{array}{ccc} B_1 & B_2 & B_3 \end{array} \\ \begin{bmatrix} 4 & 2 & 4 \\ 2 & 4 & 0 \\ 4 & 0 & 8 \end{bmatrix} \end{array}$$

17.
$$\begin{array}{c} \\ A_1 \\ A_2 \\ A_3 \end{array} \begin{array}{c} \begin{array}{ccc} B_1 & B_2 & B_3 \end{array} \\ \begin{bmatrix} 3 & -2 & 4 \\ -1 & 4 & 2 \\ 2 & 2 & 6 \end{bmatrix} \end{array}$$

18. $\begin{array}{c} \\ A_1 \\ A_2 \\ A_3 \end{array}\begin{array}{c} \begin{array}{ccc} B_1 & B_2 & B_3 \end{array} \\ \begin{bmatrix} 1 & -1 & 0 \\ -1 & 1 & 1 \\ 2 & -1 & -2 \end{bmatrix} \end{array}$

19. $\begin{array}{c} \\ A_1 \\ A_2 \\ A_3 \end{array}\begin{array}{c} \begin{array}{ccc} B_1 & B_2 & B_3 \end{array} \\ \begin{bmatrix} 1 & -1 & -1 \\ -1 & -1 & 3 \\ -1 & 2 & -1 \end{bmatrix} \end{array}$

Solve the following games using the Simplex method:

20. $\begin{array}{c} \\ A_1 \\ A_2 \\ A_3 \end{array}\begin{array}{c} \begin{array}{ccc} B_1 & B_2 & B_3 \end{array} \\ \begin{bmatrix} -1 & 1 & 1 \\ 2 & -2 & 2 \\ 3 & 3 & -3 \end{bmatrix} \end{array}$

21. $\begin{array}{c} \\ A_1 \\ A_2 \\ A_3 \end{array}\begin{array}{c} \begin{array}{ccc} B_1 & B_2 & B_3 \end{array} \\ \begin{bmatrix} -1 & 2 & 1 \\ 1 & -2 & 2 \\ 3 & 4 & -3 \end{bmatrix} \end{array}$

22. $\begin{array}{c} \\ A_1 \\ A_2 \\ A_3 \end{array}\begin{array}{c} \begin{array}{ccc} B_1 & B_2 & B_3 \end{array} \\ \begin{bmatrix} 3 & -1 & -3 \\ -3 & 3 & -1 \\ -4 & -3 & 3 \end{bmatrix} \end{array}$

Solve the following games approximately performing 10 iterations

23. $\begin{array}{c} \\ A_1 \\ A_2 \\ A_3 \end{array}\begin{array}{c} \begin{array}{ccc} B_1 & B_2 & B_3 \end{array} \\ \begin{bmatrix} 1 & -1 & -1 \\ -1 & -1 & 3 \\ -1 & 2 & -1 \end{bmatrix} \end{array}$

24. $\begin{array}{c} \\ A_1 \\ A_2 \\ A_3 \end{array}\begin{array}{c} \begin{array}{ccc} B_1 & B_2 & B_3 \end{array} \\ \begin{bmatrix} 2 & 0 & 1 \\ 0 & 2 & 1 \\ 3 & 0 & 0 \end{bmatrix} \end{array}$

25. $\begin{array}{c} \\ A_1 \\ A_2 \\ A_3 \end{array}\begin{array}{c} \begin{array}{cccc} B_1 & B_2 & B_3 & B_4 \end{array} \\ \begin{bmatrix} 2 & 3 & -1 & 0 \\ 5 & 4 & 2 & -2 \\ 1 & 3 & 8 & 2 \end{bmatrix} \end{array}$

ANSWERS

Exercise 10(A)

Part A

7. $S_A = \begin{pmatrix} A_1 & A_2 \\ \frac{c+d}{\sum a} & \frac{a+b}{\sum a} \end{pmatrix}; S_B = \begin{pmatrix} B_1 & B_2 \\ \frac{b+d}{\sum a} & \frac{a+c}{\sum a} \end{pmatrix}$ and $V = \frac{ad-bc}{\sum a}$

8. $S_A = (2, 0, 2)$; $S_B = (-2, 0, -2)$ and $V = 0$

9. $S_A = (1, 3, 1)$; $S_B = (1, 0, 1)$ or $(3, -4, 5)$ and $V = 1$

10. $S_A = (6, 8, 6)$; $S_B = (6, 2)$ and $V = 6$

Part B

11. $S_A = (3, 2, 3, 4, 3)$; $S_B = (0, -3, 2, 1, 0)$ and $V = 2$;

12. $-1 \leq \lambda \leq 2$.

13. $S_A = \begin{pmatrix} A_1 & A_2 \\ 4/7 & 3/7 \end{pmatrix}; S_B = \begin{pmatrix} B_1 & B_2 \\ 2/7 & 5/7 \end{pmatrix}$ and $V = \frac{29}{7}$.

14. $S_A = \begin{pmatrix} A_1 & A_2 \\ \frac{1}{3} & \frac{2}{3} \end{pmatrix}; S_B = \begin{pmatrix} B_1 & B_2 \\ \frac{3}{5} & \frac{2}{5} \end{pmatrix}$ and $V = 0$;

15. $S_A = S_B \begin{pmatrix} H & T \\ \frac{1}{2} & \frac{1}{2} \end{pmatrix}$ and $V = 0$;

16. $S_A = \begin{pmatrix} A_1 & A_2 & A_3 \\ \frac{2}{5} & \frac{3}{5} & 0 \end{pmatrix}; S_B = \begin{pmatrix} B_1 & B_2 & B_3 \\ \frac{1}{2} & \frac{1}{2} & 0 \end{pmatrix}$ and $V = 4$.

17. $S_A = \begin{pmatrix} A_1 & A_2 & A_3 \\ \frac{2}{5} & 0 & \frac{3}{5} \end{pmatrix}; S_B = \begin{pmatrix} B_1 & B_2 & B_3 \\ \frac{3}{5} & \frac{2}{5} & 0 \end{pmatrix}$ and $V = \frac{17}{5}$

18. $S_A = \begin{pmatrix} A_1, A_2, A_3 \\ 0\ ,\ 0\ ,\ 1 \end{pmatrix}; S_B = \begin{pmatrix} B_1, B_2, B_3 \\ \frac{2}{5}, \frac{3}{5}, 0 \end{pmatrix}$ and $V = 2$.

19. $S_A = \begin{pmatrix} A_1, A_2, A_3, A_4 \\ 0\ ,\ 0\ ,\ \frac{1}{2}, \frac{1}{2} \end{pmatrix}; S_B = \begin{pmatrix} B_1, B_2, B_3, B_4 \\ \frac{3}{8}, 0, \frac{5}{8}, 0 \end{pmatrix}$ and $V = \frac{11}{2}$

20. $S_A = \begin{pmatrix} A_1, A_2, A_3, A_4, A_5 \\ 0\ ,\ \frac{6}{7}, \frac{1}{7}, 0, 0 \end{pmatrix}; S_B = \begin{pmatrix} B_1, B_2, B_3, B_4, B_5, B_6 \\ 0\ ,\ 0\ ,\ \frac{4}{7}, \frac{3}{7}, 0\ ,\ 0 \end{pmatrix}$ and $V = \frac{13}{7}$

Exercise 10(B)

Part B

6. $S_A = \left(\frac{3}{11}, \frac{8}{11}\right); S_B = \left(0, \frac{9}{11}, \frac{2}{11}\right); V = \frac{49}{11}$

7. $S_A = \left(\frac{2}{5}, \frac{3}{5}\right); S_B = \left(0, \frac{4}{5}, 0, \frac{1}{5}\right); V = \frac{2}{5}$

8. $S_A = \left(\frac{1}{2}, \frac{1}{2}\right); S_B = \left(0, 0, \frac{13}{20}, \frac{7}{20}\right); V = \frac{1}{2}$

9. $S_A = \left(\frac{1}{2}, \frac{1}{2}\right); S_B = \left(0, \frac{1}{2}, \frac{1}{2}, 0\right); V = \frac{5}{2};$

   or $S_A = \left(\frac{1}{2}, \frac{1}{2}\right); S_B = \left(0, 0, \frac{7}{8}, \frac{1}{8}\right)$

10. $S_A = \left(\frac{9}{14}, \frac{5}{14}\right); S_B = \left(\frac{1}{14}, \frac{13}{14}, 0, 0\right); V = -\frac{19}{14}$

11. $S_A = \left(\frac{1}{2}, \frac{1}{2}\right); S_B = \left(0, 0, 0, \frac{5}{8}, \frac{3}{8}\right); V = -\frac{3}{2}$

12. $S_A = \left(\frac{9}{14}, 0, \frac{5}{14}\right); S_B = \left(\frac{5}{14}, \frac{9}{14}\right); V = \frac{73}{14}$

13. $S_A = \left(\frac{1}{6}, 0, 0, \frac{5}{6}\right); S_B = \left(\frac{5}{6}, \frac{1}{6}\right); V = \frac{13}{6}$

14. $S_A = \left(0, \frac{3}{5}, 0, \frac{2}{5}, 0\right); S_B = \left(\frac{4}{5}, \frac{1}{5}\right); V = \frac{17}{5}$

15. $S_A = \left(\frac{17}{46}, \frac{20}{46}, \frac{9}{46}\right); S_B = \left(\frac{14}{46}, \frac{12}{46}, \frac{20}{46}\right); V = \frac{30}{46}$

16. $S_A = \left(0, \frac{2}{3}, \frac{1}{3}\right); S_B = \left(0, \frac{2}{3}, \frac{1}{3}\right); V = \frac{8}{3}$

17. $S_A = (0, 0, 1); S_B = \left(\frac{2}{5}, \frac{3}{5}, 0\right); V = 2$

18. $S_A = \left(\frac{2}{9}, \frac{5}{9}, \frac{2}{9}\right); S_B = \left(\frac{4}{9}, \frac{3}{9}, \frac{2}{9}\right); V = \frac{1}{9}$

19. $S_A = \left(\frac{6}{13}, \frac{3}{13}, \frac{4}{13}\right); S_B = \left(\frac{6}{13}, \frac{4}{13}, \frac{3}{13}\right); V = -\frac{1}{13}$

20. $S_A = \left(\frac{6}{11}, \frac{3}{11}, \frac{2}{11}\right); S_B = \left(\frac{5}{22}, \frac{8}{22}, \frac{9}{22}\right); V = \frac{6}{11}$

21. $S_A = \left(\frac{17}{46}, \frac{20}{46}, \frac{9}{46}\right); S_B = \left(\frac{7}{23}, \frac{6}{23}, \frac{10}{23}\right); V = \frac{15}{23}$

22. $S_A = \left(\frac{20}{45}, \frac{11}{45}, \frac{14}{45}\right); S_B = \left(\frac{14}{45}, \frac{11}{45}, \frac{20}{45}\right); V = -\frac{29}{45}$

23. $S_A \simeq \left(\frac{4}{10}, \frac{3}{10}, \frac{3}{10}\right); S_B \simeq \left(\frac{4}{10}, \frac{2}{10}, \frac{4}{10}\right);$

    $-\frac{2}{10} \le V \le \frac{2}{10}$

24. $S_A \simeq \left(\frac{3}{10}, \frac{6}{10}, \frac{1}{10}\right); S_B \simeq \left(\frac{2}{10}, \frac{4}{10}, \frac{4}{10}\right);$

    $\frac{9}{10} \le V \le \frac{12}{10}$

25. $S_A \simeq \left(0, \frac{1}{10}, \frac{9}{10}\right); S_B \simeq \left(\frac{4}{10}, 0, 0, \frac{6}{10}\right);$

    $\frac{14}{10} \le V \le \frac{16}{10}$

11 INVENTORY CONTROL

11.0 INTRODUCTION

Inventory means a physical stock of goods such as raw materials for manufacturing a product or finished products waiting for sale in the future. Since holding inventories involves considerable investment, decisions regarding the amount of inventory are important. The objective of an inventory control problem is to minimise the total or expected cost of holding inventories or to maximise the total or expected profit of the industry or the company. Inventory models and the mathematical analysis of inventory systems provide a basis for inventory management problems.

11.1 ADVANTAGES AND DISADVANTAGES OF KEEPING INVENTORIES

Advantages

1. Having inventories avoids reduction in profit due to loss of sales and also due to loss of good will of the customers, viz., it provides adequate service to customers and promotes smooth and efficient running of business.
2. It takes advantage of price discounts by bulk purchasing, viz., it avoids purchase of raw materials at high prices due to piece-meal buying.
3. It takes advantage of reduction in production costs due to batching, viz., due to production with large run sizes or longer production plans.
4. It serves as a buffer in case of delayed deliveries of raw materials.
5. It provides flexibility in production plans due to changes in demands.
6. It provides economies in transportation, clearing and forwarding costs.

Disadvantages

1. Money is tied up in unnecessary inventory, viz., it results in lock-up capital associated with interest.
2. It costs money by way of warehouse rent, equipment, personal insurance, etc.
3. It may result in stocked items getting depreciated, deteriorated or obsolete.

11.2 TECHNIQUES OF INVENTORY CONTROL

Now that we know the advantages and disadvantages of holding inventories, we understand that though inventory of materials is essential, excess inventories are undesirable. This necessitates the control of inventories in the most profitable way. The basic inventory control problem deals with deciding

(i) when to order for raw materials,
(ii) how much to order and
(iii) how much shortage to allow.

***Note*:** Orders which cannot be fulfilled as soon as it is received (due to their receipt after the inventory level has reached zero) are called *shortages* or *back orders*

We shall discuss the inventory control problems for two categories of inventory models separately – deterministic and stochastic models. *Deterministic models* are inventory models in which the demand is assumed to be fixed and known for a subsequent period of time. *Stochastic (or probabilistic)* models are inventory models in which the future demand is a random variable having a known probability distribution formed by the knowledge of data collected in the past.

11.3 EOQ FOR THE PURCHASING MODEL WITH NO SHORTAGE (MODEL I)

Before we proceed to derive a formula for EOQ, let us familiarise ourselves with the meanings of the terms associated with this model.

Economic order quantity, abbreviated as EOQ is the optimum quantity of finished items or spare parts to order for sales in future so as to minimise the total annual inventory cost.

In manufacturing situations, EOQ is referred to as *economic lot size* which means the optimum quality of raw material required to produce in each production run.

Purchasing model is the inventory model for which delivery rate is infinite, viz., the supply of complete orders (in a sales situation) or the production of the complete requirement (in a production situation) is instantaneous.

No shortages means that as soon as the level of inventory reaches zero, the inventory is replenished, if it is assumed that the lead time is zero.

Lead time is the time gap between placing an order (or starting of the production) and the actual time of arrival (or delivery of manufactured product) at the inventory. If the lead time is a constant, not equal to zero, and also if the demand is known, then one may place an order in advance by an amount of time equal to the lead time.

Let D denote the (fixed) number of units required for sale or produced for supply in one time period. If the time period is taken as 1 year, D denotes annual demand.

Let Q denote the number of units to be ordered per order or to be produced per production run.

Let C_0 be the cost of placing an order (or) C_s be the set up cost per production run.

Let C_1 be the holding cost per unit in inventory for one time period, viz., one year.

The purchasing model with no shortages can be graphically represented by an inventory–time diagram as shown here:

We assume that Q units are ordered or produced after each time t.

$\therefore$ The total inventory during the period of duration t = Area of the triangle OQA $= \frac{1}{2}Qt$.

$\therefore$ The average inventory level on any day during the period t and so during the year $= \frac{1}{2}Qt/t = \frac{1}{2}Q$

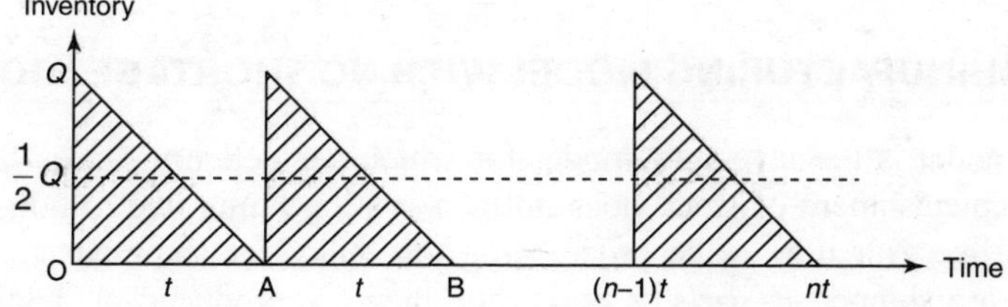

$\therefore$ Annual inventory holding cost $= \frac{1}{2}QC_1$.

Annual ordering cost (or set up cost) $= \frac{D}{Q}C_0 \left(\text{or } \frac{D}{Q}C_s \right)$

$\therefore$ The total annual (inventory) cost C is given by

$$C = \frac{1}{2}QC_1 + \frac{D}{Q}C_0 \tag{1}$$

EOQ is the optimum value of Q for which C is minimum.

$$\frac{dC}{dq} = \frac{1}{2}C_1 - \frac{D}{Q^2}C_0 \tag{2}$$

and
$$\frac{d^2C}{dQ^2} = \frac{2DC_0}{Q^3} \tag{3}$$

Solving
$$\frac{dC}{dQ} = 0,$$

we get
$$Q = \sqrt{\frac{2DC_0}{C_1}} = Q^0, \text{ say,}$$

$$\left(\frac{d_2C}{dQ^2} \right)_{Q=Q^0} = \frac{2DC_0}{\left(Q^0\right)^3} > 0$$

$\therefore$
$$\text{EOQ } (= Q^0) = \sqrt{\frac{2DC_0}{C_1}}$$

The minimum total annual (inventory) cost C is given by

$$C^0 = \frac{1}{2}Q^0C_1 + \frac{DC_0}{Q^0}$$

$$= \frac{1}{2}C_1\sqrt{\frac{2DC_0}{C_1}} + DC_0\sqrt{\frac{C_1}{2DC_0}} = \sqrt{2DC_0C_1}$$

The optimum time interval between consecutive orders is given by $t^0 = \frac{Q^0}{D}$, since $nt = 1$ and $nQ = D$

$$= \sqrt{\frac{2C_0}{C_1D}}$$

11.4 EOQ FOR MANUFACTURING MODEL WITH NO SHORTAGE (MODEL II)

The manufacturing model is the inventory model for which the delivery rate is finite. Very often a complete order for replenishment of stock does not arrive at one time due to non-availability of the complete order of delivery or due to the order being too bulky for shipment at one time. In such situations, the items are shipped in parts as and when they are produced or obtained from another source. We assume that the ordered items arrive at a certain fixed rate, viz., finite delivery rate, until the complete order is received

Also we assume that the arrival rate is greater than the use (or sale) rate so that no shortage occurs.

11.4.1 Graphical Representation of Manufacturing Model with No Shortage

The manufacturing model with no shortages can be represented graphically as shown below.

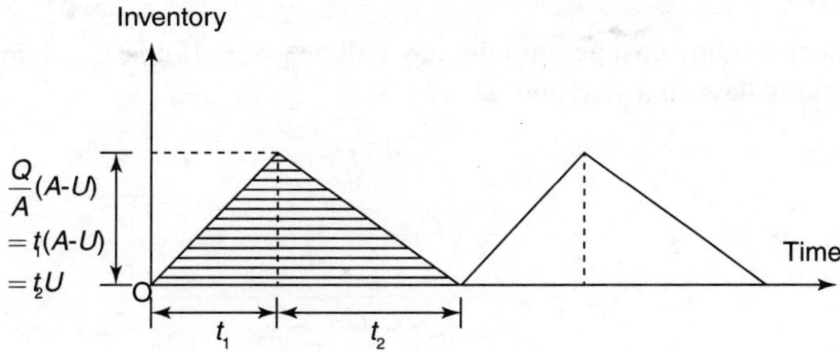

D, Q, C_0 and C_1 are assumed as in model I.

Let t be the time interval (in days) between two consecutive orders.

Let A be the arrival rate of an order per day, viz., the number of units received per day.

Let U be the use (or sale) rate per day, viz., the number of units used per day. At the start of the time period t, the inventory level is 0.

Let $t = t_1 + t_2$, where t_1 is the period when the inventory is building up at the rate of $(A–U)$ units per day and t_2 is the period when there is no replenishment and the inventory is decreasing at the rate of U units per day.

The number of days required to receive the entire order $Q = t_1 = \frac{Q}{A}$.

$$\text{The number of units used during the arrival period} = \frac{Q}{A}.U$$

$$\therefore \quad \text{Maximum inventory level which can accumulate} = Q - t_1U = Q\left(1-\frac{U}{A}\right)$$

$$\therefore \quad \text{Average inventory level} = \frac{1}{2}Q\left(1-\frac{U}{A}\right)$$

Now the total annual inventory cost (TAIC) is given by

$$C = \text{Annual ordering cost (AOC)} + \text{Annual holding cost (AHC)}$$

$$= \frac{D}{Q}C_0 + \frac{1}{2}Q\left(1-\frac{U}{A}\right)C_1$$

$$\frac{dC}{dQ} = -\frac{D}{Q^2}C_0 + \frac{1}{2}C_1\left(1-\frac{U}{A}\right) = 0$$

gives $$Q^2 = \frac{2DC_0}{C_1\left(1-\frac{U}{A}\right)}$$

and hence
$$Q^0 = \sqrt{\frac{2DC_0}{C_1\left(1-\frac{U}{A}\right)}}$$

since
$$\frac{d^2C}{dQ^2} = \frac{2DC_0}{Q^3} > 0.$$

Note Often the inventory holding cost per unit per day will be given. If it is C_1', then $C_1' = C_1/x$, where x is the number of working days in a year and $D = U \cdot x$.

In such problems,
$$Q^0 = \sqrt{\frac{2C_0Ux}{C_1'x\left(1-\frac{U}{A}\right)}}$$

$$= \sqrt{\frac{2C_0UA}{C_1'(A-U)}}$$

and
$$C^0 = \frac{DC_0}{Q^0} + \frac{1}{2}C_1\left(1-\frac{U}{A}\right)Q^0$$

$\therefore$ The minimum total daily inventory cost is given by:

$$\left(C'\right)^0 = \left(\frac{C^0}{x}\right) = \frac{UC_0}{Q^0} + \frac{C_1'}{2}\left(1-\frac{U}{A}\right)Q^0$$

$$= UC_0\sqrt{\frac{C_1'(A-U)}{2C_0UA}} + \frac{C_1'}{2}\left(\frac{A-U}{A}\right)\sqrt{\frac{2C_0UA}{C_1'(A-U)}}$$

$$= \sqrt{C_1'(A-U)}\left\{\sqrt{\frac{UC_0}{2A}} + \sqrt{\frac{UC_0}{2A}}\right\}$$

$$= \sqrt{2C_0C_1'U(A-U)/A}$$

$$t^0 = \frac{Q^0}{U} = \frac{1}{U}\sqrt{\frac{2C_0UA}{C_1'(A-U)}}$$

$$= \sqrt{\frac{2C_0A}{C_1'U(A-U)}}$$

11.5 EOQ FOR PURCHASING MODEL (FOR WHICH THE DELIVERY RATE IS INFINITE) WITH SHORTAGE (OR BACK ORDERING) (MODEL III)

In the previous two models, it was assumed that unfilled demands were lost and hence the inventory was replenished as soon as the level of inventory reached zero. Quite often, customers will be prepared to wait for an ordered item to arrive. In such situations, an unfilled demand that can be filled at a later date is known as a *back order* or as a *shortage*. If back orders are allowed, they are first satisfied as soon as the inventory is replenished. Of course, the back orders are associated with costs, viz., costs due to loss of goodwill, additional book–keeping, etc.

The purchasing model with back ordering can be represented graphically as shown here:

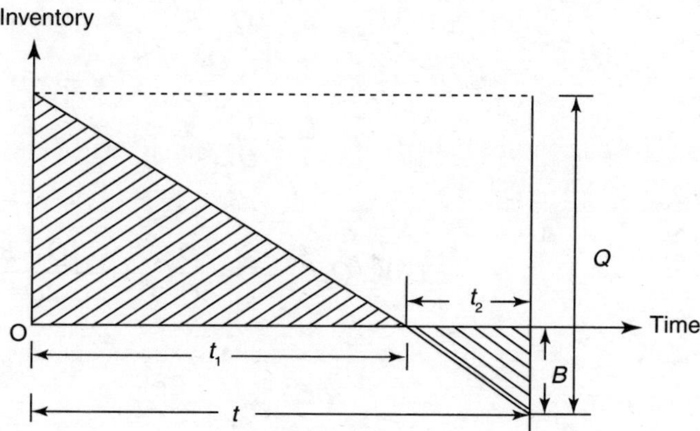

D, Q, C_0 and C_1 are assumed as in the previous models I and II.

Let B be the number of back orders allowed before replenishing the inventory.

Let C_2 be the annual back ordering (shortage) cost per unit.

Let t_1 be the time from the receipt of an order until the inventory level becomes zero.

Let t_2 be the time from zero inventory level until a new order is received, viz., the time during which orders are registered for future supply.

$$\text{Now total inventory over the time period } t = \frac{1}{2}(Q-B)t_1$$

$$\therefore \quad \text{Average inventory at any time over the time period } t = \frac{1}{2}(Q-B)\frac{t_1}{t}$$

$$\therefore \quad \text{Annual inventory holding cost} = \text{AIHC}(Q,B) = \frac{C_1}{2}(Q-B)\frac{t_1}{t}.$$

$$\text{Total amount of back orders over the time period } t = \frac{1}{2}Bt_2.$$

$$\therefore \quad \text{Average back orders (shortage) over the time period } t = \frac{1}{2}\frac{Bt_2}{t}.$$

$$\therefore \quad \text{Annual back ordering cost} = \text{ABC}(Q,B) = \frac{C_2}{2}\frac{Bt_2}{t}$$

As in the previous models, annual ordering cost

$$(\text{AOC}) = \frac{D}{Q}C_0.$$

$$\therefore \quad \text{TAIC} = \text{AOC} + \text{AIHC} + \text{ABC}$$

viz.,

$$\text{TAIC}(Q,B) = \frac{D}{Q}C_0 + \frac{C_1}{2}(Q-B)\frac{t_1}{t} + \frac{C_2}{2}\frac{Bt_2}{t} \qquad (1)$$

From similar triangles in the graphic representation,

$$\frac{t_1}{Q-B} = \frac{t_2}{B} = \frac{t_1+t_2}{Q} \text{ or } \frac{t}{Q}$$

$\therefore$

$$\frac{t_1}{t} = \frac{Q-B}{Q}$$

and

$$\frac{t_2}{t} = \frac{B}{Q}. \tag{2}$$

Using (2) in (1),

$$\text{TAIC}(Q,B) = C = \frac{D}{Q}C_0 + \frac{C_1(Q-B)^2}{2Q} + \frac{C_2B^2}{2Q}. \tag{3}$$

C is minimum,

when

$$\frac{\partial C}{\partial Q} = 0$$

and

$$\frac{\partial C}{\partial B} = 0$$

i.e., when

$$-\frac{DC_0}{Q^2} + \frac{C_1}{2}\left[\frac{Q\times 2(Q-B)-(Q-B)^2}{Q^2}\right] - \frac{C_2B^2}{2Q^2} = 0. \tag{4}$$

and

$$\frac{-C_1(Q-B)}{Q} + \frac{C_2B}{Q} = 0. \tag{5}$$

From (5), we get

$$B^0 = \frac{C_1Q}{C_1+C_2}. \tag{6}$$

Simplifying (4), we get $\quad C_1(Q^2 - B^2) - C_2B^2 = 2DC_0 \qquad (4)'$

Using (6) in (4)′;

$$C_1\left[Q^2 - \frac{C_1^2Q^2}{(C_1+C_2)^2}\right] - C_2\frac{C_1^2Q^2}{(C_1+C_2)^2} = 2DC_0$$

viz.,

$$\frac{Q^2}{(C_1+C_2)^2}\left[C_1\left\{(C_1+C_2)^2 - C_1^2\right\} - C_1^2C_2\right] = 2DC_0$$

viz.,

$$\frac{Q^2}{(C_1+C_2)^2}\left[C_1^2C_2 + C_1C_2^2\right] = 2DC_0$$

viz.,

$$Q^2 = \frac{2DC_0(C_1+C_2)}{C_1C_2}$$

or
$$Q^0 = \sqrt{\frac{2DC_0(C_1+C_2)}{C_1C_2}}. \tag{7}$$

$\therefore$ The optimum values of Q and B are given by (7) and (6)′, where B^0 is given by

$$B^0 = \frac{C_1Q^0}{C_1+C_2} \tag{6'}$$

Using the values of Q^0 and B^0 in (3), the minimum value of TAIC is given by.

$$C^0 = \frac{1}{2Q^0}\left[2DC_0 + C_1\left(Q^0 - B^0\right)^2 + C_2(B^0)^2\right]$$

$$= \frac{1}{2Q^0}\left[2DC_0 + C_1\left(\frac{C_2}{C_1+C_2}\right)^2\left(Q^0\right)^2 + C_2\left(\frac{C_1}{C_1+C_2}\right)^2\left(Q^0\right)^2\right]$$

$$= \frac{1}{2Q^0}\left[2DC_0 + \frac{C_1C_2}{C_1+C_2}\left(Q^0\right)^2\right]$$

$$= \frac{1}{2Q^0}\left[2DC_0 + 2DC_0\right] \quad \text{using (7)}$$

$$= 2DC_0\sqrt{\frac{C_1C_2}{2DC_0(C_1+C_2)}} \text{ or } \sqrt{\frac{2DC_0C_1C_2}{C_1+C_2}} \tag{8}$$

Note: When shortages are prohibited, viz., when $C_2 \to \infty$, (7) and (8) become $Q^0 = \sqrt{\frac{2DC_0}{C_1}}$ and $C^0 = \sqrt{2DC_0C_1}$, which have been derived in model I.

11.6 EOQ FOR MANUFACTURING MODEL (FOR WHICH THE DELIVERY RATE IS FINITE) WITH SHORTAGE (MODEL IV)

The manufacturing model with shortages can be represented graphically as shown here:

$D, Q, B, C_0, C_1, C_2, A$ and U are assumed as in models II and III.

Let t_1 be the time when inventory is building up from zero level at a constant rate of $(A - U)$ units per day.

Let t_2 be the time when there is no replenishment and the inventory is decreasing at the rate of U units per day until the inventory level reaches zero again.

Let t_3 be the time when shortage starts at the rate of U units per day until a new order starts coming in.

Let t_4 be the time when shortages are satisfied at the rate of $(A - U)$ units per day until the inventory level become zero again.

During the arrival period $(t_1 + t_4)$ when $(Q - B)$ are received, the number of units used $= (t_1 + t_4)\,U$.

$\therefore$ Maximum inventory level that can accumulate over the time $(t_1 + t_2) = Q - B - (t_1 + t_4)$ $U = Q_M$, say

$\therefore$ Total inventory over the time period $t =$ Area of triangle OXY

$$= \frac{1}{2}\left[Q - B - (t_1 + t_4)U\right](t_1 + t_2)$$

or
$$\frac{Q_M}{2}(t_1 + t_2)$$

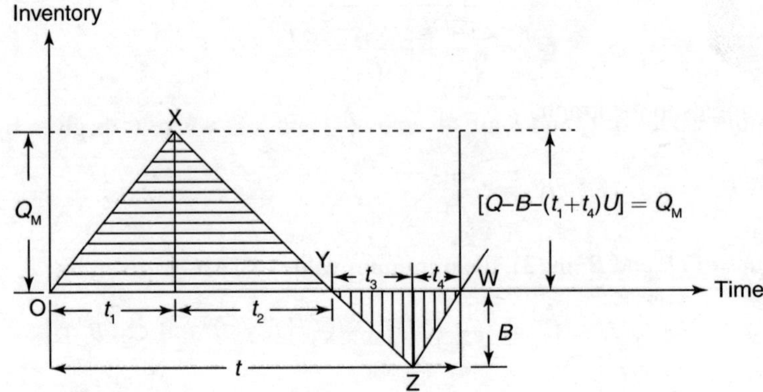

∴ Average inventory level at any time over the time period t

$$= \frac{1}{2}\left[Q - B - (t_1 + t_4)U\right]\frac{(t_1 + t_2)}{t}$$

or

$$\frac{Q_M}{2}\frac{(t_1 + t_2)}{t}$$

∴

$$\text{AIHC }(Q, B) = \frac{C_1}{2}\left[Q - B - (t_1 + t_4)U\right]\frac{(t_1 + t_2)}{t}$$

or

$$\frac{C_1 Q_M}{2}\frac{(t_1 + t_2)}{t} \tag{1}$$

Total back orders (shortages) over the time period $t = \frac{1}{2}B(t_3 + t_4)$

∴ Average back orders at any time over the time period $t = \frac{B}{2}\frac{t_3 + t_4}{t}$

∴

$$\text{ABC}(Q, B) = \frac{C_2}{2}\frac{B(t_3 + t_4)}{t} \tag{2}$$

As usual

$$\text{AOC} = \frac{D}{Q}C_0 \tag{3}$$

Now

$$Q_M = Q - B - \left(\frac{Q_M}{A-U} + \frac{B}{A-U}\right)U,$$

since

$$\frac{Q_M}{t_1} = \text{slope of } OX = \text{slope of } ZW = \frac{B}{t_4}$$

∴

$$Q_M\left(1 + \frac{U}{A-U}\right) = Q - B\left(1 + \frac{U}{A-U}\right)$$

or

$$\frac{1}{\lambda}Q_M = Q - \frac{1}{\lambda}B$$

or

$$Q_M = \lambda Q - B,$$

where $$\lambda = \frac{A-U}{A}$$

From similar triangles in the graph,

$$\frac{t_1+t_2}{Q_M} = \frac{t_3+t_4}{B} = \frac{t}{Q_M+B}$$

$\therefore$ $$\frac{t_1+t_2}{t} = \frac{Q_M}{Q_M+B}$$

and $$\frac{t_3+t_4}{t} = \frac{B}{Q_M+B}$$

viz., $$\frac{t_1+t_2}{t} = \frac{\lambda Q - B}{\lambda Q}$$

and $$\frac{t_3+t_4}{t} = \frac{B}{\lambda Q} \qquad (4)$$

Now TAIC (*Q*, *B*) = AOC + AIHC + ABC

viz., $$C\,(Q, B) = \frac{D}{Q}C_0 + \frac{C_1}{2}\cdot\frac{(\lambda Q - B)^2}{\lambda Q} + \frac{C_2}{2}\cdot\frac{B^2}{\lambda Q},$$

using (1), (2), (3) and (4)

C is minimum,

when $$\frac{\partial C}{\partial Q} = 0$$

and $$\frac{\partial C}{\partial B} = 0$$

viz.,

when $$-\frac{DC_0}{Q^2} + \frac{C_1}{2\lambda}\left\{\frac{Q\cdot 2\lambda(\lambda Q - B) - (\lambda Q - B)^2}{Q^2}\right\} - \frac{C_2 B^2}{2\lambda Q^2} = 0 \qquad (5)$$

and $$-\frac{C_1}{2\lambda Q}\cdot 2(\lambda Q - B) + \frac{C_2}{2\lambda Q}\cdot 2B = 0 \qquad (6)$$

viz., when $$C_1\left(\lambda^2 Q^2 - B^2\right) - C_2 B^2 = 2\lambda D C_0. \qquad (5)'$$

and $$B = \frac{\lambda C_1 Q}{C_1 + C_2} \qquad (6)'$$

Using (6)′ in (5)′, we get $$C_1 \lambda^2 Q^2 = 2\lambda D C_0 + \frac{\lambda^2 C_1^2 Q^2}{C_1 + C_2}$$

viz., $$\frac{\lambda^2 C_1 C_2}{C_1 + C_2}\cdot Q^2 = 2\lambda C_0 D$$

Thus, $$Q^0 = \sqrt{\frac{2DC_0(C_1+C_2)}{\lambda C_1 C_2}}$$

and $$B^0 = \frac{\lambda C_1}{C_1+C_2} Q^0,$$

where $$\lambda = \frac{A-U}{A}.$$

Note: When the delivery rate is infinite, viz., when $A \to \infty$, we see that $\lambda \to 1$. In this case, $Q^0 = \sqrt{\frac{2DC_0(C_1+C_2)}{C_1 C_2}}$ and $B^0 = \frac{C_1}{C_1+C_2} Q^0$ which are the same as the results (7) and (6)′ obtained for model III.

WORKED EXAMPLES 11(A)

Example 1

The demand for a certain item is 40,000 units per year. The inventory holding cost per unit is ₹ 5/- per year and the cost of one procurement is ₹ 640/-. No shortages are allowed and the replacement rate is instantaneous. Determine (i) the economic order quantity, (ii) the number of orders per year, (iii) the time between orders and (iv) total cost per year when the cost of one unit is ₹ 2.5.

Solution

$$D = 40{,}000 \text{ units/year}$$
$$C_0 = \text{ordering (procurement) cost/order} = 640$$
$$C_1 = 5\text{/unit/year}$$

(i) EOQ is given by $Q^0 = \sqrt{\frac{2DC_0}{C_1}}$.

$$= \sqrt{\frac{2 \times 40{,}000 \times 640}{5}} = 3200 \text{ units}$$

(ii) n = the number of orders per year $= \frac{D}{Q^0} = \frac{40{,}000}{3200} = 12.5$

(iii) $t^0 = \frac{Q^0}{D} = \frac{3200}{40{,}000}$ or 0.08 year or 29.2 days

(iv) Total cost per year = Cost of the items used + minimum TAIC

$$= 40{,}000 \times 2.5 + \sqrt{2DC_0C_1}$$
$$= 1{,}00{,}000 + \sqrt{2 \times 40{,}000 \times 640 \times 5}$$
$$= 1{,}00{,}000 + 16{,}000$$
$$= ₹1{,}16{,}000/-$$

Example 2

The production is instantaneous for a certain item. The storage cost of one item is ₹ 4/- per month and the set up cost is ₹ 250/- per run. If the demand of the item is 500 units per month, find the optimum number of items to be produced per set-up and hence, determine the total cost of storage and set up per month.

Solution

We will assume that one entire period of time is 1 month.

$$D = 500 \text{ units/month;}$$

$$C_0 = \text{set-up cost} = 250/\text{run.}$$

$$C_1 = \text{inventory cost/unit/month} = 4$$

$$\text{EOQ} = Q^0 = \sqrt{\frac{2DC_0}{C_1}}$$

$$= \sqrt{\frac{2 \times 500 \times 250}{4}} = 250$$

$$t^0 = \frac{Q^0}{D}$$

$$= \frac{250}{500} = \frac{1}{2} \text{ month.}$$

viz., the production must be done twice a month

The minimum total cost of storage and set up

per month $= C^0 = \sqrt{2DC_0C_1} = \sqrt{2 \times 500 \times 250 \times 4} = ₹\,1000/-$

Example 3

A company can manufacture and market 2500 machine (spare) parts per year. Raw materials are available when required. The raw materials for each spare part cost ₹ 500/-. It is estimated that the annual inventory holding cost of the raw materials for each spare part is 20% of the cost of the raw materials for each spare part. In addition, a cost of ₹ 200/- is incurred each time an order for raw materials is placed. What are EOQ and TAIC? Suppose the supplier of raw materials agrees to give the company a 5% discount, if it buys a 6 months' supply of raw materials each time an order is placed. Should the company accept the offer?

Solution

$$D = 2500/\text{year;}$$

$$C_0 = 200/\text{order;}$$

$$C_1 = 20\% \text{ of cost of one part}$$

$$= \frac{1}{5} \times 500 = 100 \text{ /unit/year}$$

$$\text{EOQ} = Q^0 = \sqrt{\frac{2DC_0}{C_1}}$$

$$= \sqrt{\frac{2\times 2500\times 200}{100}} = 100$$

Total amount of cost corresponding to EOQ ordering policy is given by $(\text{TC})_1$ = Cost of materials for 2500 spare parts + TAIC

$$= 2500\times 500 + \left(\frac{1}{2}Q^0C_1 + \frac{DC_0}{Q^0}\right)$$

$$= 12,50,000 + \frac{1}{2}\times 100\times 100 + \frac{2500\times 200}{100}$$

$$= 12,50,500 + 5000 + 5000$$

$$= ₹12,60,000/-$$

Total amount of cost corresponding to discount offer acceptance

$$= (\text{TC})_2 = 2500\times 475 + \frac{1}{2}\times 1250\times\frac{1}{5}\times 475 + \frac{2500\times 200}{1250}$$

$$= 11,87,500 + 59,375 + 400$$

$$= ₹\,12,47,275/-$$

Since $(\text{TC})_2 < (\text{TC}_1)_1$, the proposed discount offer should be accepted by the company, as it results in a profit of ₹ 12,725/- .per year.

Example 4

A company purchases 9000 parts of a machine for its annual requirements, ordering a one month usage at a time. Each part costs ₹ 200/-. The ordering cost per order is ₹ 150/- and the inventory holding charges are 15% of the cost of each part per year. Can the company follow a more economical purchasing policy? Why?

Solution

Present purchase policy;

$Q_P = \dfrac{D}{n_P}$, where n_p is the number of orders made in one year

$$= \frac{9000}{12} = 750 \text{ units.}$$

$$\text{Annual ordering cost (AOC)} = 12\times 150 = ₹\,1800/-$$

C_1 = 15% of the cost of one part $\dfrac{15}{100}\times 200 = ₹\,30/-$;

$$\text{AIHC} = \frac{1}{2}Q_P \times C_1 = ₹\,11{,}250/\text{-}$$

$\therefore$

$$\text{TAIC} = \text{AOC} + \text{AIHC} = ₹\ 13{,}050/\text{-}$$

EOQ purchase policy;

$$Q^0 = \sqrt{\frac{2DC_0}{C_1}}$$

$$= \sqrt{\frac{2 \times 9000 \times 150}{30}} = 300$$

$$t^0 = \frac{Q^0}{D}$$

$$= \frac{300}{9000} = \frac{1}{30}\text{ year;}$$

$$n = 30 \text{ per year}$$

$$\text{Min. TAIC} = C^0 = \sqrt{2DC_0C_1}$$

$$= \sqrt{2 \times 9000 \times 150 \times 30} = ₹\,9000/\text{-}$$

If the EOQ purchasing policy is adopted, the company has to place 30 orders per year, but the TAIC is only ₹ 9000/-

$\therefore$ The company should follow the EOQ policy, thereby saving ₹ 13,050/- – ₹ 9000/- = ₹ 4,050/- per year.

Example 5

In the purchasing model with no shortages, if the set up cost consists of a fixed part C_0 and a variable part QC_0', where C_0' is the set up cost per item produced, show that there is no change in the economic lot size produced due to the revised set up cost.

SOLUTION

The revised set up C_0^N is given by $C_0N = C_0 + QC_0'$

The annual inventory holding cost (AIHC) is given by

$$C = \frac{1}{2}QC_1 + \frac{D}{Q}C_0^N$$

or

$$\frac{1}{2}QC_1 + \frac{D}{Q}\left(C_0 + QC_0'\right)$$

For minimum C,

$$\frac{dC}{dQ} = 0$$

viz., $$\frac{1}{2}C_1 - \frac{DC_0}{Q^2} + 0 = 0$$

viz., $$Q = Q^0 = \sqrt{\frac{2DC_0}{C_1}}$$

and $$\left(\frac{d^2C}{dQ^2}\right)_{Q=Q^0} = \frac{2DC_0}{\left(Q^0\right)^3} > 0.$$

$\therefore$ There is no change in the value of Q^0 due to the revised set up cost C_0^N.

Example 6

A contractor has to supply 8000 bearings per day to an automobile manufacturer. When he starts a production run, he finds that he can produce 20,000 bearings per day. The cost of holding a bearing in stock for one year is ₹ 1/- and the set up cost of a production run is ₹ 200/-. Find the optimum number of bearings to be produced per run. How frequently should the production run be made?

Solution

$$A = \text{arrival rate per day} = 20{,}000$$

$$U = \text{use rate per day} = 8000$$

$$C_S = ₹\ 200;\ C_1 = ₹\ 1/\text{- per year};$$

$$\therefore \quad C_0' = ₹\left(\frac{1}{365}\right) \text{per day.}$$

$$Q^0 = \sqrt{\frac{2C_SUA}{C_1'(A-U)}}$$

$$= \sqrt{\frac{2 \times 200 \times 8000 \times 20{,}000 \times 365}{(20{,}000 - 8000)}}$$

$$= \sqrt{\frac{400 \times 8000 \times 20{,}000}{12{,}000} \times 365}$$

$$= \sqrt{\frac{20{,}000 \times 3200 \times 365}{12}}$$

$$= 4000 \times \sqrt{121.6} \simeq 44{,}000 \text{ bearings}$$

$$t^0 = \frac{Q^0}{U}$$

$$= \frac{44{,}000}{8000} = 5.5 \text{ days.}$$

Example 7

An item is produced at the rate of 100 per day. The demand occurs at the rate of 75 items per day. If the set up cost is ₹ 900/- per run and the holding cost is ₹ 1.5 per unit of the item per day, find the economic lot size for one run, assuming that shortages are not permitted. Also find the time of the cycle and minimum total cost for one run.

SOLUTION

$$C_S = ₹\ 900/\text{-};$$

$$C_1' = ₹\ 1.5;$$

$$A = 100;\ U = 75$$

This is a problem following the manufacturing model with no shortage. EOQ is given by

$$Q^0 = \sqrt{\frac{2C_0UA}{C_1'(A-U)}}$$

$$= \sqrt{\frac{2\times900\times75\times100}{1.5\times(100-75)}} = 600 \text{ items}$$

$$t^0 = \sqrt{\frac{2C_0A}{C_1'U(A-U)}}$$

$$= \sqrt{\frac{2\times900\times100}{1.5\times75\times25}} = 8 \text{ days}$$

viz., 600 items must be produced every 8 days. Minimum total inventory holding cost per day

$$= \sqrt{2C_0C_1'U(A-U)/A}$$

$$= \sqrt{2\times900\times\frac{3}{2}\times75\times25\times\frac{1}{100}} = ₹\ 225/\text{-}$$

∴ Minimum total holding cost for one cycle = ₹ 2000/-

Example 8

The demand for an item in a company is 24,300 units per year and the company can produce the item at a rate of 2500 per month. The cost of one step up is ₹ 800/- and the inventory holding 1 unit per month is ₹ 1.25. Determine the optimal manufacturing quantity and the total cost per year, assuming the cost of 1 unit as ₹ 5/-. Also find the maximum inventory level.

SOLUTION

Let us take 1 month as the unit of time (instead of 1 day); U = 2025/month; A = 2500/month; C_0 = ₹ 800/-; C_1' = ₹ 1.25/unit/month.

$$Q^0 = \sqrt{\frac{2C_0UA}{C_1'(A-U)}}$$

$$= \sqrt{\frac{2\times 800\times 2025\times 2500}{(5/4)\times 500}}$$

$$= 80\times 45 = 3600 \text{ units per month}$$

$$\text{Total annual cost} = 24,300\times 5 + 12\times\sqrt{2C_0C_1'U\left(1-\frac{U}{A}\right)}$$

$$= 1,21,500 + 12\times\sqrt{2\times 800\times\frac{5}{4}\times 2025\times\frac{475}{2500}}$$

$$= 1,21,500 + 12\times 877.5$$

$$= 1,21,500 + 10,530$$

$$= \text{Rs } 1,32,030/-$$

$$\text{Maximum inventory level} = Q^0 = \left(1-\frac{U}{A}\right)$$

$$= 3600\times\left(1-\frac{2025}{2500}\right) = 2916 \text{ units}$$

Example 9

The demand of an item is uniform at a rate of 300 units per month. The fixed cost is ₹ 200/- each time a production run is made and the variable cost of production is ₹ 10/- per unit. If the inventory carrying cost is ₹ 4/- per month and the shortage cost is ₹ 12/- per month, determine how often a production is to be made and of what size it should be. Find also the optimum back orders to be allowed.

Solution

$$D = 300/\text{month};$$

$$C_0 = ₹\ 200/- \text{ per production};$$

$$C_0' = ₹\ 10 \text{ per unit.}$$

C_0' has no relevance as far as the computation of Q^0 is concerned (Refer to Worked Example 5)

$$C_1 = ₹\ 4/\text{-per month};$$

$$C_2 = ₹\ 12/- \text{ per month}$$

$$Q^0 = \sqrt{\frac{2DC_0(C_1+C_2)}{C_1C_2}}$$

$$= \sqrt{\frac{2\times 300\times 200\times 16}{48}}$$

$$= \sqrt{2 \times 100 \times 200} = 200 \text{ units}$$

$$t^0 = \frac{Q^0}{D} = \frac{200}{300} \text{ of a month or 20 days}$$

$$B^0 = \frac{C_1}{C_1 + C_2} Q^0 \text{ units.}$$

$$= \frac{4}{4+12} \times 200 = 50$$

Example 10

The demand for an item is 15,000 units per year. The holding cost is ₹ 2/- per unit per year and the cost of shortages is ₹ 6/- per unit per year. The production cost is ₹ 750 per production.

Assuming that the replacement rate is instantaneous, determine the optimum order quantity and the number of back orders to be allowed.

Solution

$D = 15{,}000$ per year;

$C_0 =$ ₹ 750/- per production

$C_1 =$ ₹ 2/- per year;

$C_2 =$ ₹ 8/- per year.

$$Q^0 = \sqrt{\frac{2DC_0(C_1 + C_2)}{C_1C_2}}$$

$$= \sqrt{\frac{2 \times 15000 \times 750 \times 10}{16}}$$

$$= \frac{15000}{4} = 3750 \text{ units.}$$

$$B^0 = \frac{C_1}{C_1 + C_2} Q^0$$

$$= \frac{2 \times 3750}{10} = 750 \text{ units.}$$

Example 11

The demand for an item in a company is 19,200 units per year and the company can produce the item at a rate of 3200 per month. The cost of one set up is ₹ 500/- and the holding cost per unit is ₹ 10/- per year and the shortage cost of one unit is ₹ 20/- per year. Determine the optimum manufacturing quantity and the number of shortages. Also determine the time interval between consecutive production runs.

Solution

This is a problem following the manufacturing model with shortage.

$$D = 19{,}200 \text{ per year or } 1600 \text{ per month, viz.,}$$

$$U = 1600/\text{month}$$

$$A = 3200/\text{month};$$

$$C_0 = ₹\ 500/\text{- per set-up}$$

$$C_1 = ₹\ 10/\text{- per year};$$

$$C_2 = ₹\ 20/\text{- per year}$$

$$Q^0 = \sqrt{\frac{2C_0D(C_1+C_2)}{C_1C_2\lambda}},$$

where $\lambda = 1 - \dfrac{U}{A}$

$$= \sqrt{\frac{2\times 500\times 19{,}200\times 30}{200\times\left(1-\dfrac{1600}{3200}\right)}}$$

$$= \sqrt{2\times 5\times 19{,}200\times 30}$$

$$= \sqrt{64\times 300^2} = 2400 \text{ units}$$

$$B^0 = \frac{\lambda C_1 Q^0}{C_1+C_2}$$

$$= \frac{1}{2}\times\frac{10\times 2400}{(10+20)} = 400 \text{ units}$$

$$t^0 = \frac{Q^0}{D}$$

$$= \frac{2400}{19200} = \frac{1}{8} \text{ year or 1.5 months.}$$

Example 12

Solve Example 11 with the following data:

Demand = 16,000/year;
Production = 32,000/year;
Set up cost = ₹ 900/-;
Holding cost = ₹ 3/- per year;
Shortage cost = ₹ 15/- year.

Solution

Given:

$$D = 16{,}000 \text{ or } U = 1333.3/\text{month};$$

$$A = 32{,}000/\text{year or } 2666.6/\text{month}.$$

$$C_0 = ₹\ 900/\text{- per production run;}$$

$$C_1 = ₹\ 3/\text{- year; } C_2 = ₹\ 15/\text{- year}$$

$$Q^0 = \sqrt{\frac{2C_0D(C_1 + C_2)}{C_1C_2\left(1 - \frac{U}{A}\right)}}$$

$$= \sqrt{\frac{2 \times 900 \times 16{,}000 \times 18}{45 \times \left(1 - \frac{1}{2}\right)}}$$

$$= \sqrt{4 \times 20 \times 16{,}000 \times 18}$$

$$= \sqrt{14400 \times 1600} = 4800 \text{ units.}$$

$$B^0 = \frac{\lambda C_1 Q^0}{C_1 + C_2}$$

$$= \frac{1}{2} \times \frac{3 \times 4800}{18} = 400 \text{ units}$$

$$t^0 = \frac{Q^0}{D} = \frac{4800}{16000} \text{ or } \frac{3}{10} \text{ year.}$$

EXERCISE 11(A)

Part A (Short answer questions)

1. What is the objective of inventory control?
2. State the advantages and disadvantages of keeping inventories.
3. What is meant by shortage in inventory theory?
4. Distinguish between deterministic and stochastic inventory models.
5. What do you mean by EOQ?
6. What is meant by purchasing model and manufacturing model in inventory theory?
7. What is 'lead time' in the context of inventory control?
8. Define the various costs associated with inventory management.

Part B

9. A manufacturer has to supply his customers 1200 units of his product per year. Shortages are not allowed and the storage cost amounts to ₹ 6/- per unit per year. The set up cost per run is ₹ 400/-. Find the optimum run size and the minimum average yearly cost.
10. An oil engine manufacturer purchases lubricants at the rate of ₹ 200/- per piece from a vendor. The requirement of these lubricants is 2000 per year. What should be the order quantity per order, if the cost per placement of an order is ₹ 100/- and the inventory carrying charges per year is only 20 paise

per rupee. Find also the optimum total inventory cost and the interval between two consecutive orders.

11. A company can manufacture and market 6400 spare parts for a particular brand of machine per year. The raw materials that are required for production are available in plenty, and cost ₹ 50/- for each part. The annual inventory holding cost of the raw material for each spare part is 20% of its cost. A cost of ₹ 180/- is incurred each time an order is placed for raw materials. What are EOQ and TAIC? Suppose the supplier of raw materials agrees to give the company a 10% discount, if it buys a 6 month's supply of raw materials each time an order is placed, should the company accept the offer?
12. A company purchases 9000 items per year for use in its production shop. The unit cost is ₹ 50/-, the holding cost is ₹ 4/- per month and the cost of making a purchase is ₹ 240/-. If no shortages are allowed, determine the following:

    (a) optimal order quantity,
    (b) optimal total annual cost,
    (c) number of orders per year and
    (d) time between orders

13. A company uses 24,000 units of a raw material annually, which costs ₹ 5/- per unit. Placing each order costs ₹ 120/- and the carrying cost is 5% per year of the average inventory. Find the economic lot size and the total annual inventory cost.
14. A contractor has to supply 15,000 units per day. He can produce 18,000 units per day. The cost of holding a unit in stock is 10 paise per day and the set up cost is ₹ 200/-. How frequently and of what size should the production runs be made?
15. A contractor has to supply 10,000 bearings per day to an automobile manufacturer. He finds that when he starts a production run, he can produce 25,000 bearings per day. The cost of holding a bearing in stock for one year is 20 paise and the set up cost of production is ₹ 180/-. Find the optimal number of bearings to be produced per run. How frequently should the production run be made?
16. Find the optimal production lot size and the optimal cycle length when production rate is 8000 units per year, annual demand is 2000 units, set up cost is ₹ 300/- per run and holding cost is ₹ 1.60 per unit year.
17. A manufacturer has to supply his customer 48,000 units of his product per year. This demand is fixed and known. The customer has no storage space and so the manufacturer has to ship a day's supply each day. If the manufacturer fails to supply, the penalty is ₹ 1.5/- per unit per month. The inventory holding cost amounts to ₹ 1/-per unit per month and the set up cost is ₹ 300/- per production run. Find (a) the optimum lot size for the manufacturer, (b) the optimum shortage for the customer and (c) the total monthly inventory cost.
18. A contractor undertakes to supply diesel engines to a truck manufacturer at a rate of 25 per day. He finds that the cost of holding a completed engine in stock is ₹ 75/- per month and there is a clause in the contract penalizing him ₹ 10/-per engine (unit) per day for missing the scheduled delivery rate. Production of engines is in batches and each time a new batch is started, there are set up costs of ₹ 10,000/-. How frequently should the batches be started and what should be the initial inventory level at the time each batch is completed?
19. The demand for an item in a company is 720 units per year. The company can produce the item at a rate of 120 per month. The cost of one set up is ₹ 100/- and the holding cost of 1 unit per year is ₹ 6/-. The shortage cost of one unit is ₹ 30/- per year. Determine the optimum manufacturing quantity and the number of shortages. Find also the time between set ups.

20. The demand for an item in a company is 10,000 units per year. The company can produce the item at a rate of 20,000 units per year. The cost of one production run is ₹ 400/-. The holding and shortage costs of one unit are ₹ 5/- and ₹ 20/- respectively per year. Determine the optimum production quantity and optimum number of back orders. Find also the time between consecutive production runs.

11.7 INVENTORY MODEL WITH PRICE BREAKS (MODEL V)

11.7.1 Introduction

In the purchasing model with no shortages (viz., Model I), the unit cost of the item delivered or produced was assumed to be a constant, irrespective of the number of items delivered or produced. In real situations, often price discounts in the form of price breaks are offered for bulk purchases with a view to maximise sales. This results in the cost of the item varying based on the quantity purchased.

For inventory models with price breaks, the following assumptions are taken to hold good:

1. Demand is known and constant.
2. Shortages are not permitted.
3. Production or delivery rate is instantaneous.

The computational procedure for the optimum purchase quantity in the case of a purchasing model with *n* price breaks (which is described as follows) is given here:

Range of quantity	Purchase cost per unit
$0 \le Q_1 < b_1$	K_1
$b_1 \le Q_2 < b_2$	K_2
-	-
$b_{n-2} \le Q_{n-1} < b_{n-1}$	K_{n-1}
$b_{n-1} \le Q_n$	K_n

where $b_1, b_2, \ldots b_{n-1}$ are the quantities corresponding to which price breaks occur.

Step 1: Compute $Q_n^0 = \sqrt{\dfrac{2DC_0}{FK_n}}$, where *F* is the fraction of the unit cost (*F* will be usually given).

If $Q_n^0 \ge b_{n-1}$, the optimum purchase quantity is Q_n.

Step 2: If $Q_n^0 < b_{n-1}$, compute $Q_{n-1}^0 = \sqrt{\dfrac{2DC_0}{FK_{n-1}}}$

Step 3: If $Q_{n-1}^0 < b_{n-2}$, compute $Q_{n-2}^0 = \sqrt{\dfrac{2DC_0}{FK_{n-2}}}$

Step 4: Proceed in this way, until $Q_{n-j}^0 \ge b_{n-(j+1)}$ $\left(\text{Using } Q_{n-j}^0 = \sqrt{\dfrac{2DC_0}{FK_{n-j}}}\right)$ for $0 \le j \le n-1$, where $b_0 = 0$

Step 5: Compare TC (Q_{n-j}^0), TC (b_{n-j}), TC (b_{n-j+1}), … , TC (b_{n-1})

Step 6: Then, the optimum order quantity (OOQ) $= Q^0$ is that value of *b*, for which TC is the least.

11.7.2 Inventory Control with Uncertain Demand

In practice, the demand of the item cannot be definitely assessed beforehand and hence is uncertain due to its fluctuations on both sides. Moreover, in many situations, it is observed that neither the sale rate of the item nor the lead time is constant throughout the year. To tackle these uncertainties, an extra stock, called *buffer stock* or *safety stock* is maintained. In other words, the buffer stock is the extra stock maintained in excess of the demand during the lead time. Since the lead time is also uncertain, the buffer stock is calculated as follows:

Optimum buffer stock = (Maximum lead time – Normal lead time) × Sale rate r of the item

On the other hand, if the buffer stock is not maintained, we proceed as follows: The inventory requirement during the lead time L is computed as Lr. As soon as the stock reaches the level Lr, a fresh order for the quantity Q is placed. This level of the inventory, when a fresh order for Q is placed, is called *re-order level* (ROL). This way of determining the ROL results in shortages for nearly 50% of the time. To avoid this situation, an ad hoc buffer stock (B) is maintained and the fresh order for Q is placed, when the stock reaches ($B + Lr$) level. Thus, ROL $= B + Lr$

viz., Re-order level = Buffer stock + Normal lead time sale or consumption.

If EOQ system of ordering (producing) is followed, the situation may be explained graphically as follows:

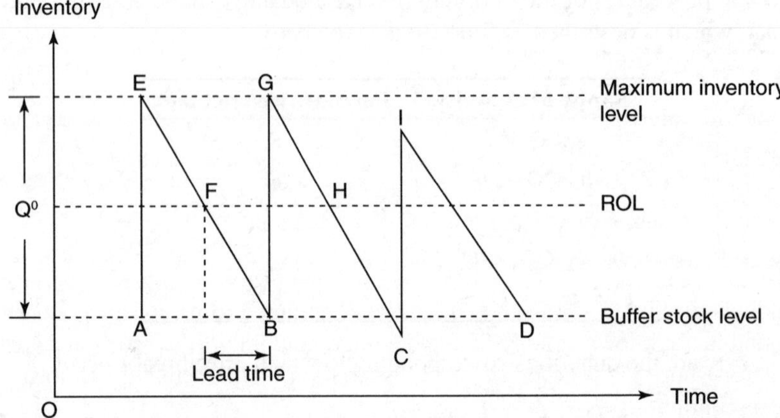

From the graph, we note that a supply of the items equal to EOQ ($= Q^0$) is received at the time point A and the inventory level reaches the maximum at E. The items are then sold or used and at time point F, when the stock reaches the ROL, a fresh order for quantity EOQ ($= Q^0$) is placed, while sale or use is continued. The supply of items for which order was placed at F is received at the point B and the stock reaches the maximum level at G. At the point C, the supply of items ordered at H has not been received and hence, the items available in the buffer stock are sold or used. As the delayed supply of items will be used to replenish the buffer stock partly, the inventory level will not be maximum at this stage.

11.7.3 Multi-item Deterministic Inventory Model

Sometimes, the inventory has to accommodate n different items, each with infinite delivery (instantaneous production) rate and no lead time, subject to the constraint that the sum of the average numbers of all items in the inventory does not exceed K.

We assume that, for the ith item ($i = 1, 2, \ldots, n$), the uniform annual demand (or demand per unit period) is D_i, the annual inventory holding cost is C_{1i} per unit, the ordering cost per order is C_{0i} and shortages are not allowed.

Then, proceeding as in Model I, the total annual inventory holding cost for all the n items is given by

$$C = \sum_{i=1}^{n} \frac{D_i}{Q_i} C_{0i} + \frac{1}{2} Q_i \, C_{1i},$$

where Q_i is the number of the ith item ordered each time whenever the order is placed.

Also, as in Model I, Q_i^0 (viz., the optimum value of Q_i) is given by $Q_i^0 = \sqrt{\frac{2D_i C_{0i}}{c_{1i}}}$ $(i = 1, 2, \ldots, n)$,

but subject to the constraint $\frac{1}{2}\sum_{i=1}^{n} Q_i^o \le k$.

If this constraint is satisfied by the values of Q_i^0 obtained , they are the required EOQ values.

If $\frac{1}{2}\sum_{i=1}^{n} Q_i^0 > k$, we make use of the Lagrange multiplier method to find the minimum values of Q_i, using the Lagrange function

$$\mathrm{L} = \sum_{i=1}^{n} \left(\frac{Di}{Qi} C_{0i} + \frac{1}{2} Q_i c_{1i} \right) + 2\lambda \left\{ \frac{1}{2} \sum_{i=1}^{n} Q_i - k \right\}$$

The optimum values of Q_i are given by $\frac{\partial L}{\partial Q_i} = 0$ and $\frac{\partial L}{\partial \lambda} = 0$ viz., $Q_i^0 = \sqrt{\frac{2D_i C_{0i}}{C_{1i} + 2\lambda}}$ and $\sum_{i=1}^{n} Q_i^0 = 2k$. To find the value of λ in the formula for Q_i^0 , we use the second condition, proceeding by trial and error.

11.7.4 Inventory Models with Random Demands

In the inventory models discussed so far, the demand was assumed to be constant and known. In real situations, the demand is not known exactly and is obtained randomly from previous experience of the company or production shop, viz., the probability distribution of the demand is known, whether the demand is discrete or continuous.

11.7.4.1 Single Period Model

Let us derive an expression for the optimal order quantity Q^0 at the beginning of a single period in order to minimise the expected value of the total cost (i) when the demand X is a discrete random variable.

The following assumptions are made:

$p(x)$ = the probability function of X,

viz., $P(X = x) = p(x)$; $x = 1, 2, \ldots, \infty$

Q = order quantity

C = cost per item produced or purchased

C_0 = ordering cost

C_1 = inventory holding cost per unit (or) cost of oversupplying per unit.

C_2 = shortage cost per unit (or) cost of undersupplying per unit.

$$TC(Q) = \begin{cases} C_0 + CQ + C_1(Q - X), \text{ if } Q \geq X \\ C_0 + CQ + C_2(X - Q), \text{ if } Q < X \end{cases}$$

$$\therefore \qquad E\{TC(Q)\} = C_0 + CQ + C_1 \sum_{X=1}^{Q} (Q - x)\, p(x) + C_2 \sum_{x=Q+1}^{\infty} (x - Q) p(x) \qquad (1)$$

Now $Q = Q^0$ is the required OOQ, if it is the local minimum,
viz., if $E\{\text{TC}(Q)\} < E\{\text{TC}(Q + 1)\}$ and $E\{\text{TC}(Q)\} < E\{\text{TC}(Q - 1)\}$

$$E\{TC(Q+1)\} = C_0 + C(Q+1) + C_1 \sum_{X=1}^{Q+1} (Q + 1 - x)\, p(x) + C_2 \sum_{x=Q+2}^{\infty} (x - Q - 1) p(x), \text{ by (1)}$$

$$= C_0 + C(Q+1) + C_1 \sum_{X=1}^{Q} (Q + 1 - x)\, p(x) + C_2 \sum_{x=Q+1}^{\infty} (x - Q - 1) p(x),$$

on omitting the last zero term in the first Σ, including the first Σ and including the first zero term in the second Σ.

$$= E\{TC(Q)\} + C + C_1 \sum_{x=1}^{Q} p(x) - C_2 \sum_{x=Q+1}^{\infty} p(x)$$

$$= E\{TC(Q)\} + C + C_1 \sum_{x=1}^{Q} p(x) - C_2 \left\{1 - \sum_{x=1}^{\infty} p(x)\right\}$$

$$= E\{TC(Q)\} + (C_1 + C_2) \sum_{x=1}^{Q} p(x) + C - C_2 \qquad (2)$$

Proceeding similarly,

$$E\{TC(Q-1)\} = E\{TC(Q)\} - (C_1 + C_2) \sum_{x=1}^{Q-1} p(x) - C + C_2 \qquad (3)$$

From (2), $E\{\text{TC}(Q)\} < E\{\text{TC}(Q + 1)\}$, if $(C_1 + C_2) \sum_{x=1}^{Q} p(x) + (C - C_2) > 0$

viz., if

$$\sum_{x=1}^{Q} p(x) > \frac{C_2 - C}{C_1 + C_2} \qquad (4)$$

From (3), $E\{\text{TC}(Q)\} < E\{\text{TC}(Q - 1)\}$, if $(C_1 + C_2) \sum_{x=1}^{Q-1} p(x) + C - C_2 < 0$

viz., if

$$\sum_{x=1}^{Q-1} p(x) < \frac{C_2 - C}{C_1 + C_2} \qquad (5)$$

$\therefore$ Q is the OOQ, if $\sum_{x=1}^{Q-1} p(x) < \frac{C_2 - C}{C_1 + C_2} < \sum_{x=1}^{Q} p(x)$

viz., The OOQ is the value of Q which satisfies the condition $P\{X \le Q-1\} < \dfrac{C_2 - C}{C_1 + C_2} < P\{X \le Q\}$

Note: If we are interested in minimising only TAIC (instead of TC), the OOQ is given by $P\{X \le Q-1\} < \dfrac{C_2}{C_1 + C_2} < P\{X \le Q\}$

(ii) *When the demand X is a continuous random variable with pdf f(x); $1 \le x \le \infty$.*

In this case, Step 1 of the discrete demand case would become

$$E\{TC(Q)\} = C_0 + CQ + C_1 \int_1^Q (Q-x) f(x)\, dx + C_2 \int_Q^\infty (x-Q) f(x)\, dx \tag{1}$$

Noting that $\dfrac{d}{dt}\displaystyle\int_{a(t)}^{b(t)} \phi(x,t)\, dx = \int_{a(t)}^{b(t)} \frac{\partial \phi}{\partial t} dx + \phi(b,t)\frac{db}{dt} - \phi(a,t)\frac{da}{dt}$ and differentiating (1) w r t Q,

$$\frac{dE}{dQ} = C + C_1\left[\int_1^Q f(x)\, dx + (Q-Q) f(Q)\frac{dQ}{dQ} - (Q-1) f(1)\frac{d(1)}{dQ}\right]$$

$$+ C_2\left[\int_Q^\infty -f(x)\, dx + \left\{(x-Q) f(x)\frac{dx}{dQ}\right\}_{x \to Q} - (Q-Q) f(Q)\frac{dQ}{dQ}\right]$$

$$= C + C_1 \int_1^Q f(x)\, dx - C_2 \int_Q^\infty f(x)\, dx$$

$$= C + C_1 \int_1^Q f(x)\, dx - C_2\left[1 - \int_1^Q f(x)\, dx\right]$$

$$= (C - C_2) + (C_1 + C_2)\int_1^Q f(x)\, dx \tag{2}$$

$$\frac{d^2E}{dQ^2} = (C_1 + C_2)\left[\int_1^Q \frac{d}{dQ} f(x)\, dx + f(Q)\frac{dQ}{dQ} - f(1)\frac{d(1)}{dQ}\right]$$

$$= (C_1 + C_2) f(Q) \tag{3}$$

E is minimum, when $\dfrac{dE}{dQ} = 0$, i.e., $\displaystyle\int_1^Q f(x)\, dx = \frac{C_2 - C}{C_1 + C_2}$ and when $\dfrac{d^2E}{dQ^2} > 0$ which is true.

$\therefore$ The OOQ is the value of Q which satisfies the condition

$$\int_1^Q f(x)\, dx \text{ or } P\{X \le Q\} = \frac{C_2 - C}{C_1 + C_2}$$

Note: If the minimum of only TAIC is required, then the OOQ is given by $P\{X \le Q\} = \dfrac{C_2}{C_1 + C_2}$

WORKED EXAMPLES 11(B)

Example 1

The annual demand of a product is 10,000 units. Each unit costs ₹ 100/- if orders are placed in quantities below 400 units, but for orders of 400 or above, the cost per unit is ₹ 90/-. The annual inventory holding cost is 10% of the value of the item and the ordering cost is ₹ 45/- per order. Find the economic lot size.

Solution

Quantity	Unit cost in ₹
$0 \le Q_1 < 400$	$k_1 =$ ₹ 100/-
$400 \le Q_2$	$k_2 =$ ₹ 90/-

viz., the price break occurs at $b = 400$.

$$Q_2^0 = \sqrt{\frac{2DC_0}{C_1}} = \sqrt{\frac{2\times 10{,}000\times 45}{90\times 0.1}} = 316.5 \simeq 317 \text{ units}$$

As $Q_2^0 < b(=400)$, we compute Q_1^0

$$Q_1^0 = \sqrt{\frac{2\times 10{,}000\times 45}{100\times 0.1}} = 300$$

Now
$$\text{TC}(Q_1^0) = k_1 D + \frac{1}{2}Q_1^0 C_1 + \frac{DC_0}{Q_1^0}$$

$$= 100\times 10{,}000 + \frac{1}{2}\times 300\times 10 + \frac{10{,}000\times 45}{300}$$

$$= 10{,}00{,}000 + 1500 + 1500$$

$$= ₹\ 10{,}03{,}000/-$$

$$\text{TC}(b) = \text{TC}(400)$$

$$= 90\times 10{,}000 + \frac{1}{2}\times 400\times 9 + \frac{10{,}000\times 45}{400}$$

$$= 9{,}00{,}000 + 1800 + 1125$$

$$= ₹\ 9{,}02{,}925/-$$

Since $\text{TC}(b) < \text{TC}(Q_1^0)$, the optimum order quantity (OOQ) = 400 units.

Note: If $Q_2^0 > 400$, then the OOQ is Q_2^0. If $\text{TC}(b) > \text{TC}(Q_1^0)$, then the $\text{OOQ} = Q_1^0$

Example 2

Find the optimal order quantity when the demand is 200 units per month, the ordering cost is ₹ 100/- per order and the cost of storage is 2% of the unit cost per month. The price breaks are as follows:

Quantity	Unit cost in ₹
$0 \le Q_1 < 500$	$k_1 = 10.00$
$500 \le Q_2 < 750$	$k_2 = 9.25$
$750 \le Q_3$	$k_3 = 8.75$

Solution

The price breaks occur at $b_1 = 500$ and $b_2 = 750$.

$$Q_3^0 = \sqrt{\frac{2DC_0}{C_1}}$$

$$= \sqrt{\frac{2\times 200\times 100}{8.75\times 0.02}} \simeq \frac{200}{0.42} \simeq 476$$

Since $Q_3^0 < b_2$, we compute Q_2^0.

$$Q_2^0 = \sqrt{\frac{2\times 200\times 100}{9.25\times 0.02}} \simeq \frac{200}{0.43} \simeq 465$$

Since $Q_2^0 < b_1$, we compute Q_1^0.

$$Q_1^0 = \sqrt{\frac{2\times 200\times 100}{10\times 0.02}} \simeq \frac{200}{0.45} \simeq 444$$

Now
$$\text{TC}(Q_1^0) = \text{TC}(444) = k_1 D + \frac{1}{2}Q_1^0 C_1 + \frac{DC_0}{Q_1^0}$$

$$= 10\times 200 + \frac{1}{2}\times 444\times 0.2 + \frac{200\times 100}{444} = ₹\ 2089 \text{ nearly}$$

$$\text{TC}(b_1) = \text{TC}(500)$$

$$= 9.25\times 200 + \frac{1}{2}\times 500\times 0.185 + \frac{200\times 100}{500} = ₹1936 \text{ nearly}$$

$$\text{TC}(b_2) = \text{TC}(750)$$

$$= 8.75\times 200 + \frac{1}{2}\times 750\times 0.175 + \frac{200\times 100}{750} = ₹1842 \text{ nearly}$$

Since TC(b_2) is the least, the optimum order quantity is equal to $b_2 = 750$ units.

Example 3

Find the optimal order quantity for a product where the annual demand for the product is 500 units, the cost of storage per unit is 10% of the unit cost and the ordering cost per order is ₹ 200/-. The unit costs are given here:

Quantity	Unit cost in ₹
$0 \le Q_1 < 500$	$k_1 = 25$
$500 \le Q_2 < 1000$	$k_2 = 24$
$1000 \le Q_3 < 1500$	$k_3 = 23$
$1500 \le Q_4$	$k_4 = 22$

$b_1 = 500,\ b_2 = 1000,\ b_3 = 1500$

$$Q_4{}^0 = \sqrt{\frac{2DC_0}{C_1}} = \sqrt{\frac{2\times 500\times 200}{0.1\times 22}} \simeq 302;$$

$$Q_3{}^0 = \sqrt{\frac{2\times 500\times 200}{0.1\times 23}} \simeq 295;$$

$$Q_2{}^0 = \sqrt{\frac{2\times 500\times 200}{0.1\times 24}} \simeq 289;$$

$$Q_1{}^0 = \sqrt{\frac{2\times 500\times 200}{0.1\times 25}} \simeq 283.$$

Now
$$\mathrm{TC}\left(Q_1{}^0\right) = TC(283) = K_1 D + \frac{1}{2}Q_1{}^0 C_1 + \frac{DC_0}{Q_1{}^0}$$

$$= 500\times 25 + \frac{1}{2}\times 283\times 2.5 + \frac{500\times 200}{283} = ₹13,207 \text{ nearly}$$

$$\mathrm{TC}(b_1) = TC(500)$$

$$= 500\times 24 + \frac{1}{2}\times 500\times 2.4 + \frac{500\times 200}{500} = ₹12,800 \text{ nearly}$$

$$TC(b_2) = TC(1000)$$

$$= 500\times 23 + \frac{1}{2}\times 1000\times 2.3 + \frac{500\times 200}{1000} = ₹11,715 \text{ nearly}$$

$$TC(b_3) = TC(1500)$$

$$= 500\times 22 + \frac{1}{2}\times 1500\times 2.2 + \frac{500\times 200}{1500} = ₹11,292 \text{ nearly}$$

Since TC (b_3) is the least, the optimum order quantity is equal to $b_3 = 1500$ units.

Example 4

In a central cement godown, it takes about 12 days to get the stock after placing the order. Cement (600 tonnes) are despatched to the sales depots daily. Ten days consumption is to kept as ad hoc buffer stock. Calculate the inventory level when a fresh order is to be placed.

SOLUTION

Normal lead time = 12 days

Buffer stock = 10 days' consumption = 10×600 = 6000 tonnes.

ROL = $B + Lr$ = $6000 + 12 \times 600$ = 13,200 tonnes.

viz., when the inventory is 13,200 tonnes, a fresh order is to placed.

Example 5

A company uses 50,000 units of an item annually, each costing ₹ 10/-. Each order costs ₹ 90/- and inventory carrying costs 10% of the average annual value. Find the economic order quantity. If the company operates 300 days in a year, the procurement (lead) time is 15 days and the buffer stock is 750 units, find the re-order level, maximum, minimum and average inventories.

SOLUTION

C_0 = ₹ 90/- per order

C_1 = 10% of the value of each item

$= \frac{10}{100} \times 10$ = ₹ 1/- per year.

Total demand per year = 50,000.

$$\text{EOQ} = \sqrt{\frac{2DC_0}{C_1}}$$

$$= \sqrt{\frac{2 \times 50,000 \times 90}{1}} = 3000 \text{ units}$$

Lead time (L) = 15 days; Buffer stock (B) = 750 units

$$\text{Re-order level (ROL)} = L \times \frac{50,000}{300} + B \text{ [viz., Lead time use + } B\text{]}$$

$$= 15 \times \frac{50,000}{300} + 750 = 3250 \text{ units}$$

∴ Maximum inventory = EOQ + Buffer stock

= 3000 + 750 = 3750 units

Minimum inventory = Buffer stock = 750 units

$$\text{Average inventory} = \left[\frac{1}{2}(0+\text{EOQ})+\text{Buffer stock}\right] \text{or} [\text{Average of max. and min. inv.}]$$

$$\frac{1}{2}\times 3000+750 = 2250 \text{ units.}$$

Example 6

A company that produces three different items has a limited storage space for nearly 300 items of all varieties on the average. Determine the optimal production quantities for the three items separately, using the following information:

Item produced	:	1	2	3
Holding cost per unit per year (₹)	:	1.5	0.5	1.0
Set up cost per production run (₹)	:	25	20	30
Demand of the item per year	:	1000	1200	800

SOLUTION

This is a problem following the constrained multi-item deterministic model. The optimum production quantity of the *i*th item is given by

$$Q_i^0 = \sqrt{\frac{2C_{0i}D_i}{C_{1i}+2\lambda}},$$

where $$\sum_{i=1}^{3} Q_i^0 = 2K \text{ or } \frac{1}{2}\left(Q_1^0+Q_2^0+Q_3^0\right) = 300$$

Iteration 0:

Taking $\lambda = 0$,

$$Q_1^0 = \sqrt{\frac{2\times 25\times 1000}{1.5}} \simeq 183$$

$$Q_2^0 = \sqrt{\frac{2\times 20\times 1200}{0.5}} = \sqrt{96{,}000} \simeq 310$$

and $$Q_3^0 = \sqrt{\frac{2\times 30\times 800}{1}} = \sqrt{48{,}000} \simeq 219$$

Now $$\frac{1}{2}(Q_1^0+Q_2^0+Q_3^0) = \frac{1}{2}\times 712 = 356 > 300$$

We increase the value of λ. Taking $\lambda = 0.1$, the improved values of Q_i^0 are obtained as follows:

Iteration 1:

$$Q_1^0 = \sqrt{\frac{2\times 25\times 1000}{1.7}} = 100\times\sqrt{2.94} \simeq 172;$$

$$Q_2^0 = \sqrt{\frac{2\times 20\times 1200}{0.7}} = 100\times\sqrt{6.86} \simeq 262$$

$$Q_3^0 = \sqrt{\frac{2\times 30\times 800}{1.2}} = 100\times\sqrt{4} \simeq 200\,;$$

$$\frac{1}{2}(Q_1^0 + Q_2^0 + Q_3^0) = \frac{1}{2}\times 634 = 317 > 300$$

We further increase the values of λ. Let us take $\lambda = 0.15$.

Iteration 2:

$$Q_1^0 = \sqrt{\frac{2\times 25\times 1000}{1.8}} = 100\times\sqrt{2.78} \simeq 167;$$

$$Q_2^0 = \sqrt{\frac{2\times 20\times 1200}{0.8}} = 100\times\sqrt{6} \simeq 245;$$

$$Q_3^0 = \sqrt{\frac{2\times 30\times 800}{1.3}} = 100\times\sqrt{3.69} \simeq 192\,.$$

Now, $\frac{1}{2}(Q_1^0 + Q_2^0 + Q_3^0) \simeq \frac{1}{2}\times 604 = 302$

This sum of average inventory values (302) is nearly equal to 300, viz., the sum of the given inventory values.

∴ The optimal production quantities of the three products are

$$Q_1^0 = 167,\ Q_2^0 = 245 \text{ and } Q_3^0 = 192\,.$$

Example 7

For the following data, determine approximately the economic order quantities when the total value of the average of inventory levels of all the three products is ₹ 15,000/-.

Products:	1	2	3
Holding cost per unit per year (₹):	20	30	40
Cost of ordinary per order (₹):	50	40	60
Annual demand:	10,000	12,000	9000
Cost per unit (₹):	60	70	50

Solution

The EOQ of the ith item is given by

$$Q_i^0 = \sqrt{\frac{2D_iCo_i}{C_{1i}+2\lambda}}\text{, where } \sum_{i=1}^{3} k_i\cdot Q_i^0 = 10{,}000\text{; where } k_i \text{ is the cost per unit of the } i\text{th item.}$$

Iteration 0: $\lambda = 0$

$$Q_1^0 = \sqrt{\frac{2\times 10{,}000\times 50}{20}} = 100\times 2.24 \simeq 224$$

$$Q_2^0 = \sqrt{\frac{2\times 12,000\times 40}{30}} = 100\times 1.79 \simeq 179$$

$$Q_3^0 = \sqrt{\frac{2\times 9000\times 60}{40}} = 100\times 1.64 \simeq 164$$

$$(\text{TCI})^0 = \frac{1}{2}(224\times 60 + 179\times 70 + 164\times 50) = ₹\ 17,085/\text{-}$$

Iteration 1: $\lambda = 5$

$$Q_1^0 = \sqrt{\frac{2\times 10,000\times 50}{30}} = 100\times \sqrt{3.33} \simeq 183$$

$$Q_2^0 = \sqrt{\frac{2\times 12,000\times 40}{40}} = 100\times \sqrt{2.4} \simeq 153$$

$$Q_3^0 = \sqrt{\frac{2\times 9000\times 60}{50}} = 100\times \sqrt{2.16} \simeq 147$$

$\therefore$ $$(\text{TCI})^0 = \frac{1}{2}(183\times 60 + 153\times 70 + 147\times 50) = ₹\ 14,590/\text{-}$$

Iteration 2: $\lambda = 4.5$

$$Q_1^0 = \sqrt{\frac{2\times 10,000\times 50}{29}} = 100\times \sqrt{3.45} \simeq 186$$

$$Q_2^0 = \sqrt{\frac{2\times 12,000\times 40}{39}} = 100\times \sqrt{2.46} \simeq 157$$

$$Q_3^0 = \sqrt{\frac{2\times 9000\times 60}{49}} = 100\times \sqrt{2.20} \simeq 148$$

$\therefore$ $$(\text{TCI})^0 = \frac{1}{2}(186\times 60 + 157\times 70 + 148\times 50) = ₹\ 14\ 775/\text{-}$$

Iteration 3: $\lambda = 4$

$$Q_1^0 = \sqrt{\frac{2\times 10,000\times 50}{28}} = 100\times \sqrt{3.57} \simeq 189$$

$$Q_2^0 = \sqrt{\frac{2\times 12,000\times 40}{38}} = 100\times \sqrt{2.53} \simeq 159$$

$$Q_3^0 = \sqrt{\frac{2\times 9000\times 60}{48}} = 100\times \sqrt{2.25} \simeq 150$$

$\therefore$ $$(\text{TCI})^0 = \frac{1}{2}(189\times 60 + 159\times 70 + 150\times 50) = ₹\ 14\ 985/\text{-}$$

₹ 14985/- is nearly equal to ₹ 15,000, EOQ values have been obtained.

The EOQ values of the three items are given by

$$Q_1^0 = 189,\ Q_2^0 = 159 \text{ and } Q_3^0 = 150$$

Example 8

A newspaper boy buys papers for ₹ 3.50 each and sells them for ₹ 4.00 each. He cannot return unsold papers. Daily demand follows the discrete distribution:

No. of customers:	41	42	43	44	45	46	47	48	49	50
Probability:	0.01	0.03	0.06	0.10	0.20	0.25	0.15	0.10	0.05	0.05

If each day's demand is independent of the previous day's, how many paper's should be ordered each day?

SOLUTION

Let Q be the number of newspapers ordered per day and X be the daily demand, viz., the number of papers that are actually sold per day.

The amount the paper boy pays for each paper is taken as the inventory holding cost per unit per day $= c_1 =$ ₹ 3.50 [viz., the cost of over-supplying per unit per day]

The cost of undersupply per unit per day = daily shortage cost $= C_2 =$ ₹ 4.00 – ₹ 3.50 = ₹ 0.50

The cumulative probability distribution corresponding to the given discrete distribution is

$\le X$:	41	42	43	44	45	46	47	48	49	50
$p(x)$	.01	.03	.06	.10	.20	.25	.15	.10	.05	.05
$\sum_{x=1}^{X} p(x)$	.01	.04	.10	.20	.40	.65	.80	.90	.95	1.00

The optimum value of Q is given by

$$P(X \le Q-1) < \frac{C_2}{C_1 + C_2} < P(X \le Q)$$

or

$$\sum_{x=1}^{Q-1} p(x) < \frac{0.5}{3.5 + 0.5} < \sum_{x=1}^{Q} p(x)$$

viz.,

$$P(X \le 43) \ < 0.125 \ < P(X \le 44)$$

$P(X \le 44)$ exceeds 0.125 level.

$\therefore$ 44 papers must be ordered daily.

Example 9

The demand of an item in a season follows a discrete probability distribution given here:

X:	17	18	19	20	21	22	23	24
p(x):	.02	.06	.12	.34	.20	.14	.08	.04
$P(x)=\sum_{x=1}^{X} p(x)$	.02	.08	.20	.54	.74	.88	.96	1.00

The inventory holding cost is ₹ 10/- per year minus the end of the season discount price of ₹ 60/-. The purchasing cost per unit is ₹ 100/- and the stock-out cost is ₹ 150/- per unit. How many units of the item should be stocked at the start of the season to minimise the total cost?

SOLUTION

$$C = ₹\ 100/\text{-};$$
$$C_1 = ₹\ 10 - ₹\ 60 = ₹\ (-50)/\text{-};$$
$$C_2 = ₹\ 150/\text{-}$$

The optimal value of Q is given by

$$P(X \le Q-1) < \frac{C_2 - C}{C_2 + C_1} < P(X \le Q)$$

viz., $$P(X \le Q-1) < \frac{150-100}{150-50} < P(X \le Q)$$

viz., $P(X \le 19) < 0.5 < P(X \le 20)$, since $.20 < 0.5 < 0.54$

$\therefore$ $Q^0 = 20$, viz., 20 items should be stocked at the start of the season.

Example 10

The demand of a certain item X during a day is found to follow a uniform distribution in $0 \le x \le 12$. The unit inventory carrying cost is ₹ 7/- per day and the shortage cost is ₹ 10/- per day. If the purchasing cost per unit is ₹ 3/-, determine the optimum order level for the inventory.

SOLUTION

X follows a uniform distribution in $0 \le x \le 12$.

$$\therefore \quad f(x) = \begin{cases} \frac{1}{12}, & \text{in } 0 \le x \le 12 \\ 0, & \text{elsewhere} \end{cases}$$

The optimum order quantity Q^0 is given by

$$P(X \le Q^0) = \frac{C_2 - C}{C_2 + C_1}$$

Here $C = ₹\ 3/\text{-}$, $C_1 = ₹\ 5/\text{-}$ and $C_2 = ₹\ 7/\text{-}$

$$\therefore \quad P(X \le Q^0) = \frac{7-3}{7+5} = \frac{1}{3}$$

viz.,
$$\int_0^{Q^0} \frac{1}{12} dx = \frac{1}{3};$$

viz.,
$$\frac{Q^0}{12} = \frac{1}{3} \text{ or } Q^0 = 4 \text{ units.}$$

Example 11

A fish stall sells a variety of fish at the rate of ₹ 140/- per kg on the day of the catch. If the stall fails to sell the catch on the same day, it pays for storage at the rate of ₹ 40/- per kg and the next day, the price fetched for the fish is ₹ 120/-. Past record shows that there is an unlimited demand for one day old fish. Determine how much fish should be procured every day, so that the total expected cost may be minimum. Assume that the daily demand of fish follows an exponential distribution with parameter $\lambda = 0.012$.

SOLUTION

The cost of inventory holding 1 kg of fish for a day is given by

$$C_1 = \text{(Actual storage cost + loss due to over purchasing)/kg}$$

$$= ₹\ 40/- + ₹\ (140 - 120) = ₹\ 60/-$$

$$C_2 = ₹\ 140/- \text{ (loss due to under purchasing/kg)}$$

The daily demand follows the exponential distribution with pdf $f(x) = 0.012e^{-0.012x}; x \geq 0$

The optimum order quantity Q^0 is given by $P(X \leq Q^0) = \dfrac{C_2}{C_2 + C_1}$

viz.,
$$P(X \leq Q^0) = \frac{140}{140 + 60} = 0.7$$

viz.,
$$\int_0^{Q^0} 0.012\, e^{-0.012x} dx = 0.7$$

viz.,
$$1 - e^{-0.012Q^0} = 0.7$$

viz.,
$$e^{-.012Q^0} = 0.3,$$

∴
$$-.012Q^0 = 1.2$$

∴
$$Q^0 = \frac{1.2}{0.012} = 100 \text{ kg}.$$

Example 12

The demand of an item is normally distributed with mean 20 and variance 25. The inventory holding cost is ₹ 10/- per year minus the end of the season discount price of ₹ 50/-, the purchasing cost per unit is ₹ 55/- and the stock out cost per unit is ₹ 100/-. How many units of the item should be kept in stock at the start of the season to minimise the total cost?

$$C = ₹\ 55/-;$$

$$C_1 = ₹\ 10/- - ₹\ 50/- = -(₹\ 40/-);$$

$$C_2 = ₹\ 100/-$$

The optimum purchasing quantity Q^0 is given by

$$P(X \leq Q^0) = \frac{C_2 - C}{C_2 + C_1} \text{ or } \frac{100 - 55}{100 - 40} \text{ or } 0.75 \quad (1)$$

Since X follows a normal distribution with $\mu = 20$ and $\sigma = 5$, (1) is equivalent to $\int_{-\infty}^{Q^0} f(x)\,dx = 0.75$

viz.,
$$\int_{-\infty}^{(Q^0-m)/\sigma} \phi(z)\,dz = 0.75,$$

on putting
$$z = \frac{Q^0 - m}{\sigma}$$

From the table of normal areas,

$$\int_{-\infty}^{-0.72} \phi(z)\,dz = 0.75, \text{ nearly}$$

$\therefore$
$$\frac{Q^0 - m}{\sigma} = 0.72, \text{ nearly}$$

viz.,
$$\frac{Q^0 - 20}{5} = 0.72, \text{ nearly}$$

$\therefore\ Q^0 = 23.6$ or $Q^0 = 24$ nearly.

EXERCISE 11(B)

Part A (Short answer questions)

1. What are the assumptions made for the inventory models with price breaks?
2. Give the step-by-step procedure to find the OOQ ($= Q^0$) in the case of purchasing model with 'n' price breaks.
3. What is buffer stock? Why it is necessary to maintain buffer stock in inventory control?
4. Define re-order level. How is it connected to buffer stock?
5. Explain briefly how the formula for the OOQ ($= Q_i^0$) is obtained for the *i*th item in a multi-item deterministic inventory model with no shortage and lead time, when $\frac{1}{2}\sum Q_1^0 \leq k$.
6. How is the formula for Q_i^0 in Question 5 modified when $\frac{1}{2}\sum Q_1^0 > k$?
7. State the formula for Q^0 to minimise the total cost of inventory in a single period inventory model when the demand is a discrete RV.

8. State the formula for Q^0 to minimise the total cost of inventory in a single period inventory model when the demand is a continuous RV.

Part B

9. Find the optimum order quantity for a product for which the price breaks are as follows:

Quantity	Unit cost in ₹
$0 \leq Q_1 < 5$	10.00
$500 \leq Q_2$	9.25

The monthly demand for the product is 200 units, the cost of storage is 2% of the unit cost and the cost of ordering is ₹ 100/-.

10. Find the optimum order quantity for a product for which the price breaks are as follows:

Quantity	Unit cost in ₹
$0 \leq Q_1 < 50$	10/-
$50 \leq Q_2 < 100$	9/-
$100 \leq Q_3$	8/-

The monthly demand for the product is 200 units, the cost of storage is 25% of the unit cost and the ordering cost is ₹ 20/- per order.

11. Find the optimum order quantity for a product for which the price breaks are as follows:

Quantity	Unit cost in ₹
$0 \leq Q_1 < 100$	20
$100 \leq Q_2 < 200$	18
$200 \leq Q_3$	16

The monthly demand for the product is 400 units. The storage cost is 20% of the unit cost of the product and the cost of ordering is ₹ 25/- per order.

12. A company uses 20,000 units of an item annually, each costing ₹ 20/-. Each order costs ₹ 100/- and the inventory carrying costs 20% of the average annual value. Find the EOQ. If the company functions 250 days in a year, the lead time is 10 days and the buffer stock is 400 units, find the ROL, maximum, minimum and average inventories.
13. A factory uses 2500 units of a raw material annually which costs ₹ 12.5 per unit. Placing each order costs ₹ 75/- and the carrying cost is 3% of the average inventory per year. Find the EOQ and the total inventory cost (including the cost of the material). The factory works for 250 days in a year. If the normal lead time is 8 days and the safety stock is 250 units, find the ROL, maximum, minimum and average inventories.
14. A company producing three different items has a limited storage space for 700 items of all varieties on the average. Determine the optimal production quantities for the items separately, using the following data:

Item produced:	1	2	3
Holding cost per year per unit (₹):	5	2	4
Set up cost per production run (₹):	500	400	600
Demand of the item per year:	1000	1200	750

15. Using the following information to determine approximately the economic order quantities when the total value of the average of inventory levels of all the three products is ₹ 10,000/-. The carrying charge on each product is 20% of the average inventory valuation per annum.

Product:	1	2	3
Annual demand (in units):	1000	500	2000
Variable cost (₹ per unit):	20	100	50
Set up cost per lot (₹):	50	75	100

16. For a single period stochastic model, the inventory holding cost is ₹ 5/- per unit and the shortage cost is ₹ 10/-. Determine the optimal order level of the inventory, given the probability distribution of demand of the item is as follows:

$X = x$:	0	1	2	3	4	5
$p(x)$:	0.10	0.20	0.25	0.20	0.15	0.10

17. The probability distribution of monthly sales of a certain item is as follows:

$X = x$:	0	1	2	3	4	5	6
$p(x)$:	.02	.05	0.30	0.27	0.20	0.10	0.06

The cost of carrying inventory is ₹ 10/- per unit per month. The current policy is to maintain a stock of 4 items at the beginning of each month. Assuming that the cost of storage is proportional to both time and quantity short, obtain the imputed cost of a shortage of one item for one time unit.

18. If the demand for a certain item follows a uniform distribution between 4000 and 5000, find the optimal order quantity, if storage cost is ₹ 2/- per unit and the shortage cost is ₹ 6/- per unit.
19. A bakery sells one of its types of cakes by weight. If the product is not sold on the day it is baked, it can be sold at a loss of ₹ 20/- per unit weight the next day and there is unlimited market for one day old cake. The company makes a profit of ₹ 100/- on a unit weight sold on the day it is baked. Past daily orders follow a triangular distribution with probability density function $f(x) = 0.02 - 0.0002x$; $0 \leq x \leq 100$. Determine the optimal unit weight of cake baked.
20. If a variety of ice-cream prepared by a company is not sold on the day of preparation, it can be sold the next day at a loss of ₹ 10/- per $\frac{1}{4}$ kg. However, there is an unlimited market for one day old ice-cream. On the other hand, the company makes a profit of ₹ 30/- per $\frac{1}{4}$ kg ice-cream sold on the day of preparation. Past daily sales follow a distribution with probability density function $f(x) = 0.04 - 0.0004x$; $0 \leq x \leq 50$. How many $\frac{1}{4}$ kg packets of ice-cream should the company prepare every day?

ANSWERS

Exercise 11(A)

Part B

9. 400; ₹ 2400/-
10. 100; ₹ 4000/-; $\frac{1}{20}$ year.
11. 480; ₹ 4800/-; $(TC)_1 =$ ₹ 3,24,800/-; $(TC)_2 =$ ₹ 2,94,760/-; The discount offer should be accepted.
12. 300; ₹ 4,64,400/-; 30; $\frac{1}{30}$ year.
13. 4800; ₹ 1200/-.
14. $\frac{8}{5}$ day; 6000
15. 1,05,000 bearings; 10.5 days.
16. 1000 units; 0.5 year.
17. 2000; 800; ₹ 1200/-.
18. 20 days; 400.
19. 240; 20; 4 months.
20. 2000; 200; $\frac{1}{5}$ year.

Exercise 11(B)

Part B

9. 500.
10. 100.
11. 100.
12. 1000; 1200; 1400; 400; 900.
13. 1000; ₹ 31,625/-; 330; 250; 290.
14. $Q_1^0 = 408$; $Q_2^0 = 566$; $Q_3^0 = 424$.
15. $Q_1^0 = 183$; $Q_2^0 = 63$; $Q_3^0 = 202$.
16. $Q^0 = 3$.
17. $17.7 < C_2 < 52.2$.
18. 4750.
19. 59 units.
20. 21 quarter kg packets.

12 REPLACEMENT AND MAINTENANCE PROBLEMS

12.0 INTRODUCTION

The study of replacement and maintenance has become necessary in industrial and domestic situations where machinery, electric light bulbs, men, etc. are used. The replacement of these items or equipment is inevitable due to deterioration in their efficiency, failure or breakdown, which may be either gradual or sudden, resulting in the whole system getting affected. Moreover, the maintenance (or running) and service costs of an equipment also increases with time. Often it becomes more economical to replace an old equipment with a new one than to take some remedial action to restore the efficiency of deteriorating units.

We shall deal with the most economical replacement policy in the following situations:

1. Replacement of items that deteriorate with time.
2. Replacement of items in anticipation of complete breakdown or failure, the probability of which increases with time.
3. Problems in mortality and staffing.

12.1 REPLACEMENT OF ITEMS THAT DETERIORATE WITH TIME

Maintenance cost of items, such as machines, usually increases with time and a stage is reached when the maintenance cost is so large that it is economical to replace the machine with a new one. Our interest in this case is to find the optimal age of the machine at which the old machine is to be replaced by the new one.

Let C be the capital cost of the item (a machine), S be the scrap (re-sale) value of the item, assumed to be constant and M_r be the maintenance cost of the item during the rth year.

$\therefore$ The total maintenance cost incurred on the item over a period of n years $=\sum_{r=1}^{n} M_r$.

$\therefore$ The total cost of the item at the end of the nth year is given by

$$T(n)=C-S+\sum_{r=1}^{n} M_r$$

$\therefore$ The average cost of the item per year over a period of n years is given by

$$A(n) = \frac{1}{n}T(n) = \frac{C-S}{n} + \frac{1}{n}\sum_{r=1}^{n} M_r$$

The replacement of the old item by a new one should be done at the end of the nth year, provided $A(n)$ is minimum

viz., when $A_n < A_{n+1}$ (1)

and $A_n < A_{n-1}$ (2)

(1) is true,

when $$\left(\frac{C-S}{n} + \frac{1}{n}\sum_{r=1}^{n} M_r\right) - \left(\frac{C-S}{n+1} + \frac{1}{n+1}\sum_{r=1}^{n+1} M_r\right) < 0$$

viz., $$\frac{C-S}{n(n+1)} + \frac{1}{n(n+1)}\sum_{r=1}^{n} M_r - \frac{1}{n+1}M_{n+1} < 0$$

viz., $$\frac{1}{n+1}\left\{\frac{C-S}{n} + \frac{1}{n}\sum_{r=1}^{n} \mathrm{M}_r\right\} - \frac{1}{n+1}\mathrm{M}_{n+1} < 0$$

viz., $$A_n < M_{n+1}$$

Similarly (2) is true when $A_n > M_n$

$\therefore$ Replacement of the old item should be done at the end of the nth year if $M_n < A_n < M_{n+1}$ or equivalently if

$$A(1) > A(2) > \ldots > A(n-1) > A(n) < A(n+1) < A(n+2) < \ldots$$

Change in the Value of Money With Time

If we borrow ₹ 100/- at an interest rate of 10% per year, then we will have to return ₹ 110/- after one year.

Thus, ₹ 110/- after one year, is equivalent to ₹ 100/- today, viz., ₹ 1/- spent after one year is equivalent to ₹$\frac{100}{110}$ or ₹ $(1.1)^{-1}$ today at the rate of 10%.

₹ 1/- spent after n years from today is equivalent to $(1.1)^{-n}$ today at the rate of 10%.

$(1.1)^{-1}$ is called the *present worth factor* (pwf) of ₹ 1/- spent after n years from today at 10% rate.

$(1+0\cdot1)^{-1}$ is called the *discount rate* or *depreciation factor* at 10% rate of interest.

In general, if r is the rate of interest per year on ₹ 1/-, the *pwf* is $=(1+r)^{-n}$. The depreciation factor is $\frac{1}{1+r}$ and is denoted by γ.

Replacement of Items Whose Maintenance Costs Increase and When the Value of Money Changes with Time

Let C be the capital cost of the item and S be the scrap (re-sale of salvage) value of the item.

Let M_r be the maintenance cost during the rth year and $\gamma = \frac{1}{1+r}$ be the depreciation factor of money.

Present value of the total maintenance cost over a period of n years

$$= M_1 + M_2\gamma + M_3\gamma^2 + \cdots + M_n\gamma^{n-1} = \sum_{r=1}^{n} M_r\gamma^{r-1}$$

Present value of the total cost of the item at the end of the nth year is

$$T(n) = C - S + \sum_{r=1}^{n} M_r \gamma^{r-1} \tag{1}$$

Let $A(n)$ be the average cost of the item per year over a period of n years.
Then, the present value of the total cost of the item at the end of the nth year can also be taken as

$$\begin{aligned} T(n) &= A(n) + A(n)\gamma + A(n)\gamma^2 + \cdots + A(n)\gamma^{n-1} \\ &= A(n)\left[1 + \gamma + \gamma^2 + \cdots \gamma^{n-1}\right] \\ &= A(n)\left(\frac{1-\gamma^n}{1-\gamma}\right) \end{aligned} \tag{2}$$

Equating (1) and (2), we get

$$A_n \text{ or } A(n) = \frac{(1-\gamma)}{(1-\gamma^n)}\left\{C - S + \sum_{r=1}^{n} M_r \gamma^{r-1}\right\} \tag{3}$$

Proceeding as in the discussion of the previous section, replacement of the old item should be done at the end of the nth year, if $M_n < A(n) < M_{n+1}$ or equivalent by

$$A_{n-1} > A_n < A_{n+1}, \text{ where } A_n \text{ is given by (3).}$$

12.2 REPLACEMENT OF ITEMS THAT FAIL COMPLETELY

The failure of a certain item may occur all of a sudden, instead of gradual deterioration (as in the case of electric bulbs), resulting in complete breakdown of the system. Apart from the cost of the failed item, the complete breakdown of the system results in loss in production, idle labour, wastage and other damages. If it is possible to know in advance when an item is going to fail, we can consider *preventive replacement*, viz., replacing the item before its sudden breakdown and hence, eliminating or reducing the indirect cost (due to loss in production etc.). However, in many cases the failure time cannot be predicted accurately. Instead, the probability distribution of failure time may be available, based on past experience.

In case of complete failure of items, the following two replacement policies are usually followed.

1. *Individual Replacement Policy*
   According to this policy, an old item is replaced by the new one immediately after its failure.

2. *Group Replacement Policy*
   Under this policy, we have to decide as to when all the items must be replaced, irrespective of the fact that the items have failed or not, with provision that if any single item fails before the optimal (group) replacement time, it may be replaced individually.

12.3 PROBLEMS IN STAFFING

The problems related to recruitment and promotion of staff in any organisation (system) are known as *staffing problems*. Such problems can also be treated as replacement problems where the members of the staff are considered as machine parts. The problem of staff replacement arises due to resignation, retirement or deaths (mortality) of the staff members from time to time. Hence, there is a need for some useful replacement (recruitment or promotion) policy, in order to maintain the required strength of the staff in a system. The procedure to be used in such problems can be understood with the help of a few worked examples.

WORKED EXAMPLES 12

Example 1

A small-scale industry is considering replacement of a machine whose cost price is ₹ 35,000/-. The scrap value is ₹ 5000/-. The maintenance costs are found from past experience as follows:

Year(r)	1	2	3	4	5	6	7	8	9
M_r (in hundreds of ₹)	5.0	12.5	20.0	30.0	45.0	62.5	80.0	100.0	150.0

When should the machine be replaced?

SOLUTION

$$C = ₹\ 35,000/-$$

and

$$S = ₹\ 5,000/-$$

Year (r)	M_r (₹)	$\sum_{r=1}^{r} M_r$ (₹)	$T(r)$ (₹) $= (C - S) + \sum M_r$	$A(r) = \frac{T(r)}{r}$ (₹)
1	500	500	30,500	30,500
2	1250	1,750	31,750	15,875
3	2000	3,750	33,750	11,250
4	3000	6,750	36,750	9,188.5
5	4500	11,250	41,250	8,250
6	6250	17,500	47,500	7,916.7⊛
7	8000	25,500	55,500	7,928.6
8	10,000	35,500	65,500	8,187.5
9	15,000	50,500	80,500	8,944.4

The values of A_r in the last column of the table decrease, gradually upto 7,916.7⊛ and afterwards, increase gradually.

⊛ Since A_r is the least when $r = 6$, replacement of the machine must be done at the end of the 6th year.

Example 2

The following table gives the running costs per year and the re-sale value of a certain equipment, the purchase cost of which is ₹ 80,000/-

Year	1	2	3	4	5	6	7	8
Running cost in 1000's of ₹	24.0	25.6	28.8	33.6	40.0	46.4	54.4	64.0
Re-sale value in 1000's of ₹	56.0	40.0	27.2	19.2	9.6	8.0	8.0	8.0

Find the optimal time of replacement of the equipment.

SOLUTION

$$\text{Capital cost } (C) = 80 \times 10^3 \text{ ₹}$$

Year Yr	Re-sale value ($S_r \times 10^{-3}$) (₹)	$(C - S_r)\ 10^{-3}$ (₹)	Running cost ($M_r \times 10^{-3}$) (₹)	$\sum_{r=1}^{r} M_r \times 10^{-3}$ (₹)	Average cost $A_r = 10^{-3} \times \frac{1}{r}\{C - S + \sum M_r\}$ (₹)
1	56.0	24.0	24.0	24.0	48.0/1 = 48.0
2	40.0	40.0	25.6	49.6	89.6/2 = 44.8
3	27.2	52.8	28.8	78.4	131.2/3 = 43.733
4	19.2	60.8	33.6	112.0	172.8/4 = 43.2⊛
5	9.6	70.4	40.0	152.0	222.4/5 = 44.48
6	8.0	72.0	46.4	198.4	270.4/6 = 45.067
7	8.0	72.0	54.4	252.8	324.8/7 = 46.4
8	8.0	72.0	64.0	316.8	388.8/8 = 48.6

⊛ Since A_r is least when $r = 4$, the replacement of the equipment should be done at the end of the 4th year.

Example 3

Machine A costs ₹ 75,000/-. Annual operating costs are ₹ 5000/- for the first year and then increase by ₹ 20,000/- every year. If the optimum replacement is followed, find the best age at which the machine may be replaced.

Another machine B costs ₹ 1,00,000/-. Annual operating costs for this machine are ₹ 4000/- for the first year and then increase by ₹ 12,000/- every year. If there is a one-year old machine of type A, should it be replaced with a machine of type B and if so, when? Assume that both types of machines have no re-sale value.

SOLUTION

$$\textit{Machine A}: C = \text{₹ } 75{,}000/\text{-};$$

$$S = \text{Nil};$$

$$C - S = \text{₹ } 75{,}000/\text{-}$$

Year (r)	$C-S$ (₹)	M_r (₹)	$\sum_{r=1}^{r} M_r$ (₹)	$T(r)=C-S+\sum_{r=1}^{r} M_r$ (₹)	$A(r)=\frac{1}{r}T(r)$ (₹)
1	75,000	5,000	5,000	80,000	80,000
2	75,000	25,000	30,000	1,05,000	52,500
3	75,000	45,000	75,000	1,50,000	50,000⊛
4	75,000	65,000	1,40,000	2,15,000	53,750
5	75,000	85,000	2,25,000	3,00,000	60,000
6	75,000	1,05,000	3,30,000	4,05,000	67,500

⊛ Since A_r is the least when $r = 3$, machine A should be replaced at the end of the third year.

Machine B: $C =$ ₹ 1,00,000/-;

$S =$ Nil;

$C - S =$ ₹ 1,00,000/-

Year (r)	$C-S$(₹)	M_r (₹)	$\sum_{r=1}^{r} M_r$ (₹)	$T(r)=C-S+\sum_{r=1}^{r} M_r$ (₹)	$A(r)=\frac{1}{r}T(r)$ (₹)
1	1,00,000	4,000	4,000	1,04,000	1,04,000
2	1,00,000	16,000	20,000	1,20,000	60,000
3	1,00,000	28,000	48,000	1,48,000	49,333
43	1,00,000	40,000	88,000	1,88,000	47,000⊛
5	1,00,000	52,000	1,40,000	2,40,000	48,000
6	1,00,000	64,000	2,04,000	3,04,000	50,667

⊛ Since A_r is the least when $r = 4$, machine B should be replaced at the end of the fourth year.

Comparison: The average total cost of a one-year old machine A is ₹ 52,500/- during the first year, ₹ 50,000/- during the second year and ₹ 53,750/- during the third year of operation. However, the average total cost of a new machine B is only ₹ 49,333/- during the third year of operation. Hence, a one-year old machine A must be replaced by machine B at the end of the third year.

Example 4

The capital cost of an equipment is ₹ 60,000/- and the present value of ₹ 1/- to be spent in a year's time is ₹ 0.9. The maintenance costs of the equipment are given in the following table. When should the equipment be replaced? Assume that the re-sale value is nil.

Year	1	2	3	4	5	6	7
Maintenance cost (in thousands of ₹s)	10	12	16	20	26	32	60

SOLUTION

Year (r)	M_r (₹)	γ^{-1} ($\gamma = 0.9$)	$\sum_{r=1}^{r} \gamma^{r-1}$	$M_r \cdot \gamma^{r-1}$	$\sum_{r=1}^{r} M_r \gamma^{r-1}$	$C + \sum_{r=1}^{r} M_r \gamma^{r-1}$	$A(r) = [C + \sum M_r \gamma^{r-1}] \div \sum \gamma^{r-1}$
1	10,000	1	1	10,000	10,000	70,000	70,000
2	12,000	0.9	1.9	10,800	20,800	80,800	42,526
3	16,000	0.81	2.71	12,960	33,760	93,760	34,597
4	20,000	0.729	3.439	14,588	48,348	1,08,348	31,506
5	26,000	0.6561	4.0951	17,059	65,407	1,25,407	30,624⊛
6	32,000	0.5905	4.6856	18,896	84,303	1,44,303	30,797
7	60,000	0.5314	5.2170	31,884	1,16,187	1,76,187	33.772

⊛ Since $26,000 < 30,624 < 32,000$ [*viz.,*, $M_5 < A(5) < M_6$], the equipment should be replaced at the end of the 5th year

Example 5

The cost of a new machine is ₹ 1,20,000/-. The maintenance cost during the nth year is given by $M_n = 10,000\,(n - 1)$ $n = 1, 2, 3, \ldots$. If the discount rate per year is 0.5, when will it be economical to replace the machine by a new one.

SOLUTION

To find the year of optimal replacement, the computations required are given in the following table, with $C =$ ₹ 1,20,000/- $\gamma = 0.5$ and $M_r = 10,000(r - 1)$ $r = 1, 2, 3, \ldots$

Year (r)	M_r (₹)	γ^{-1} ($\gamma = 0.5$)	$M_r \gamma^{r-1}$	$\sum_{r=1}^{r} \gamma^{r-1}$	$C + \sum_{r=1}^{r} M_r \gamma^{r-1}$	$A(r) = \left[C + \sum_{r=1}^{r} M_r \gamma^{r-1}\right] \div (\sum \gamma^{r-1})$
1	0	1	0	1	1,20,000	1,20,000
2	10,000	0.5	5,000	1.5	1,25,000	83,333
3	20,000	0.25	5,000	1.75	1,30,000	74,286
4	30,000	0.125	3,750	1.875	1,33,750	71,333
5	40,000	0.0625	2,500	1.9375	1,36,250	70,322
6	50,000	0.0313	1,565	1.9688	1,37,815	69,999
7	60,000	0.0156	936	1.9844	1,38,751	69,920⊛
8	70,000	0.0078	546	1.9922	1,39,297	69,921

⊛ Since $60,000 < 69,920 < 70,000$ [viz., $M_7 < A(7) < M_8$], the machine must be replaced at the end of the 7th year.

Example 6

A manufactures is offered two machines A and B. A is priced at ₹ 1lakh and running costs are estimated at ₹ 15000/- for each of the first 5 years, increasing by ₹ 3000/- per year in the 6th and subsequent years.

Machine B, which has the same capacity as A is priced at ₹ 80,000/-, but will incur running costs of ₹ 18,000/- for each of the first 6 years, increasing by ₹ 4000/- per year thereafter. If money is worth 10% per year, which machine can the manufacturer prefer for purchase? The ultimate scrap values of both machines are negligible.

SOLUTION

Since the money is worth 10% per year, the discount rate for both machines is given by

$$\gamma = \frac{1}{1+0\cdot 1} = 0\cdot 9091$$

The optimal age of replacement of machine A is found by using the following calculation with capital cost $C =$ ₹ 1/- lakh.

Year (r)	M_r(₹)	γ^{r-1} ($\gamma = 0.9091$)	$M_r \cdot \gamma^{r-1}$ (₹)	$C + \sum_{r=1}^{r} M_r \gamma^{r-1}$	$\sum_{r=1}^{r} \gamma^{r-1}$	$A(r) = [C + \sum M_r \gamma^{r-1}] \div (\sum \gamma^{r-1})$
1	15,000	1.0000	15,000	1,15,000	1.0000	1,15,000
2	15,000	0.9091	13,637	1,28,637	1.9091	67,381
3	15,000	0.8264	12,396	1,41,033	2.7355	51,557
4	15,000	0.7513	11,270	1,52,302	3.4868	43,680
5	15,000	0.6830	10,245	1,62,547	4.1698	38,982
6	18,000	0.6209	11,176	1,73,723	4.7907	36,263
7	21,000	0.5645	11,855	1,85,578	5.3552	34,654
8	24,000	0.5132	12,317	1,97,895	5.8684	33,722
9	27,000	0.4665	12,596	2,10,491	6.3349	33,227
10	30,000	0.4241	12,723	2,23,214	6.7590	33,025
11	33,000	0.3855	12,722	2,35,936	7.1445	33,023⊛
12	36,000	0.3505	12,618	2,48,554	7.4950	33,163

⊛ Since 33,000 < 33,023 < 36,000 [viz., $M_{11} < A(11) < M_{12}$], machine A must be replaced at the end of the 11th year.

The optimal age of replacement of machine B is found from the following calculations with capital cost $C =$ ₹ 80,000/-

Year (r)	M_r (₹)	γ^{r-1} $(\gamma = 0.9091)$	$M_r\gamma^{r-1}$ (₹)	$C + \sum_{r=1}^{r} M_r\gamma^{r-1}$	$\sum_{r=1}^{r} \gamma^{r-1}$	$A(r) = [C + \sum M_r\gamma^{r-1}] \div (\sum \gamma^{r-1})$
1	18,000	1.0000	18,000	98,000	1.0000	98,000
2	18,000	0.9091	16,364	1,14,364	1.9091	59,985
3	18,000	0.8264	14,875	1,29,239	2.7355	47,245
4	18,000	0.7513	13,523	1,42,762	3.4868	40,944
5	18,000	0.6830	12,308	1,55,070	4.1698	37,189
6	18,000	0.6209	11,176	1,66,246	4.7907	34,702
7	22,000	0.5645	12,419	1,78,665	5.3552	33,363
8	26,000	0.5132	13,343	1,92,008	5.8684	32,719
9	30,000	0.4665	13,995	2,06,003	6.3349	32,519⊛
10	34,000	0.4241	14,419	2,20,422	6.7590	32,612
11	38,000	0.3855	14,649	2,35,071	7.1445	32,902
12	42,000	0.3505	14,721	2,49,792	7.4950	33,328

⊛ Since 30,000 < 32,519 < 34,000 [viz., $M_9 < A(9) < M_{10}$], machine B must be replaced at the end of the 9th year.

Comparison: Since $A(9)$ for machine B, viz., ₹ 32,519/- is less than $A(11)$ for machine A, viz., ₹ 33,023/-, the manufacturer can prefer machine B.

Example 7

The following failure rates have been observed for a certain type of light bulb:

End of week	1	2	3	4	5
% of failure by end of week	10	25	50	80	100

Find the cost per week of individual replacement policy of an installation of 300 bulbs, based on the information given, if the cost of replacing a bulb is ₹ 20/-

SOLUTION

Week No. (r)	1	2	3	4	5
Cumulative probability of failure (Σp_r)	0.10	0.25	0.50	0.80	1.00
Probability (p_r)	0.10	0.15	0.25	0.30	0.20

Expected life of a bulb $= \sum_{r=1}^{5} r p_r$

$$= 0.10 + 0.30 + 0.75 + 1.20 + 1.00$$

$$= 3.35 \text{ weeks.}$$

$\therefore$ Expected number of bulbs to be replaced in a week $= \dfrac{300}{3\cdot35} \simeq 90$

$\therefore$ Cost of individual replacement per week = ₹ 20/- × 90 = ₹ 1800/-

Example 8

There is a large number of light bulbs in a system, all of which must be kept in working condition. If a bulb fails in service, it costs ₹ 18/- to replace it, but if all the bulbs are replaced simultaneously, it costs ₹ 10/- only per bulb.

The following failure distribution of bulbs is known:

Week (r)	1	2	3	4	5	6
Probability during the week (p_r)	0.09	0.16	0.24	0.36	0.12	0.03

Decide the best replacement policy.

SOLUTION

Excepted life of a bulb

$$= \sum r p_r = 0.09 + 0.32 + 0.72 + 1.44 + 0.60 + 0.18 = 3.35$$

Let us assume that the system consists of 1000 bulbs.

Then, the average number of bulbs to be replaced in a week $= \dfrac{1000}{3\cdot5} \simeq 299$.

$\therefore$ Cost of individual replacement per week = ₹ (299 × 18) = ₹ 5382/-

Let N_r denote the number of replacements made at the end of the rth week.

Then, N_0 = number of bulbs in the beginning = 1000

$$N_1 = N_0 \cdot p_1$$
$$= 1000 \times 0.09 = 90$$

$$N_2 = N_0 \cdot p_2 + N_1 \cdot p_1$$
$$= 160 + 8.1 \simeq 168$$

$$N_3 = N_0 p_3 + N_1 p_2 + N_2 p_1$$
$$= 240 + 14.4 + 15.12 \simeq 270$$

$$N_4 = N_0 p_4 + N_1 p_3 + N_2 p_2 + N_3 p_1$$
$$= 360 + 21.6 + 26.88 + 24.3 \simeq 433$$

$$N_5 = N_0 p_5 + N_1 p_4 + N_2 p_3 + N_3 p_2 + N_4 p_1$$
$$= 120 + 32.4 + 40.32 + 43.2 + 38.97 \simeq 275$$

$$N_6 = N_0 p_6 + N_1 p_5 + \cdots + N_5 p_1$$
$$= 30 + 10.8 + 60.48 + 64.8 + 69.28 + 24.75 \simeq 260$$

End of week	Total cost of group replacement	Average cost per week of group replacement
1	1000 × 10 + 90 × 18 = ₹ 11,620/-	₹ 11,620/-
2	10,000 + (90 + 168) × 18 = ₹ 14,644	₹ 7,322/-
3	10,000 + (258 + 270) × 18 = ₹ 19,504/-	₹ 6,501/-
4	10,000 + (528 + 433) × 18 = ₹ 27,298/-	₹ 6825/-
5	10,000 + (961 + 275) × 18 = ₹ 32,248/-	₹ 6450/-
6	10,000 + (1236 +260) × 18 = ₹ 36,928/-	₹ 6155/-

Minimum average cost per week of group replacement = ₹ 6155/-, that is greater than the cost of individual replacement (viz., ₹ 5382/-)

∴ The individual replacement policy is the best replacement policy.

Example 9

All the items in a system are new to start with. The probability that each item fails just before the end of the first month of life is p and that before the end of the second month of life is q, where $p + q = 1$(viz., all items fail at the end of the second month). If all items are replaced as they fail, show that the expected number of failures at the end of the rth month is given by

$$F(r) = \frac{N}{1+q}\left\{1-(-q)^{r+1}\right\},$$

where N is the number of items in the system.

If the cost per item of individual replacement is C_1 and the cost per item of the group replacement is C_2, find the condition under which

(a) a group replacement policy at the end of each month is most profitable.
(b) a group replacement policy at the end of every other month is the most profitable.
(c) a pure individual replacement policy is better than the other two.

SOLUTION

Let N_0 be the number of items in the system in the beginning (= N)

Let N_r be the number of items expected to fail just before the end of the rth month.

Then, $N_1 = N_0 p$ or Np or $N(1 - q)$

$$N_2 = N_0 q + N_1 p = Nq + N(1 - q)^2$$

or $N(1 - q + q^2)$

$$N_3 = N_1 q + N_2 p$$
$$= Npq + N(1 - q + q^2)p$$
$$= N(1 - q + q^2 - q^3)$$

and so on.

In general,

$$N_r = N[1 - q + q^2 - q^3 + \cdots + (-q)^r]$$

$$N_{r+1} = N_{r-1}q + N_r p = N[1 - q + q^2 - \cdots + (-q)^{r-1}]q + N[1 - q + q^2 - \cdots + (-q)^r](1 - q)$$

$$= N[1 - q + q^2 - \cdots + (-q)^{r+1}]$$

$\therefore$ By mathematical induction, we get

$$F(r) = N\left[1 - q + q^2 - \cdots + (-q)^r\right]$$

$$= \frac{N[1-(-q)^{r+1}}{1+q}$$

$$\text{Expected life of an item } = \sum_{r=1}^{2} rp_r = 1 \times p + 2 \times q = 1 - q + 2q = (1+q) \text{ months}$$

$$\therefore \text{ Expected no of items to be replaced in a month } = \frac{N}{1+q}$$

$$\therefore \text{ Average cost of individual replacement per month } = \frac{N \times C_1}{1+q} \tag{1}$$

If group replacement policy at the end of each month is adopted, the average cost incurred per month

$$N \times C_2 \tag{2}$$

If group replacement policy at the end of every other month (viz., at the end of the second month) is adopted, the average cost incurred per month

$$= \frac{1}{2}(N_1C_1 + NC_2) = \frac{1}{2}(Np \cdot C_1 + NC_2) \tag{3}$$

(a) The group replacement policy at the end of each month is the most profitable if

$$N \times C_2 < \frac{N \times C_1}{1+q}$$

and $$N \times C_2 < \frac{1}{2}(Np \cdot C_1 + N \cdot C_2)$$

viz., if $$C_2 < \frac{1}{1+q} C_1$$

and $$C_2 < (1-q)C_1$$

viz., if $$C_2 < (1-q)C_1,$$

since $$1 - q < \frac{1}{1+q},$$

as $$1-q^2<1$$

(b) The group replacement policy at the end of every other month is the most profitable if

$$\frac{1}{2}(p\cdot C_1+C_2)<\frac{C_1}{1+q}$$

and $$\frac{1}{2}(p\cdot C_1+C_2)<C_2$$

viz., if $$(1-q^2)C_1+(1+q)C_2<2C_1$$

and $$(1-q)C_1+C_2<2C_2$$

viz., if $$(1+q)C_2<(1+q^2)C_1$$

and $$C_2>(1-q)C_1$$

viz., if $$C_2<\left(\frac{1+q^2}{1+q}\right)C_1$$

and $$C_2>(1-q)C_1$$

viz., if $$(1-q)C_1<C_2<\left(\frac{1+q^2}{1+q}\right)C_1$$

(c) A pure individual replacement policy is better than both the group replacement policies if

$$\frac{C_1}{1+q}<C_2$$

and $$\frac{C_1}{1+q}<\frac{1}{2}(p\cdot C_1+C_2)$$

viz., if $$C_2>\frac{C_1}{1+q}$$

and $$2C_1<(1-q^2)C_1+(1+q)C_2$$

or $$C_2>\left(\frac{1+q^2}{1+q}\right)C_1$$

viz., if $$C_2>\left(\frac{1+q^2}{1+q}\right)C_1,$$

since $$\frac{1+q^2}{1+q}>\frac{1}{1+q}$$

Example 10

Calculate the probability of a staff leaving the present position (either due to resignation, promotion, retirement or death) in each year from the following survival table:

Year (r)	0	1	2	3	4	5	6	7	8	9	10
No, of staff in the same position at the end of year r	1000	940	820	580	400	280	190	130	70	30	0

SOLUTION

Let p_r denote the probability that a staff leaves the present position in the rth year.

Then, $$p_r = \frac{(N_{r-1} - N_r)}{N},$$

where N is the original staff in the system

$\therefore$ $$p_1 = \frac{N_0 - N_1}{N} = \frac{1000 - 940}{1000} = 0.06;$$

$$p_2 = \frac{N_1 - N_2}{N} = \frac{940 - 820}{1000} = 0.12;$$

Similarly,

$$p_3 = 0.24;$$
$$p_4 = 0.18;$$
$$p_5 = 0.12;$$
$$p_6 = 0.9;$$
$$p_7 = 0.06;$$
$$p_8 = 0.06;$$
$$p_9 = 0.04$$

and $$p_{10} = 0.03.$$

It can be verified that $\sum_{r=1}^{10} p_r = 1$

Example 11

A pharmaceutical company is planning to gradually build up over a period of 10 years and ultimately maintain a strength of 50 chemists in its research team. The serving chemists are terminated based on their experience and efficiency. The following data relate to the recruitment of 100 chemists in the beginning every year: No recruit is allowed to continue in service after 10 years.

Year (r)	0	1	2	3	4	5	6	7	8	9	10
No, of chemists who leave at the end of the year(r)	0	5	36	56	63	68	73	79	87	97	100

What is the recruitment necessary per year to maintain the required strength? There are 8 senior posts for which the length of service is the main criterion. What is the average length of service after which a new entrant may expect his promotion to one of these senior posts?

SOLUTION

The probability of a chemist continuing in service at the end of the rth year is computed as follows:

Year (r)	0	1	2	3	4	5	6	7	8	9	10
No of chemists leaving at the end of r	0	5	36	56	63	68	73	79	87	97	100
No of chemists in service at the end of r	100	95	64	44	37	32	27	21	13	3	0 = 436
Probability that a chemist is in service at the end of r (p_r)	1,	0.95,	0.64,	0.44,	0.37,	0.32,	0.27,	0.21,	0.13,	0.03,	0

Thus if we assume that if 100 persons are recruited at the begining of every year, the total number of chemists serving in the company would have been 436 over the period of 10 years viz., to maintain 436 chemists, the number to be recruited every year is 100.

∴ To maintain 50 chemists, the number to be recruited every year $= \frac{100}{436} \times 50 = 11 \cdot 47 \simeq 12$.

If 12 chemists are recruited every year, the number N_r of chemists continuing in service at the end of the rth year is given by $(12 \times p_r)$. The following table gives the number of chemists in service at the end of each of 10 years.

r	0	1	2	3	4	5	6	7	8	9	10
N_r	12	11.4	7.68	5.28	4.44	3.84	3.24	2.52	1.56	0.36	0
N_r(rounded)	12	11	8	5	4	4	3	2	1	0	0

As there are 8 senior posts for which the length of service is the criterion, we have to consider 8 or less senior persons in service.

We note that there are 3, 2 and 1 senior persons in service at the end of the 6th ,7th and 8th years.

∴ Promotions of the recruits to senior posts should be given to those who have completed 5 years of service.

EXERCISE 12

Part A (Short answer questions)

1. Why is replacement of items necessary in industrial situations?
2. Explain briefly some important replacement situations.
3. Are we justified in treating a staffing problem as a replacement problem?
4. Explain the terms: individual, preventive and group replacements.
5. Derive the condition for the optimal replacement age of items whose maintenance costs increase with time, assuming that money value remains constant.
6. Define the depreciation factor and present worth factor.
7. Derive the condition for the optimal replacement age of items whose maintenance costs increase with time, assuming that money value changes with time.

Part B

8. The cost of a machine is ₹ 18,000/- and its scrap value is ₹ 3000/-. The maintenance costs are found from experience as follows:

Year	1	2	3	4	5	6	7	8
Maintenance cost in ₹	300	750	1200	1800	2700	3750	4800	6000

When should the machine be replaced?

9. The cost of a machine is ₹ 90,000/-. The running cost and the re-sale value of the machine are as given here. Find the optimal age of replacement.

Year	1	2	3	4	5	6	7	8
Running cost (in 100's of ₹ s)	270	288	324	378	450	522	612	720
Resale value (in 100's of ₹ s)	630	450	306	216	144	90	90	90

10. A manufacturer wants to purchase one of the two machines A and B. The capital cost of either machine is ₹ 1, 00,000/-. The maintenance costs over the first 6 years for both machines are identical as given here:

Year (*r*)	1	2	3	4	5	6
M_r (₹)	12,500	15,000	18,750	25,000	37,500	52,500

However the scrap values of the machines vary slightly as given here:

End of year (*r*)	1	2	3	4	5	6
S_r for *A*(*r*)	62,500	41,250	25,000	13,750	11,500	10,250
S_r for *B*(*r*)	67,500	42,500	29,500	19,750	13,750	11,750

Find the optimal ages of replacement for the machines. Which machine would you suggest for purchase? Why?

11. A truck is priced at ₹ 1,50,000/- and running costs are estimated at ₹ 15,000/- for each of the first four years, increasing by ₹ 5000/- per year in the fifth and subsequent years. If the discount rate is $\gamma = 0.9$ per year, when should the truck be replaced? Assume that the truck will eventually be sold for scrap at a negligible price.

12. The capital cost of an equipment is ₹ 45,000/-. If money is worth 10% per year, find when the equipment should be replaced. The scrap value is negligible and the maintenance costs are given here:

Year	1	2	3	4	5	6	7
Maintenance cost	7500	9000	12000	15000	18500	24000	30,000

13. The costs of two new machines A and B are respectively ₹ 1.2 lakhs and ₹ 1.3 lakhs. The maintenance costs for A and B during the *n*th year are given by $M_n = 7000(n - 1)$ and $M_n = 6000(n - 1)$; $n =$ 1, 2, 3 ... respectively. If the discount rate is 0.8/ year, when will it be economical to replace the machines by new ones? Which machine can be preferred for purchase?

14. The failure pattern of bulbs is as follows:

Life in months	1	2	3	4	5	6
Proportion of bulbs failing	0.08	0.12	0.20	0.30	0.20	0.10

The cost of replacing an individual bulb when it fails is ₹ 15/-, while the cost when replaced on a group basis is ₹ 10/- per bulb. The company has 1000 bulbs of this type installed. Determine the optimum replacement policy.

15. The following mortality rates have been observed for a certain type of light bulb:

Week	1	2	3	4	5
Percentage failing by end of week	10	25	50	80	100

There are 1000 bulbs in use and it costs ₹ 12/- to replace an individual bulb which has burnt out. If all the bulbs are replaced simultaneously, it would cost ₹ 5/- per bulb. It is proposed to replace all bulbs at fixed intervals, whether or not they have burnt out and to continue replacing burnt out bulbs as they fail. At what intervals should all the bulbs be replaced?

16. The probability p_r of failure just before age r months are given here: If individual replacements costs ₹ 12/- and group replacement costs ₹ 5/- per item, find the optimal group replacement policy.

r	1	2	3	4	5	6	7	8	9	10	11
p_r	.01,	.03,	.05,	.07,	.10,	.15,	.20,	.15,	.11,	.08,	.05

17. A computer contains 10,000 resistors. When any resistor fails, it is replaced. The cost of replacing a resistor individually is ₹ 5/- only. If all the resistors are replaced at the same time, the cost of the resistor would be reduced to ₹ 2/- only. The probability p_r of failure of the resistor during the month r is given here. What is the optimum replacement plan?

Month (r)	1	2	3	4	5	6
p_r	0.03	0.07	0.20	0.40	0.15	0.15

18. A process control panel has 1000 LED's the failure rates of which are given here:

Just before week's (r)	1	2	3	4	5
Probability of failure (p_r)	0.15	0.20	0.25	0.20	0.20

The cost of replacing an individual LED is 3 times that of an LED in group replacement. Determine the optimum replacement policy.

19. The following failure rates have been observed for a certain item:

End of month (r)	1	2	3	4	5
Probability of failure to date (Σp_r)	0.10	0.30	0.55	0.85	1.00

The cost of replacing an individual item is ₹ 12.50. The decision is made to replace all items simultaneously at fixed intervals and also to replace individual items as they fail. If the cost of group replacement is ₹ 5/- per time, what is the best interval of group replacement? At what group replacement price per item, would a policy of strictly individual replacement become preferable to the adopted policy?

20. A big industry maintains a strength of 25 engineering graduates in its R & D wing .The serving graduates are terminated based on their experience and efficiency. The following data relate to the recruitment of 100 graduates in the beginning. No recruit is allowed to continue in the R & D wing after 10 years.

Year (r)	0	1	2	3	4	5	6	7	8	9	10
No. of graduates who leave at the end of year r	0	3	7	13	19	28	40	58	82	94	100

How many engineering graduates should be recruited every year to maintain the required strength? There are 5 posts of team heads for which the length of service is the main criterion. What is the average length of service after which a new entrant may expect his elevation to one of the posts of team heads? [see worked example (11)]

ANSWERS

Exercise 12

Part B

8. At the end of the 6th year; minimum average cost per year = ₹ 4250/-;
9. At the end of the 4th year; min. ave. cost per year = ₹ 26,550;
10. At the end of the 4th year for both A and B; Since $A(4)$ for B < $A(4)$ for A, machine B must be preferred.
11. Since M_{10} (= ₹ 45,000) < $A(10)$ (= ₹ 45,408) < M_{11}(= ₹ 50,000), the truck should be replaced at the end of the 10th year.
12. Since M_5 (= ₹ 18,500) < $A(5)$ (= ₹ 22,664) < M_6(= ₹ 22,837), the equipment should be replaced at the end of the 5th year.
13. For machine A, M_5 (= ₹ 28,000) < $A(5)$ (= ₹ 31,356) < M_6(= ₹ 35,000). Machine A can be replaced at the end of the 5th year.
    For machine B, M_6 (= ₹ 24,000) < $A(6)$ (= ₹ 28,560) < M_7(= ₹ 30,000)
    Machine B can be replaced at the end of the 6th year. Since $A(6)$ for $B < A(5)$ for A and the average costs for B in subsequent years are less, machine B can be preferred for purchase.
14. Average cost of IR (= ₹ 4035) < minimum average cost of GR (= ₹ 5029)
    It is economical to have individual replacement of bulbs.
15. Minimum average cost of GR at the end of 5 years (= ₹ 3396) < average cost of IR (= ₹ 3588); All bulbs should be replaced at the end of every 5 years.
16. Minimum average cost of GR at the end of 5 months (= ₹ 1641) < average cost of IR (= ₹ 1836); the optimal replacement policy is the GR policy at the end of 5 months.
17. Minimum average cost of GR at the end of 3 months (= ₹ 11,752) < average cost of IR (= ₹ 12,440); the optimal replacement policy is the GR policy at the end of 3 months.
18. Average cost of IR (= ₹ 969k) < minimum average cost of GR at the end of 4 months (= ₹ 1011k); the optimal replacement policy is the IR policy.
19. Optimal time of GR is the end of the third month, since the average cost of GR is the minimum then, equal to ₹ 4171. If the cost per item of GR ≥ ₹ 4·23, I.R. policy will be preferable.
20. 6; 5 years.

13 NETWORK SCHEDULING OF PROJECTS BY PERT/ CPM TECHNIQUES

13.0 INTRODUCTION

A network is a graphic representation of operations or activities and nodes of large projects such as construction of buildings, production, marketing, etc., which consume resources and time. Networks play an important role in determining an optimum solution to a problem, as it is easy to select the best sequence of operations in a network out of a finite number of available alternatives that been represented in the network.

The following terms are used in network analysis of projects:

Activity in a project is a job that requires time and resource for its completion and that is represented by an *arrow* with its head indicating the direction of progress in the project. An *event* in a project represents a point in time that signifies the completion of a certain activity and the beginning of a new activity. The point is called a *node*, and is denoted by a small circle, rectangle or other simple geometrical shapes. Thus, an activity can also be described by a pair of nodes, usually knows as the tail node and head node respectively.

The following rules for construction of networks (or arrow diagrams) are to be followed:

1. Only one activity should connect any two nodes. This rule in necessary so that an activity may be specified by giving the numbers of the start and end node.
2. There should be no loops in the network. In other words, no continuous path may go through one node twice.
3. Every node must have at least one activity preceding it and at least one activity following it, except for the node at the beginning and at the end of the network. In other words, the beginning node has no activity before it and the end node has no activity following it.
4. A dummy activity may be introduced to keep the continuity of sequence of jobs which represents the precedence relationship and does not consume time and resources. The dummy activity is represented by a dotted line in the network.

13.1 PERT/CPM

There are two basic scheduling techniques that make use of networks to analyse and complete a project. These are

1. Critical path method (CPM)
2. Program evaluation and review technique (PERT)

13.2 CRITICAL ACTIVITIES AND CRITICAL PATH

The manager or the planner of projects has to estimate how long it will take to complete projects. The time for completing an entire project is not necessarily the sum of all the individual activity times, since some activities can be done simultaneously. In fact, the total time for completion of a project will be governed by a particular set of jobs, knows as *critical activities*. The path formed by an unbroken chain of critical activities (jobs) from the start event to the end event of the project is called *the critical path* (or the path that takes the longest time to complete the project) of the network. Using the critical path, we can identify the jobs whose duration times must be shortened in order to shorten the time to complete the project.

An efficient method to find the critical jobs in a project depends on the computation of the earliest and the latest times for each activity which is defined here:

The starting time of the project is assumed to be zero. This is also taken as the *earliest start time* of jobs emerging from the start node 1 and denoted as ES $(1, \beta)$ $[= 0]$ for all β.

If t_{ij} represents the duration of the job (i, j) then $\text{ES}(j, \beta) = \max_i \{\text{ES}(i, j) + t_{ij}\}$, for all β, where ES (j, β) and ES(i, j) are the earliest start times of jobs starting and ending with the node j respectively.

If we have fixed a target time for completing the project, it is also taken as the *latest finish time* of jobs ending with the end node n of the project and denoted as LF(α, n), for all α

If t_{ij} represents the duration of the job (i, j), then $\text{LF}(\alpha, i) = \min_j \{\text{LF}(i, j) - t_{ij}\}$, for all α

The *earliest finish time* of an activity (i, j) is then found by using the formula $\text{EF}(i,j) = \text{ES}(i, j) + t_{ij}$.

The *latest start time* of the an activity (i, j) is then found by using the formula $\text{LS}(i,j) = \text{LF}(i, j) - t_{ij}$

Now the critical activities can be easily identified as explained here:

Any activity (i, j) will be a critical activity and hence, will lie on the critical path, provided it satisfies the condition (i) $\text{ES}(i, j) = \text{LS}(i, j)$ or (ii) $\text{EF}(i, j) = \text{LF}(i, j)$

Note: These conditions actually mean that there is no slack (or float or spare) time between the earliest start and latest start times of a critical activity.

Total and free slacks

There are two types of slack, namely, total slack (TS) and free slack (FS) for any activity (i, j), that are defined as follows:

The total slack of an activity is the maximum time by which it can be delayed before starting without affecting the project completion time. It is computed by using the formula

$$\text{TS}(i, j) = \text{LS}(i, j) - \text{ES}(i, j) \text{ or } \text{LF}(i, j) - \text{EF}(i, j)$$

Note: From this formula, it is clear that the total slack of a critical activity is zero.

The free slack of any activity represents the time by which the activity can be delayed before starting without affecting the staring time of subsequent activities

It is completed by using the formula

$$\text{FS}(i, j) = \text{TS}(i, j) - \min_{\beta}\{\text{TS}(j, \beta)\}$$

If j is the final node of the project, TS(j, β) is taken as zero.

Note: Total slack and free slack will be positive for non-critical activities and will be zero for critical activities.

13.3 MINIMUM COST SCHEDULING OF PROJECT BY CPM

Generally, the cost of a project includes direct costs and indirect costs. Direct costs are associated with the individual activities of the project such as manpower employed, equipment utilised, materials consumed, etc. The Indirect costs are associated with the whole project such as overhead items, administrative and establishment charges. Both these costs are affected by the length of the project. Direct costs increase if durations of activities are reduced by using extra resources, whereas indirect costs increase if durations of activities are increased. The total project cost is the sum of the direct and indirect costs.

Though one may be prepared to use extra resources to reduce the direct cost of an activity, the activity requires a minimum duration beyond which it cannot be further reduced. This minimum duration of the activity is called *the crash time* of the activity.

Since direct costs increase and indirect costs decrease as the durations of the various activities of the project are reduced, we intuitively feel that there must be some optimum project length corresponding to which the total project cost will be optimum (minimum). The CPM can be used to find such a minimum cost schedule. It is explained in the following procedure.

Step-by-step procedure

Step 1: Draw the network 1of the given project and identify the critical path corresponding to the normal durations of activities of the project. Note down the direct, indirect and total costs corresponding to this maximum length schedule of the project

Step 2: Generate new schedules by crashing critical jobs (one at a time) which will result in reduction of total project cost. For this, examine the cost–time slope (defined in the following equation) of each of the critical jobs and find the one having the least slope. This will be the job that can be shortened with minimum extra cost.

$$\text{Cost slope} = \frac{\text{Crash cost} - \text{Normal cost}}{\text{Normal time} - \text{Crash time}}$$

Step 3: Continue the process of crashing critical jobs by unit time (month, day, hour, etc.) at each step, until

(a) The total cost of the project gets reduced or
(b) The duration of the job (being crashed) reaches the crash point or
(c) Some other job becomes critical along a parallel path.

Step 4: Examine the remaining critical jobs on the original critical path and proceed as in Step 2.

Step 5: If there are parallel critical paths, then a combination of jobs, one on each of the critical paths, which results in a minimum sum of crashing costs, must be considered for crashing.

Step 6: Repeat the process until no further crashing of critical jobs is possible or the total cost of the project starts increasing. The schedule of the project at this final stage is the minimum cost schedule of the project.

13.4 PERT METHOD

If CPM can be considered as activity-oriented scheduling technique, the PERT can be regarded as event (node)-oriented scheduling technique. The events (nodes) of the networks are numbered in such a way that their ascending order indicates the direction of progress in the network.

In PERT terminology, ES(i, β)is called *the earliest expected occurrence time of the ith event* and denoted as $(\text{ET})_i$ and LF(α, j) is called *the latest allowable occurrence of the jth event* and denoted as $(\text{LT})_j$. $(\text{ET})_i$ and $(\text{LT})_j$ are computed using the same formulas for computing ES(i, β) and LF(α , j). Sometimes, in PERT-oriented project network, the following notation in respect to an event is used.

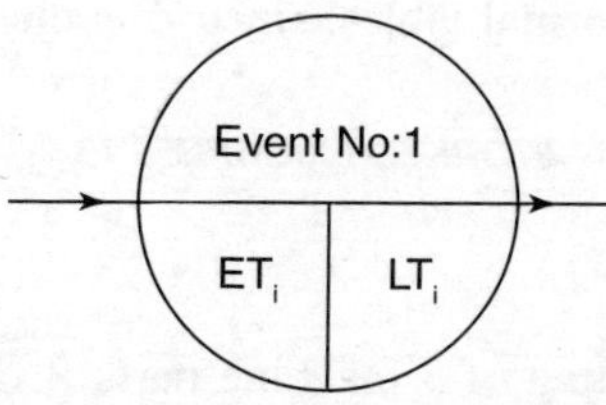

Except for these changes, the identification of the critical path and activity slacks are common features of both CPM and PERT.

The primary difference between CPM and PERT is that CPM uses single value deterministic time (duration) estimates for each of the activities, whereas PERT uses statistically related time estimates for each of the activities.

Accordingly, CPM is more useful for projects such as construction of houses, highways, bridges and in the field of manufacturing for which the time estimates of activities are available with some degree of certainty. PERT can be used in research and development projects and design of new machines where there is uncertainty regarding the durations of various activities.

In PERT approach, the expected (approximate) duration for each activity is computed based on the following time estimates.

(i) *Optimistic time estimate* (t_o): It is the shortest time required for completing an activity, if everything goes well.

(ii) *Pessimistic time estimate* (t_p): It is the maximum time that an activity could take under adverse conditions

(iii) *Most likely time estimate* (t_m): It is the normal time that an activity would take for completion under normal working conditions.

In PERT method, the duration of an activity t under uncertain conditions is regarded as a random variable that follows a beta distribution approximately with and

and
$$t_e = E(t) = \frac{1}{6}(t_o + 4t_m + t_p)$$

$$V(t) = \sigma_t^2 = \left(\frac{t_p - t_o}{6}\right)^2$$

Based on the expected times (t_e) of completion of various activities of the project, the critical path is determined.

If T denotes the total time necessary for completing the whole project, then T is also a random variable with

$$E(T) = \text{Expected value of } T = \Sigma t_e$$

and
$$V(T) = \text{Variance of } T = \Sigma \text{V}(t),$$

where the summation is done with respect to critical activities.

Since T is the sum of independent random variables, by central limit theorem, T follows a normal distribution with mean $E(T)$ and variance $V(T)$

Using this concept, the probability with which a project can be completed on or before a given scheduled time (ST) can be computed as follows:

$$P(T \leq ST) = P\left[\left\{\frac{T - E(T)}{\sigma_T}\right\} \leq \left\{\frac{ST - E(T)}{\sigma_T}\right\}\right]$$
$$= P[Z \leq A],$$

which is obtained from standard normal tables, since Z is the standard normal variable with mean zero and variance 1.

WORKED EXAMPLES 13

Example 1

The following project consists of casting of 3 machine parts P, Q, R in a foundry, taking them to the assembly shop through a lathe and/or milling machine and finally sending the assembled machine parts to stores.

Activity name	Description of the activity	Time required in days
a	Casting part P and taking it to the assembly shop (AS) from the foundry	4
b	Casting parts Q and R and taking them from foundry to the lathe	3
c	Turning part Q and taking it from lathe to the AS	6
d	Turning part R and taking it from lathe to the milling machine	2
e	Milling part R and taking it from the milling machine to the AS	4
f	Assembling parts P,Q and R and delivering them to the stores	3

Draw the network diagram for the project and number the events according to Fulkerson's rule. Name the various paths connecting the start node and the end node and hence, find the critical path and the shortest possible time for completing the project.

SOLUTION

The network consists of 6 activities (directed arrows) and 5 nodes corresponding to the places, such as foundry, lathe, etc.

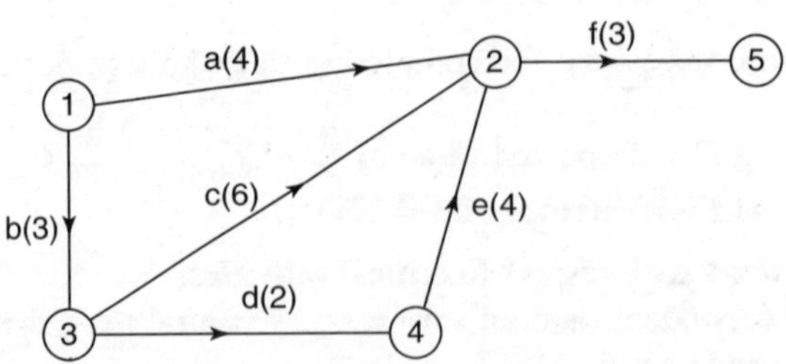

Fulkerson's rule states that numbering the events should be done in such a way that their ascending order indicates the direction of progress in the network. Accordingly, the events (nodes) have been numbered.

There are 3 paths from the start node to the end node, namely, (i) $a–f$ (1–2–5), (ii) $b–c–f$ (1–3–2–5) and $b–d–e–f$ (1–3–4–2–5) with a total time of 7, 12 and 12 days respectively.

The shortest possible time for completing the project is 12 days. Hence, there are two (parallel) critical paths for the project namely *b–c–f* and *b–d–e–f*

Example 2

Activities A, B, C, …, H, I constitute a project. The notation X < Y means that the activity X must be completed before Y can start. With this notation, draw the network diagram of the project satisfying the following constraints.

A < D, A < E, B < F, D < F, C < G, C < H, F < I, G < I

The duration (in days) of each job is as given here:

Activity	A	B	C	D	E	F	G	H	I
Duration	8	10	8	10	17	16	18	14	9

Number the nodes according to Fulkerson's rule. Find the critical path of the project and hence, the shortest possible time for completing the project.

Solution

The constraints on the activities of the project can be re-written as follows:

A < D, E; B, D < F; C < G, H; F, G < I

Note: A < D, E means that jobs D and E can start only after job A is completed. B, D < F means that F can start only after jobs B and D are completed.

From the modified form of constraints, we note that jobs A, B and C have no proceeding jobs and hence, they can emanate from the same starting node.

Jobs E, H and I have no succeeding jobs and hence, they can convergence at the same terminating node. Accordingly, the network diagram of the project is drawn and nodes numbered as shown here:

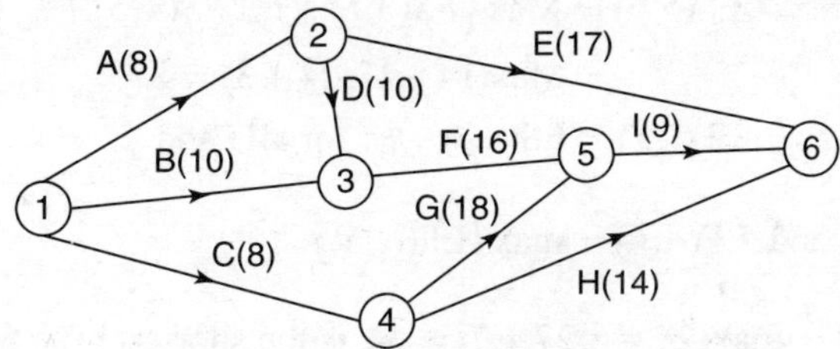

The various paths of the project are 1–4–6, 1–2–6, 1–3–5–6 and 1–2–3–5–6 with respective length 22, 25, 35 and 43 days.

Hence, the critical path is 1–2–3–5–6 and the critical jobs are A, D, F and I. The shortest possible time for completing the project is 43 days.

Example 3

Draw the network diagram for the following project whose activities and corresponding durations are given as follows.

Activity(*i, j*)	(1, 2)	1, 3)	(2, 3)	(2, 4)	(3, 4)	(3, 5)	(4, 5)	(4, 6)	(5,6)
Duration(days)	5	4	9	8	10	13	3	6	7

Compute the early start, early finish, late start, late finish, total slack and free slack of various activities and display them in a table. Find also the critical path of the project and hence, the shortest possible time for completing the project.

SOLUTION

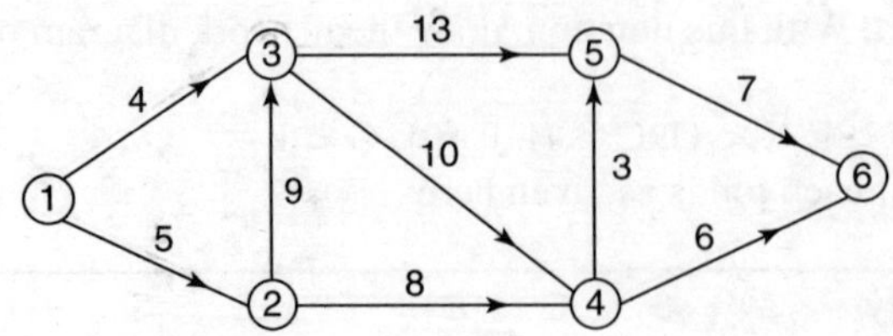

Since the jobs (1, 2) and (1, 3) originate from the start node 1,

$$ES(1, 2) = ES(1, 3) = 0$$

Since $$ES(j, \beta) = \max \{ ES(i, j) + t_{ij} \} \text{ for all } \beta,$$

we have

$$ES(2,3) = ES(1,2) + t_{12} = 0 + 5 = 5 = ES(2,4)$$

$$\begin{aligned} ES(3,4) &= \text{Max}\{ES(1,3) + t_{13}, ES(2,3) + t_{23}\} \\ &= \text{Max}(0 + 4, 5 + 9) = 14 \\ &= ES(3,5) \end{aligned}$$

$$\begin{aligned} ES(4,5) &= \text{Max}\{ES(2,4) + t_{24}, ES(3,4) + t_{34}\} \\ &= \text{Max}(5 + 8, 14 + 10) = 24 \\ &= ES(4,6) \end{aligned}$$

$$\begin{aligned} ES(5,6) &= \text{Max}\{ES(3,5) + t_{35}, ES(4,5) + t_{45}\} \\ &= \text{Max}(14 + 13, 24 + 3) = 27 \end{aligned}$$

$$EF(i, j) = ES(i, j) + t_{ij}, \text{ for all } i \text{ and } j$$

Since 6 is the final node and $LF(\alpha,6) = \max_{\alpha}\{EF(\alpha,6)\}$,

$\text{Max}\{EF(4,6), EF(5,6)\} = \max(24 + 6, 27 + 7) = 34$ is the shortest time for completing the project, viz., $LF(4,6) = 34 = LF(5,6)$

Since $$LF(\alpha, i) = \min_{j}\{LF(i, j) - t_{i,j}\} \text{ for all } \alpha,$$

we have

$$\begin{aligned} LF(4,5) &= LF(5,6) - t_{56} \\ &= 34 - 7 = 27 = LF(3,5) \end{aligned}$$

$$\begin{aligned} LF(3,4) &= \text{Min}\{LF(4,6) - t_{46}, LF(4,5) - t_{45}\} \\ &= \text{Min}(34 - 6, 27 - 3) \\ &= 24 = LF(2,4) \end{aligned}$$

$$LF(2.3) = \text{Min}\{LF(3,4) - t_{34}, LF(3,5) - t_{35}\}$$
$$= \text{Min}(24 - 10, 27 - 13)$$
$$= 14 = LF(1,3)$$
$$LF(1,2) = \text{Min}\{LF(2,3) - t_{23}, LF(2,4) - t_{24}\}$$
$$= \text{Min}(14 - 9, 24 - 8) = 5$$

Critical jobs are those for which ES(*i, j*) = LS(*i, j*) or EF(*i, j*) = LF(*i, j*). Accordingly, there are two critical paths, namely, 1–2–3–5–6 and 1–2–3–4–5–6

Total slackness (TS) of various jobs is computed by using the formula $TS(i, j) = LS(i, j) - ES(i, j)$ $or\ LF(i, j) - EF(i, j)$.

Free slackness (FS) of various jobs is computed by using the formula $FS(i, j) = TS(i, j) - \underset{\dagger}{\text{Min}}\, TS(j, \beta)$ where $TS(j, n) = 0$, and n is the final node.

The CPM calculations for the project are given in the following table

Activity (*i*, *j*)	Duration t_{ij}	ES(*i*, *j*)	EF(*i*, *j*)	LF(*i*, *j*)	LS(*i*, *j*)	TS(*i*, *j*)	FS(*i*, *j*)
(1, 2)	5	0	5	5	0	0	0
(1, 3)	4	0	4	14	10	10	10
(2, 3)	9	5	14	14	5	0	0
(2, 4)	8	5	13	24	16	11	11
(3, 4)	10	14	24	24	14	0	0
(3, 5)	13	14	27	27	14	0	0
(4, 5)	3	24	27	27	24	0	0
(4, 6)	6	24	30	34	28	4	4
(5, 6)	7	27	34	34	27	0	0

Example 4

Draw the network diagram for the following project:

Activity	A	B	C	D	E	F	G	H
Immediate predecessor	-	-	-	A	B	C	D, E	F, G
Duration (days)	3	5	4	3	4	9	8	4

Number the events according to Fulkerson's rule. Compute the earliest occurrence time and the latest occurrence time of various nodes of the project and hence, find the critical path of the project.

SOLUTION

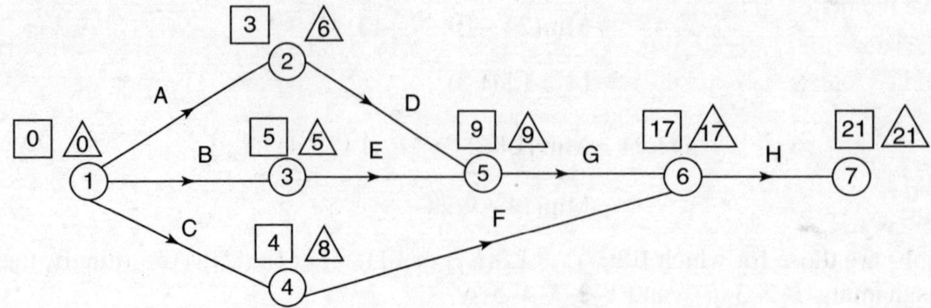

Proceeding as in example 3, we can get the following results:

$$ES(1,2) = ES(1,3) = ES(1,4) = 0;$$

$$ES(2,5) = 3; ES(3,5) = 5;$$

$$ES(4,6) = 4;$$

$$ES(5,6) = 9$$

and $$ES(6,7) = 17$$

Using $EF(i, j) = ES(i, j) + t_{ij}$, we get the following result:

$$EF(1,2) = EF(1,3) = 5; EF(1,4) = 3;$$

$$EF(1,4) = 4; EF(2,5) = 6;$$

$$EF(3,5) = 9;$$

$$EF(4,6) = 13; EF(5,6) = 17$$

and $$EF(6,7) = 21$$

Since (6, 7) is the last activity, LF(6, 7) is assumed to be EF(6, 7) = 21
Proceeding as in example 3, we compute $LF(5,6) = 17; LF(4,6) = 17;$

$$LF(3,5) = 9; LF(2,5) = 9;$$

$$LF(1,4) = 8; LF(1,3) = 5;.$$

$$LF(1,2) = 6$$

Using $LS(i, j) = LF(i, j) - t_{ij}$, we get the following results

$$LS(6,7) = 17; LS(5,6) = 9;$$

$$LS(4,6) = 8; LS(3,5) = 5;$$

$$LS(2,5) = 6; LS(1,4) = 4;$$

$$LS(1,3) = 0; LS(1,2) = 0$$

Note: ES(i, j) is placed inside a □ and written above the node i and LS(i, j) is placed inside a Δ and written above the node i in the network diagram.

The critical activities (i, j) are those for which ES(i, j) = LS(i, j), viz., those connecting the nodes for which □ and Δ entries are equal. Hence, the critical path of the project is B–E–G–H.

Since, ES(i, β) = ET, the ET's of the nodes 1, 2,…, 6 are 0, 3, 5, 4, 9 and 17.

Since, LF(α, j) = LT(j), the LT's of the nodes 2, 3, 7, …, are 6, 5, 8, 8, 17 and 21.

Example 5

For the project whose details are given here, find the maximum length of the schedule and the minimum cost schedule when the overhead cost is Rs 100/- per day

Event No	Successor event No	Normal duration (days)	Normal cost (Rs)	Minimum (crash) duration (days)	Minimum (crash) cost (Rs)
1	2	6	280	2	440
1	3	7	450	5	510
1	4	5	245	3	275
2	5	5	225	3	325
5	6	4	245	2	365
3	6	4	310	2	380
4	6	7	160	3	240

Solution

The paths of the project are A–D–G, B–E and C–F with respective lengths 15, 11 and 12 days. Hence, the critical path is A–D–G, corresponding to the maximum length of the schedule.

The network diagram of the given project is as follows:

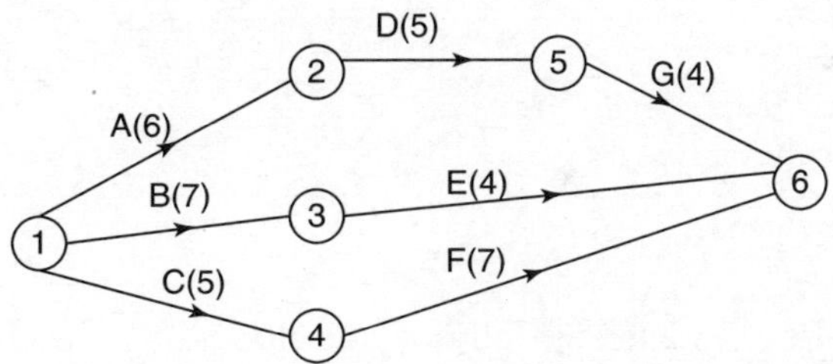

Initial direct cost of the project = sum of the costs of all the activities of the project at normal cost = 280 (A) + 450 (B) + 245 (C) + 225 (D) + 310 (E) + 160 (F) + 245 (G) = Rs 1915/-

Cost slopes (= costs of crashing by 1 day) of the activities of the project are computed by using the formula.

$$\text{Cost slope(crashing cost by 1 day)} = \frac{\text{Crash cost} - \text{Normal cost}}{\text{Normal time} - \text{Crash time}}$$

They are tabulated here:

Activity	A	B	C	D	E	F	G
Cost slope	$\frac{440-280}{6-2}$	$\frac{510-450}{7-5}$	$\frac{275-245}{5-3}$	$\frac{325-225}{5-3}$	$\frac{380-310}{4-2}$	$\frac{240-160}{7-3}$	$\frac{365-245}{2}$
	or 40	30	15	50	35	20	60

Computations of cost scheduling / shortening of the project are tabulated here:

Stage No:	Jobs considered for crashing with cost / day	Jobs selected for crashing with days	Direct cost(Rs)	Critical jobs with durations in days after crashing	Indirect cost (Rs)	Total cost(Rs)
0	-	-	1915	A(6), D(5), G(4)	$15 \times 100 = 1500$	3415
1	(i) A(40)					
	(ii) D(50)					
	(iii) G(60)	A(3)	$1915 + 120 = 2035$	(i) A(3), D(5),G(4)	$12 \times 100 = 1200$	3235
				(ii) C(5), F(7)		
2	(i) A(40) & C(15)√					
	(ii) A(40) & F(20)	A(1) & C(1)	$2035 + 55 = 2090$	(i) A(2), D(5),G(4)		
				(ii) C(4), F(7)	$11\times 100 = 1100$	3190
				(iii) B(7),E(4)		
				(A cannot be crashed further)		
3	(i) D(50), B(30), C(15) √					
	(ii) D(50), B(30), F(20)					
	(iii) D(50), E(35),C(15)					
	(iv) D(50), E(35), F(20)					
	(v) G(60), B(30), C(15)			(i) A(2), D(4), G (4)		
	(vi) G(60), B(30), F(20)			(ii) B(6), E(4)		
	(vii) G(60), E(35), C(15)			(iii) C(3), F(7)		

(Contd)

Table (*Contd*)

Stage No:	Jobs considered for crashing with cost / day	Jobs selected for crashing with days	Direct cost(Rs)	Critical jobs with durations in days after crashing	Indirect cost (Rs)	Total cost(Rs)
	(viii) G(60), E(35), F(20)	D(1), B(1) & C(1)	2090 + 95 = 2185	(C cannot be further crashed)	10 × 100 = 1000	3185
4	(i) D(50), B(30), F(20) √			(i) A(2), D(3),G(4)		
	(ii) D(50), E(35), F(20)			(ii) B(5), E(4)		
	(iii) G(60), B(30), F(20)			(iii) C(3), F(6)	9 × 100 = 900	3185
	(iv) G(60), E(35), F(20)	D(1), B(1) & F(1)	2185 + 100 = 2285	(B and D cannot be further crashed)		
5	G(60), E(35), F(20)	G(1), E(1) & F(1)	2285 + 115 = 2400	(i) A(2), D(3),G (3)		
				(ii) B(5), E(3)		
				(iii) C(3), F(5)	8 × 100 = 800	3200
6	G(60), E(35), F(20)	G(1), E(1) & F(1)	2400 + 115 = 2515	(i) A(2), D(3),G(2)		
				(ii) B(5), E(2)		
				(iii) C(3),F(4)	7× 100 = 700	3215

Further crashing is not possible after Stage 6, since none of the jobs in the paths A–D–G and B–E can be further crashed.

From the tabulated computations, we note that there are two alternative optimal duration schedules (10 or 9 days), namely,

(i) A(2)–D(4)–G(4), B(6)–E(4) and C(3)–F(7) and

(ii)A(2)–D(3)–G(4), B(5)–E(4) and C(3)–F(6)

The minimum total cost of each of the schedules = Rs 3185/-

Example 6

The data on normal time and normal cost and crash time and crash cost for a project are given in the following table:

Activity	Normal		Crash	
	Time (day)	Cost (Rs)	Time (day)	Cost(Rs)
A(1, 2)	8	100	6	200
B(1, 3)	4	150	2	350

(*Contd*)

Table (*Contd*)

Activity	Normal		Crash	
	Time (day)	Cost (Rs)	Time (day)	Cost(Rs)
C(2, 4)	2	50	1	90
D(2, 5)	10	100	5	400
E(3, 4)	5	120	1	200
F(4, 5)	3	80	1	100

The indirect cost of the project is Rs 300/- per day

Draw the network, crash the activities systematically and find the optimum project completion time and the associated total cost of the project.

SOLUTION

The network of the given project is as follows:

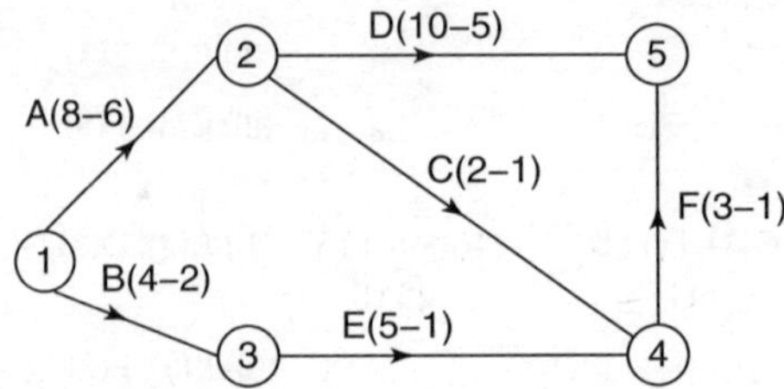

Initial direct cost of the project = sum of the normal costs of all the activities =Rs 100 + Rs 150 + Rs 50 + Rs 100 + Rs 120 + Rs 80 = Rs 600/-

Cost of crashing the various activities by 1 day are computed using the formula.

$$\text{Cost of crashing per day (cost slope)} = \frac{\text{Crash cost} - \text{Normal cost}}{\text{Normal time} - \text{Crash time}}$$

They are tabulated as follows:

Activity	A	B	C	D	E	F
Crash cost/day	$\frac{200-100}{8-6}$ $=50$	$\frac{350-150}{4-2}$ $=100$	$\frac{90-50}{2-1}$ $=40$	$\frac{400-100}{10-5}$ $=60$	$\frac{200-120}{5-1}$ $=20$	$\frac{100-80}{3-1}$ $=10$

Computations of systematic cost crashing and the associated total costs are tabulated below:

The paths of the project are A–D, A–C–F and B–E–F with respective lengths (based on normal times of jobs) ... 18, 13 and 12 days. Hence, the critical path is A–D corresponding to the maximum length.

Stage No:	Jobs considered for crashing with cost/day	Jobs selected for crashing with days	Direct Cost(Rs)	Critical jobs with times after crashing	Indirect Cost(Rs)	Total Cost(Rs)
0		–	600	A(8), D(10)	18×300 $= 5400$	6000
1	(i) A(50) ✓ (ii) D(60)	A(2)	$600 + 100 = 700$	(i) A(6), D(10) (A cannot be further crashed)	16×300 $= 4800$	5500
2	D(60)	D(4)	$700 + 240 = 940$	(i) A(6),D(6) (ii) B(4),E(5),F(3)	12×300 $= 3600$	4540
3	(i) D(60), B(100) (ii) D(60), E(20) (iii) D(60),F(10) ✓	D(1) and F(1)	$940 + 70 = 1010$	(i) A(6), D(5) (ii) B(4), E(5), F(2) (D cannot be further crashed)	11×300 $= 3300$	4310

As further crossing of any job on the path (A–D) is not possible, the minimum time required to complete the project is 11days and the optimum total cost Rs 4310/-

Example 7

A small marketing project consists of 6 jobs as given in the following table. The normal time, the crash time and the cost in Rs of crashing per day of each job are also given:

Job	Normal duration (days)	Minimum (crash) duration (days)	Cost of crashing (per day)
A(1–2)	9	6	40
B(1–3)	8	5	50
C(1–4)	15	10	60
D(2–4)	5	3	20
E(3–4)	10	6	30
F(4–5)	2	1	80

Find the minimum cost schedule by systematically crashing the jobs from their normal project length to the minimum project length, assuming that the overhead cost total is Rs 120/- per day.

SOLUTION

Since the normal costs of the activities are not given in the problem, we will assume that the initial direct cost of the project is assumed as a constant k.

The paths of the project are A–D–F, B–E–F and C–F with respective normal lengths 16, 20 and 17 days.

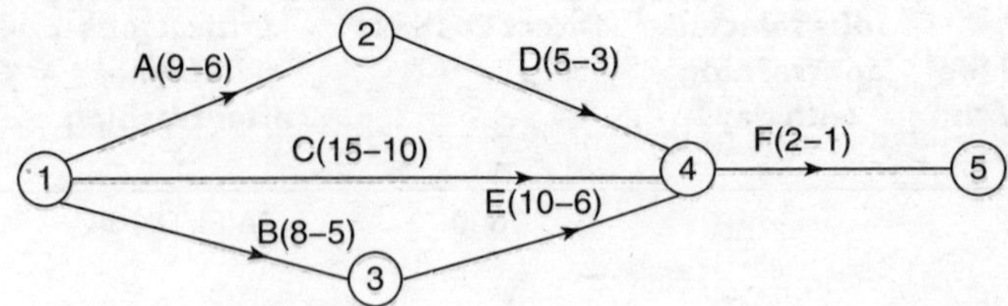

Computations of systematic cost crashing and the associated total costs are tabulated here:

Stage No	Jobs considered for crashing with cost / day	Jobs selected for crashing with days	Direct cost (Rs)	Critical jobs with times after crashing	Indirect cost (Rs)	Total cost(Rs)
0	-	-	k	B(8), E(10), F(2)	$20 \times 120 = 2400$	$(k + 2400)$
1	(i) B(50)	E(3)	$k + 90$	(i) B(8), E(7), F(2)	17×120	$(k + 2130)$
	(ii) E(30) √			(ii) C(15), F(2)		
	(iii) F(80)					
2	(i) B(50), C(60)	C(1) & E(1)	$k + 180$	(i) B(8), E(6), F(2)	16×120	$(k + 2100)$
	(ii) B(50), F(80)			(ii) A(9), D(5), F(2)		
	(iii) C(60), E(30) √			(iii) C(14), F(2)		
	(iv) E(30), F(80)			(E cannot be crashed further)		
3	(i) A(40), B(50), C(60)	B(2), C(2) & D(2)	$k + 440$	(i) B(6), E(6), F(2)	14×120	$(k + 2120)$
	(ii) A(40), B(50), F(80)			(ii) A(9), D(3), F(2)		
	(iii) B(50), C(60), D(20)√			(iii) C(12), F(2)		
	(iv) B(50), D(20), F(80)			(D cannot be crashed further)		
	(v) B(50), C(60), F(80)					
4	(i) A(40), B(50), F(80)	A(1), B(1), C(1)	$k + 590$	(i) B(5), E(6), F(2)	13×120	$k + 2150$
	(ii) A(40), B(50), C(60)√			(ii) A(8), D(3), F(2)		
	(iii) A(40), C(60), F(80)			(iii) C(11), F(2)		
	(iv) B(50), C(60), F(80)			(B cannot be further crashed)		
5	F(80)	F(1)	$k + 670$	(i) B(5), E(6), F(1)	12×120	$k + 2110$
				(ii) A(8), D(3), F(1)		
				(iii) C(11), F(1)		

The minimum cost schedule of Rs (k +2100) occurs when the project length is 16 days.

Example 8

The following table gives the jobs of a project along with their time estimates:

Job (i–j)	Duration (days)			Duration (days)			
	Optimistic (t_0)	Most likely (t_m)	Pessimistic (t_p)	Job (i, j)	t_o	t_m	t_p
1–2	1	7	13	-	-	-	-
1–6	2	5	14	4–5	5	5	17
2–3	2	14	26	6–7	5	8	29
2–4	2	5	8	5–8	3	3	9
3–5	7	10	19	7–8	8	17	32

(i) Draw the project network
(ii) Calculate the length and variance of the critical path

Find the probability that the project will be completed in 40 days

Solution

First we compute $t_e = E(t) = \frac{1}{6}(t_0 + 4t_m + t_p)$ and $V(t) = \left(\frac{t_p - t_0}{6}\right)^2$ for the activities of the project.

Job	1–2	1–6	2–3	2–4	3–5	4–5	6–7	5–8	7–8
E(t)	7	6	14	5	11	7	11	4	18
V(t)	4	4	16	1	4	4	16	1	16

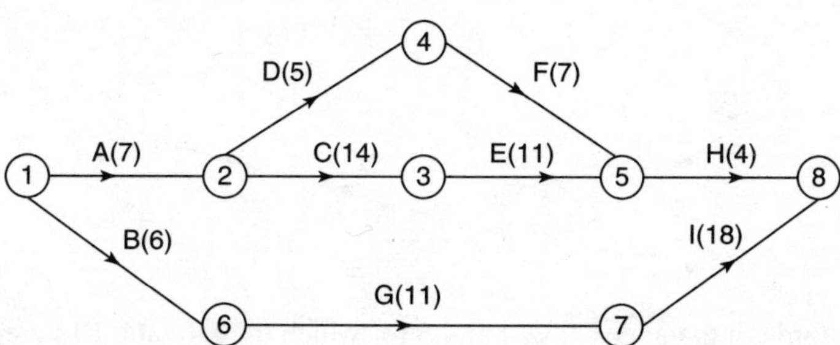

The paths of the project are A–D–F–H, A–C–E–H and B–G–I with respective lengths based on the expected activity times 23, 36 and 35 days

If T is the critical length of the project, $E(T) = 36$ and $V(T) = \Sigma V(\text{critical jobs}) = 4 + 16 + 4 + 1 = 25$

T follows a normal distribution with mean 36 and variance 25.

Now
$$P(T \leq 40) = P\left\{\frac{T - E(T)}{\sigma(T)} \leq \frac{40 - 36}{5}\right\}$$

$= P(z \leq 0.8)$, where z is the standard normal variable

$= 0.5 + P(0 \leq z \leq 0.8)$

$= 0.7882$ from the normal tables.

Example 9

The following table gives the jobs of a small project along with their time estimates:

Activity	1–2	1–3	2–4	3–4	4–5	3–5
t_o	2	9	5	2	6	8
t_m	5	12	14	5	9	17
t_p	14	15	17	8	12	20

(i) Draw the PERT network.
(ii) Find the earliest and latest occurrence times of each node and hence identify the critical path.
(iii) Calculate the length and variance of the critical path
(iv) Find the time by which the project can be completed with a probability of 0.9

Solution

Job(i, j)	1–2	1–3	2–4	3–4	4–5	3–5
E(t)	6	12	17	5	9	16
V(t)	4	1	4	1	1	4

We now draw the network and using the $E(t)$ values of the jobs, we compute the earliest (ET) and the latest (LT) occurrence times of various nodes

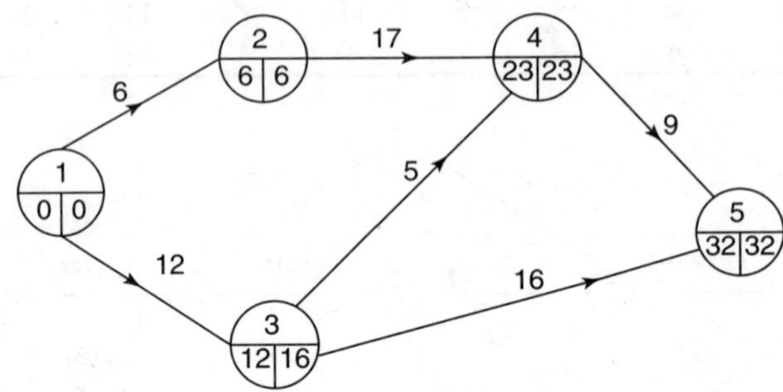

The path passing through the nodes 1, 2, 4 and 5 for which the ETs and LTs are equal is the critical path.

$\therefore$ $E(T)$ length of the critical path $= 6 + 17 + 9 = 32$

$$V(T) = \sum V(\text{critical jobs})$$

$$= 4 + 4 + 1 = 9$$

T follows a normal distribution with mean 32 and variance 9.
Let T_R be the required time of completion of the project.

Then, $$P(T \leq T_R) = 0.9$$

$\therefore$ $$P\left\{\frac{T-32}{\sqrt{9}} \leq \frac{T_R - 32}{\sqrt{9}}\right\} = 0.9$$

viz., $$P\left\{z \le \frac{T_R - 32}{3}\right\} = 0.9$$

or $$P\left\{0 \le z \le \frac{T_R - 32}{3}\right\} = 0.4$$

From the normal tables, $\frac{T_R - 32}{3} \simeq 1.28$

$\therefore \quad T_R \simeq 32 + 3.84 \simeq 36$ days.

Example 10

Assuming that the expected times of the activities are normally distributed, find the probability of meeting the scheduled date of 30 days. Also find the date on which the project manager can complete the project with a probability of 0.9. The required data given here may be used.

Activity	Days		
	Optimistic	Most likely	Pessimistic
1–2	2	5	14
1–3	9	12	15
2–4	5	14	17
3–4	2	5	8
4–5	6	9	12
3–5	8	17	20

SOLUTION

Job (*i–j*)	1–2	1–3	2–4	3–4	4–5	3–5
E(*t*)	6	12	13	5	9	16
V(*t*)	4	1	4	1	1	4

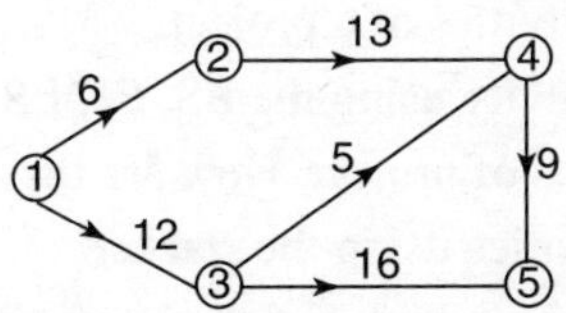

From the network of the project given here, we see that the paths of the network are 1–2–4–5, 1–3–4–5 and 1–3–5 with respective lengths based on expected activity times 28, 26 and 28 days. Hence, there are two alternative critical paths, namely, 1–2–4–5 and 1–3–5 with $E(T) = 28$, where T is the critical length of the project.

$V(T)$ corresponding to the path 1–2–4–5 = 9

and $V(T)$ corresponding to the path 1–3–5 = 5

Important note: Whenever there is more than one critical path with the same value of $E(T)$, the path having the largest value of variance has to be taken as the critical path for further PERT analysis.

Hence, the critical path for further PERT analysis is 1–2–4–5 with $E(T) = 28$ and $V(T) = 9$

$$P(T \le 30) = P\left\{\frac{T - E(T)}{\sqrt{V(T)}} \le \frac{30 - 28}{3}\right\}$$

$$= P\{z \le 0.6667\}$$

$$= 0.5 + P\{0 \le z \le 0.6667\}$$

$$= 0.5 + 0.2486, \text{ from the normal tables}$$

viz., $P(T \le 30) = 0.75$

Let T_R be the completion time of the project with probability 0.9

viz., $P\{T \le T_R\} = 0.9$

viz., $P\left\{\frac{T-28}{3} \le \frac{T_R - 28}{3}\right\} \le 0.9$

viz., $P\left\{0 \le z \le \frac{T_R - 28}{3}\right\} \le 0.4$

From the normal tables, $\frac{T_R - 28}{3} = 1.28$

$\therefore$ $T_R = 28 + 3.84 \simeq 32$ days.

EXERCISE 13

Part A (Short answer questions)

1. Define activity and event in a project. How are they represented in the network of the project?
2. State the rule to be followed in the construction of networks of a project.
3. Define the earliest start time and the latest finish time of activities of a project. How are they computed?
4. Define the earliest finish time and the latest start time of activities of a project. How are they computed?
5. Define critical path and critical activities of a project.
6. How will you identify a critical activity using the ES, EF, LS and LS times of activities?
7. Define total and free slack activities of project. How are they computed?
8. How will you identify critical activities using the slacks?
9. Define the earliest expected occurrence time and the latest allowable occurrence time of an event of a project.
10. How are ES times of activities and ET times of events related?
11. How are LF times of activities and LT times of events related?
12. In what ways do CPM and PERT techniques differ from each other?
13. Define normal and crash durations and normal cost and crash cost of an activity. How are they used to find the crashing cost per unit time?
14. Define optimistic, pessimistic and most likely time estimates of an activity.
15. How are the mean and variance of the duration of an activity computed?
16. How are the mean and variance of the completion time of a project computed?
17. What are the probability distributions followed by the duration of an activity (t) and the project completion time (T)?

27. The following table gives information regarding the mean duration and variance of the activities of a project. Find the probability of completing the project within 28 days.

Job	E(t)	V(t)	Job	E(t)	V(t)
1–2	7	2	3–5	4	0
1–3	8	3	4–6	9	1
2–4	5	2	5–6	10	4
2–6	6	2	-	-	-

28. A project consists of 8 activities with the time estimate given in the following table:

Activity	A	B	C	D	E	F	G	H
Immediate predecessor	–	–	–	A	B	C	D,E	F,G
Optimistic time (days)	1	1	2	1	2	2	3	2
Most likely time (days)	1	4	2	1	5	5	6	2
Pessimistic time (days)	7	7	8	1	14	8	15	8

Find the time by which the project can be completed with a probability of 0.95.

29. The following table gives the jobs of a project with their time estimates in days:

Job (i–j)	1–2	1–6	2–3	2–4	3–5	4–5	6–7	5–8	7–8
t_0	3	2	6	2	5	3	3	1	9
t_m	6	5	12	5	11	6	9	4	12
t_p	15	14	30	8	17	15	27	7	15

Find approximately the probability that the jobs on the critical path can be completed in 41 days.

30. The following table gives the jobs of a project with their time estimate in days:

Job (i–j)	1–2	2–3	2–4	3–4	3–5	3–7	4–5	4–6	5–6	5–7	6–7
t_0	1	1	2	0	2	6	4	3	0.5	5	3
t_m	2	4	4	0	3	8	6	5	1.0	7	5
t_p	9	7	12	0	4	16	8	7	1.5	15	13

Find the time by which the project can be completed with a probability of 0.85.

ANSWERS

Exercise 13

Part B

18.

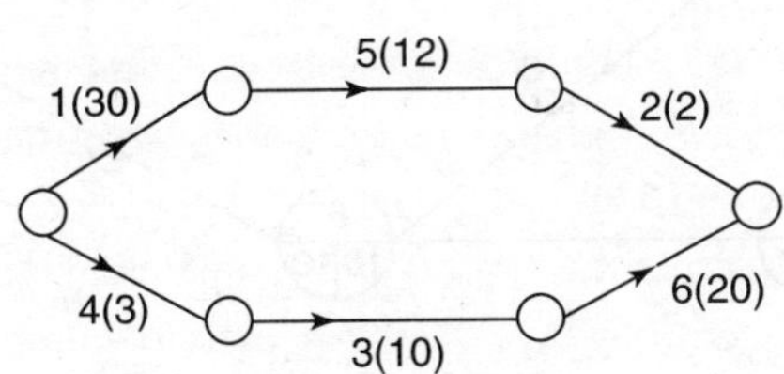

Shortest time for the completion of the project = 44 days

19.

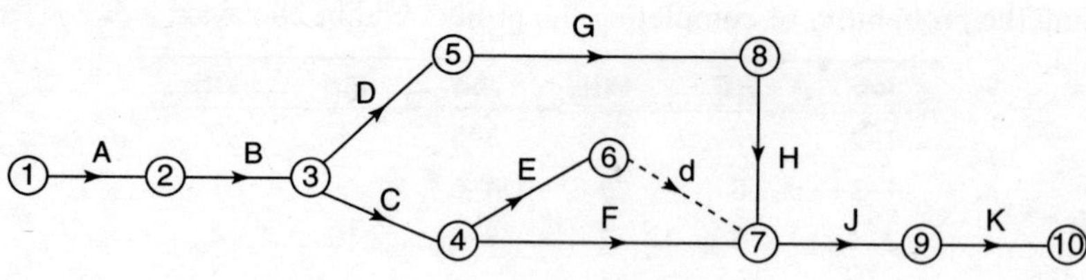

20.

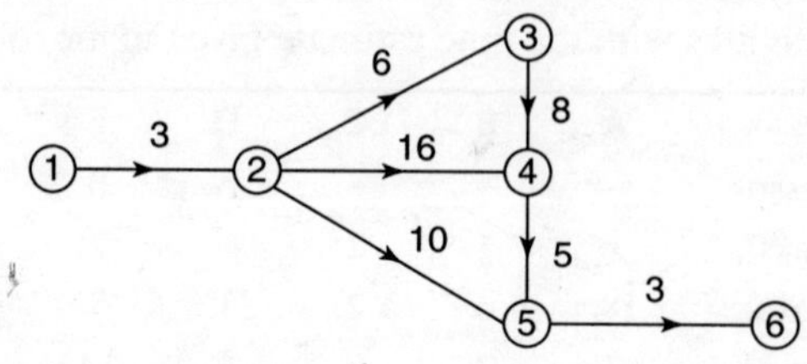

Job(*i, j*)	t_{ij} (days)	ES	EF	LF	LS	TS	LS
1–2	3	0	3	3	0	0	0
2–4	16	3	19	19	3	0	0
2–3	6	3	9	11	5	2	0
3–4	8	9	17	19	11	2	2
2–5	10	3	13	24	14	11	11
4–5	5	19	24	24	19	0	0
5–6	3	24	27	27	24	0	0

Critical path: 1–2–4–5–6

Length of the critical path =27 days

21.

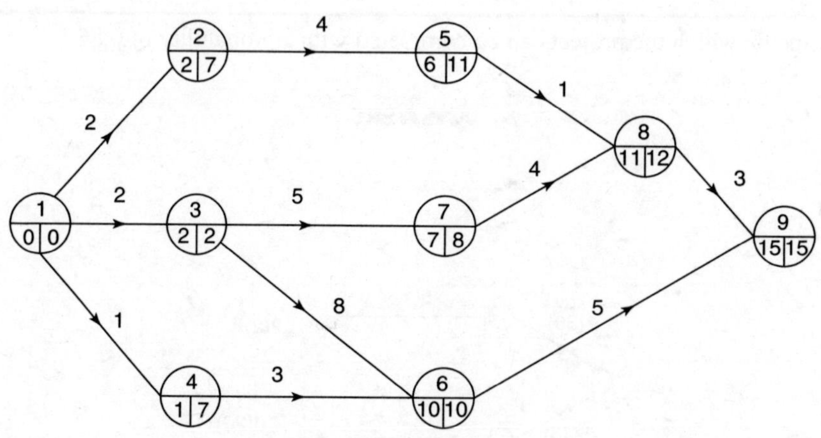

Critical path: 1–3–6–9; length of the critical path = 15 days

22.

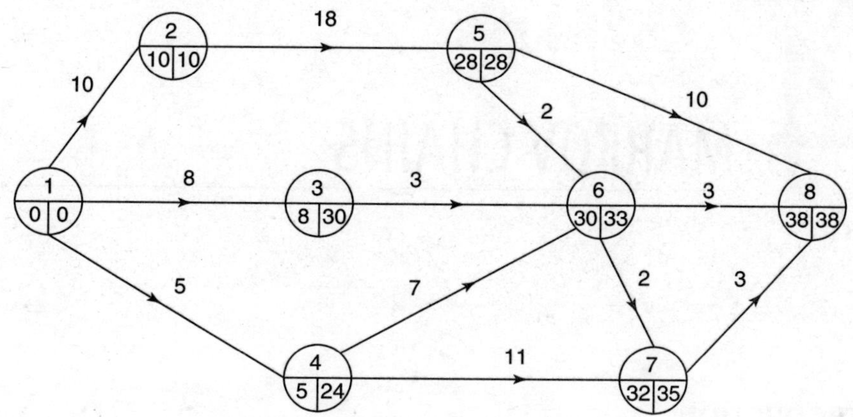

Critical path: 1–2–5–8; length of the critical path = 38 days

23.

No of days:	48	43	38	37	36
Total cost (Rs)	58,300	53,900	49,600	48,800	48200

24. Optimum duration =16 days

    Minimum total cost =Rs 15,000/- corresponding to 17 days

25. Minimum sum =Rs 2200/- corresponding to 42 days

26. Optimum length of the schedule = 9 days

    Minimum total cost = Rs 880/-

27. 0.91;

28. 24 days;

29. 0.86 ;

30. 25 days

Chapter 14 MARKOV CHAINS

14.0 INTRODUCTION

The reader is assumed to have a basic idea about Probability theory, Random experiment and Random variable (RV). A random variable is a rule (or function) that assigns a real number to every outcome of a random experiment, while a random process is a rule (or function) that assigns a time function to every outcome of a random experiment.

For example, consider the random experiment of tossing a die at $t = 0$ and observing the number on the top face. The sample space of this experiment consists of the outcomes $\{1, 2, 3, 4, 5, 6\}$. For each outcome of the experiment, let us arbitrarily assign a function of $(0 \le t < \infty)$ in the following manner:

Outcome:	1	2	3	4	5	6
Function of time:	$x_1(t) = -4$	$x_2(t) = -2$	$x_3(t) = 2$	$x_4(t) = -4$	$x_5(t) = -\frac{b}{2}$	$x_6(t) = \frac{b}{2}$

The set of functions $\{x_1(t), x_2(t), \ldots, x_6(t)\}$ represents a random process.

Definition

A random process is a collection (or ensemble of R.V. is $\{X(s,t)\}$ that are functions of a real variable, namely time t, where $s \in$ **the sample space** S and $t \in$ **the parameter or index set** T.

In electrical systems, voltage or current wave-forms are used as signals for collecting, transmitting or processing information as well as for controlling and providing power to a variety of devices. These signals are functions of time and are of two classes – deterministic and random. Deterministic signals can be described by the usual mathematical functions with time t as the independent variable. But a random signal always has some element of uncertainity associated with it and hence it is not possible to determine its value exactly at any given point of time. However we may be able to describe the random signal in terms of its average properties such as the average power in the random signal, its spectral distribution and the probability that the signal amplitude exceeds a given value. The probabilistic is an example of a random process or stochastic process.

We have to note that

(i) if s and t are fixed, $\{X(s,t)\}$ is a mere number

(ii) if s is fixed, $\{X(s,t)\}$ is simply a time function

(iii) if t is fixed, $\{X(s,t)\}$ is a random variable and

(iv) if s and t are variables, $\{X(s,t)\}$ is a set or collection of random variables that are time functions, viz., $\{X(s,t)\}$ is a random process.

As the dependence of a random process on s is obvious, s will be usually omitted in the notation of a random process. If the parameter set T is discrete, the random process will be denoted by $\{X(n)\}$ or $\{X_n\}$. If the parameter set T is continuous, the process will be denoted by $X(n)$.

Depending on the discrete or continuous nature of the state space s and the parameter set T, a random process is classified into four types. If T is discrete (irrespective of the nature of s), $\{X(t,s)\}$ is called a *random sequence*. If T is continuous (irrespective of the nature of s), $\{X(t,s)\}$ is called a *random process*. Sometimes a random sequence is also referred to as a random process.

14.1 MARKOV PROPERTY AND MARKOV PROCESS

An important feature of a random process is the relationship among the members of the set $\{X(s,t)\}$. Usually the nature of relationship is understood by the joint distribution function of the member R.V.'s. A random process is said to be specified, only when the parameter set, the state space and the nature of dependence relationship existing among the members of the set $\{X(s,t)\}$ are specified.

Based on the dependence relationship among the members of the process, random processes are classified broadly into a few special types, of which an interesting one is the *Markov process*.

Markov process is a random process possessing the property, called Markovian property, which is stated below: The value of a random process depends only upon the most recent previous value and is independent of all values in the more distant past. In other words, a Markov process is one in which the future value is independent of the past values, given the present value. Alternatively, if for $t_1 < t_2 < t_3 < \cdots t_n < t$, $P\{X(t) \le x / x(t_1) = x_1, x(t_2) = x_2, \cdots, x(t_n) = x_n\} = P\{X(t) \le x / x(t_n) = x_n\}$, then the process $\{X(t)\}$ is called a Markov process.

14.2 APPLICATIONS OF MARKOV ANALYSIS

Marlov analysis is a way of analysing the current movement of some variable in an attempt to forecast the future movement of the same variable. This method has come into use as a marketing aid, for examining and predicting the behaviour of customers from the stand point of their loyalty to one brand and their witching patterns to other brands. Although this is a leading use of the Markov technique, there are also er areas in which its application has made significant contribution.

onsider the experiment of tossing a fair coin a number of times. The possible outcomes at each trial wo – 'head' with probability $\frac{1}{2}$ and 'tail' with probability $\frac{1}{2}$. If we denote the outcome of the nth which is a RV, by X_n and the outcomes 'head' and 'tail' by 1 and 0 respectively, then

$$P\{X_n = 1\} = \frac{1}{2}$$

$$P\{X_n = 0\} = \frac{1}{2}; n = 1, 2, \cdots$$

Thus, we have a sequence of independent RVs $X_1, X_2, \cdots$, since the trials are independent and hence the outcome of the nth trial does not depend in any way on the previous trials.

Consider now the RV that represents the total number of heads in the first n trials and is given by $S_n = X_1 + \cdots + X_n$. The possible values of S_n are 0, 1, 2, ... n. If $S_n = k(k = 0,1,\cdots n)$, then the $RVS_{n+1}(= S_n + X_{n+1})$ can assume only 2 possible values, namely $k+1$ [if the $(n+1)th$ trial results in a head] and k [if the $(n+1)$ th trial results in a tail].

Thus
$$P\left\{S_{n+1} = \frac{k+1}{S_n} = k\right\} = \frac{1}{2}$$

$$P\left\{S_{n+1} = \frac{k}{S_n} = k\right\} = \frac{1}{2}$$

These probabilities re not at all affected by the values of the $RVs\ S_1, S_2, \cdots, S_{n-1}$. Also the conditional probability of S_{n+1} given S_n depends on the value of S_n and not on the manner in which the value of S_n was reached. This is a simple example of a Markov chain. Random processes $\{X(t)\}$ (with Markov property) which take discrete values, whether t is discrete or continuous, are called *Markov chains*. In this section, we will discuss time Markov chains.

14.3 DEFINITION OF A MARKOV CHAIN

If, for all n, $P\left\{X_n = \frac{a_n}{X_n - 1} = a_{n+1}, X_{n-2} = a_{n-2}, \ldots, X_0 = a_0\right\} = P\left\{X_n = \frac{a_n}{X_{n-1}} = a_{n-1}\right\}$, then the process $\{X_n\}$, $n = 0, 1, 2, \ldots$, is called a Markov chain.

$(a_1, a_2, \cdots, a_n, \cdots)$ are called the states of the Markov chain. The conditional probability $P\left\{x_n = \frac{a_j}{X_{n-1}} = a_i\right\}$ is called the ***one-step transition probability*** from state a_i to state a_j at the nth step (trial) and is denoted by $P_{ij}(n-1, n)$.

If the one-step transition probability does not spend on the step, i.e., $P_{ij}(n-1, n) = P_{ij}(m-1, m)$ the Markov chain is called a ***homogeneous Markov chain*** or the chain is said to have stationary transition probabilities. The use of the word 'stationary' does not imply a stationary random sequence.

When the Markov chain is homogeneous, the one-step transition probability is denoted by P_{ij}. The matrix $P = \{p_{ij}\}$ is called (one-step) ***transition probability matrix***, shortly, tpm.

Note: The tpm of a Markov chain is a stochastic matrix, since $p_{ij} \geq 0$ and $\sum_j P_{ij} = 1$ for all i. i.e. sum of all the elements of any row of the tpm is 1. This is obvious because the transition from state any one of the states (including a_i itself) is a certain event.

The conditional probability that the process is in state a_j at step n, given that it was in state step 0, i.e., $P\left\{X_n = \frac{a_j}{X_0} = a_i\right\}$ is called the ***n-step transition probability*** and denoted by $P_{ij}(n)$

Note:

$$P_{ij}^{(1)} = P_{ij}.$$

Let us consider an example in which we explain how the tpm is formed for a Markov chain. Assume that a man is at an integral point of the x-axis between the origin and the point $x = 3$. He takes a unit step either to the right with probability 0.7 or to the left with probability 0.3, unless he is at the origin when he takes a step to the right to reach $x = 1$ or he is at the point $x = 3$, when he takes a step to the left to reach $x = 2$. The chain is called 'Random walk with reflecting barriers'. The tpm is given below:

$$\text{States of } X_{n-1} \begin{array}{c} \\ 0 \\ 1 \\ 2 \\ 3 \end{array} \overset{\text{States of } X_n}{\begin{array}{cccc} 0 & 1 & 2 & 3 \end{array}} \begin{pmatrix} 0 & 1 & 0 & 0 \\ 0.3 & 0 & 0.7 & 0 \\ 0 & 0.3 & 0 & 0.7 \\ 0 & 0 & 1 & 0 \end{pmatrix}$$

Note P_{23} = is the element in the 2nd row, 3rd column of this tpm = 0.7. This means that, if the process is at state 2 at step $(n - 1)$, the probability that it moves to state 3 at step $n = 0.7$, where n is any positive integer.

Definition

If the probability that the process is in state a_i is P_i $(i = 1, 2, \cdots, k)$ at any arbitrary step, then the row vector $p = (p_1, p_2, \ldots, p_k)$ is called ***the probability distribution of the process*** at that time. In particular, $p^{(0)} = \{p_1^{(0)}, p_2^{(0)}, \cdots, P_k^{(0)}\}$ is the initial probability distribution.

Remark: The transition probability matrix together with the initial probability distribution completely specifies a Markov chain $\{X_n\}$. In the example given above, let us assume that the initial probability distribution of the chain is $p^{(0)} = \left(\frac{1}{4}, \frac{1}{4}, \frac{1}{4}, \frac{1}{4}\right)$.

i.e.,

$$P\{X_0 = i\} = \frac{1}{4}, \; i = 0, 1, 2, 3$$

Then we have, for the example given above,

$$P\left\{X_1 = \frac{2}{X_0} - 1\right\} = 0.7; \quad P\left\{X_2 = \frac{1}{X_1} = 2\right\} = 0.3,$$

$$P\left\{X_2 - 1, X_1 = \frac{2}{X_0} = 1\right\}$$

$$= P\left\{X_2 = \frac{1}{X_1} = 2\right\} \times P\left\{X_1 = \frac{2}{X_0} = 1\right\}$$

$$= 0.3 \times 0.7 = 0.21 \tag{1}$$

$$P\{X_2 = 1, X_1 = 2, X_0 = 1\}$$

$$= P\{X_0 = 1\} \times P\left\{X_2 = 1,\ X_1 = \frac{2}{X_0} = 1\right\}$$

$$= \frac{1}{4} \times 0.21 = 0.0525 \text{ [by (1)]} \tag{2}$$

$$P\{X_3 = 3, X_2 = 1, X_1 = 2, X_0 = 1\}$$

$$= P\{X_2 = 1,\ X_1 = 2, X_0 = 1\}$$

$$\times P\left\{X_{3.} = \frac{3}{X_2} = 1, X_1 = 2, X_0 = 1\right\}$$

$$= 0.0525P\left\{X_3 = \frac{3}{X_2} = 1\right\} \text{ (Markov property) [by (2)]}$$

$$= 0.0525 \times 0 = 0$$

14.4 CHAPMAN–KOLMOGOROV THEOREM

If P is the tpm of a homogenous Markov chain, then the n-step tpm $P^{(n)}$ is equal top P^n.

i.e.,
$$= \left[P_{ij}^{(n)}\right] = \left[P_{ij}\right]^n$$

Proof

$$P_{ij}^{(2)} = P\left\{X_2 = \frac{j}{X_0} = i\right\}, \text{ since the chain is homogeneous.}$$

The state j can be reached from the state i in 2 steps through some intermediate state k.

Now
$$P_{ij}^{(2)} = P\left\{X_2 = \frac{j}{X_0} = i\right\}$$

$$= P\left\{X_2 = j, X_1 = \frac{k}{X_0} = i\right\}$$

$$= P\left\{X_2 = \frac{j}{X_1} = k, X_0 = i\right\}$$

$$P\left\{X_1 = \frac{k}{X_0} = i\right\}$$

$$= P_{kj}^{(1)} P_{ik}^{(1)}$$

$$= P_{ik} P_{kj}$$

Since the transition from state i to state j in 2 steps can take place through any one of the intermediate states, k can assume the values 1, 2, 3, …. The transitions through various intermediate states are mutually exclusive.

Hence

$$P_{ij}^{(2)} = \sum_k P_{ik} P_{kj}$$

i.e., the ij-th element of 2 step tpm = the ij-th element of the product of the 2 one-step tpm's

i.e., $$P^{(2)} = P^2$$

Now $$Pij^{(3)} = P\left\{X_3 = \frac{j}{X_0} = i\right\}$$

$$= \sum_k P\left\{X_3 = \frac{j}{X_2} = k\right\} P\left\{X_2 = \frac{k}{X_0} = i\right\}$$

$$= \sum_k P_{kj} P_{ik}^{(2)}$$

$$= \sum_k P_{ik}^{(2)} P_{kj}$$

Similarly $$P_{ij}^{(3)} = \sum_k P_{ik} P_{kj}^{(2)}$$

i.e., $$P^{(3)} = P^2 P = P \ P^2 = P^3$$

Proceeding further in a similar way, we get

$$P^{(n)} = P^n$$

For example, consider the problem of Random walk with reflecting barriers, discussed above, for which the tpm is

$$P = \begin{pmatrix} 0 & 1 & 0 & 0 \\ 0.3 & 0 & 0.7 & 0 \\ 0 & 0.3 & 0 & 0.7 \\ 0 & 0 & 1 & 0 \end{pmatrix}$$

$$P^2 = \begin{pmatrix} 0.3 & 0 & 0.7 & 0 \\ 0 & 0.51 & 0 & 0.49 \\ 0.09 & 0 & 0.91 & 0 \\ 0 & 0.3 & 0 & 0.7 \end{pmatrix}$$

From this matrix, we see that $P_{11}^{(2)} = 0.51$. This is so, because

$$\begin{aligned} P_{11}^{(2)} &= P_{10}P_{01} + P_{11}P_{11} + P_{12}P_{21} + P_{13}P_{31} \\ &= (0.3)(1) + (0)(0) + (0.7)(0.3) + (0)(0) \\ &= 0.51 \end{aligned}$$

Definition

A stochastic matrix P is said to be a regular matrix, if all the entries of P^m (for some integer m) are positive. A homogeneous Markov chain is said to be *regular* if its tpm is regular.

We state below two theorems without proof:

1. If $p=\{p_i\}$ is the state probability distribution of the process at an arbitrary time, then that after one step is pP, where P is the tpm of the chain and that after n steps in pP^n.
2. If a homogenous Markov chain is regular, then every sequence of state probability distributions approaches a unique fixed probability distribution called the ***stationary (state) distribution*** or ***steady-state distribution*** of the Markov chain.

   That is, $\lim_{n\to\infty}\{p^{(n)}\}=\pi$, where the state probability distribution at step n, $p^{(n)}=\left(p_1^{(n)},p_2^{(n)},\cdots p_k^{(n)}\right)$ and the stationary distribution $\pi=(\pi_1,\pi_2,\cdots\pi_k)$ are row vectors.
3. Moreover, if P is the tpm of the regular chain, hen $\pi P=\pi$ (π is a row vector). Using this property of π, it can be found out, as in the worked examples given below:

14.5 CLASSIFICATION OF STATES OF A MARKOV CHAIN

If $P_{ij}^{(n)}>0$ for some n and for all i and j, then every state can be reached from every other state . When this condition is satisfied, the Markov chain is said to be ***irreducible***. The tpm of an irreducible chain is an irreducible matrix. Otherwise, the chain is said to be ***nonirreducible*** or ***reducible***.

State i of a Markov chain is called a **return state**, if $p_{ii}(n)>0$ for some $n>1$.

The period d_i of a return state i is defined as the greatest common divisor of all m such that $P_{ii}^{(m)}>0$, i.e., $d_i=GCD\{M:p_{ii}(m)>0\}$. State i is said to be **periodic** with period d_i if $d_i>1$ and aperiodic if $d_i=1$.

Obviously state i is aperiodic if $p_{ii}\neq 0$. The probability that the chain returns to state i, having started from state i, for the first time at the nth step (or after n transitions) is denoted by $f_{ii}(n)$ and *called the **first return time probability*** or *the **recurrence time probability***. $\{n,f_{ii}^{(n)}\},n=1,2,3,\ldots,$ is the distribution of recurrence times of the state i.

If $F_{ii}=\sum_{n=1}^{\infty}f_{ii}^{(n)}=1$, the return to state i is certain.

$\mu_{ii}=\sum_{n=1}^{\infty}nf_{ii}^{(n)}$ is called *the **mean recurrence time*** of the state i.

A state i is said to be ***persistent*** or ***recurrent*** if the return to state i is certain, i.e., if $F_{ii}=1$. The state i is said to be ***transient*** if the return to state i is uncertain, i.e., if $F_{ii}<1$. The state i is said to be ***nonnull persistent*** if its mean recurrence time μ_{ii} is finite and ***null persistent***, if $\mu_{ii}=\propto$.

A nonnull persistent and aperiodic state is called ***ergodic***.

We give below two theorems, without proof, which will be helpful to classify the states of a Markov chain.

1. If a Markov chain is irreducible, all its states are of the same type. They are all transient, all null persistent or all nonnull persistent. All its states are either aperiodic or periodic with the same period.
2. If a Markov chain is finite irreducible, all its states are nonnull persistent.

WORKED EXAMPLES 14

Example 1

The transition probability of a Markov chain $\{X_n\}$, $n = 1, 2, 3, \ldots$, having 3 states 1, 2 and 3 is

$$P = \begin{pmatrix} 0.1 & 0.5 & 0.4 \\ 0.6 & 0.2 & 0.2 \\ 0.3 & 0.4 & 0.3 \end{pmatrix}$$

and the initial distribution is $p^{(0)} = (0.7, 0.2, 0.1)$.

Find (i) $P\{X_2 = 3\}$ and (ii) $P\{X_3 = 2, X_2 = 3, X_1 = 3, X_0 = 2\}$.

$$P^{(2)} = P^2 = \begin{pmatrix} 0.1 & 0.5 & 0.4 \\ 0.6 & 0.2 & 0.2 \\ 0.3 & 0.4 & 0.3 \end{pmatrix} \begin{pmatrix} 0.1 & 0.5 & 0.4 \\ 0.6 & 0.2 & 0.2 \\ 0.3 & 0.4 & 0.3 \end{pmatrix}$$

$$= \begin{pmatrix} 0.43 & 0.31 & 0.26 \\ 0.24 & 0.42 & 0.34 \\ 0.36 & 0.35 & 0.29 \end{pmatrix}$$

SOLUTION

(i)
$$P\{X_2 = 3\} = \sum_{i=1}^{3} P\left\{X_2 = \frac{3}{X_0} = i\right\} \times P\{X_0 = i\}$$

$$= p_{13}^{(2)} P\{(X_0 = 1) + p_{23}^{(2)} P(X_0 = 2) + p_{33}^{(2)} P(X_0 = 3)\}$$

$$= 0.26 \times 0.7 + 0.34 \times 0.2 + 0.29 \times 0.1$$

$$= 0.182 + 0.068 + 0.029$$

$$= 0.279$$

(ii)
$$P\left\{X_1 = \frac{3}{X_0} = 2\right\} = p_{23} = 0.2 \quad (1)$$

$$P\{X_1 = 3, X\ 0 = 2\} = P\left\{X_1 = \frac{3}{X_0} = 2\right\} \times P\{X_0 = 2\}$$

$$= 0.2 \times 0.2 = 0.04 \text{ [by (1)} \quad (2)$$

$$P\{X_2 = 3, X_1 = 3, X_0 = 2\} = P\left\{X_2 = \frac{3}{X_1} = 3, X_0 = 2\right\} \times P\{X_1 = 3, X_0 = 2\}$$

$$= P\left\{X_2 = \frac{3}{X_1} = 3\right\} \times P\{X_1 = 3, X_0 = 2\}$$

(by Markov property)

$$= 0.3 \times 0.04 \text{ [by (2)]}$$

$$= 0.012 \quad (3)$$

$$P\{X_3 = 2, X_2 = 3, X_1 = 3, X_0 = 2\}$$

$$P\left\{X_3 = \frac{2}{X_2} = 3, X_1 = 3, X_0 = 2\right\} \times P\{X_2 = 3, X_1 = 3, X_0 = 2\}$$

$$= P\left\{X_3 = \frac{2}{X_2} = 3\right\} \times P\{X_2 = 3, X_1 = 3, X_0 = 2\}$$

(by Markov process)

$$= 0.4 \times 0.012 \text{ [by(3)]}$$
$$= 0.0048$$

Example 2

A fair die is tossed repeatedly. If X_n denotes the maximum of the numbers occuring in the first n tosses, find the transition probability matrix P of the Markov chain X_n.

Find also P^2 and $P(X_2 = 6)$

State space: $\{1, 2, 3, 4, 5, 6\}$

Solution

The tpm is formed using the following analysis.

Let X_n = the maximum of the numbers occuring in the first n trials = 3, say

Then $X_{n+1} = 3$, if the $(n + 1)$th trial results in 1, 2 or 3

= 4, if the $(n + 1)$th trial results in 4

= 5, if the $(n + 1)$th trial results in 5

= 6, if the $(n + 1)$th trial results in 6

$$\therefore \quad P\left\{X_{n+1} = \frac{3}{X_n} = 3\right\}$$

$$= \frac{1}{6} + \frac{1}{6} + \frac{1}{6} = \frac{3}{6}$$

$$P\left\{X_{n+1} = \frac{i}{X_n} = 3\right\} = \frac{1}{6}, \text{ when } i = 4, 5, 6$$

Therefore, the transition probability matrix of the chain is

$$P = \begin{bmatrix} \frac{1}{6} & \frac{1}{6} & \frac{1}{6} & \frac{1}{6} & \frac{1}{6} & \frac{1}{6} \\ 0 & \frac{2}{6} & \frac{1}{6} & \frac{1}{6} & \frac{1}{6} & \frac{1}{6} \\ 0 & 0 & \frac{3}{6} & \frac{1}{6} & \frac{1}{6} & \frac{1}{6} \\ 0 & 0 & 0 & \frac{4}{6} & \frac{1}{6} & \frac{1}{6} \\ 0 & 0 & 0 & 0 & \frac{5}{6} & \frac{1}{6} \\ 0 & 0 & 0 & 0 & 0 & 1 \end{bmatrix}$$

$$P^2 = \frac{1}{36}\begin{bmatrix} 1 & 3 & 5 & 7 & 9 & 11 \\ 0 & 4 & 5 & 7 & 9 & 11 \\ 0 & 0 & 9 & 7 & 9 & 11 \\ 0 & 0 & 0 & 16 & 9 & 11 \\ 0 & 0 & 0 & 0 & 25 & 11 \\ 0 & 0 & 0 & 0 & 0 & 36 \end{bmatrix}$$

Initial state probability is $p(0) = \left(\frac{1}{6}, \frac{1}{6}, \frac{1}{6}, \frac{1}{6}, \frac{1}{6}, \frac{1}{6}\right)$ since all the values 1, 2, …, 6 are equally likely.

$$P\{X_2 = 6\} = \sum_{i=1}^{6} P\left\{X_2 = \frac{6}{X_0} = i\right\} \times P\{X_0 = i\}$$

$$= \frac{1}{6}\sum_{i=1}^{6} p_{i6}^{(2)}$$

$$= \frac{1}{6} \times \frac{1}{36} \times (11+11+11+11+11+36)$$

$$= \frac{91}{216}$$

Example 3

A man either drives a car or catches a train to go to office each day. He never goes 2 days in a row by train but if he drives one day, then the next day he is just as likely to drive again as he is to travel by train. Now suppose that on the first day of the week, the man tossed a fair die and drove to work if and only if a 6 appeared. Find (i) the probability that he makes a train on the third day and (ii) the probability that he drives to work in the long run.

SOLUTION

The travel pattern is a Markov chain, with state space = (train, car)

The tpm of the chain is

$$P = \begin{matrix} & \begin{matrix} T & C \end{matrix} \\ \begin{matrix} T \\ C \end{matrix} & \begin{pmatrix} 0 & 1 \\ \frac{1}{2} & \frac{1}{2} \end{pmatrix} \end{matrix}$$

The initial state probability distribution is $p^{(1)} = \left(\frac{5}{6}, \frac{1}{6}\right)$.

since P(travelling by car) = P(getting 6 in the toss of the dice)

$$= \frac{1}{6}$$

and $$P(\text{travelling by train}) = \frac{5}{6}$$

$$p^{(2)} = p^{(1)}P$$

$$= \left(\frac{5}{6}, \frac{1}{6}\right)\begin{pmatrix} 0 & 1 \\ \frac{1}{2} & \frac{1}{2} \end{pmatrix} = \left(\frac{1}{12}, \frac{11}{12}\right)$$

$$p^{(3)} = p^{(2)}P$$

$$= \left(\frac{1}{12}, \frac{11}{12}\right)\begin{pmatrix} 0 & 1 \\ \frac{1}{2} & \frac{1}{2} \end{pmatrix} = \left(\frac{11}{24}, \frac{13}{24}\right)$$

$\therefore P$(the man travels by train on the third day) $= \dfrac{11}{24}$

Let $\pi = (\pi_{1,}\pi_2)$ be the limiting form of the state probability distribution or stationary state distribution of the Markov chain.

By the property of π, $pP = p$

i.e., $$(\pi_1, \pi_2)\begin{pmatrix} 0 & 1 \\ \frac{1}{2} & \frac{1}{2} \end{pmatrix} = (\pi_1, \pi_2)$$

i.e., $$\frac{1}{2}\pi_2 = \pi_1$$

and $$\pi_1 + \frac{1}{2}\pi_2 = \pi_2$$

Equations (1) and (2) are one and the same.
Therefore, consider (1) or (2) with $\pi_1 + \pi_2 = 1$, since π is a probability distribution.

Solving, $\pi_1 = \dfrac{1}{3}$ and $\pi_2 = \dfrac{2}{3}$

$\therefore P\{\text{the man travels by car in the long run}\} = \dfrac{2}{3}$.

Example 4

Consider a communication system which transmits the 2 digits 0 and 1 through several stages. Let $X_n (n \geq 1)$ be the digit leaving the nth stage if the system and X_0 be the digit entering the first stage (or leaving the 0th stage). At each stage there is a constant probability q that the digit which enters will be transmitted unchanged (i.e., the digit changes when it leaves), where $p + q = 1$. Write down the tpm P of the homogeneous two-state Markov chain $\{X_n\}$. Find P^m, $P^\propto$ and the conditional probability that the digit entering the first stage is 0, given that the digit leaving the mth stage is 0. Assume that the initial state probability distribution is $p(0) = (a, 1-a)$.

Solution

State space = (0, 1);

$$P \equiv \begin{array}{c} \\ \text{State of } X_n \end{array} \begin{array}{c} \text{State of } X_{n+1} \\ \begin{array}{cc} 0 & 1 \end{array} \end{array}$$

$$\begin{array}{c} 0 \\ 1 \end{array}\begin{pmatrix} q & p \\ p & q \end{pmatrix}$$

Now $$P^2 = \begin{pmatrix} q & p \\ p & q \end{pmatrix}\begin{pmatrix} q & p \\ p & q \end{pmatrix}$$

$$= \begin{pmatrix} p^2 + q^2 & 2pq \\ 2pq & p^2 + q^2 \end{pmatrix}$$

$$= \begin{pmatrix} \frac{1}{2}\left[(q+p)^2 + (q-p)^2\right] & \frac{1}{2}\left[(q+p)^2 - (q-p)^2\right] \\ \frac{1}{2}\left[(q+p)^2 - (q-p)^2\right] & \frac{1}{2}\left[(q+p)^2 + (q-p)^2\right] \end{pmatrix}$$

$$= \begin{pmatrix} \frac{1}{2} + \frac{1}{2}r^2 & \frac{1}{2} - \frac{1}{2}r^2 \\ \frac{1}{2} - \frac{1}{2}r^2 & \frac{1}{2} + \frac{1}{2}r^2 \end{pmatrix}, \text{ where } q - p = r$$

$$P^3 = \begin{pmatrix} \frac{1}{2} + \frac{1}{2}r^3 & \frac{1}{2} - \frac{1}{2}r^3 \\ \frac{1}{2} - \frac{1}{2}r^3 & \frac{1}{2} + \frac{1}{2}r^3 \end{pmatrix}$$

The values of P^2 and P^3 make us guess that

$$P^m = \begin{pmatrix} \frac{1}{2}+\frac{1}{2}r^m & \frac{1}{2}-\frac{1}{2}r^m \\ \frac{1}{2}-\frac{1}{2}r^m & \frac{1}{2}+\frac{1}{2}r^m \end{pmatrix}$$

It is correct as can be proved by induction as follows:

$$P^{m+1} = \begin{pmatrix} q & p \\ p & q \end{pmatrix}\begin{pmatrix} \frac{1}{2}+\frac{1}{2}r^m & \frac{1}{2}-\frac{1}{2}r^m \\ \frac{1}{2}-\frac{1}{2}r^m & \frac{1}{2}+\frac{1}{2}r^m \end{pmatrix}$$

$$= \begin{pmatrix} \frac{q}{2}+\frac{q}{2}r^m+\frac{p}{2}-\frac{p}{2}r^m & \frac{q}{2}-\frac{q}{2}r^m+\frac{p}{2}+\frac{p}{2}r^m \\ \frac{p}{2}+\frac{p}{2}r^m+\frac{q}{2}-\frac{q}{2}r^m & \frac{p}{2}-\frac{p}{2}r^m+\frac{q}{2}+\frac{q}{2}r^m \end{pmatrix}$$

$$= \begin{pmatrix} \frac{1}{2}+\frac{1}{2}r^{m+1} & \frac{1}{2}-\frac{1}{2}r^{m+1} \\ \frac{1}{2}-\frac{1}{2}r^{m+1} & \frac{1}{2}+\frac{1}{2}r^{m+1} \end{pmatrix}$$

$$P^m = \begin{pmatrix} \frac{1}{2}+\frac{1}{2}r^m & \frac{1}{2}-\frac{1}{2}r^m \\ \frac{1}{2}-\frac{1}{2}r^m & \frac{1}{2}+\frac{1}{2}r^m \end{pmatrix},$$

where m is a positive integer ≥ 1.

$$p^{\infty} = \lim_{m\to\infty}\left(P^m\right)$$

$$= \begin{pmatrix} \frac{1}{2} & \frac{1}{2} \\ \frac{1}{2} & \frac{1}{2} \end{pmatrix} \quad \text{since } |r| < 1$$

Now $$P\{X_m = 0, X_0 = 0\} = P\left\{X_m = \frac{0}{X_0} = 0\right\} \times P\{X_0 = 0\}$$

$$= ap_{00}{}^{(m)}$$

and $$P\{X_m = 0, X_0 = 0\} = bp_{10}{}^{(m)}$$

where $b = 1 - a$

Now $$P\left\{X_0 = \frac{0}{Xm} = 0\right\} = \frac{p\{X_0 = 0\} \times P\left\{X_m = \frac{0}{X_0} = 0\right\}}{p\{X_0 = 0\} \times P_{00}{}^{(m)} + p\{X_0 = 1\} \times P_{10}{}^{(m)}} \text{ (by Baye's theorem)}$$

$$= \frac{a\left\{\frac{1}{2} + \frac{1}{2}r^m\right\}}{a\left\{\frac{1}{2} + \frac{1}{2}r^m\right\} + b\left\{\frac{1}{2} - \frac{1}{2}r^m\right\}}$$

$$= \frac{a(1+r^m)}{1+(a-b)r^m},$$

where $b = 1 - a$

Example 5

A gambler has Rs 2/-. He bets Re.1 at a time and wins Re.1 with probability $\frac{1}{2}$. He stops playing if he loses Rs 2 or wins Rs.4(a) What is the tpm of the related Markov chain? (b) What is the probability that he has lost his money at the end of 5 plays(c) what is the probability that the game lasts more than 7 plays?

SOLUTION

Let X_n represent the amount with the player at the end of the nth round of the play.

Since space of $\{X_n\} = (0,1,2,3,4,5,6)$, as the game ends, if the player loses all the money $(X_n = 0)$ or wins Rs. 4, ie.e., has Rs. 6 $(X_n = 6)$. The tpm of the Markov chain is

$$P = \begin{array}{c} \\ 0 \\ 1 \\ 2 \\ 3 \\ 4 \\ 5 \\ 6 \end{array}\begin{array}{c} \begin{array}{ccccccc} 0 & 1 & 2 & 3 & 4 & 5 & 6 \end{array} \\ \begin{pmatrix} 1 & 0 & 0 & 0 & 0 & 0 & 0 \\ \frac{1}{2} & 0 & \frac{1}{2} & 0 & 0 & 0 & 0 \\ 0 & \frac{1}{2} & 0 & \frac{1}{2} & 0 & 0 & 0 \\ 0 & 0 & \frac{1}{2} & 0 & \frac{1}{2} & 0 & 0 \\ 0 & 0 & 0 & \frac{1}{2} & 0 & \frac{1}{2} & 0 \\ 0 & 0 & 0 & 0 & \frac{1}{2} & 0 & \frac{1}{2} \\ 0 & 0 & 0 & 0 & 0 & 0 & 1 \end{pmatrix} \end{array}$$

Note: This is called a random walk with ***absorbing barriers*** at 0 and 6, since the chain cannot come out of the states 0 and 6, once it has entered them.

The initial probability distribution of $\{X_n\}$ is $p(0) = (0,0,1,0,0,0,0)$, as the player has got Rs.2/- to start with.

$$p^{(1)} = p^{(0)} P = \left(0, \frac{1}{2}, 0, \frac{1}{2}, 0, 0, 0\right)$$

$$p^{(2)} = p^{(1)} P = \left(\frac{1}{4}, 0, \frac{1}{2}, 0, \frac{1}{4}, 0, 0\right)$$

$$p^{(3)} = p^{(2)} P = \left(\frac{1}{4}, \frac{1}{4}, 0, \frac{3}{8}, 0, \frac{1}{8}, 0\right)$$

$$p^{(4)} = p^{(3)} P = \left(\frac{3}{8}, 0, \frac{5}{16} .0, \frac{1}{4}, 0, \frac{1}{16}\right)$$

$$p^{(5)} = p^{(4)} P = \left(\frac{3}{8}, \frac{5}{32}, 0, \frac{9}{32}, 0, \frac{1}{8}, \frac{1}{16}\right)$$

$P\{$the man has lost his money at the end of 5 plays$\}$

$= P\{X_5 = 0\}$

= the entry corresponding to state 0 in $p^{(5)}$

$$= \frac{3}{8}$$

Again $$p^{(6)} = p^{(5)} P = \left(\frac{29}{64}, 0, \frac{7}{32}, 0, \frac{13}{64}, 0, \frac{1}{8}\right)$$

$$p^{(7)} = p^{(6)} P = \left(\frac{29}{64}, \frac{7}{654}, 0, \frac{27}{128}, 0, \frac{13}{128}, \frac{1}{8}\right)$$

$P\{$the game lasts more than 7rounds$\}$ = $P\{$the system is neither in state 0 nor in 6 at the end of the seventh round$\}$

$$= P\{X_7 = 1, 2, 3, 4 \text{ or } 5\}$$

$$= \frac{7}{64} + 0 + \frac{27}{128} + 0 + \frac{13}{128} = \frac{27}{64}$$

Example 6

There are 2 white marbles in urn A and 3 red marbles in urn B. At each step of the process, a marble is selected from each urn and the 2 marbles selected are inter-changed. Let the state a_i of the system be the number of red marbles in A after i changes. What is the probability that there are 2 red marbles in A after 3 steps? In the long run, what is the probability that there are 2 red marbles in urn A?

SOLUTION

State space of the chain $\{X_n\} = (0, 1, 2)$, since the number of balls in the urn A is always 2.

Let the tpm of the chain $\{X_n\}$ be

$$P = \begin{array}{c} \\ 0 \\ 1 \\ 2 \end{array} \begin{array}{c} \begin{array}{ccc} 0 & 1 & 2 \end{array} \\ \begin{pmatrix} p_{00} & p_{01} & p_{02} \\ p_{10} & p_{11} & p_{12} \\ p_{20} & p_{21} & p_{22} \end{pmatrix} \end{array}$$

$p_{00} = 0$ (since the state cannot remain at 0 after interchange of marbles)

$p_{02} = p_{20} = 0$ (since the number of red marbles in urn cannot increase or decrease by 2 in one interchange)

To start with, A contains 0 red marble (and 1 white marble) and B contains 1 white and 2 red marbles.

Then $X_{n+1} = 0$, if A contains 0 red marble (and 2 white marbles) and B contains 3 red marbles, i.e., if 1 red marble is chosen from A and 1 white marble is chosen from B and interchanged.

$$\therefore \qquad P\left\{X_{n+1} = \frac{0}{X_n} = 1\right\}$$

$$= p_{10} = \frac{1}{2} \times \frac{1}{3} = \frac{1}{6}$$

Similarly, we can find $p_{12} = \frac{1}{2} \times \frac{1}{3} = \frac{1}{3}$

Since P is a stochastic matrix,

$$P_{10} + P_{11} + P_{12} = 1$$

$$\therefore \qquad P_{11} = \frac{1}{2}$$

Similarly, $$p_{21} = \frac{2}{3}$$

and $$p_{22} = 1 - (p_{20} + p_{21}) = \frac{1}{3}$$

$$\therefore \qquad P = \begin{pmatrix} 0 & 1 & 0 \\ \frac{1}{6} & \frac{1}{2} & \frac{1}{3} \\ 0 & \frac{2}{3} & \frac{1}{3} \end{pmatrix}$$

Now $p^{(0)} = (1,0,0)$, as there is no red marble in A in the beginning.

$$p^{(1)} = p^{(0)} P = (0,1,0)$$

$$p^{(2)} = p^{(1)} P = \left(\frac{1}{6}, \frac{1}{2}, \frac{1}{3}\right)$$

$$p^{(3)} = p^{(1)} P = \left(\frac{1}{12}, \frac{23}{26}, \frac{5}{18}\right)$$

$\therefore P\{$there are 2 red marbles in A after 3 steps$\}$

$$= P\{X_3 = 2\} = p_2^{(3)} = \frac{5}{18}$$

Let the stationary probability distribution of the chain be $\pi = (\pi_0, \pi_1, \pi_2)$.

By the property of π, $p\mathrm{P} = \pi$ and $\pi_0 + \pi_1 + \pi_2 = 1$

i.e., $$(\pi_0, \pi_1, \pi_2)\begin{pmatrix} 0 & 1 & 0 \\ \frac{1}{6} & \frac{1}{2} & \frac{1}{3} \\ 0 & \frac{2}{3} & \frac{1}{3} \end{pmatrix} = (\pi_0, \pi_1, \pi_2)$$

i.e., $$\frac{1}{6}\pi_1 = \pi_0$$

$$= \pi_0 + \frac{1}{2}\pi_1 + \frac{2}{3}\pi_2 = \pi_1$$

$$= \frac{1}{3}\pi_1 + \frac{1}{3}\pi_2 = \pi_2$$

and $$\pi_1 + \pi_2 + \pi_3 = 1$$

Solving, $$\pi_0 = \frac{1}{10}, \pi_1 = \frac{6}{10}, \pi_2 = \frac{3}{10}$$

$\therefore P\{\text{there are 2 red marbles in A in the long run}\} = 0.3.$

Example 7

Find the nature of the states of the Markov chain with the tpm

$$P = \begin{array}{c} \\ 0 \\ 1 \\ 2 \end{array}\begin{array}{c} \begin{matrix} 0 & 1 & 2 \end{matrix} \\ \begin{pmatrix} 0 & 1 & 0 \\ \frac{1}{2} & 0 & \frac{1}{2} \\ 0 & 1 & 0 \end{pmatrix} \end{array}$$

Solution

$$P^2 = \begin{pmatrix} \frac{1}{2} & 0 & \frac{1}{2} \\ 0 & 1 & 0 \\ \frac{1}{0} & 0 & \frac{1}{2} \end{pmatrix}; P^3 = P$$

$\therefore$ $$P^4 = P^2$$

and so on. In general, $P^{2n} = P^2, P^{2n+1} = P$

We know that $$p_{00}^{(2)} > 0, p_{01}^{(1),} p_{02}^{(2)} > 0$$

$$p_{10}^{(1)} > 0, p_{11}^{(2),} p_{12}^{(1)} > 0$$

$$p_{20}^{(2)} > 0, p_{21}^{(1),} p_{22}^{(2)} > 0$$

Therefore, the Markov chain is irreducible.

Also $p_{ii}^{(2)} = p_{ii}^{(4)} = p_{ii}^{(6)} > 0$, for all i, all the states of the chain are periodic, with period 2.
Since the chain is finite and irreducible, all its states are nonnull persistent.
All states are not ergodic.

Example 8

Three boys A, B, and C are throwing a ball to each other. A always throws the ball to B and B always throws the ball to C, but C is just as likely to throw the ball to B as to A. Show that the process is Markovian. Find the transition matrix and classify the states. (MKU—Nov. 96)

Solution

The transition probability matrix of the process $\{X_n\}$ is given below:

$$\text{State of } X_{\text{n+1}} \begin{matrix} \\ A \\ B \\ C \end{matrix} \overset{\text{State of } X_n}{\begin{matrix} A & B & C \\ \end{matrix}} \begin{pmatrix} 0 & 1 & 0 \\ 0 & 0 & 1 \\ \frac{1}{2} & \frac{1}{2} & 0 \end{pmatrix} \equiv P, \text{ say}$$

States of X_n, depend on states of X_{n-1}, but not on states of X_{n-2}, X_{n-3}, ..., or earlier states. Therefore, $\{X_n\}$ is a Markov chain.

Now $$P^2 = \begin{pmatrix} 0 & 1 & 2 \\ \frac{1}{2} & \frac{1}{2} & 0 \\ 0 & \frac{1}{2} & \frac{1}{2} \end{pmatrix}; P^3 = \begin{pmatrix} \frac{1}{2} & \frac{1}{2} & 0 \\ 0 & \frac{1}{2} & \frac{1}{2} \\ \frac{1}{4} & \frac{1}{4} & \frac{1}{2} \end{pmatrix}$$

$p_{11}^{(3)} > 0, p_{13}^{(2)} > 0, p_{21}^{(2)} > 0, p_{33}^{(2)} > 0$ and all other $p_{ij}^{(1)} > 0$.

Therefore the chain is irreducible.

$$P^4 = \begin{pmatrix} 0 & \frac{1}{2} & \frac{1}{2} \\ \frac{1}{4} & \frac{1}{4} & \frac{1}{2} \\ \frac{1}{4} & \frac{1}{2} & \frac{1}{4} \end{pmatrix};$$

$$P^5 = \begin{pmatrix} \frac{1}{4} & \frac{1}{4} & \frac{1}{2} \\ \frac{1}{4} & \frac{1}{2} & \frac{1}{4} \\ \frac{1}{8} & \frac{3}{8} & \frac{1}{2} \end{pmatrix};$$

$$P^6 = \begin{pmatrix} \frac{1}{4} & \frac{1}{2} & \frac{1}{4} \\ \frac{1}{4} & \frac{3}{8} & \frac{1}{2} \\ \frac{1}{8} & \frac{3}{8} & \frac{3}{8} \end{pmatrix};$$

and so on.

We note that $p_{ii}^{(2)}, p_{ii}^{(3)}, p_{ii}^{(5)}; p_{ii}^{(6)}$ etc are > 0 for i = 2, 3, and GCD of 2, 3, 5, 6, … =1.

Therefore, the states 2 and 3 (i.e., *B* and *C*) are periodic with period 1. i.e., aperiodic.

We note that $p_{11}^{(3)}, p_{11}^{(5)}, p_{11}^{(6)}$ etc. are > 0 and GCD of 3, 5, 6, … =1.

Therefore, the state 1(i.e., state *A*) is periodic with period 1, i.e., aperiodic.

Since the chain is finite and irreducible, all its states and nonnull peristent. Morover all the states are ergodic.

EXERCISE 14

Part A (Short answer questions)

1. Define a Markov process.
2. Define a Markov chain ad give an example of a Markov chain.
3. Prove that the Poisson process is a Markov process.
4. When is a Markov chain called homogeneous?
5. When is a homogeneous Markov chain said to be regular?
6. Define transition probability matrix of a Markov chain.
7. What is a stochastic matrix? When is it said to be regular?
8. Prove that the tpm of a Markov chain is a stochastic matrix.
9. Define *n*-step transition probability in a Markov chain.
10. State Chapman–Kolmogorov theorem.
11. What do you mean by probability distribution of a Markov chain?
12. When is a Markov chain completely specified?
13. What is meant by steady-state distribution of a Markov chain?
14. Write down the relation satisfied by the steady-state distribution and the tpm of a regular Markov chain.
15. If the tpm of a Markov chain is $\begin{pmatrix} 0 & 1 \\ \frac{1}{2} & \frac{1}{2} \end{pmatrix}$ find the steady-state distribution of the chain.
16. When is a Markov chain said to be irreducible or ergodic?
17. Prove that t matrix $\begin{pmatrix} 0 & 1 & 0 \\ 0 & 0 & 1 \\ \frac{1}{2} & \frac{1}{2} & 0 \end{pmatrix}$ is the tpm of an irreducible Markov chain.
18. What do you mean by an absorbing Markov chain. Give an example.

19. If the initial state probability distribution of a Markov chain is $p^{(0)} = \left(\frac{5}{6}, \frac{1}{6}\right)$ and the tpm of the chain is $\begin{pmatrix} 0 & 1 \\ \frac{1}{2} & \frac{1}{2} \end{pmatrix}$, find the probability distribution of the chain after 2 steps.

Part B

20. The tpm of a Markov chain with three states 0, 1, 2 is

$$P = \begin{pmatrix} \frac{3}{4} & \frac{1}{4} & 0 \\ \frac{1}{4} & \frac{1}{2} & \frac{1}{4} \\ 0 & \frac{3}{4} & \frac{1}{4} \end{pmatrix}$$

and the initial state distribution of the chain is $P\{X_0 = i\} = \frac{1}{3}$, $i = 0, 1, 2$.

Find (i) $P\{X_2 = 2\}$ and

(ii) $P\{X_3 = 1, X_2 = 2, X_1 = 1, X_0 = 2\}$.

21. A man is at an integral point on the *x*-axis between the origin and the point 3. He takes a unit step to the right with probability $\frac{1}{3}$ or to the left with probability $\frac{2}{3}$, unless he is at the origin, where he takes a step to the right to reach the point 1 or is at the point 3, where he takes a step to the left to reach the point 2. What is the probability that (i) he is at the point 1 after three walks? and (ii) he is at the point 1 in the long run?
22. Suppose that the probability of a dry day (state 0) following a rainy day (state 1) is $\frac{1}{3}$ and that the probability of a rainy day following a dry day is $\frac{1}{2}$. Given that May 1 is a dry day and (ii) May 5 is also a dry day.
23. A gambler has Rs 3/-. At each play of the game, he loses Re 1 with probability $\frac{3}{4}$, but wins Rs 2/- with probability $\frac{1}{4}$. He stops playing if he has lost his initial amount of Rs.3/-or he has won at least Rs3/-. Write down the tpm of the associated Markov chain. Find the probability that there are at least 4 rounds to the game.
24. A communication source can generate 1 of 3 possible messages 1, 2 and 3. Assume that the generation can be described by a homogeneous Markov chain withthe following transition probability matrix

Current message	**Next message**		
	1	2	3
1	0.5	0.3	0.2
2	0.4	0.2	0.4
3	0.3	0.3	0.4

and the initial state probability distribution $p^{(0)} = (0.3, 0.3, 0.4)$. Find $p^{(3)}$.

25. Assume that the weather in a certain locality can be modeled as the homogeneous Markov chain whose transition probability matrix is given below

Today's weather	Tomorrow's weather		
	Fair	Cloudy	Rainy
Fair	0.8	0.15	0.05
Cloudy	0.5	0.3	0.2
Rainy	0.6	0.3	0.1

If the initial state distribution is given by $p^{(0)} = (0.7, 0.2, 0.1)$, find $p^{(2)}$ and $\lim_{n\to\infty} p^{(n)}$.

26. A fair coin is tossed until 3 heads occur in a row. Let X_n be the sequence of heads ending at the nth trial. What is the probability that there are at least 8 tosses of the coin?
27. There are 2 white marbles in urn A and 4 red marbles in urn *B*. At each step of the process, a marble is selected from each urn and the 2 marbles selected are interchanged. The state of the related Markov chain is the number of the red balls in *A* after the interchange. What is the probability that there are 2 red balls in urn *A* (i) after 3 steps and (ii) in the long run?
28. A student's study habits are as follows: If he studies one night, he is 70% sure not to study the next night. On the other hand, if he does not study one night, he is 60% sure not to study the next night as well. In the long run, how often does he study?
29. A salesman's territory consists of 3 cities *A*, *B*, and *C*. He never sells in the same city on successive days. If he sells in city *A*, then the next day he sells in *B*. However, if he sells either in *B* or *C*, then the next day he is twice as likely to sell in city *A* as in the other city. How often does he sell in each of the cities in the steady-state?
30. A housewife buys 3 kinds of cereals, *A*, *B* and *C*. She never buys the same cereal in successive weeks. If she buys cereal *A*, the next week she buys cereal *B*. However if she buys *B* or *C*, the next week she is 3 times as likely to buy *A* as the other cereal. In the long run, how often does she buy each of the three cereals?
31. Two boys B_1, B_2 and two girls G_1, G_2 are throwing a ball from one to another. Each boy throws the ball to the other boy with probability $\frac{1}{2}$ and to each girl with probability $\frac{1}{4}$. On the other hand, each girl throws the ball to each boy with probability $\frac{1}{2}$ and never to the other girl. In the long run, how often does each receive the ball?
32. A gambler's luck follows a pattern. If he wins the game, the probability of his winning the next game is 0.6. However if he loses a game, the probability of his losing the next game is 0.7. There is an even chance that the gambler wins the first game. What is the probability that he wins (i) the second game, (ii) the third game and (iii) in the long run?
33. The three-state Markov chain is given by the tpm

$$P = \begin{pmatrix} 0 & \frac{2}{3} & \frac{1}{3} \\ \frac{1}{2} & 0 & \frac{1}{2} \\ \frac{1}{2} & \frac{1}{2} & 0 \end{pmatrix}$$

Prove that the chain is irreducible and all the states are aperiodic and non-null persistent. Find also the steady –state distribution of the chain.

34. A man tosses a fair coin until 3 heads occur in a row. Let $X_n = K$, if at the nth trial, the last trail occurred at the $(n-k)$th trial; i.e., X_n denotes the longest string of heads ending at the nth trial. Show that the process is Markovian. Find the transition matrix and classify the states.

ANSWERS

Exercise 14

Part A

7. A square matrix in which the sum of all the elements of each row is 1, is called a stochastic matrix. A stochastic matrix P is said to be regular if all the entries of P^m (for some positive integer m) are positive.

12. A Markov chain is completely specified when n the initial probability distribution and the tpm are given.

14. If $\pi = (\pi_1, \pi_2, \cdots \pi_n)$ is the steady-state distribution of the chain whose tpm is the nth order square matrix P, then $\pi P = \pi$.

15. If (π_1, π_2) is the steady – state distribution of the chain, $(\pi_1, \pi_2)\begin{pmatrix} 0 & 1 \\ \frac{1}{2} & \frac{1}{2} \end{pmatrix} = (\pi_1, \pi_2)$; $\frac{1}{2}\pi_2 = \pi_1$ and $\pi_1 + \frac{1}{2}\pi_2 = \pi_1$, i.e., $2\pi_1 = \pi_2$. Also $\pi_1 + \pi_2 = 1$

$$\therefore \quad \pi_1 = \frac{1}{3}, \pi_2 = \frac{2}{3}.$$

18. A state i of a Markov chain is said to be an absorbing state if $p_{ii} = 1$, i.e., if it is impossible to leave it.

A Markov chain is said to be absorbing if it has at least one absorbing state.

$$\text{Example: } \begin{array}{c} \\ 1 \\ 2 \\ 3 \\ 4 \end{array}\begin{array}{c} \begin{array}{cccc} 1 & 2 & 3 & 4 \end{array} \\ \begin{pmatrix} 1 & 0 & 0 & 0 \\ \frac{1}{2} & 0 & \frac{1}{2} & 0 \\ 0 & \frac{1}{2} & 0 & \frac{1}{2} \\ 0 & 0 & 0 & 1 \end{pmatrix} \end{array}$$

19. $p^{(1)} = P^{(0)}P; p^{(2)} = P^{(1)}P = \left(\frac{11}{14}, \frac{13}{24}\right)$.

Part B

20. $\frac{1}{6};\frac{3}{64}$

21. $P=\begin{pmatrix}0 & 1 & 0 & 0\\ \frac{2}{3} & 0 & \frac{1}{3} & 0\\ 0 & \frac{2}{3} & 0 & \frac{1}{3}\\ 0 & 0 & 1 & 0\end{pmatrix}$

(i) $\frac{22}{27}$ (ii) $\frac{3}{7}$

22. (i) $\frac{5}{12}$; (ii) $\frac{173}{432}$

23. $\frac{27}{64}$

24. (.4083, .2727, .3190)

25. (.7245, .1920, .0835); $\left(\frac{111}{157},\frac{30}{157},\frac{13}{157}\right)$

26. $\frac{81}{128}$

27. $\frac{3}{8};\frac{2}{5}$

28. $\frac{4}{11}$ of the nights

29. $P=\begin{pmatrix}0 & 1 & 0\\ \frac{2}{3} & 0 & \frac{1}{3}\\ \frac{2}{3} & \frac{1}{3} & 0\end{pmatrix}$; (40%, 45%, 15%)

30. $P=\begin{pmatrix}0 & 1 & 0\\ \frac{3}{4} & 0 & \frac{1}{4}\\ \frac{3}{4} & \frac{1}{4} & 0\end{pmatrix};\left(\frac{15}{35},\frac{16}{35},\frac{4}{35}\right)$

31. $\left(\frac{1}{3},\frac{1}{3},\frac{1}{6},\frac{1}{6}\right)$

32. $\frac{9}{20},\frac{87}{200},\frac{3}{7}$

33. $\left(\frac{9}{27},\frac{10}{27},\frac{8}{27}\right)$

34. $$X_n \begin{matrix} \\ 0\\ 1\\ 2\\ 3\end{matrix}\overset{X_{n+1}}{\begin{pmatrix}0 & 1 & 2 & 3\\ \frac{1}{2} & \frac{1}{2} & 0 & 0\\ \frac{1}{2} & 0 & \frac{1}{2} & 0\\ \frac{1}{2} & 0 & 0 & \frac{1}{2}\\ 0 & 0 & 0 & 1\end{pmatrix}}$$

The chain is not irreducible. State 3 is absorbing and other states are aperiodic.

Chapter 15 Queueing Models

15.0 INTRODUCITON

There are many situations in daily life when a queue is formed. For example, machines waiting to be repaired, patients waiting in a doctor's room, cars waiting at a traffic signal and passengers waiting to buy tickets in counters form queues. A queue is formed if the service required by the customer (machine, patient, car, etc.) is not immediately available, that is, if the current demand for a particular service exceeds the capacity to provide the service.

Queues may be decreased in size or prevented from forming by providing additional service facilities which results in a drop in profits. On the other hand, excessively long queues may result in lost sales and lost customers. Hence, the problem is how to achieve a balance between the cost associated with long waiting (queues) and the cost associated with the prevention of waiting in order to maximise the profits. As queueing theory provides an answer to this problem, it has become a topic of interest. Before we consider the **Solutions** of queueing problems, we shall consider the general framework of a queueing system.

Although there are many types of queueing systems, all of them can be classified and described according to the following characteristics:

15.1 INPUT (OR ARRIVAL) PATTERN

The input describes the manner in which the customers arrive and join the queueing system. It is not possible to observe and control the actual moment of arrival of a customer for service. Hence, the number of arrivals in one time period or the interval between successive arrivals is not treated as a constant, but a random variable. Therefore, the mode of arrival of customers is expressed by means of the probability distribution of the number of arrivals per unit of time or of the inter-arrival time.

In this chapter, we shall mostly deal with those queueing systems in which the numbers of arrivals per unit of time has a Poisson distribution with mean λ. In this case it is known that the time between consecutive arrivals has an exponential distribution with mean $\frac{1}{\lambda}$.

Further, the input process should specify the number of queues that are permitted to form, the maximum queue length and the maximum number of customers requiring service, viz., the nature of the source (finite of infinite) from which the customers emanate.

15.2 SERVICE MECHANISM (OR PATTERN)

The mode of service is represented by means of the probability distribution of the number of customers serviced per unit of time or of the inter-service time. We shall mostly deal with only those queueing systems in which the number of customers serviced per unit of time has a Poisson distribution with mean μ or equivalently, the inter-service time (viz., the time to complete the service for a customer) has an exponential distribution with mean $\frac{1}{\mu}$.

Further, the service process should specify the number of servers and the arrangement of servers (in parallel, in series, etc.), as the behavior of the queueing system depends on them also. Figures 15.1–15.3 represent the framework of queueing systems in which only one queue is permitted to form.

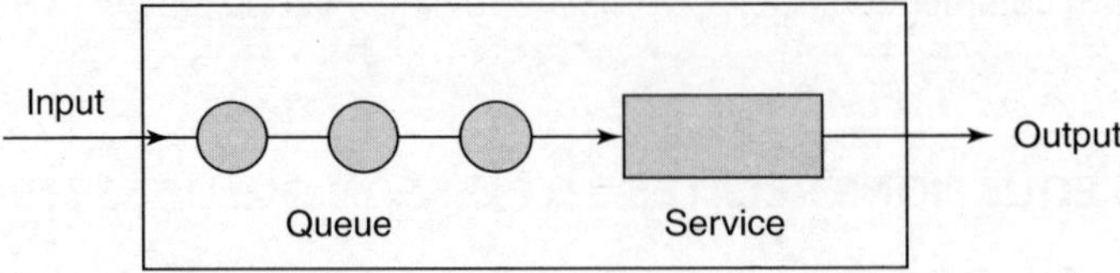

Fig. 15.1 Single server queueing system

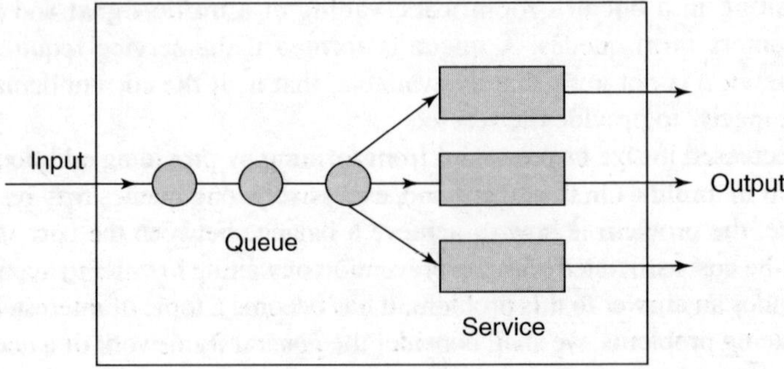

Fig. 15.2 Multiple servers (in parallel) queueing system

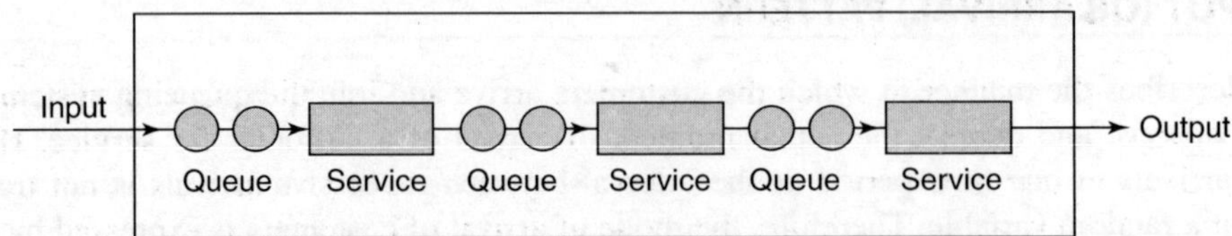

Fig. 15.3 Multiple servers (in series) queueing system

15.3 QUEUE DISCIPLINE

Queue discipline specifies the manner in which the customers form the queue or equivalently, the manner in which they are selected for service, when a queue has been formed. The most common discipline is the FCFS (first come first served) or FIFO (first in first out) according to which the customers are served in the strict order of their arrival. If the last arrival in the system is served first, we have the LCFS or LIFO (last in first out) discipline. If the service is given in random order, we have the service in random order (SIRO) discipline. In the queueing systems which we deal with, we shall assume that service is provided on the FCFS (first come first served) basis.

Symbolic Representation of a Queueing Model

Usually a queueing model is specified and represented symbolically in the form (*a*/*b*/*c*): (*d*/*e*), where *a* denotes the type of distribution of the number of arrivals per unit time, *b* the type of distribution of the service time, *c* the number of servers, *d* the capacity of the system, viz., the maximum queue size and *e* the queue discipline.

Accordingly, the first four models which we will deal with will be denoted by the symbols (M/M/1): (∞/FIFO), (M/M/*s*): (∞/FIFO), (M/M/1): (*k*/FIFO) and (M/M/*s*): (*k*/FIFO).

In the aforementioned symbols, the letter M stands for Markov, indicating that the number of arrivals in time *t* and the number of completed services in time *t* follow the Poisson process which is a continuous time Markov chain.

15.4 DIFFERENCE EQUATIONS RELATED TO POISSON QUEUE SYSTEMS

If the characteristics of a queueing system (such as the input and output parameters) are independent of time or equivalently, if the behaviour of the system is independent of time, the system is said to be in *steady state*. Otherwise, it is said to be in *transient state*.

Let $P_n(t)$ be the probability that there are n customers in the system at time t ($n > 0$). Let us first derive the differential equation satisfied by $P_n(t)$ and then deduce the difference equation satisfied by P_n (probability of n customers at any time) in the steady state.

Let λ_n be the average arrival rate and μ_n be the average service rate when there are n customers in the system (both waiting in the queue and being served).

Note: *The system being in steady state does not mean that the arrival rate and the service rate are independent of the number of customers in the system.*

Now $P_n(t + \Delta t)$ is the probability of n customers at time $t + \Delta t$.

The presence of n customers in the system at time $t + \Delta t$ can happen in any one of the following four mutually exclusive ways:

(i) Presence of n customers at t and no arrival or departure during Δt time.
(ii) Presence of $(n - 1)$ customers at t and one arrival and no departure during Δt time.
(iii) Presence of $(n + 1)$ customers at t and no arrival and one departure during Δt time.
(iv) Presence of n customers at t and one arrival and one departure during Δt time (since more than one arrival/departure during Δt is ruled out)

$$\therefore \quad \begin{aligned} P_n(t+\Delta t) = {} & P_n(t)(1-\lambda_n \Delta t)(1-\mu_n \Delta t) + P_{n-1}(t)\lambda_{n-1}\Delta t(1-\mu_{n-1}\Delta t) \\ & + P_{n+1}(t)(1-\lambda_{n+1}\Delta t)\mu_{n+1}\Delta t + P_n(t)\cdot\lambda_n \Delta t \cdot \mu_n \Delta t \end{aligned}$$

[Since P (an arrival occurs during Δt time) $= \lambda\Delta t$ etc.]

i.e., $P_n(t+\Delta t) = P_n(t) - (\lambda_n + \mu_n)P_n(t)\Delta t + \lambda_{n-1}P_{n-1}(t)\Delta t + \mu_{n+1}P_{n+1}(t)$, on omitting terms containing $(\Delta t)^2$ which is negligibly small.

$$\therefore \quad \frac{p_n(t+\Delta t) - P_n(t)}{\Delta t} = \lambda_{n-1}P_{n-1}(t) - (\lambda_n + \mu_n)P_n(t) + \mu_{n+1}P_{n+1}(t) \tag{1}$$

Taking limits on both sides of (1) as $\Delta t \to 0$, we have

$$P_n'(t) = \lambda_{n-1}P_{n-1}(t) - (\lambda_n + \mu_n)P_n(t) + \mu_{n+1}P_{n+1}(t) \tag{2}$$

Equation (2) does not hold good for $n = 0$, as $P_{-1}(t)$ does not exist. Hence, we derive the differential equation satisfied by $P_0(t)$ independently. Proceeding as before,

$$P_0(t+\Delta t) = p_0(t)(1-\lambda_0\Delta t) + P_1(t)(1-\lambda_1\Delta t)\mu_1\Delta t,$$

[by the possibilities (i) and (iii) given earlier and as no departure is possible when $n = 0$]

$$\therefore \quad \frac{P_0(t+\Delta t) - P_0(t)}{\Delta t} = -\lambda_0 P_0(t) + \mu_1 P_1(t) \tag{3}$$

Taking limits on both sides of (3) as $\Delta t \to 0$, we have

$$P_0'(t) = -\lambda_0 P_0(t) + \mu_1 P_1(t) \tag{4}$$

Now in the steady state, $P_n(t)$ and $P_0(t)$ are independent of time and hence, $P_n'(t)$ and $P_0'(t)$ become zero. Therefore, the differential equations (2) and (4) reduce to the difference equations.

$$\lambda_{n-1}P_{n-1} - (\lambda_n + \mu_n)P_n + \mu_{n+1}P_{n+1} = 0 \tag{5}$$

and

$$-\lambda_o P_0 + \mu_1 P_1 = 0 \tag{6}$$

15.5 VALUES OF P_0 AND P_N FOR POISSON QUEUE SYSTEMS

From (6) derived previously, we have

$$P_1 = \frac{\lambda_0}{\mu_1} P_0 \tag{7}$$

Putting $n = 1$ in (5) and using (7), we have

$$\mu_2 P_2 = (\lambda_1 + \mu_1)P_1 - \lambda_0 P_0$$

$$= (\lambda_1 + \mu_1)\frac{\lambda_0}{\mu_1}P_0 - \lambda_0 P_0 = \frac{\lambda_0\lambda_1}{\mu_1}P_0$$

$$\therefore \quad P_2 = \frac{\lambda_0\lambda_1}{\mu_1\mu_2}P_0 \tag{8}$$

Successively putting $n = 2, 3, \ldots$ in (5) and proceeding similarly, we can get

$$P_3 = \frac{\lambda_0\lambda_1\lambda_2}{\mu_1\mu_2\mu_3}P_0 \quad \text{etc.}$$

Finally,

$$P_n = \frac{\lambda_0\lambda_1\lambda_2\cdots\lambda_{n-1}}{\mu_1\mu_2\mu_3\cdots\mu_n}\cdot P_0 \quad \text{for } n = 1, 2, \ldots \tag{9}$$

Since the number of customers in the system can be 0 or 1 or 2 or 3 etc., which events are mutually exclusive and exhaustive, we have $\sum_{n=0}^{\infty} P_n = 1$.

i.e.,

$$P_0 + \sum_{n=1}^{\infty}\left(\frac{\lambda_0\lambda_1\cdots\lambda_{n-1}}{\mu_1\mu_2\cdots\mu_n}\cdot\right)P_o = 1$$

$$\therefore \quad P_0 = \frac{1}{1 + \sum_{n=1}^{\infty}\left(\frac{\lambda_0\lambda_1\cdots\lambda_{n-1}}{\mu_1\mu_2\cdots\mu_n}\cdot\right)} \tag{10}$$

Equations (9) and (10) will be used to derive the important characteristics of the four queueing models.

15.6 CHARACTERISTICS OF INFINITE CAPACITY, SINGLE SERVER POISSON QUEUE MODEL I

[M/M/1): (∞/FIFO) model], when $\lambda_n = \lambda$ and $\mu_n = \mu$ ($\lambda < \mu$)

1. *Average number L_s of customers in the system*: Let N denote the number of customers in the queueing system (i.e., those in the queue and the one who is being served).

   N is a discrete random variable, which can take the values 0, 1, 2, … ,∞

   such that $P(N = n) = P_n = \left(\frac{\lambda}{\mu}\right)^n P_0$, from (9) of the previous discussion.

   From (10), we have

$$P_0 = \frac{1}{1 + \sum_{n=1}^{\infty}\left(\frac{\lambda}{\mu}\right)^n}$$

$$= \frac{1}{\sum_{n=0}^{\infty}\left(\frac{\lambda}{\mu}\right)^n} = 1 - \frac{\lambda}{\mu}$$

$\therefore$
$$P_n = \left(\frac{\lambda}{\mu}\right)^n \left(1 - \frac{\lambda}{\mu}\right)$$

Now
$$L_s = E(N) = \sum_{n=0}^{\infty} n \times P_n$$

$$= \left(\frac{\lambda}{\mu}\right)\left(1 - \frac{\lambda}{\mu}\right)\sum_{n=1}^{\infty} n\left(\frac{\lambda}{\mu}\right)^{n-1}$$

$$= \frac{\lambda}{\mu}\left(1 - \frac{\lambda}{\mu}\right)\left(1 - \frac{\lambda}{\mu}\right)^{-2}, \quad \text{by binomial summation}$$

$$= \frac{\frac{\lambda}{\mu}}{1 - \frac{\lambda}{\mu}} = \frac{\lambda}{\mu - \lambda} \tag{1}$$

2. *Average number L_q of customers in the queue or average length of the queue:*
   If N is the number of customers in the system, then the number of customers in the queue is (N – 1).

$\therefore$
$$L_q = E(N - 1) = \sum_{n=1}^{\infty}(n - 1)P_n$$

$$= \left(1 - \frac{\lambda}{\mu}\right)\sum_{n=1}^{\infty}(n - 1)\left(\frac{\lambda}{\mu}\right)^n$$

$$= \left(\frac{\lambda}{\mu}\right)^2 \left(1 - \frac{\lambda}{\mu}\right)\sum_{n=2}^{\infty}(n - 1)\left(\frac{\lambda}{\mu}\right)^{n-2}$$

$$= \left(\frac{\lambda}{\mu}\right)^2 \left(1-\frac{\lambda}{\mu}\right)\left(1-\frac{\lambda}{\mu}\right)^{-2}$$

$$= \frac{\left(\frac{\lambda}{\mu}\right)^2}{1-\frac{\lambda}{\mu}} = \frac{\lambda^2}{\mu(\mu-\lambda)} \tag{2}$$

3. *Average number L_w of customers in non-empty queues*

$L_w = E\{(N-1)/(N-1) > 0\}$, since the queue is non-empty

$$= \frac{E(N-1)}{P(N-1>0)} = \frac{\lambda^2}{\mu(\mu-\lambda)} \times \frac{1}{\sum_{n=2}^{\infty} P_n}$$

$$= \frac{\lambda^2}{\mu(\mu-\lambda)} \times \frac{1}{\sum_{n=2}^{\infty}\left(\frac{\lambda}{\mu}\right)^n\left(1-\frac{\lambda}{\mu}\right)}$$

$$= \frac{\lambda^2}{\mu(\mu-\lambda)} \times \frac{1}{\left(\frac{\lambda}{\mu}\right)^2\left(1-\frac{\lambda}{\mu}\right)\sum_{n=0}^{\infty}\left(\frac{\lambda}{\mu}\right)^n}$$

$$= \frac{\mu}{\mu-\lambda} \times \frac{1}{\left(1-\frac{\lambda}{\mu}\right)\left(1-\frac{\lambda}{\mu}\right)^{-1}} = \frac{\mu}{\mu-\lambda} \tag{3}$$

4. *Probability that the number of customers in the system exceeds k*

$$P(N>k) = \sum_{n=k+1}^{\infty} P_n = \sum_{n=k+1}^{\infty}\left(\frac{\lambda}{\mu}\right)^n\left(1-\frac{\lambda}{\mu}\right)$$

$$= \left(\frac{\lambda}{\mu}\right)^{k+1}\left(1-\frac{\lambda}{\mu}\right)\sum_{n=k+1}^{\infty}\left(\frac{\lambda}{\mu}\right)^{n-(k+1)}$$

$$= \left(\frac{\lambda}{\mu}\right)^{k+1}\cdot\left(1-\frac{\lambda}{\mu}\right)\sum_{n=0}^{\infty}\left(\frac{\lambda}{\mu}\right)^n$$

$$= \left(\frac{\lambda}{\mu}\right)^{k+1}\cdot\left(1-\frac{\lambda}{\mu}\right)\left(1-\frac{\lambda}{\mu}\right)^{-1} = \left(\frac{\lambda}{\mu}\right)^{k+1} \tag{4}$$

5. *Probability density function of the waiting time in the system*

Let W_s be the continuous random variable that represents the waiting time of a customer in the system, viz., the time between arrival and completion of service.

Let its probability density function (pdf) be $f(w)$ and let $f(w/n)$ be the density function of W_s subject to the condition that there are n customers in the queueing system when the customer arrives,

Then, $$f(w)=\sum_{n=0}^{\infty} f(w/n)P_n \tag{5}$$

Now $f(w/n)$ = pdf of sum of $(n + 1)$ service times (one part-service time of the customer being served + n complete service times) = pdf of sum of $(n + 1)$ independent random variables, each of which is exponentially distributed with parameter μ

$$=\frac{\mu^{n+1}}{n!}e^{-\mu w}w^n; w>0$$

which is the pdf of an Erlang distribution. [$\because$ The moment generating function (mgf) of the exponential distribution (μ) is $\left(1-\frac{t}{\mu}\right)^{-1}$ and hence, the mgf of the sum of $(n + 1)$ independent exponential (μ) variables is $\left(1-\frac{t}{\mu}\right)^{n+1}$, which is the mgf of an Erlang distribution with parameters μ and $(n + 1)$] by the additive property of the exponential distribution.

$$\therefore \quad f(w)=\sum_{n=0}^{\infty}\frac{\mu^{n+1}}{n!}e^{-\mu w}w^n\left(\frac{\lambda}{\mu}\right)^n\left(1-\frac{\lambda}{\mu}\right), \text{ by (5)}$$

$$=\mu e^{-\mu w}\left(1-\frac{\lambda}{\mu}\right)\sum_{n=0}^{\infty}\frac{1}{n!}(\lambda w)^n$$

$$=\mu\left(1-\frac{\lambda}{\mu}\right)e^{-\mu w}e^{\lambda w}, \text{ by exponential summation}$$

$$=(\mu-\lambda)e^{-(\mu-\lambda)w} \tag{6}$$

which is the pdf of an exponential distribution with parameter $(\mu - \lambda)$.

6. *Average waiting time of a customer in the system*
   W_s follows an exponential distribution with parameter $(\mu - \lambda)$.

$$\therefore \quad E(W_s)=\frac{1}{\mu-\lambda} \tag{7}$$

($\because$ the mean of an exponential distribution is the reciprocal of its parameter).

7. *Probability that the waiting time of a customer in the system exceeds t*

$$P(W_s>t)=\int_t^{\infty} f(w)\,dw$$

$$=\int_t^{\infty}(\mu-\lambda)\,e^{-(\mu-\lambda)w}\,dw$$

$$=[-e^{-(\mu-\lambda)w}]_t^{\infty}=e^{-(\mu-\lambda)t} \tag{8}$$

8. *Probability density function of the waiting time W_q in the queue*
   W_q represents the time between arrival and reach of service point. Let the pdf of W_q be $g(w)$ and let $g(w/n)$ be the density function of W_q subject to the condition that there are n customers in the system or

there are $(n-1)$ customers in the queue apart from one customer receiving service. Now, $g(w/n)$ = pdf of the sum of n service times [one residual service time + $(n-1)$ full service times]

$$=\frac{\mu^n}{(n-1)!}e^{-\mu w}w^{n-1};w>0$$

$$\therefore \quad g(w)=\sum_{n=1}^{\infty}\frac{\mu^n}{(n-1)!}e^{-\mu w}w^{n-1}\left(\frac{\lambda}{\mu}\right)^n\left(1-\frac{\lambda}{\mu}\right)$$

$$=\lambda\left(1-\frac{\lambda}{\mu}\right)e^{-\mu w}\sum_{n=1}^{\infty}\frac{1}{(n-1)!}(\lambda w)^{n-1}$$

$$=\frac{\lambda}{\mu}(\mu-\lambda)e^{-\mu w}e^{\lambda w}$$

$$=\frac{\lambda}{\mu}(\mu-\lambda)e^{-(\mu-\lambda)w};w>0 \tag{9}$$

and $$g(w)=1-\frac{\lambda}{\mu},\text{when } w=0$$

Note:

1. W_q *is a continuous random variable in* $\omega > 0$ *and it takes the value 0 with a non-zero probability.*
2. W_q *does not follow an exponential distribution.*

9. *Average waiting time of a customer in the queue*

$$E\left(W_q\right)=\frac{\lambda}{\mu}(\mu-\lambda)\int_0^{\infty}we^{-(\mu-\lambda)w}dw$$

$$=\frac{\lambda}{\mu}\int_0^{\infty}xe^{-x}\frac{dx}{\mu-\lambda},\text{on putting } (\mu-\lambda)\omega=x$$

$$=\frac{\lambda}{\mu(\mu-\lambda)}[x\left(-e^{-x}\right)-e^{-x}]_0^{\infty}$$

$$=\frac{\lambda}{\mu(\mu-\lambda)} \tag{10}$$

10. *Average waiting time of a customer in the queue, if he has to wait*

$$E(w_q/W_q>0)=\frac{E\left(W_q\right)}{P\left(W_q>0\right)}$$

$$=\frac{E\left(W_q\right)}{1-P\left(W_q=0\right)}$$

$$=\frac{E\left(W_q\right)}{1-P(\text{no customer in the queue})}$$

$$= \frac{E(W_q)}{1-P_0}$$

$$= \frac{\lambda}{\mu(\mu-\lambda)} \times \frac{\mu}{\lambda} \quad \because \left(P_0 = 1 - \frac{\mu}{\lambda}\right)$$

$$= \frac{1}{\mu-\lambda} \tag{11}$$

Relationship between $E(N_s)$, $E(N_q)$, $E(W_s)$ and $E(W_q)$

(i) $$E(N_s) = \frac{\lambda}{\mu-\lambda} = \lambda E(W_s) \quad \text{where } [E(N_s) = L_s]$$

(ii) $$E(N_q) = \frac{\lambda^2}{\mu(\mu-\lambda)} = \lambda E(W_q) \quad \because [E(N_q) = L_q]$$

(iii) $$E(W_s) = E(W_q) + \frac{1}{\mu}$$

(iv) $$E(N_s) = E(N_q) + \frac{\lambda}{\mu}$$

Note:

1. *If any one of the quantities $E(N_s)$, $E(N_q)$, $E(W_s)$ and $E(W_q)$ is known, the other three can be found out using the relations given here.*
2. *These relations, called Little's formulas hold good for models with infinite capacity; a slight modification is needed for models with finite capacity.*

15.7 CHARACTERISTICS OF INFINITE CAPACITY, MULTIPLE SERVER POISSON QUEUE MODEL II

[M/M/s): (∞/FIFO) model], When $\lambda_n = \lambda$ for all n($\lambda < s\mu$)

1. *Values of P_0 and P_n:*

   For the Poisson queue system, P_n is given by

$$P_n = \frac{\lambda_0\lambda_1\lambda_2\cdots\lambda_{n-1}}{\mu_1\mu_2\mu_3\cdots\mu_n} \times P_0, n \geq 1, \tag{1}$$

where
$$P_0 = \left[1 + \sum_{n=1}^{\infty}\left(\frac{\lambda_0\lambda_1\cdots\lambda_{n-1}}{\mu_1\mu_2\cdots\mu_n}\right)\right]^{-1} \tag{2}$$

If there is a single server, $\mu_n = \mu$ for all n. However, there are s servers working independently of each other.

If there be less than **s** customers, i.e., if $n < s$, only n of the s servers will be busy and the others idle. Hence, the mean service rate will be $n\mu$.

If $n \geq s$, all the s servers will be busy and hence, the mean service rate $= s\mu$.

Hence, $$\mu_n = \begin{cases} n\mu, & \text{if } 0 \le n < s \\ s\mu, & \text{if } n \ge s \end{cases} \tag{3}$$

Using (3) in (1) and (2), we have

$$P_n = \frac{\lambda^n}{1\mu \cdot 2\mu \cdot 3\mu \ldots n\mu} P_0, \text{ if } 0 \le n < s$$

$$= \frac{1}{n!}\left(\frac{\lambda}{\mu}\right)^n P_0, \text{ if } 0 \le n < s \tag{4}$$

and $$P_n = \frac{\lambda^n}{\{1\mu \cdot 2\mu \cdots (s-1)\mu\}\{s\mu \cdot s\mu \cdots (n-s+1) \text{ times}\}} P_0$$

$$= \frac{\lambda^n}{(s-1)!\,\mu^{s-1}(s\mu)^{n-s+1}} \cdot P_0$$

$$= \frac{1}{s!\,s^{n-s}}\left(\frac{\lambda}{\mu}\right)^n P_0, \text{ if } n \ge s \tag{5}$$

Now P_0 is given by $$\sum_{n=0}^{\infty} P_n = 1$$

i.e., $$\left[\sum_{n=0}^{s-1} \frac{1}{n!}\left(\frac{\lambda}{\mu}\right)^n + \sum_{n=s}^{\infty} \frac{1}{s!\,s^{n-s}}\left(\frac{\lambda}{\mu}\right)^n\right] P_0 = 1$$

i.e., $$\left[\sum_{n=0}^{s-1} \frac{1}{n!}\left(\frac{\lambda}{\mu}\right)^n + \frac{s^s}{s!}\sum_{n=s}^{\infty}\left(\frac{\lambda}{\mu s}\right)^n\right] P_0 = 1$$

i.e., $$\left[\sum_{n=0}^{s-1} \frac{1}{n!}\left(\frac{\lambda}{\mu}\right)^n + \frac{s^s}{s!}\left(\frac{\lambda}{\mu s}\right)^s \frac{1}{1-\dfrac{\lambda}{\mu s}}\right] P_0 = 1$$

i.e., $$\left[\left\{\sum_{n=0}^{s-1} \frac{1}{n!}\left(\frac{\lambda}{\mu}\right)^n\right\} + \left\{\frac{1}{s!\left(1-\dfrac{\lambda}{\mu s}\right)}\left(\frac{\lambda}{\mu}\right)^s\right\}\right] P_0 = 1$$

or
$$P_0 = \frac{1}{\left\{\sum_{n=0}^{s-1} \frac{1}{n!}\left(\frac{\lambda}{\mu}\right)^n\right\} + \left\{\frac{1}{s!\left(1-\frac{\lambda}{\mu s}\right)}\left(\frac{\lambda}{\mu}\right)^s\right\}} \tag{6}$$

2. *Average number of customers in the queue or average queue length*

$$\begin{aligned}
L_q &= E(N_q) \\
&= E(n-s) = \sum_{n=s}^{\infty}(n-s)P_n \\
&= \sum_{x=0}^{\infty} xP_{x+s} \\
&= \sum_{x=0}^{\infty} x \times \frac{1}{s!s^x}\left(\frac{\lambda}{\mu}\right)^{s+x} \cdot P_0 \\
&= \frac{1}{s!}\left(\frac{\lambda}{\mu}\right)^s P_0 \sum_{x=0}^{\infty} x\left(\frac{\lambda}{\mu s}\right)^x \\
&= \frac{1}{s!}\left(\frac{\lambda}{\mu}\right)^s \frac{\lambda}{\mu s} \cdot P_0 \frac{1}{\left(1-\frac{\lambda}{\mu s}\right)^2} \\
&= \frac{1}{s \cdot s!} \cdot \frac{\left(\frac{\lambda}{\mu}\right)^{s+1}}{\left(1-\frac{\lambda}{\mu s}\right)^2} P_0
\end{aligned} \tag{7}$$

3. *Average number of customers in the system*
   By Little's formula (iv),

$$\begin{aligned}
E(N_s) &= E(N_q) + \frac{\lambda}{\mu} \\
&= \frac{1}{s \cdot s!} \cdot \frac{\left(\frac{\lambda}{\mu}\right)^{s+1}}{\left(1-\frac{\lambda}{\mu s}\right)^2} P_0 + \frac{\lambda}{\mu}
\end{aligned} \tag{8}$$

Result (8) can also be directly derived by using the definition $E(N_s) = \sum_{n=0}^{\infty} n\, P_n$

4. *Average time a customer has to spend in the system*
   By Little's formula (i),

$$E(W_s) = \frac{1}{\lambda} E(N_s)$$

$$= \frac{1}{\mu} + \frac{1}{\mu} \cdot \frac{1}{s \cdot s!} \frac{\left(\frac{\lambda}{\mu}\right)^s}{\left(1 - \frac{\lambda}{\mu s}\right)^2} \cdot P_0 \tag{9}$$

5. *Average time a customer has to spend in the queue*
   By Little's formula (ii),

$$E(W_q) = \frac{1}{\lambda} E(N_q)$$

$$= \frac{1}{\mu} \cdot \frac{1}{s \cdot s!} \frac{\left(\frac{\lambda}{\mu}\right)^s}{\left(1 - \frac{\lambda}{\mu s}\right)^2} \cdot P_0 \tag{10}$$

6. *Probability that an arrival has to wait*

   Required probability = Probability that there are s or more customers in the system

   i.e.,
$$P(W_s > 0) = P(N \geq s)$$

$$= \sum_{n=s}^{\infty} P_n = \sum_{n=s}^{\infty} \frac{1}{s! s^{n-s}} \left(\frac{\lambda}{\mu}\right)^n P_0$$

$$= \frac{1}{s!}\left(\frac{\lambda}{\mu}\right)^s \cdot P_0 \sum_{n=s}^{\infty} \left(\frac{\lambda}{\mu s}\right)^{n-s}$$

$$= \frac{\left(\frac{\lambda}{\mu}\right)^s \cdot P_0}{s!\left(1 - \frac{\lambda}{\mu s}\right)} \tag{11}$$

7. *Probability that an arrival enters the service without waiting*
   Required probability = 1 − P(an arrival has to wait)

$$= 1 - \frac{\left(\frac{\lambda}{\mu}\right)^s \cdot P_0}{s!\left(1 - \frac{\lambda}{\mu s}\right)} \tag{12}$$

8. *Mean waiting time in the queue for those who actually wait*

$$E(W_q / W_s > 0) = \frac{E(W_q)}{P(W_s > 0)}$$

$$= \frac{1}{\mu} \cdot \frac{1}{s \cdot s!} \frac{\left(\frac{\lambda}{\mu}\right)^s}{\left(1 - \frac{\lambda}{\mu s}\right)^2} P_0 \times \frac{s!\left(1 - \frac{\lambda}{\mu s}\right)}{\left(\frac{\lambda}{\mu}\right)^s \cdot P_0} \quad \text{[using (10) and (11)]}$$

$$= \frac{1}{\mu s\left(1 - \frac{\lambda}{\mu s}\right)} = \frac{1}{\mu s - \lambda} \tag{13}$$

9. *Probability that there will be someone waiting*

Required probability $= P(N \geq s + 1)$

$$= \sum_{n=s+1}^{\infty} P_n = \sum_{n=s}^{\infty} P_n - P(N = s)$$

$$= \frac{\left(\frac{\lambda}{\mu}\right)^s \cdot P_0}{s!\left(1 - \frac{\lambda}{\mu s}\right)} - \frac{\left(\frac{\lambda}{\mu}\right)^s \cdot P_0}{s!} \quad \text{[using (10) and (5)]}$$

$$= \frac{\left(\frac{\lambda}{\mu}\right)^s P_0}{s!} \cdot \frac{\left(\frac{\lambda}{\mu s}\right)}{1 - \frac{\lambda}{\mu s}} \tag{14}$$

10. *Average number of customers (in non-empty queues), who have to actually wait.*

$$L_w = E(N_q / N_q \geq 1)$$

$$= E(N_q) / P(N \geq s)$$

$$= \frac{1}{s \cdot s!} \frac{\left(\frac{\lambda}{\mu}\right)^{s+1} \cdot P_0}{\left(1 - \frac{\lambda}{\mu s}\right)^2} \cdot \frac{s!\left(1 - \frac{\lambda}{\mu s}\right)}{\left(\frac{\lambda}{\mu}\right)^s P_0}$$

$$= \frac{\left(\frac{\lambda}{\mu s}\right)}{1 - \frac{\lambda}{\mu s}} \tag{15}$$

15.8 CHARACTERISTICS OF FINITE CAPACITY, SINGLE SERVER POISSON QUEUE MODEL III

[(M/M/1): (*k*/FIFO) Model]

1. *Values of P_0 and P_n*

For the Poisson queue system, $P_n = P(N = n)$ in the steady state is given by the difference equations

$$\lambda_{n-1}P_{n-1} - (\lambda_n + \mu_n)P_n + \mu_{n+1}P_{n+1} = 0;\ n > 0$$

and
$$-\lambda_0 p_0 + \mu_1 P_1 = 0; n = 0$$

This model represents the situation in which the system can accommodate only a finite number *k* of arrivals. If a customer arrives and the queue is full, the customer leaves without joining the queue.

Therefore, for this model,

$$\mu_n = \mu, n = 1,\ 2,\ 3,\ \ldots$$

and
$$\lambda_n = \begin{cases} \lambda, \text{ for } n = 0,1,2,\cdots(k-1) \\ 0, \text{ for } n = k, k+1,\cdots \end{cases}$$

Using these values in the difference equations given earlier, we have

$$\mu P_1 = \lambda P_0 \tag{1}$$

$$\mu P_{n+1} = (\lambda + \mu)P_n - \lambda P_{n-1}, \text{for } 1 \le n \le k-1 \tag{2}$$

and
$$\mu P_k = \lambda P_{k-1}, \text{for } n = k\ \ (\because P_{k+1} \text{ has no meaning}) \tag{3}$$

From (1),
$$P_1 = \frac{\lambda}{\mu} P_0$$

From (2),
$$\mu P_2 = (\lambda + \mu)\frac{\lambda}{\mu} \cdot P_0 - \lambda P_0$$

$\therefore$
$$P_2 = \left(\frac{\lambda}{\mu}\right)^2 P_0 \text{ and so on}$$

In general,
$$P_n = \left(\frac{\lambda}{\mu}\right)^n P_0, \text{ true for } 0 \le n \le k - 1$$

From (3),
$$P_k = \frac{\lambda}{\mu} \cdot \left(\frac{\lambda}{\mu}\right)^{k-1} P_0 = \left(\frac{\lambda}{\mu}\right)^k P_0$$

Now
$$\sum_{n=0}^{k} P_n = 1$$

i.e.,
$$P_0 \sum_{n=0}^{k} \left(\frac{\lambda}{\mu}\right)^n = 1$$

i.e., $$P_0 \frac{\left\{1-\left(\frac{\lambda}{\mu}\right)^{k+1}\right\}}{1-\frac{\lambda}{\mu}} = 1,$$

which is valid even for $\lambda > \mu$

$$\therefore \qquad P_0 = \begin{cases} \dfrac{1-\frac{\lambda}{\mu}}{1-\left(\frac{\lambda}{\mu}\right)^{k+1}}, \text{ If } \lambda \neq \mu & (4) \end{cases}$$

$$\text{and} \qquad P_0 = \begin{cases} \dfrac{1}{k+1}, \text{ if } \lambda = \mu, \text{ since } \lim\limits_{\frac{\lambda}{\mu}\to 1} \left[\dfrac{1-\frac{\lambda}{\mu}}{1-\left(\frac{\lambda}{\mu}\right)^{k+1}}\right] = \dfrac{1}{k+1} & (5) \end{cases}$$

$$\therefore \qquad P_n = \begin{cases} \left(\dfrac{\lambda}{\mu}\right)^n \cdot \left[\dfrac{1-\frac{\lambda}{\mu}}{1-\left(\frac{\lambda}{\mu}\right)^{k+1}}\right], \text{ if } \lambda \neq \mu & (6) \\[2ex] \dfrac{1}{k+1}, \quad \text{if } \lambda = \mu & (7) \end{cases}$$

2. *Average number of customers in the system*

$$E(N) = \sum_{n=0}^{k} nP_n = \frac{\left(1-\frac{\lambda}{\mu}\right)}{1-\left(\frac{\lambda}{\mu}\right)^{k+1}} \cdot \sum_{n=0}^{k} n\left(\frac{\lambda}{\mu}\right)^n$$

$$= \frac{\left(1-\frac{\lambda}{\mu}\right)\cdot\frac{\lambda}{\mu}}{1-\left(\frac{\lambda}{\mu}\right)^{k+1}} \cdot \sum_{n=0}^{k} \frac{d}{dx}\left(x^n\right), \text{ where } x = \frac{\lambda}{\mu}$$

$$= \frac{\left(1-\frac{\lambda}{\mu}\right)\cdot\frac{\lambda}{\mu}}{1-\left(\frac{\lambda}{\mu}\right)^{k+1}} \frac{d}{dx}\left(\frac{1-x^{k+1}}{1-x}\right)$$

$$= \frac{(1-x)x}{1-x^{k+1}} \cdot \left[\frac{(1-x)\{-(k+1)x^k\} + (1-x^{k+1})}{(1-x)^2} \right]$$

$$= \frac{x\left[1-(k+1)x^k + kx^{k+1}\right]}{(1-x)(1-x^{k+1})}$$

$$= \frac{x(1-x^{k+1}) - (k+1)(1-x)x^{k+1}}{(1-x)(1-x^{k+1})}$$

$$= \frac{x}{1-x} - \frac{(k+1)x^{k+1}}{1-x^{k+1}}$$

$$= \frac{\lambda}{\mu-\lambda} - \frac{(k+1)\left(\frac{\lambda}{\mu}\right)^{k+1}}{1-\left(\frac{\lambda}{\mu}\right)^{k+1}}, \text{ if } \lambda \neq \mu \tag{8}$$

and $$E(N) = \sum_{n=0}^{k} \frac{n}{k+1} = \frac{k}{2}, \text{ if } \lambda = \mu \tag{9}$$

3. *Average number of customers in queue*

$$E(N_q) = E(N-1) = \sum_{n-1}^{k} (n-1) P_n$$

$$= \sum_{n=0}^{k} nP_n - \sum_{n=1}^{K} P_n$$

$$= E\,(N) - (1 - P_0) \tag{10}$$

According to Little's formula (iv),

$$E(N_q) = E(N) - \frac{\lambda}{\mu},$$

which is true when the average arrival rate is λ throughout. Now we see that in Step 8, $1 - P_0 \neq \frac{\lambda}{\mu}$, because the average arrival rate is λ as long as there is a vacancy in the queue; it is zero when the system is full.

Hence, we define *the overall effective arrival rate*, denoted by λ' or λ_{eff}, by using Step 8 and Little's formula as

$$\frac{\lambda'}{\mu} = 1 - P_0 \text{ or } \lambda' = \mu(1 - P_0) \tag{11}$$

Thus, Step 8 can be rewritten as

$$E(N_q) = E(N) - \frac{\lambda'}{\mu}, \tag{12}$$

which is the modified Little's formula for this model.

4. *Average waiting times in the system and in the queue*:
By the modified Little's formulas,

$$E(W_s) = \frac{1}{\lambda'} E(N) \tag{13}$$

and
$$E(W_q) = \frac{1}{\lambda'} E(N_q) \tag{14}$$

where λ' is the effective arrival rate, given by Step 9.

15.9 CHARACTERISTICS OF FINITE QUEUE, MULTIPLE SERVER POISSON QUEUE MODEL IV

[M/M/s): (*k*/FIFO) model]

1. *Values of P_0 and P_n*
For the Poisson queue system, P_n is given by

$$P_n = \frac{\lambda_0 \lambda_1 \cdots \lambda_{n-1}}{\mu_1 \mu_2 \cdots \mu_n} \cdot P_0, \; n \geq 1, \tag{1}$$

where
$$P_0 = \left\{ 1 + \sum_{n=1}^{k} \left(\frac{\lambda_0 \lambda_1 \cdots \lambda_{n-1}}{\mu_1 \mu_2 \cdots \mu_n} \right) \right\}^{-1} \tag{2}$$

For this (*M*/*M*/*s*): (*k*/*FIFO*) model,

$$\lambda_n = \begin{cases} \lambda, \text{ for } n = 0, 1, 2, \ldots k-1 \\ 0 \text{ for } n = k, k+1, \ldots \end{cases}$$

$$\mu_n = \begin{cases} n\mu, \; for \; n = 0, 1, 2, \ldots s-1 \\ s\mu \text{ for } n = s, s+1, \ldots \end{cases}$$

Using these values of λ_n and μ_n in (2) and noting that $1 < s < k$, we get

$$P_0^{-1} = \left\{ 1 + \frac{\lambda}{1!\mu} + \frac{\lambda^2}{2!\mu^2} + \cdots + \frac{\lambda^{s-1}}{(s-1)!\mu^{s-1}} \right\} + \left\{ \begin{array}{l} \dfrac{\lambda^s}{(s-1)!\mu^{s-1} \cdot \mu s} + \dfrac{\lambda^{s+1}}{(s-1)!\mu^{s-1} \cdot (\mu s)^2} \\ + \cdots + \dfrac{\lambda^k}{(s-1)!\mu^{s-1} \cdot (\mu s)^{k-s+1}} \end{array} \right\}$$

$$= \sum_{n=0}^{s-1} \frac{1}{n!} \left(\frac{\lambda}{\mu} \right)^n + \frac{\lambda^s}{s!\mu^s} \left[1 + \frac{\lambda}{\mu s} + \left(\frac{\lambda}{\mu s} \right)^2 + \cdots \left(\frac{\lambda}{\mu s} \right)^{k-s} \right]$$

$$= \sum_{n=0}^{s-1} \frac{1}{n!} \left(\frac{\lambda}{\mu} \right)^n + \frac{1}{s!} \left(\frac{\lambda}{\mu} \right)^s \sum_{n=s}^{k} \left(\frac{\lambda}{\mu s} \right)^{n-s} \tag{3}$$

$$P_n = \left\{ \frac{1}{n!} \left(\frac{\lambda}{\mu} \right)^n P_0, \text{ for } n \leq s \right\}$$

and
$$P_n = \begin{cases} \dfrac{1}{s!s^{n-s}} \cdot \left(\dfrac{\lambda}{\mu}\right)^n P_0, & \text{for } s < n \le k \\ 0, & \text{for } n > k \end{cases} \tag{4}$$

2. *Average queue length or average number of customers in the queue*

$$E(N_q) = E(N-s) = \sum_{n=s}^{k}(n-s)P_n$$

$$= \frac{P_0}{s!}\sum_{n=s}^{k}(n-s)\left(\frac{\lambda}{\mu}\right)^n / s^{n-s} \quad \text{[using (4)]}$$

$$= \frac{\left(\dfrac{\lambda}{\mu}\right)^s \cdot P_0}{s!}\sum_{x=0}^{k-s} x \cdot \left(\frac{\lambda}{\mu s}\right)^x$$

$$= \frac{\left(\dfrac{\lambda}{\mu}\right)^s \cdot P_0\rho}{s!}\sum_{x=0}^{k-s} x\rho^{x-1} \quad \text{where } \rho = \frac{\lambda}{\mu s}$$

$$= \left(\frac{\lambda}{\mu}\right)^s \frac{P_0\rho}{s!}\sum_{x=0}^{k-s}\frac{\mathrm{d}}{\mathrm{d}\rho}(\rho^x)$$

$$= \left(\frac{\lambda}{\mu}\right)^s \frac{P_0\rho}{s!}\frac{\mathrm{d}}{\mathrm{d}\rho}\left\{\frac{1-\rho^{k-s+1}}{1-\rho}\right\}$$

$$= \left(\frac{\lambda}{\mu}\right)^s \frac{P_0\rho}{s!}\left[\frac{-(1-\rho)(k-s+1)\rho^{k-s} + (1-\rho^{k-s+1})}{(1-\rho)^2}\right]$$

$$= \left(\frac{\lambda}{\mu}\right)^s \frac{P_0\rho}{s!}\left[\frac{-(k-s)(1-\rho)\rho^{k-s} - (1-\rho)\rho^{k-s} + 1 - \rho^{k-s+1}}{(1-\rho)^2}\right]$$

$$= P_0\left(\frac{\lambda}{\mu}\right)^s \frac{\rho}{s!}\left[\frac{-(k-s)(1-\rho)\rho^{k-s} + 1 - \rho^{k-s}(1-\rho+\rho)}{(1-\rho)^2}\right]$$

$$= P_0\left(\frac{\lambda}{\mu}\right)^s \frac{\rho}{s!(1-\rho)^2}[1-\rho^{k-s} - (k-s)(1-\rho)\rho^{k-s}],$$

where
$$\rho = \frac{\lambda}{\mu s}$$

3. *Average number of customers in the system*

$$E(N) = \sum_{n=0}^{k} nP_n = \sum_{n=0}^{s-1} nP_n + \sum_{n=0}^{k} nP_n$$

$$= \sum_{n=0}^{s-1} nP_n + \sum_{n=0}^{k} (n-s)P_n + \sum_{n=0}^{k} sP_n$$

$$= \sum_{n=0}^{s-1} nP_n + E(N_q) + s\left\{\sum_{n=0}^{k} P_n - \sum_{n=0}^{s-1} P_n\right\}$$

$$= E(N_q) + \text{s} - \sum_{n=0}^{s-1} (s-n)P_n \left(\because \sum_{n=0}^{k} P_n = 1\right) \tag{6}$$

Obviously $\left\{\text{s} - \sum_{n=0}^{s-1} (s-n)P_n\right\} \neq \frac{\lambda}{\mu}$, so that step (6) represents Little's formula.

In order to make (6) assume the form of Little's formula, we define the *overall effective arrival rate* λ' or λ_{eff} as follows:

$$\frac{\lambda'}{\mu} = s - \sum_{n=0}^{s-1} (s-n)P_n$$

i.e.,

$$\lambda' = \mu\left[s - \sum_{n=0}^{s-1} (s-n)P_n\right] \tag{7}$$

With this definition of λ', Step 6 becomes

$$E(N) = E(N_q) + \frac{\lambda'}{\mu} \tag{8}$$

which is the modified Little's formula for this model.

4. *Average waiting time in the system and in the queue*:
By the modified Little's formulas,

$$E(W_s) = \frac{1}{\lambda'} E(N) \tag{9}$$

and

$$E(W_q) = \frac{1}{\lambda'} E(N_q) \tag{10}$$

where λ' is the effective arrival rate, given by Step 7.

15.10 NON-MARKOVIAN QUEUEING MODEL V

[(M/G/1): (∞/GD) Model]

So far we have discussed Markovian queue models in which the inter-arrival and inter-service times were assumed to follow exponential distributions with parameters λ and μ. When the arrivals and the departures do not follow Poisson distributions, the discussion of the queueing models is tedious. However, we can derive the characteristics of a particular non-Markovian model (M/G/1): (∞/GD), where M indicates that the number of arrivals in time t follows a Poisson process, G indicates that the service time follows a general (arbitrary) distribution and GD indicates general queue discipline (viz., any kind of queue discipline).

The average number L of customers in the M/G/1 queueing system is given by a formula known as the Pollaczek–Khinchine formula.

15.11 POLLACZEK–KHINCHINE FORMULA

Let N and N' be the number of customers in the system at times t and $t + T$, when two consecutive customers have just left the system after getting service.

Thus, T is the random service time, which is a continuous random variable. Let $f(t)$, $E(T)$, Var (T) be the pdf, mean and variance of T. Also, let M be the number of customers arriving in the system during the service time T.

Hence, $$N' = \begin{cases} M, \text{ if } N = 0 \\ N - 1 + M, \text{ if } N > 0 \end{cases}$$

where M is a discrete random variable, taking the values 0, 1, 2…

Equivalently, $$N' = N - 1 + M + \delta \quad (1)$$

where $$\delta = \begin{cases} 1, \text{ if } N = 0 \\ 0, \text{ if } N > 0 \end{cases}$$

$$\therefore \quad E(N') = E(N) - 1 + E(M) + E(\delta) \quad (2)$$

When the system has reached the steady state, the probability of the number of customers in the system will be a constant.

Hence, $$E(N) = E(N') \text{ and } E(N^2) = E(N'^2) \quad (3)$$

Using this in (2), we get $E(\delta) = 1 - E(M)$ (4)

Squaring both sides of (1), we have

$$(N')^2 = N^2 + (M - 1)^2 + \delta^2 + 2N(M - 1) + 2(M - 1)\,\delta + 2N\delta \quad (5)$$

Now $$\delta^2 = \delta$$

($\because \delta^2 = 0$ or 1, according as $\delta = 0$ or 1)

and $$N\delta = \begin{cases} 0 \times 1, \text{ if } N = 0 \\ N \times 0, \text{ if } N > 0 \end{cases}$$

$$= 0$$

Using these values in (5), we have

$$N'^2 = N^2 + M^2 + 2N(M - 1) + (2M - 1)\delta - 2M + 1$$

i.e., $$2N(1 - M) = N^2 - N'^2 + M^2 + (2M - 1)\delta - 2M + 1$$

$$\therefore \quad 2E\{N(1 - M)\} = E(N^2) - E(N'^2) + E(M^2) + E\{(2M - 1)\delta\} - 2E(M) + 1$$

i.e., $$2E(N)\{1 - E(M)\} = E(M^2) + \{2E(M) - 1\}E(\delta) - 2E(M) + 1$$

[by independence and by (3)]

$$\therefore \quad E(N) = \frac{E(M^2) + \{2E(M) - 1\}\{1 - E(M)\} - 2E(M) + 1}{2\{1 - E(M)\}}, \text{ by (4)}$$

$$= \frac{E(M^2) - 2E^2(M) + E(M)}{2\{1 - E(M)\}} \quad (5)$$

Since the number M of arrivals in time T follows a Poisson process with parameter λ, say, then $E(M) = \lambda T$ and Var $(M) = \lambda T$ or $E(M^2) = (\lambda T)^2 + \lambda T$

Now
$$E(M) = E\{E(M/T)\}$$
$$= E(\lambda T) = \lambda E(T) \tag{6}$$
$$E(M^2) = E\{E(M^2/T)\} = E\{\lambda^2 T^2 + \lambda T\}$$
$$= \lambda^2\{\text{Var}(T) + E^2(T)\} + \lambda E(T) \tag{7}$$

Using (6) and (7) in (5), we have

$$L_s = E(N) = \frac{\lambda^2 V(T) + \lambda^2 E^2(T) + \lambda E(T) - 2\lambda^2 E^2(T) + \lambda E(T)}{2\{1 - \lambda E(T)\}}$$
$$= \lambda E(T) + \frac{\lambda^2\{V(T) + E^2(T)\}}{2\{1 - \lambda E(T)\}}$$

Note:

1. *The other characteristics $L_q = E(N_q)$, $E(W_s)$ and $E(W_q)$ of this model can be obtained by using Little's formulas.*
2. *$\lambda E(T)$ must be less than 1, otherwise L_s becomes negative, which is meaningless.*
3. *In this M/G/1 model, if $G \equiv M$, viz., the service time T follows an exponential distribution with parameter μ, then*

$$E(T) = \frac{1}{\mu}$$

and
$$V(T) = \frac{1}{\mu^2}$$

and hence
$$L_s = \frac{\lambda}{\mu} + \frac{\lambda^2\left\{\frac{1}{\mu^2} + \frac{1}{\mu^2}\right\}}{1\left(1 - \frac{\lambda}{\mu}\right)} = \frac{\lambda}{\mu - \lambda},$$

which has already been derived in the M/M/1 model.

WORKED EXAMPLES 15

Example 1

Arrivals at a telephone booth are considered to be Poisson with an average time of 12 min. between one arrival and the next. The length of a phone call is assumed to be distributed exponentially with mean 4 min.

(a) Find the average number of persons waiting in the system.
(b) What is the probability that it will take him more than 10 min. altogether to wait in the queue?
(c) What is the probability that it will take him more than 10 min. altogether to wait for a phone and complete his call?

(d) Estimate the fraction of the day when the phone will be in use.
(e) The telephone department will install a second booth, when convinced that an arrival has to wait on the average for at least 3 min. for the phone. By how much should the flow of arrivals increase in order to justify a second booth?
(f) What is the average length of the queue that forms from time to time?

SOLUTION

Mean inter-arrival time $= \frac{1}{\lambda} = 12 \text{ min}$

Therefore, mean arrival rate $= \lambda = \frac{1}{12}$ per minute

Mean service time $= \frac{1}{\mu} = 4 \text{ min}$

Therefore, mean service rate $= \mu = \frac{1}{4}$ per minute

(a)

$$E(N) = \frac{\lambda}{\mu - \lambda} \quad \text{(by formula (1) of model I)}$$

$$= \frac{\frac{1}{12}}{\frac{1}{4} - \frac{1}{12}} = 0.5 \text{ customer}$$

(b)

$$P(W > 0) = 1 - P(W = 0)$$

$$= 1 - P \text{ (no customer in the system)}$$

$$= 1 - P_0$$

$$= 1 - \left(1 - \frac{\lambda}{\mu}\right) \quad \text{(by formula for } P_0 \text{ of model I)}$$

$$= \frac{\lambda}{\mu} = \frac{1/12}{1/4} = \frac{1}{3}$$

(c)

$$P(W > 10) = e^{-(\mu-\lambda)} \times 10 \text{ [by formula (8) of model I]}$$

$$e^{-\left(\frac{1}{4} - \frac{1}{12}\right) \times 10}$$

$$= e^{-\frac{5}{3}} = 0.1889$$

(d) P (the phone will be idle) $= P(N = 0) = P_0$

$$= 1 - \frac{\lambda}{\mu} = \frac{2}{3}$$

$\therefore$ P (the phone will be in use) $= 1 - \frac{2}{3} = \frac{1}{3}$

or the fraction of the day when the phone will be in use $= \frac{1}{3}$.

(e) The second phone will be installed, if $E(W_q) > 3$.

(f) i.e., if $\frac{\lambda}{\mu(\mu - \lambda)} > 3$ [by formula (10) of model I]

(g) i.e., if $\frac{\lambda_R}{\frac{1}{4}\left(\frac{1}{4} - \lambda_R\right)} > 3$,

where λ_R is required arrival rate.

i.e., if $$\lambda_R > \frac{3}{4}\left(\frac{1}{4} - \lambda_R\right)$$

i.e., if $$\lambda_R > \frac{3}{28}$$

Hence, the arrival rate should increase by $\frac{3}{28} - \frac{1}{12} = \frac{1}{42}$ per minute to justify a second phone.

(f) $E(N_q /$ the queue is always available)

$$= E(N_q / N_q > 0)$$

$$= E(N_q / N > 1)$$

$$= \frac{E(N_q)}{P(N > 1)} = \frac{E(N_q)}{1 - P_0 - P_1}$$

$$= \frac{\lambda^2}{\mu(\mu - \lambda)} \times \frac{1}{1 - \left(1 + \frac{\lambda}{\mu}\right) P_0}$$

[by formula (2) of model 1]

$$= \frac{\lambda^2}{\mu(\mu - \lambda)} \cdot \frac{1}{1 - \left(1 + \frac{\lambda}{\mu}\right)\left(1 - \frac{\lambda}{\mu}\right)}$$

$$= \frac{\lambda^2}{\mu(\mu-\lambda)} \cdot \frac{\mu^2}{\lambda^2} = \frac{\mu}{\mu-\lambda}$$

$$= \frac{1/4}{1/4-1/12} = 1.5 \text{ persons}$$

Example 2

Customers arrive at a one-man barber shop according to a Poisson process with a mean inter-arrival time of 12 min. They spend an average of 10 min. in the barber's chair.

(a) What is the expected number of customers in the barber shop and in the queue?
(b) Calculate the percentage of time an arrival can walk straight into the barber's chair without having to wait.
(c) How much time can a customer expect to spend in the barber's shop?
(d) Management will provide another chair and hire another barber when a customer's waiting time in the shop exceeds 1.25 h. How much must the average rate of arrivals increase to warrant a second barber?
(e) What is the average time customers spend in the queue?
(f) What is the probability that the waiting time in the system is greater than 30 min.?
(g) Calculate the percentage of customers who have to wait prior to getting into the barber's chair.
(h) What is the probability that more than 3 customers are in the system?

SOLUTION

$$\frac{1}{\lambda} = 12$$

$\therefore$ $$\lambda = \frac{1}{12} \text{ per minute}$$

$$\frac{1}{\mu} = 10$$

$\therefore$ $$\mu = \frac{1}{10} \text{ per minute}$$

(a) $$E(N_s) = \frac{\lambda}{\mu-\lambda}$$

$$= \frac{1/12}{1/10-1/12} = 5 \text{ customers [by formula(1) of model I]}$$

(b) $$E(N_q) = \frac{\lambda^2}{\mu(\mu-\lambda)} \text{ [by formula (2) of model I]}$$

(c) $$= \frac{\frac{1}{144}}{\frac{1}{10}\left(\frac{1}{10}-\frac{1}{12}\right)} = 4.17 \text{ customers}$$

(b) P(a customer goes straight to the barber's chair)

$= P$(No customer in the system)

$$= P_0 = 1 - \frac{\lambda}{\mu}$$

$$= 1 - \frac{\frac{1}{12}}{\frac{1}{10}} = \frac{1}{6}$$

Therefore, percentage of time an arrival need not wait = 16.7

(c) $$E(W) = \frac{1}{\mu - \lambda} \text{ [by formula (7) of mode I]}$$

$$= \frac{1}{\frac{1}{10} - \frac{1}{12}} = 60 \text{ min or 1h}$$

(d) $$E(W) > 75, \text{ if } \frac{1}{\mu - \lambda_r} > 75$$

i.e., if $$\lambda_r > \mu - \frac{1}{75}$$

i.e., if $$\lambda_r > \frac{1}{10} - \frac{1}{75}$$

i.e., if $$\lambda_r > \frac{13}{150}$$

Hence, to warrant a second barber, the average arrival rate must increase by $\frac{13}{150} - \frac{1}{12} = \frac{1}{300}$ per minute

(e) $$E(W_q) = \frac{\lambda}{\mu(\mu - \lambda)}$$

[by formula (10) of model I]

$$= \frac{\frac{1}{12}}{\frac{1}{10}\left(\frac{1}{10} - \frac{1}{12}\right)} = 50 \text{ min}$$

(f) $$P(W > t) = e^{-(\mu - \lambda)t}, \text{ [by formula (8) of model I]}$$

$\therefore$ $$P(W > 30) = e^{-\left(\frac{1}{10} - \frac{1}{12}\right) \times 30}$$

$$= e^{-0.5} = 0.6065$$

(g) $$P(\text{a customer has to wait}) = P(W > 0)$$

$$= 1 - P(W = 0)$$

$$= 1 - P(N = 0) = 1 - P_0$$

$$= \frac{\lambda}{\mu} = \frac{1/12}{1/10} = \frac{5}{6}$$

∴ Percentage of customer who have to wait

$$= \frac{5}{6} \times 100 = 83.33$$

(h) $$P(N > 3) = P_4 + P_5 + P_6 + \ldots$$

$$= 1 - \{P_0 + P_1 + P_2 + P_3\}$$

$$= 1 - \left(1 - \frac{\lambda}{\mu}\right)\left\{1 + \frac{\lambda}{\mu} + \left(\frac{\lambda}{\mu}\right)^2 + \left(\frac{\lambda}{\mu}\right)^3\right\}$$

since $$P_n = \left(\frac{\lambda}{\mu}\right)^n \left(1 - \frac{\lambda}{\mu}\right), \text{ for } n \geq 0, \text{ for model I]}$$

$$= \left(\frac{\lambda}{\mu}\right)^4 = \left(\frac{5}{6}\right)^4$$

$$= 0.4823$$

Example 3

At what average rate must a clerk in a supermarket work in order to ensure a probability of 0.90 that the customer will not wait longer than 12 min? It is assumed that there is only one counter at which customers arrive in a Poisson fashion at an average rate of 15 per hour and that the length of the service by the clerk has an exponential distribution.

Solution

$$\lambda = 15/\text{ hour};$$

$$\mu = \mu_R/\text{hour}$$

$$P\left(Wq \leq \frac{1}{5}\right) = 0.90$$

i.e., $$P\left(Wq > \frac{1}{5}\right) = 0.10$$

i.e., $$\int_{0.2}^{\infty} g(w)dw = 0.10$$

i.e., $$\int_{0.2}^{\infty} \frac{\lambda}{\mu}(\mu - \lambda)e^{-(\mu-\lambda)w} dw = 0.10 \text{ [by formula (9) of model I]}$$

i.e., $$\left[-\frac{\lambda}{\mu} e^{-(\mu-\lambda)w}\right]_{0.2}^{\infty} = 0.1$$

i.e., $$\frac{15}{\mu_R} e^{-(\mu_R - 15)\times 0.2} = 0.1$$

i.e., $$(15 - \mu_R) \times 0.2 = \log(0.1) - \log 15 + \log \mu_R$$

i.e., $$0.2 \cdot \mu_R + \log \mu_R = 3 + \log 150 = 80$$

Solving (1), we get $\mu_R = 24$ approximately.

That is, the clerk must serve at the rate of 24 customers per hour.

Example 4

If people arrive to purchase cinema tickets at the average rate of 6 per minute, it takes an average of 7.5 seconds to purchase a ticket. If a person arrives 2 min. before the picture starts and if it takes exactly 1.5 min. to reach the correct seat after purchasing the ticket,

(a) Can he expect to be seated for the start of the picture?
(b) What is the probability that he will be seated for the start of the picture?
(c) How early must he arrive in order to be 99% sure of being seated for the start of the picture?

SOLUTION

$$\lambda = 6\text{/minute;}$$

$$\mu = 8\text{/minute}$$

(a)
$$E(W) = \frac{1}{\mu - \lambda} \text{ [by formula (7) of model I]}$$

$$= \frac{1}{8-6} = \frac{1}{2} \text{ min}$$

$\therefore$ E(total time required to purchase the ticket and to reach the seat)

$$= \frac{1}{2} + 1\frac{1}{2} = 2 \text{ min}$$

Hence, he can just be seated for the start of the picture.

(b) P(total time < 2)

$$= P\left(W < \frac{1}{2}\right) = 1 - P\left(W > \frac{1}{2}\right)$$

$$= 1 - e^{-\mu\left(1 - \frac{\lambda}{\mu}\right) \times \frac{1}{2}} \text{ [by formula 8 of model I]}$$

$$= 1 - e^{-1} = 0.63$$

(c)
$$P(W < t) = 0.99$$

i.e.,
$$P(W > t) = 0.01$$

i.e.,
$$e^{-(\mu-\lambda)t} = 0.1$$

i.e.,
$$-2t = \log(0.1) = -2.3$$

$\therefore$
$$t = 1.15 \text{ min}$$

i.e., P(ticket purchasing time <1.15) = 0.99

$\therefore$ P [total time to get ticket and go to the seat < (1.15 + 1.5)] = 0.99

Therefore, the person must arrive at least 2.65 min early so as to be 99% sure of seeing the start of the picture

Example 5

A duplicating machine maintained for office use is operated by an office assistant who earns Rs. 5 per hour. The time to complete each job varies according to an exponential distribution with mean 6 min. Assume a Poisson input with an average arrival rate of 5 jobs per hour. If an 8-h day is used as a base, determine

(a) the percentage of idle time of the machine
(b) the average time a job is in the system and
(c) the average earning per day of the assistant.

SOLUTION

$$\lambda = 5/\text{ hour};$$

$$\mu = \frac{60}{6} = 10\text{ /hour}$$

$$P(\text{the machine is idle}) = P(N=0) = P_0$$

$$= 1 - \frac{\lambda}{\mu} \text{ [by the formula for } P_0 \text{ in model I)}$$

$$= 1 - \frac{5}{10} = \frac{1}{2}$$

$\therefore$ Percentage of idle time of the machine = 50

$$E(W) = \frac{1}{\mu - \lambda} \text{ [by formula (7) of model I)}$$

$$= \frac{1}{10-5} = \frac{1}{5}\text{ h or 12 min}$$

E(earning per day)

$$= E(\text{number of jobs done/day}) \times \text{earning per job}$$
$$= E(\text{number of jobs done/day}) \times E(\text{time in hours/job}) \times \text{earning/hour}$$

$$= (8 \times 5) \times \frac{1}{5} \times 5 = \text{Rs. } 40$$

Example 6

The mean rate of arrival of planes at an airport during the peak period is 20 per hour, but the actual number of arrivals in any hour follows a Poisson distribution. The airport can land 60 planes per hour on an average in good weather or 30 planes per hour in bad weather. However, the actual number landed in any hour follows a Poisson distribution with respective averages. When there is congestion, the planes are forced to fly over the field in the stack waiting for their turn to land after the other planes that had arrived earlier.

(a) How many planes would be flying over the field in the stack on an average in good weather and in bad weather?
(b) How long would a plane be in the stack and in the process of landing in good and bad weather?

(c) How much stack and landing time is to be allowed so that priority to land out of order will have to be requested only 1 in 20 times.

SOLUTION

$$\lambda = 20 \text{ per hour}$$

$$\mu = \begin{cases} 60 \text{ per hour in good weather} \\ 30 \text{ per hour in bad weather} \end{cases}$$

Note: *Landing time is service time; the planes flying over the field in the stack are assumed to form the queue.*

$E(N_q)$ = Average number of planes flying over the field $= \dfrac{\lambda^2}{\mu(\mu-\lambda)}$ [by formula (2) of model I]

$$= \begin{cases} \dfrac{20^2}{60(60-20)}, & \text{in good weather} \\ \dfrac{20^2}{30(30-20)}, & \text{in bad weather} \end{cases}$$

$$= \begin{cases} \dfrac{1}{6}, & \text{in good weather} \\ \dfrac{4}{3}, & \text{in bad weather} \end{cases}$$

$E(W)$ = Average time for flying in the stack and for landing

$$= \frac{1}{\mu-\lambda} \text{ [by formula (7) of model I]}$$

$$= \begin{cases} \dfrac{1}{40}\text{h} & \text{or 1.5 min in good weather} \\ \dfrac{1}{10}\text{h} & \text{or 6 min in bad weather} \end{cases}$$

Let t_R be the maximum stack and landing time to be allowed beyond which priority out of order is to be requested.

Then, $$P(W > t_R) = \frac{1}{20}$$

i.e., $$e^{-\mu\left(1-\frac{\lambda}{\mu}\right)t_R} = 0.05 \text{ [by formula (8) of model I]}$$

i.e., $$\begin{cases} e^{-40t_R} = 0.05, & \text{for good weather} \\ e^{-10t_R} = 0.05, & \text{for bad weather} \end{cases}$$

i.e., $$\begin{cases} t_R = 0.075 \text{ h or 4.5 min for good weather} \\ \quad\;\; = 0.299 \text{ h or 18 min for bad weather} \end{cases}$$

Example 7

There are three typists in an office. Each typist can type an average of 6 letters per hour. If letters arrive for being typed at the rate of 15 letters per hour,

(a) What fraction of time will all the typists be busy?
(b) What is the average number of letters waiting to be typed?
(c) What is the average time a letter has to spend for waiting and for being typed?
(d) What is the probability that a letter will take longer than 20 min. waiting to be typed and being typed?

SOLUTION

$$\lambda = 15/\text{hour};$$

$$\mu = 6/\text{hour};$$

$$s = 3.$$

Hence, this is a problem in multiple server [(M/M/s): (∞/FIFO)] model, i.e., model II.

(a) P (all the typists are busy) $= P(N \geq 3)$

$$= \frac{\left(\frac{\lambda}{\mu}\right)^3 \cdot P_0}{3!\left(1 - \frac{\lambda}{3\mu}\right)} \quad \text{[by formula (11) of model II]}$$

$$= \frac{(2.5)^3 P_0}{6 \times \left(1 - \frac{2.5}{3}\right)} \tag{1}$$

Now

$$P_0 = \frac{1}{\left\{\sum_{n=0}^{s-1} \frac{1}{n!}\left(\frac{\lambda}{\mu}\right)^n\right\} + \left\{\frac{1}{s!\left(1 - \frac{\lambda}{\mu s}\right)} \cdot \left(\frac{\lambda}{\mu}\right)^s\right\}} \quad \text{[by formula (6) of model II]}$$

$$= \frac{1}{\left\{1 + 2.5 + \frac{1}{2} \times (2.5)^2\right\} + \left\{\frac{1}{6 \times \left(1 - \frac{5}{6}\right)} \times (2.5)^3\right\}}$$

$$= \frac{1}{22.25} = 0.0449 \tag{2}$$

Using (2) in (1), we have $P(N \geq 3) = 0.7016$.
Hence, the fraction of the time all the typists will be busy $= 0.7016$.

(b) $$E(N_q) = \frac{1}{s.s!} \cdot \frac{\left(\frac{\lambda}{\mu}\right)^{s+1} \cdot P_0}{\left(1 - \frac{\lambda}{\mu s}\right)^2} \text{ [by formula (7) of model II]}$$

$$= \frac{1}{3 \times 6} \times \frac{(2.5)^4}{\left(1 - \frac{2.5}{3}\right)^2} \times 0.0449 = 3.5078$$

(c) $$E(W) = \frac{1}{\lambda} E(N) \text{ [by Little's formula (i)]}$$

$$= \frac{1}{\lambda}\left\{E(N_q) + \frac{\lambda}{\mu}\right\}, \text{ [by Little's formula (iv)]}$$

$$= \frac{1}{15}\{3.5078 + 2.5\}$$

$$= 0.4005 \text{ h or 24 min., nearly}$$

(d) $$P(W > t) = e^{-\mu t}\left\{1 + \frac{\left(\frac{\lambda}{\mu}\right)^s \left[1 - e^{-\mu t\left(s-1-\frac{\lambda}{\mu}\right)}\right] P_0}{s!\left(1 - \frac{\lambda}{\mu s}\right)\left(s - 1 - \frac{\lambda}{\mu}\right)}\right\}$$

(This formula has not been derived; it may be assumed.)

$$\therefore \quad P\left(W > \frac{1}{3}\right) = e^{-6 \times \frac{1}{3}} = \left[1 + \frac{(2.5)^3\{1 - e^{(-2 \times -0.5)}\} \times 0.0449}{6\left(1 - \frac{2.5}{3}\right)(-0.5)}\right]$$

$$= e^{-2}\left[1 + \frac{0.7016\,(1 - e)}{(-0.5)}\right]$$

$$= 0.4616$$

Example 8

Given an average arrival rate of 20 per hour, is it better for a customer to get service at a single channel with a mean service rate of 22 customers per hour or at one of two channels in parallel with a mean service rate of 11 customers per hour for each of the two channels. Assume both queues to be of Poisson type.

Solution

For the single channel service,

$$\lambda = 20/\text{hour};$$

$$\mu = 22/\text{hour.}$$

$$E(W) = \frac{1}{\mu - \lambda} \quad \text{[by formula (7) of model I]}$$

$$= \frac{1}{2}\text{h}$$

For the two channel service,

$$\lambda = 20/\text{hour;}$$

$$\mu = 11/\text{hour}$$

$$E(W) = \frac{1}{\mu} + \frac{1}{\mu} \cdot \frac{1}{s.s!} \cdot \frac{\left(\frac{\lambda}{\mu}\right)^s}{\left(1 - \frac{\lambda}{\mu s}\right)^2} \times P_0 \qquad \text{[by formula (9) of model II]}$$

$$= \frac{1}{11} + \frac{1}{11 \times 2 \times 2} \times \frac{\left(\frac{20}{11}\right)^2}{\left(1 - \frac{20}{22}\right)^2} \times P_0$$

$$= 0.0909 + 9.0909 \times P_0 \qquad (1)$$

Now,

$$P_0^{-1} = \sum_{n=0}^{s-1} \frac{1}{n!} \cdot \left(\frac{\lambda}{\mu}\right)^n + \left\{ \frac{1}{s!\left(1 - \frac{\lambda}{\mu s}\right)} \cdot \left(\frac{\lambda}{\mu}\right)^s \right\} \qquad \text{[by formula (6) of model II]}$$

$$= 1 + \frac{20}{11} + \frac{1}{2 \times \frac{1}{11}} \times \left(\frac{20}{11}\right)^2$$

$$= 21$$

$\therefore$

$$P_0 = 0.0476$$

Using (2) in (1), we have

$$E(W) = 0.5236 \text{ h}$$

As the average waiting time in a single channel service is less than that in a two channel service, the customer has to prefer the former.

Example 9

A telephone company is planning to install telephone booths in a new airport. It has established the policy that a person should not have to wait more than 10% of the time he tries to use a phone. The demand for use is estimated to be a Poisson distribution with an average of 30 per hour. The average phone call has an exponential distribution with a mean time of 5 min. How many phone booths should be installed?

$$\lambda = 30/\text{hour}$$

and
$$\mu = 12/\text{hour}$$

In order that an infinite queue may not build up, the traffic intensity $\frac{\lambda}{\mu s} < 1$, for a multi-server model.

i.e.,
$$s > \frac{\lambda}{\mu}$$

i.e.,
$$s > \frac{30}{12} (= 2.5)$$

Therefore, the telephone company must install at least 3 booths.

Now we have to find the number s of telephone booths such that

$$p(w > 0) \leq 0.10 \text{ or equivalently}$$

$$p(N \geq s) \leq 0.10$$

i.e., we have to find s such that

$$\frac{\left(\frac{\lambda}{\mu}\right)^s \cdot P_0}{s!\left(1 - \frac{\lambda}{\mu s}\right)} \leq 0.10 \qquad \text{[by formula (11) model II]}$$

This inequation is not easily solvable. Hence, we proceed by trials and find the least value of s that satisfies this inequation.

Let
$$s = 3:$$

Then,
$$P(w > 0) = \frac{(2.5)^3 . P_0}{6\left(1 - \frac{2.5}{3}\right)} = 15.625 P_0$$

where
$$P_0 = \left[\left\{\sum_{n=0}^{s-1} \frac{1}{n!}\left(\frac{\lambda}{\mu}\right)^n\right\} + \frac{\left(\frac{\lambda}{\mu}\right)^s}{s!\left(1 - \frac{\lambda}{\mu s}\right)}\right]^{-1}$$

i.e.,
$$P_0 = \left[\left\{1 + 2.5 + \frac{1}{2} \times (2.5)^2\right\} + \frac{(2.5)^3}{6 \times \left(1 - \frac{2.5}{3}\right)}\right]^{-1}$$

$$= (22.25)^{-1} = 0.0449$$

$$P(w > 0) = 15.625 \times 0.0449$$

$$= 0.7022 \not< 0.10$$

Let $s = 4$:

Then, $$P(W > 0) = \frac{(2.5)^4 \cdot P_0}{24\left(1 - \frac{2.5}{4}\right)} = 4.3403\, P_0$$

where $$P_0 = \left[\left\{1 + 2.5 + \frac{1}{2}\times(2.5)^2 + \frac{1}{6}\times(2.5)^3\right\} + \frac{(2.5)^4}{24\left(1 - \frac{2.5}{4}\right)}\right]^{-1}$$

$$= 0.0737$$

$\therefore$ $$P(W > 0) = 4.3403 \times 0.0737 = 0.3199 \not< 0.10$$

Similarly, when $s = 5$,

$$P(W > 0) = 0.1304 \not< 0.10.$$

When $s = 6$,

$$P(W > 0) = 0.047 < 0.10.$$

Hence, the number of booths to be installed = 6.

Example 10

A bank has two tellers working on saving accounts. The first teller handles only withdrawals. The second teller handles only deposits. It has been found that the service time distributions for both deposits and withdrawals are exponential with a mean service time of 3 min per customer. Depositors are found to arrive in a Poisson fashion throughout the day with a mean arrival rate of 16 per hour. Withdrawers also arrive in a Poisson fashion with a mean arrival rate of 14 per hour. What would be the effect on the average waiting time for the customers if each teller could handle both withdrawals and deposits? What would be the effect, if this could only be accomplished by increasing the service time to 3.5 min?

Solution

When there is a separate channel for the depositors, $\lambda_1 = 16$/hour, $\mu = 20$/hour

$\therefore$ $$E(W_q \text{ for depositors}) = \frac{\lambda_1}{\mu(\mu - \lambda_1)}$$ [by formula (10) of model I]

$$= \frac{16}{20(20-16)} = \frac{1}{5} \text{ h or 12 min}$$

When there is a separate channel for the withdrawers, $\lambda_2 = 14$/hour, $\mu = 20$/hour.

$\therefore$ $$E(W_q \text{ for withdrawers}) = \frac{\lambda_2}{\mu(\mu - \lambda_2)}$$

$$= \frac{14}{20(20-14)} = \frac{7}{60} h \text{ or } 7 \text{ min}$$

If both tellers do both services,

$$s = 2,$$

$$\mu = 20/\text{hour},$$

$$\lambda = \lambda_1 + \lambda_2 = 30/\text{hour}$$

$$\therefore \quad E(W_q \text{ for any customer}) = \frac{1}{\mu} \cdot \frac{1}{s.s!} \cdot \frac{\left(\frac{\lambda}{\mu}\right)^s \times P_0}{\left(1 - \frac{\lambda}{\mu s}\right)^2}, \qquad \text{[by formula (10) of model II]}$$

$$= \frac{1}{20} \times \frac{1}{2 \times 2} \times \frac{(1.5)^2}{(1-.75)^2} \times P_0 \qquad (1)$$

$$= 0.45 \times P_0$$

$$\text{Now} \qquad P_0 = \left[\left\{ \sum_{n=0}^{s-1} \frac{1}{n!} \cdot \left(\frac{\lambda}{\mu}\right)^n \right\} + \frac{1}{s!\left(1 - \frac{\lambda}{\mu s}\right)} \left(\frac{\lambda}{\mu}\right)^s \right]^{-1} \qquad \text{[by formula (6) of model II]}$$

$$= \left[1 + 1.5 + \frac{(1.5)^2}{2 \times 0.25} \right]^{-1}$$

$$= \frac{1}{7} \qquad (2)$$

Using (2) in (1),

$E(W_q$ for any customer$) = 0.45 \times \frac{1}{7}$ h or 3.86 min

Hence, if both tellers do both types of service, the customers get benefited as their waiting time is considerably reduced.

Now if both tellers do both types of service but with increased service time,

$$s = 2,$$

$$\lambda = 30,$$

$$\mu = \frac{60}{3.5} = \frac{120}{7} \text{ per hour.}$$

$E(W_q$ of any customer$)$

$$= \frac{7}{120} \times \frac{1}{2 \times 2} \times \frac{(1.75)^2}{\left(1 - \frac{7}{8}\right)^2} \times P_0 = 2.86 P_0,$$

where $$P_0 = \left[1 + 1.75 + \frac{(1.75)^2}{2 \times \frac{1}{8}}\right]^{-1} = \frac{1}{15}$$

$$\therefore \quad E(W_q \text{ of any customer}) = \frac{2.86}{15} \text{ h or } 11.44 \text{ min}$$

If this arrangement is adopted, withdrawers stand to lose as their waiting time is increased considerably and depositors get slightly benefited.

Example 11

A supermarket has two girls attending to sales at the counters. If the service time for each customer is exponential with a mean of 4 min. and if people arrive in Poisson fashion at the rate of 10 per hour,

(a) What is the probability that a customer has to wait for service?
(b) What is the expected percentage of idle time for each girl?
(c) If the customer has to wait in the queue , what is the expected length of his waiting time?

SOLUTION

$$s = 2, \lambda = \frac{1}{6} \text{ per minute},$$

$$\mu = \frac{1}{4} \text{ per minute}$$

(a) P (a customer has to wait for service)

$$= P(N \geq 2) = 1 - P_0 - P_1 \tag{1}$$

$$P_0 = \left[\left\{\sum_{n=0}^{s-1} \frac{1}{n!}\left(\frac{\lambda}{\mu}\right)^n\right\} + \left\{\frac{\left(\frac{\lambda}{\mu}\right)^s}{s!\left(1 - \frac{\lambda}{\mu s}\right)}\right\}\right]^{-1} \quad \text{[by formula (6) of model II]}$$

$$= \left[1 + \frac{2}{3} + \frac{\left(\frac{2}{3}\right)^2}{2 \times \left(1 - \frac{1}{3}\right)}\right]^{-1} = \frac{1}{2} \tag{2}$$

$$P_1 = \frac{\lambda}{\mu} \cdot P_0, \quad \text{[by formula (4) of model II]}$$

$$= \frac{2}{3} \times \frac{1}{2} = \frac{1}{3} \quad (3)$$

Using (2) and (3) in (1), we have

$$P(N \geq 2) = 1 - \frac{1}{2} - \frac{1}{3} = \frac{1}{6}$$

(b) Fraction of time when the girls are busy $= \frac{\lambda}{\mu s} = \frac{1}{3}$

$\therefore$ Fraction of time when the girls are idle $= \frac{2}{3}$

$\therefore$ Expected percentage of idle time for each girl $= \frac{2}{3} \times 100 = 67$

(c)

$$E(W_q / W_s > 0) = \frac{1}{\mu s - \lambda} \quad \text{[by formula (13) of model II]}$$

$$= \frac{1}{\frac{1}{4} \times 2 - \frac{1}{6}} = 3 \text{ min}$$

Example 12

A petrol pump station has 4 pumps. The service times follow the exponential distribution with a mean of 6 min. and cars arrive for service in a Poisson process at the rate of 30 cars per hour.

(a) What is the probability that an arrival would have to wait in line?
(b) Find the average waiting time, average time spent in the system and the average number of cars in the system.
(c) For what percentage of time, would a pump be idle on an average?

$$s = 4,$$

$$\lambda = 30/\text{hour},$$

$$\mu = 10/\text{hour}$$

(a)

$$P\,(\text{an arrival has to wait}) = P(W > 0)$$

$$= \frac{\left(\frac{\lambda}{\mu}\right)^s \times P_0}{s!\left(1 - \frac{\lambda}{\mu s}\right)} \quad \text{[by formula (11) of model II]}$$

$$= \frac{3^4 \times P_0}{24 \times \left(1 - \frac{3}{4}\right)} = 13.5 \times P_0 \quad (1)$$

$$P_0 = \left[\left\{\sum_{n=0}^{s-1}\frac{1}{n!}\cdot\left(\frac{\lambda}{\mu}\right)^n\right\} + \frac{\left(\frac{\lambda}{\mu}\right)^s}{s!\left(1-\frac{\lambda}{\mu s}\right)}\right]^{-1} \quad \text{[by formula (6) of model II]}$$

$$= \left[\left(1+3+\frac{1}{2}\times 9+\frac{1}{6}\times 27\right) + \frac{3^4}{24\times\left(1-\frac{3}{4}\right)}\right]^{-1}$$

$$= 0.0377$$

Using (2) in (1),

$$P(W>0) = 0.5090$$

(b)
$$E(W_q) = \frac{1}{\mu}\cdot\frac{1}{s\times s!}\times\frac{\left(\frac{\lambda}{\mu}\right)^s}{\left(1-\frac{\lambda}{\mu s}\right)^2}\times P_0 \quad \text{[by formula (10) of model II]}$$

$$= \frac{1}{10\times 4\times 24}\times\frac{3^4}{\left(1-\frac{3}{4}\right)^2}\times 0.0377$$

$$= 0.0509 \text{ h or } 3.05 \text{ min}$$

$$E(W_s) = \frac{1}{\mu} + E(W_q) \quad \text{[by formulas (9) and (10) of model II]}$$

$$= 6 + 3.05 = 9.05 \text{ min}$$

$$E(N) = \frac{1}{s\times s!}\times\frac{\left(\frac{\lambda}{\mu}\right)^{s+1}}{\left(1-\frac{\lambda}{\mu s}\right)^2}\times P_0 + \frac{\lambda}{\mu} \quad \text{[by formula (8) of model II]}$$

$$= \frac{1}{4\times 24}\times\frac{3^5}{\left(1-\frac{3}{4}\right)^2}\times 0.0377 + 3$$

$$= 4.53 \text{ cars}$$

(c) The fraction of time when the pumps are busy = traffic intensity

$$= \frac{\lambda}{\mu s} = \frac{3}{4}$$

$\therefore$ The fraction of time when the pumps are idle $= \frac{1}{4}$

Therefore, required percentage = 25%

Example 13

In a single server queueing system with Poisson input and exponential service times, if the mean arrival rate is 3 calling units per hour, the expected service time is 0.25 h and the maximum possible number of calling units in the system is 2, find $P_n (n \geq 0)$, average number of calling units in the system and in the queue and average waiting time in the system and in the queue.

SOLUTION

The situation in this problem is one of finite capacity, single server Poisson queue models.

$$\lambda = 3,$$

$$\mu = 4$$

and $$k = 2$$

As $$\lambda \neq \mu,$$

$$P_0 = \frac{1 - \frac{\lambda}{\mu}}{1 - \left(\frac{\lambda}{\mu}\right)^{k+1}}$$ [by formula (4) of model III]

$$= \frac{1 - \frac{3}{4}}{1 - \left(\frac{3}{4}\right)^3}$$

$$= \frac{16}{37} = 0.4324$$

Since $$\lambda \neq \mu,$$

$$P_n = \left(\frac{\lambda}{\mu}\right)^n \left[\frac{1 - \frac{\lambda}{\mu}}{1 - \left(\frac{\lambda}{\mu}\right)^{k+1}}\right]$$ [by formula (6) of model III]

$$= (0.4324)(0.75)^n$$

$$E(N)=\frac{\lambda}{\mu-\lambda}-\frac{(k+1)\left(\frac{\lambda}{\mu}\right)^{k+1}}{1-\left(\frac{\lambda}{\mu}\right)^{k+1}}$$ [by formula (8) of model IV]

$$=\frac{3}{4-3}-\frac{3\times\left(\frac{3}{4}\right)^{3}}{1-\left(\frac{3}{4}\right)^{3}}=3-\frac{81}{37}$$

$$=\frac{30}{37}\simeq 0.8 \text{ calling unit}$$

$$E(N_q)=E(N)-(1-P_0)$$ [by formula (10) of model III]

$$=\frac{30}{37}-\left(1-\frac{16}{37}\right)$$

$$=\frac{9}{37}=0.24 \text{ calling unit}$$

$$E(W_s)=\frac{1}{\lambda'}E(N)$$ [by formula (13) of model III]

where $\lambda'=\mu(1-P_0)$ [by formula (11) of model III]

$$=4\left(1-\frac{16}{37}\right)=\frac{84}{37}$$

$$\therefore \quad E(W_s)=\frac{37}{84}\times\frac{30}{37}$$

$$=\frac{5}{14}\text{ h or 21.4 min}$$

$$E(W_q)=\frac{1}{\lambda'}E(N_q)$$ [by formula (14) of model III]

$$=\frac{37}{84}\times\frac{9}{37}=\frac{3}{28}\text{ h or 6.4 min}$$

Example 14

The local one-person barber shop can accommodate a maximum of 5 people at a time (4 waiting and 1 getting a hair-cut). Customers arrive according to a Poisson distribution with a mean of 5 per hour. The barber cuts hair at an average rate of 4 per hour (Exponential service time).

(a) What percentage of time is the barber idle?
(b) What fraction of the potential customers is turned away?

(c) What is the expected number of customers waiting for a hair-cut?
(d) How much time can a customer expect to spend in the barber shop?

SOLUTION

$$\lambda = 5,$$
$$\mu = 4,$$
$$k = 5$$

(a) $$P\text{ (the barber is idle)} = P(N = 0)$$

$$= P_0 = \frac{1 - \frac{\lambda}{\mu}}{1 - \left(\frac{\lambda}{\mu}\right)^{k+1}} \quad \text{[by formula (4) of model III]}$$

$$= \frac{1 - \frac{5}{4}}{1 - \left(\frac{5}{4}\right)^6} = 0.0888$$

∴ Percentage of time when the barber is idle is 9.

(b) $$P\text{ (a customer is turned away)} = P(N > 5)$$

$$= \left(\frac{\lambda}{\mu}\right)^5 \left[\frac{1 - \frac{\lambda}{\mu}}{1 - \left(\frac{\lambda}{\mu}\right)^{k+1}}\right] \quad \text{[by formula (6) of model III]}$$

$$= \left(\frac{5}{4}\right)^5 \left[\frac{1 - \frac{5}{4}}{1 - \left(\frac{5}{4}\right)^6}\right]$$

$$= \frac{3125}{11529} = 0.2711$$

Therefore, 0.2711 × potential customers are turned away.

(c) $$E(N_q) = E(N) - (1 - P_0)$$

$$= \frac{\lambda}{\lambda - \mu} - \frac{(k+1)\left(\frac{\lambda}{\mu}\right)^{k+1}}{1 - \left(\frac{\lambda}{\mu}\right)^{k+1}} - (1 - P_0),$$

[by formulas (6) and (10) of model III]

$$= \left\{ -5 - \frac{6 \times \left(\frac{5}{4}\right)^6}{1 - \left(\frac{5}{4}\right)^6} \right\} - (1 - 0.0888)$$

$$= \frac{6 \times \frac{15625}{4096}}{\frac{11529}{4096}} - 5.9112 \simeq 2.2 \text{ customers}$$

(d) $$E(W) = \frac{1}{\lambda'} E(N) \qquad \text{[by formula (13) of model III]}$$

$$= \frac{1}{\mu(1 - P_0)} \times E(N) = \frac{3.1317}{3.6448}$$

$$\simeq 0.8595 \text{ h or } \quad 51.5 \text{ min}$$

Example 15

At a railway station, only one train is handled at a time. The railway yard is big enough only for 2 trains to wait, while another is given the signal to leave the station. Trains arrive at the station at an average rate of 6 per hour and the railway station can handle them on an average of 6 per hour. Assuming Poisson arrivals and exponential service distribution, find the probabilities for the numbers of trains in the system. Also find the average waiting time of a new train coming into the yard. If the handling rate is doubled, how will the previous results get modified?

SOLUTION

(i) $$\lambda = 6 \text{ per hour},$$

$$\mu = 6 \text{ per hour},$$

$$k = 2 + 1 = 3$$

Since $\lambda = \mu$,

$$P_0 = \frac{1}{k+1}$$

$$= \frac{1}{4} \quad \text{[by formula (5) of model III]}$$

$$P_n = \frac{1}{k+1} = \frac{1}{4} \qquad \text{for } n = 1, 2, 3 \text{ [by formula (7) of model III]}$$

$$E(N) = \frac{k}{2} \qquad \text{[by formula (9) of model III]}$$

$$= 1.5 \text{ trains}$$

$$E(W) = \frac{1}{\lambda'} E(N) \qquad \text{[by formula (13) of model III]}$$

$$= \frac{1.5}{\mu(1-P_0)} = \frac{1.5}{6 \times \frac{3}{4}}$$

$$= \frac{1}{3} \text{h or 20 min}$$

(ii) $\lambda = 6;$

$\mu = 12,$

$k = 3$

Since $\lambda \neq \mu,$

$$P_0 = \frac{1 - \frac{\lambda}{\mu}}{1 - \left(\frac{\lambda}{\mu}\right)^{k+1}}$$ [by formula (4) of model III]

$$= \frac{1 - \frac{1}{2}}{1 - \left(\frac{1}{2}\right)^4} = \frac{8}{15}$$

$$P_n = \left(\frac{\lambda}{\mu}\right)^n \left\{ \frac{1 - \frac{\lambda}{\mu}}{1 - \left(\frac{\lambda}{\mu}\right)^{k+1}} \right\}$$ [by formula (6) of model III]

$$= \frac{8}{15} \cdot \left(\frac{1}{2}\right)^n, \text{ for } n = 1, 2, 3.$$

$$E(N) = \frac{\lambda}{\mu - \lambda} - \frac{(k+1)\left(\frac{\lambda}{\mu}\right)^{k+1}}{1 - \left(\frac{\lambda}{\mu}\right)^{k+1}}$$ [by formula (8) of model III]

$$= 1 - \frac{4 \times \left(\frac{1}{2}\right)^4}{1 - \left(\frac{1}{2}\right)^4} = 1 - \frac{4}{15}$$

$$= \frac{11}{15} \simeq 0.73 \text{ train}$$

$$E(W) = \frac{1}{\lambda'} E(N)$$ by formula (13) of model III]

$$= \frac{1}{\mu(1-P_0)} \times E(N)$$

$$= \frac{\frac{11}{15}}{12\left(1-\frac{8}{15}\right)} = \frac{11}{84}\text{h or 7.9 min}$$

Example 16

Patients arrive at a clinic according to Poisson's distribution at a rate of 30 patients per hour. The waiting room does not accommodate more than 14 patients. Examination time per patient is exponential with a mean rate of 20 per hour.

(a) Find the effective arrival rate at the clinic.
(b)What is the probability that an arriving patient will not wait?
(c) What is the expected waiting time until a patient is discharged from the clinic?

(a)
$$\lambda = 30 \text{ per hour,}$$
$$\mu = 20 \text{ per hour,}$$
$$k = 14 + 1 = 15$$

Since $\lambda \neq \mu$,

$$P_0 = \frac{1-\frac{\lambda}{\mu}}{1-\left(\frac{\lambda}{\mu}\right)^{k+1}}$$ [by formula (4) of model III]

$$= \frac{1-\frac{3}{2}}{1-\left(\frac{3}{2}\right)^{16}} = 0.00076$$

Effective arrival rate $\lambda' = (1 - P_0)$ [by formula (4) of model III]

$$= 20 \times (1 - 0.00076)$$

$$= 19.98 \text{ per hour}$$

(b) P (a patient will not wait)

$$P_0 = 0.00076$$

(c)
$$E(N) = \frac{\lambda}{\lambda-\mu} - \frac{(k+1)\left(\frac{\lambda}{\mu}\right)^{k+1}}{1-\left(\frac{\lambda}{\mu}\right)^{k+1}}$$

$$= -3 - \frac{16 \times \left(\frac{3}{2}\right)^{16}}{1 - \left(\frac{3}{2}\right)^{16}} = 13 \text{ patients nearly}$$

$$E(W) = \frac{E(N)}{\lambda'}$$

$$= \frac{13}{19.98} = 0.65\text{h or } 39 \text{ min}$$

Example 17

A two-person barber shop has 5 chairs to accommodate waiting customers. Potential customers, who arrive when all 5 chairs are full, leave without entering the shop. Customers arrive at the average rate of 4 per hour and spend an average of 12 min. in the barber's chair. Compute P_0, P_1, P_7, $E(N_q)$ and $E(W)$.

Solution

The situation in this problem follows the finite capacity, multi-server Poisson queue model.

$$\lambda = 4 \text{ per hour},$$

$$\mu = 5 \text{ per hour},$$

$$s = 2,$$

$$k = 2 + 5 = 7$$

(a)

$$P_0 = \left[\sum_{n=0}^{s-1} \frac{1}{n!}\left(\frac{\lambda}{\mu}\right)^n + \frac{1}{s!}\left(\frac{\lambda}{\mu}\right)^s \sum_{n=s}^{k}\left(\frac{\lambda}{\mu s}\right)^{n-s}\right]^{-1}$$

[by formula (3) of model IV]

$$= \left[\sum_{n=0}^{1} \frac{1}{n!}\left(\frac{4}{5}\right)^n + \frac{1}{2}\cdot\left(\frac{4}{5}\right)^2 \sum_{n=2}^{7}\left(\frac{2}{5}\right)^{n-2}\right]^{-1}$$

$$= \left[1 + \frac{4}{5} + \frac{8}{25}\left\{1 + \frac{2}{5} + \left(\frac{2}{5}\right)^2 + \left(\frac{2}{5}\right)^3 + \left(\frac{2}{5}\right)^4 + \left(\frac{2}{5}\right)^5\right\}\right]^{-1}$$

$$= \left[\frac{9}{5} + \frac{8}{25}\left\{\frac{1-(0.4)^7}{1-0.4}\right\}\right]^{-1} = 0.4287$$

(b)

$$P_n = \frac{1}{n!}\left(\frac{\lambda}{\mu}\right)^n P_0, \text{ for } n \le s \qquad \text{[by formula (4) of model IV]}$$

$$\therefore \quad P_1 = \left(\frac{1}{5}\right) \times 0.4287 = 0.3430$$

(c) $$P_n = \frac{1}{s!.s^{n-s}}\left(\frac{\lambda}{\mu}\right)^n \cdot P_0,$$ for $s < n \le k$ [by formula (4) of model IV]

$$P_7 = \frac{1}{2\times 2^{7-2}}\times\left(\frac{4}{5}\right)^7 \times 0.4287$$

$$= 0.0014$$

(d) $$E(N_q) = P_0\left(\frac{\lambda}{\mu}\right)^s \cdot \frac{\rho}{s!(1-\rho)^2}\left[1 - P^{k-s} - (k-s)(1-\rho)\rho^{k-s}\right],$$

where $$\rho = \frac{\lambda}{\mu s}$$ [by formula (5) of model IV]

$$= (0.4287).(0.8)^2 \frac{(0.4)}{2\times(0.6)^2}\left[1-(0.4)^5 - 5\times 0.6\times(0.4)^5\right]$$

$$= 0.15 \text{ customer}$$

(e) $$E(N) = E(N_q) + s - \sum_{n=0}^{s-1}(s-n)P_n$$ [by formula (6) of model IV]

$$= 0.1462 + 2 - \sum_{n=0}^{1}(2-n)P_n$$

$$= 2.1462 - (2 \times P_0 + 1 \times P_1)$$

$$= 2.1462 - (2 \times 0.4287 + 1 \times 0.3430)$$

$$= 0.95 \text{ customer}$$

$$E(W) = \frac{1}{\lambda'}.E(N)$$ [by formula (9) of model IV]

where $$\lambda' = \mu\left[s - \sum_{n=0}^{s-1}(s-n)P_n\right]$$ [by formula (7) of model IV]

$$= 4[2 - (2 \times 0.4287 + 1 \times 0.3430)]$$

$$= 3.1984$$

$$E(W) = \frac{0.9458}{3.1984}$$

$$= 0.2957\text{h or } 17.7 \text{ min}$$

Example 18

At a port are 6 unloading berths and 4 unloading crews. When all the berths are full, arriving ships are diverted to an overflow facility 20 kms down the river. Tankers arrive according to a Poisson process with a mean of 1 every 2 h. It takes, on the average, 10 h for an unloading crew to unload a tanker, the unloading time following an exponential distribution. Find

(a) how many tankers are at the port on the average?
(b) how long does a tanker spent at the port on the average?
(c) what is the average arrival rate at the overflow facility?

SOLUTION

$$\lambda = \frac{1}{2} \text{ per hour,}$$

$$\mu = \frac{1}{10} \text{ per hour,}$$

$$s = 4,$$

$$k = 6$$

(a) $$P_0 = \left[\sum_{n=0}^{s-1} \frac{1}{n!}\left(\frac{\lambda}{\mu}\right)^n + \frac{1}{s!}\left(\frac{\lambda}{\mu}\right)^s \sum_{n=s}^{k}\left(\frac{\lambda}{\mu s}\right)^{n-s}\right]^{-1}$$

[by formula (3) of model IV]

$$= \left[\left(1+5+\frac{1}{2}\times 5^2+\frac{1}{6}\times 5^3\right)+\frac{1}{24}\times 5^4\times\left\{\left(\frac{5}{4}\right)^0+\left(\frac{5}{4}\right)^1+\left(\frac{5}{4}\right)^2\right\}\right]^{-1}$$

$$= 0.0072$$

$$E(N_q) = P_0\left(\frac{\lambda}{\mu}\right)^s \cdot \frac{\rho}{s!(1-\rho)^2}\left[1-\rho^{k-s}-(k-s)(1-\rho)\rho^{k-s}\right],$$

where $\rho = \dfrac{\lambda}{\mu s}$ [by formula (5) of model IV]

$$= 0.0072\times 5^4\times\frac{1.25}{24\times(.25)^2}\left[1-(1.25)^2-2\times(-.25)(1.25)^2\right]$$

$$= 0.8203 \text{ tanker}$$

$$E(N) = E(N_q)+s-\sum_{n=0}^{s-1}(s-n)P_n$$ [by formula (6) of model IV]

$$= 4.8203 - (4P_0 + 3P_1 + 2P_2 + P_3)$$

$$= 4.8203 - \{4\times 0.0072 + 3\times 0.0360 + 2\times 0.09 + 0.15\}$$

$$= 4.3535 \text{ tankers.}$$

(b) $$E(W) = \frac{1}{\lambda'}E(N)$$ [by formula (9) of model IV] where $$\lambda' = \mu\left[s - \sum_{n=0}^{s-1}(s-n)P_n\right]$$ [by formula (7) of model IV]

$$= \frac{1}{10}[4 - \{4P_0 + 3P_1 + 2P_2 + P_3\}]$$

$$= \frac{1}{10}[4 - 0.4668] = 0.3533$$

$$\therefore \quad E(W) = \frac{4.3535}{0.3533} = 12.32 \text{ h}$$

(c) When $N = 6$, i.e., when the number of tankers in the port is 6, overflow occurs.

$$P(N=6) = \frac{1}{s!s^{n-s}}\left(\frac{\lambda}{\mu}\right)^n P_0$$ for $n = k$ [by formula (4) of model IV]

$$= \frac{1}{24 \times 4^2} \times 5^6 \times 0.0072$$

$$= 0.2930$$

Average arrival rate at the overflow facility = (Average arrival rate at the port) × (Possibility that overflow occurs)

$$= \frac{1}{2} \times 0.2930 = 0.586 \text{ per hour}$$

Example 19

A car servicing station has 2 bays where service can be offered simultaneously. Because of space limitation, only 4 cars are accepted for servicing. The arrival pattern is a Poisson distribution with 12 cars per day. The service time in both the bays is exponentially distributed with $\mu = 8$ cars per day per bay. Find the average number of cars in the service station, the average number of cars waiting for service and the average time a car spends in the system.

Solution

$$\lambda = 12 \text{ per day},$$

$$\mu = 8 \text{ per day},$$

$$s = 2, k = 4$$

(a) $$P_0 = \left[\sum_{n=0}^{s-1}\frac{1}{n!}\left(\frac{\lambda}{\mu}\right)^n + \frac{1}{s!}\left(\frac{\lambda}{\mu}\right)^s \sum_{n=s}^{k}\left(\frac{\lambda}{\mu s}\right)^{n-s}\right]^{-1}$$ [by formula (3) of model IV]

$$= \left[1 + \frac{1.5}{1} + \frac{1}{2} \times (1.5)^2\{1 + (.75) + (.75)^2\}\right]^{-1}$$

$$= 0.1960$$

$$E(N_q) = P_0 \left(\frac{\lambda}{\mu}\right)^s \cdot \frac{\rho}{s!(1-\rho)^2}[1-\rho^{k-s} - (k-s)(1-\rho)\rho^{k-s}]$$

where $\rho = \dfrac{\lambda}{\mu s}$ [by formula (5) of model IV]

i.e., $$E(N_q) = 0.1960 \times (1.5)^2 \times \frac{0.75}{2 \times (0.25)^2} \times [1-(0.75)^2 - 2 \times 0.25 \times (0.75)^2]$$

$= 0.4134$ car

(b) $E(N)$ = Average number of cars in the service station

$$E(N_q) = +s - \sum_{n=0}^{s-1}(s-n)P_n$$ [by formula (6) of model IV]

$$= 0.4134 + 2 - \sum_{n=0}^{1}(2-n)P_n$$

$= 2.4134 - (2P_0 + P_1)$

$= 2.4134 - (2 \times 0.1960 + 15 \times 0.1960)$

$= 1.73$ cars

(c) $$E(W) = \frac{1}{\lambda'}E(N)$$ [by formula (9) of model IV]

where $$\lambda' = \mu\left[s - \sum_{n=0}^{s-1}(s-n)P_n\right]$$ [by formula (7) of model IV]

$= 8[2 - (2P_0 + P_1)]$

$= 10.512$

$\therefore$ $$E(W) = \frac{1.73}{10.512} = 0.1646 \text{ day}$$

Example 20

A group of engineers has 2 terminals available to aid in their calculations. The average computing job requires 20 min. of terminal time and each engineer requires some computation about once every half an hour. Assume that these are distributed according to an exponential distribution. If there are 6 engineers in the group, find

(a) The expected number of engineers waiting to use one of the terminals and in the computing centre and
(b) The total time lost per day.

Solution

$\lambda = 2$ per hour,

$\mu = 3$ per hour,

$s = 2, k = 6$

$$P_0 = \left[\sum_{n=0}^{s-1} \frac{1}{n!} \left(\frac{\lambda}{\mu}\right)^n + \frac{1}{s!} \left(\frac{\lambda}{\mu}\right)^s \sum_{n=s}^{k} \left(\frac{\lambda}{\mu s}\right)^{n-s} \right]^{-1}$$ by formula (3) of model IV]

$$= \left[1 + \frac{2}{3} + \frac{1}{2} \times \left(\frac{2}{3}\right)^2 \left\{ 1 + \frac{1}{3} + \left(\frac{1}{3}\right)^2 + \left(\frac{1}{3}\right)^3 + \left(\frac{1}{3}\right)^4 \right\} \right]^{-1}$$

$$= 0.5003$$

$$E(N_q) = P_0 \times \left(\frac{\lambda}{\mu}\right)^s \times \frac{\rho}{s!(1-\rho)^2} \left[1 - \rho^{k-s} - (k-s)(1-\rho)\rho^{k-s} \right],$$

where $\rho = \dfrac{\lambda}{\mu s}$ [by formula (5) of model IV]

i.e., $$E(N_q) = 0.5003 \times \left(\frac{2}{3}\right)^2 \times \frac{\left(\frac{1}{3}\right)}{2 \times \left(\frac{2}{3}\right)^2} \left[1 - \left(\frac{1}{3}\right)^4 - 4 \times \frac{2}{3} \times \left(\frac{1}{3}\right)^4 \right]$$

$$= 0.0796$$

$$E(N) = E(N_q) + s - \sum_{n=0}^{s-1} (s-n) P_n$$ [by formula (6) of model IV]

$$= 0.0796 + 2 - \sum_{n=0}^{1} (2-n) P_n$$

$$= 2.20796 - (2P_0 + P_1)$$

$$= 2.0796 - \left(2 \times 0.5003 + \frac{2}{3} \times 0.5003 \right)$$

$$= 0.75$$

$$E(W_q) = \frac{1}{\lambda'} E(N_q)$$ [by formula (10) of model IV]

where $$\lambda' = \mu \left[s - \sum_{n=0}^{s-1} (s-n) P_n \right]$$ [by formula (7) of model IV]

$$= 3 \left[2 - \sum_{n=0}^{1} (2-n) P_n \right]$$

$$= 3[2 - (2P_0 + P_1)]$$

$$= 3 \left[2 - \left(2 \times 0.5003 + \frac{2}{3} \times 0.5003 \right) \right]$$

$$= 1.9976$$

$\therefore$ $$E(W_q) = \frac{0.0796}{1.9976} = 0.0398 \text{ h}$$

Every time an engineer approaches the computer centre, he has to lose 0.0398 h by way of waiting. If the day consists of 8 working hours, he has to approach the centre 16 times.

∴ Time lost in waiting in a day per engineer = 16 × 0.0398 = 0.6368 h

∴ Total time lost in waiting in a day by all the 6 engineers = 6 × 0.06368 = 3.82 h.

Example 21

A one-man barber shop takes exactly 25 minutes to complete one hair-cut. If customers arrive at the barber shop in a Poisson fashion at an average rate of one every 40 minutes, how long, on the average, does a customer spend in the shop? Also find the average time a customer must wait for service.

SOLUTION

The service time T is a constant = 25 min viz., T follows a distribution with $E(T) = 25$ and $V(T) = 0$.

Also $$\lambda = \frac{1}{40}$$

∴ By Pollaczek–Khinchine formula,

$$E(N_s) = \lambda E(T) + \frac{\lambda^2\{V(T) + E^2(T)\}}{2(1 - \lambda E(T)\}}$$

$$= \frac{25}{40} + \frac{\frac{1}{40^2}\{0 + 25^2\}}{2\left\{1 - \frac{1}{40} \times 25\right\}}$$

$$= \frac{5}{8} + \frac{25/64}{2 \times (3/8)} = \frac{55}{48}$$

By Little's formula,

$$E(W_s) = \frac{1}{\lambda} E(N_s)$$

$$= 40 \times \frac{55}{48} = 45.8 \text{ minutes}$$

$$E(W_q) = E(W_s) - \frac{1}{\mu}$$

$$= (E(W_s) - (E(T) = 20.8 \text{ min}$$

i.e., a customer has to spend 45.8 minutes in the shop and has to wait 20.8 minutes for service on the average.

Example 22

A patient who goes to a single doctor clinic for a general check-up has to go through 4 phases. The doctor takes on the average 4 minutes for each phase of the check-up and the time taken for each phase is exponentially distributed. If the arrivals of the patient at the clinic are approximately Poisson at the average rate of 3 per hour, what is the average time spent by a patient (i) in the examination? (ii) waiting in the clinic?

Solution

Let X_1, X_2, X_3, X_4 denote the times required for the 4 phases of the check-up.

Each X_r is exponential with mean 4 min. or with parameter $\frac{1}{4}$.

Since the X_rs are independent, $(X_1 + X_2 + X_3 + X_4)$ follows an Erlang's distribution with parameters $\lambda = \frac{1}{4}$ and $k = 4$ [by the additive property of exponential distribution]

The mean and variance of Erlang's distribution with parameters λ and k are $\frac{k}{\lambda}$ and $\frac{k}{\lambda^2}$.

Thus, if T represents the service time for a patient,

$$E(T) = \frac{k}{\lambda}$$

$$= \frac{4}{1/4} = 16$$

and

$$V(T) = \frac{k}{\lambda^2}$$

$$= \frac{4}{1/16} = 64$$

$\therefore$ Average time for examination of each patient $= 16$ min. If λ_c represents the arrival rate in the clinic, then by the P–K formula.

$$E(N_s) = \lambda_c E(T) + \frac{\lambda_c^2\{V(T) + E^2(T)\}}{2\{1 - \lambda_c E(T)\}}$$

$$= \frac{1}{20} \times 16 + \frac{\frac{1}{400}\{64 + 256\}}{2\left\{1 - \frac{1}{20} \times 16\right\}}$$

$(\because \lambda_c = 3/\text{hour or } \frac{1}{20}/\text{min})$

$$= \frac{4}{5} + \frac{\frac{4}{5}}{2 \times \frac{1}{5}} = \frac{14}{5}$$

By Little's formula

$$E(W_s) = \frac{1}{\lambda_c} E(N_s)$$

$$= 20 \times \frac{14}{5} = 56 \text{ minutes}$$

and

$$E(W_q) = E(W_s) - \frac{1}{\mu} = 56 - \frac{1}{1/E(T)}$$

$$= 56 - 16 = 40 \text{ min}$$

i.e., a patient has to wait 40 minutes for check-up in the clinic.

Example 23

A car wash facility has only one bay. Cars arrive according to a Poisson distribution with a mean of 4 cars per hour and may wait in the facility's parking lot if the bay is busy. The parking lot is large enough to accommodate any number of cars. Find the average number of cars waiting in the parking lot if the time for washing and cleaning a car follows

(a) uniform distribution between 8 and 12 minutes
(b) a normal distribution with a mean of 12 minutes and an SD of 3 minutes.
(c) a discrete distribution with values equal to 4.8 and 15 minutes and corresponding probabilities 0.2, 0.6 and 1.2.

Solution

(a) $$\lambda = 4/\text{hour or } \frac{1}{15} \text{ per minute}$$

$$E(T) = \text{mean of the uniform distribution in } (8, 12)$$

$$\frac{1}{2}(8+12) = 10 \text{minutes}$$

$$V(T) = \frac{1}{12}(12-8)^2 = \frac{4}{3}$$

By the P–K formula,

$$E(N_s) = \lambda E(T) + \frac{\lambda^2\{V(T) + E^2(T)\}}{2\{1 - \lambda \cdot E(T)\}}$$

$$= \frac{1}{15} \times 10 + \frac{\frac{1}{225}\left\{\frac{4}{3} + 100\right\}}{2\left\{1 - \frac{1}{15} \times 10\right\}}$$

$$= \frac{2}{3} + \frac{152}{225} = 1.342 \text{ cars}$$

By Little's formula,

$$E(N_q) = E(N_s) - \frac{\lambda}{\mu} = 1.342 - \frac{\left(\frac{1}{15}\right)}{\left(\frac{1}{10}\right)} \quad \because \left(\mu = \frac{1}{E(T)}\right)$$

$$= 0.675 \text{ car}$$

(b) $$\lambda = \frac{1}{15};$$

$$E(T) = 12 \text{ min}.$$

and $$V(T) = 9$$

$\therefore$ $$\mu = \frac{1}{E(T)} = \frac{1}{12}$$

By the P–K formula,

$$E(N_s) = \frac{1}{15} \times 12 + \frac{\frac{1}{225}\{9 + 144\}}{2\left\{1 - \frac{1}{15} \times 12\right\}}$$

$$= \frac{4}{5} + \frac{153}{90} = 2.5 \text{ cars}$$

By Little's formula,

$$E(N_q) = E(N_s) - \frac{\lambda}{\mu}$$

$$= 2.5 - \frac{(1/15)}{(1/12)} = 1.7 \text{ cars}$$

(c) The service time T follows the discrete distribution given here:

T:	4	8	15
$p(T)$	0.2	0.6	0.2

$$E(T) = \sum T.\, p(T)$$

$$= 0.8 + 4.8 + 3 = 8.6 \text{ minutes}$$

$$E(T^2) = \sum T^2.\, p(T)$$

$$= 3.2 + 38.4 + 45 = 86.6$$

$$V(T) = \sum T^2 - E^2(T)$$

$$= 86.6 - (8.6)^2 = 12.64$$

By the P–K formula.

$$E(N_s) = \frac{1}{15} \times 8.6 + \frac{\frac{1}{225}\{12.64 + 73.96\}}{2\left\{1 - \frac{1}{15} \times 8.6\right\}}$$

$$= 0.573 + \frac{86.6}{225} \times \frac{15}{12.8}$$

$$= 0.573 + 0.451 = 1.024 \text{ cars}$$

By Little's formula

$$E(N_q) = E(N_s) - \frac{\lambda}{\mu}$$

$$= 1.024 - \frac{(1/5)}{(1/8.6)} = 0.451 \text{ car}$$

EXERCISE 15

Part A (Short answer questions)

1. What are the characteristics of a queueing system?
2. What do the letters in the symbolic representation $(a/b/c)$: (d/e) of a queueing model represent?
3. What do you mean by transient state and steady state queueing systems?
4. Write down the difference equations that give the probability that there are n customers $(n \geq 0)$ in a Poisson queueing system in steady state.
5. Write down the formulas for P_0 and P_n in a Poisson queue system in the steady state.
6. Give the formulas for the average number of customers (i) in the system, (ii) in the queue and (iii) in the non-empty queues for the (M/M/1): (∞/FIFO) model.
7. Obtain the variance of queue length for the (M/M/1): (∞/FIFO) model.
8. In the usual notation of a (M/M/1): (∞/FIFO) queue system, find $P(N > 2)$, if $\lambda = 12$ per hour and $\mu = 30$ per hour.
9. In the usual notation of a (M/M/1): (∞/FIFO) queue system, if $\lambda = 12$ per hour and $\mu = 24$ per hour, find the average number of customers in the system and in the queue.
10. Give the formulas for the waiting time of a customer in the queue and in the system for the (M/M/1): (∞/FIFO) model.
11. If a customer has to wait in a (M/M/1): (∞/FIFO) queue system, what is his average waiting time in the queue, if $\lambda = 8$ per hour and $\mu = 12$ per hour?
12. What is the probability that a customer has to wait more than 15 min. to get his service completed in a (M/M/1): (∞/FIFO) queue system, if $\lambda = 6$ per hour and $\mu = 10$ per hour?
13. Write down the probability density function of the waiting time of a customer in the (M/M/1): (∞/FIFO) queue system.
14. Write down Little's formulas that hold good for all the Poisson queue models.
15. Write down Little's formulas that hold good for the infinite capacity Poisson queue models.
16. What is the probability that there are no customers in the (M/M/s): (∞/FIFO) queueing system?
17. Write down the formula for P_n in terms of P_0 for the (M/M/s): (∞/FIFO) queueing system.
18. Give the formula for the average number of customers in the system and in the queue for the (M/M/s): (∞/FIFO) queueing model.

19. If there are 2 servers in an infinite capacity Poisson queue system with $\lambda = 10$ per hour and $\mu = 15$ per hour, what is the percentage of idle time for each server?
20. In a 3 server infinite capacity Poisson queue model if $\lambda/s\mu = \frac{2}{3}$, find P_0
21. In a 3 server infinite capacity Poisson queue model if $\lambda/s\mu = 2/3$ and P_0 is 1/9, find the average number of customers in the queue and in the system.
22. If $\lambda/s\mu = \frac{2}{3}$ in a (M/M/s): (∞/FIFO) queue system, find the average number of customers in the non-empty queue.
23. What is the probability that an arrival to an infinite capacity 3 server Poisson queue system with $\lambda/s\mu = 2/3$ and $P_0 = 1/9$ will have to wait?
24. What is the probability that an arrival to an infinite capacity 3 server Poisson queue system with $\lambda/s\mu = 2/3$ and $P_0 = 1/9$ will enter the service without waiting?
25. Give the formulas for the average waiting time of a customer in the system and in the queue for the (M/M/s): (∞/FIFO) queueing model.
26. What is the average waiting time of a customer in the 3 server infinite capacity Poisson queue if he happens to wait, given that $\lambda = 6$ per hour and $\mu = 4$ per hour.
27. Give the probability that there are no customers in an (M/M/1): (*k*/FIFO) queueing system.
28. Write down the probability that there are n customers in an (M/M/1): (*k*/FIFO) queueing system.
29. If $\lambda = 4$ per hour and $\mu = 12$ per hour in an (M/M/1): (4/FIFO) queueing system, find the probability that there is no customer in the system. If $\lambda = \mu$, what is the value of this probability?
30. Write the formulas for the average number of customers in the (M/M/1): (*k*/FIFO) queueing system and also in the queue.
31. Define effective arrival rate with respect to an (M/M/1): (*k*/FIFO) queueing model.
32. How are N_s and N_q related in an (M/M/1): (*k*/FIFO) queueing model?
33. Write down Little's formulas for the average waiting time in the system and in the queue for the (M/M/1): (*k*/FIFO) queueing model.
34. If $\lambda = 3$ per hour and $\mu = 4$ per hour and maximum capacity $k = 7$ in an (M/M/1): (*k*/FIFO) system, find the average number of customers in the system.
35. Write the formula for the probability that there is no customer in an (M/M/*s*): (*k*/FIFO) queue system.
36. Write the formula for the probability that there are n customers in an (M/M/*s*): (*k*/FIFO) queueing system.
37. Write down the formula for the average queue length in an (M/M/*s*): (*k*/*FIFO*) queueing model.
38. Define effective arrival rate with respect to an (M/M/*s*): (*k*/FIFO) queueing model.
39. How are N_s and N_q related in an (M/M/*s*): (*k*/FIFO) queueing model?
40. Write down Little's formulas for the average waiting time in the system and in the queue for an (M/M/*s*): (*k*/FIFO) queueing model.

Part B

41. Arrivals at a telephone booth are considered to be Poisson with an average time of 10 min. between one arrival and the next. The length of a phone call is assumed to be distributed exponentially with a mean of 3 min.
    (a) Find the average number of persons waiting in the system.
    (b) What is the probability that a person arriving at the booth will have to wait in the queue?
    (c) What is the probability that it will take him more than 10 min. altogether to wait for the phone and complete his call?
    (d) Estimate the fraction of the day when the phone will be in use.
    (e) The telephone department will install a second booth when convinced that an arrival has to wait on the average for at least 3 min. for the phone. By how much should the flow of arrivals increase in order to justify a second booth?
42. Customers arrive at a one man barber shop according to a Poisson process with a mean inter-arrival time of 20 min. They spend an average of 15 min. in the barber's chair.
    (a) What is the expected number of customers in the barber's shop? in the queue?
    (b) What is the probability that a customer will not have to wait for a hair cut?
    (c) How much time can a customer expect to spend in the barber shop?
    (d) Management will put another chair and hire another barber when a customer's average waiting time in the shop exceeds 1.25 h. How much must the average rate of arrivals increase to warrant a second barber?
    (e) What is the average time customers spend in the queue?
    (f) What is the probability that the waiting time in the system is greater than 30 min?
    (g) What is the probability that there are more than 3 customers in the system?
43. If customers arrive for the service according to a Poisson distribution at the average rate of 5 per day, how fast must they be serviced on the average (assume exponential service time) in order to keep the average number of customers in the system less than 4?
44. Patients arrive at a hospital for emergency service at the rate of one every hour. Currently, only one emergency can be handled at a time. Patients spend an average of 20 min. for receiving emergency service. By how much should the average service time be decreased to keep the average time to wait and receive service less than 25 min?
45. A departmental secretary receives an average of 8 jobs per hour. Many are short jobs, while others are quite long. Assume, however, that the time to perform a job has an exponential distribution with a mean of 6 min.
    (a) What is the average elapsed time from the time the secretary receives a job until it is completed?
    (b) Calculate $E(N)$, $E(Wq)$, $P(W > 2h)$, $P(N > 5)$ and the fraction of time the secretary is busy?
46. A service station expects a customer every 4 min. on the average. Service takes, on the average, 3 min. Assume Poisson input and exponential service.
    (a) What is the average number of customers waiting for service?
    (b) How long can a customer expect to wait for service?
    (c) What is the probability that a customer will spend less than15 min. waiting for and getting services?

(d) What is the probability that a customer will spend longer than 10 min. waiting for and getting service?

47. A dress shop has 3 sales persons. Assume that arrivals follow a Poisson pattern with an average of 10 min. between arrivals. Also assume that any sales person can provide the desired service for any customer. If the time to provide service for a customer is exponentially distributed with a mean of 20 min. per customer, calculate $E(N)$, $E(Nq)$, $E(W)$, $E(Wq)$ and P_n for $n = 0$, 1 and 2.

48. If the mean arrival rate is 24 per hour, find, from the customer's point of view of the time spent in the system, whether 3 channels in parallel with a mean service rate of 10 per hour is better or worse than a single channel with a mean service rate of 30 per hour.

49. Four counters are being run on the frontier of a country to check the passports and necessary papers of the tourists. The tourists choose a counter at random. If the arrivals at the frontier follow a Poisson pattern at the rate of λ and the service time is exponential with parameter $\frac{\lambda}{2}$, what is the steady state average queue at each counter?

50. An insurance company has 3 claim registers in its branch office. People with claims against the company are found to arrive in a Poisson fashion at an average rate of 20/8-h day. The amount of time that an adjuster spends with a claimant is found to have an exponential distribution with a mean service time of 40 min. Claimants are processed in order of their appearance

(a) How many hours a week can an adjuster expect to spend with claimants?

(b) How much time, on the average, does a claimant spend in the branch office?

51. A telephone exchange has 2 long distance operators. The telephone company finds that during the peak load, long distance calls arrive in a Poisson fashion at an average rate of 15 per hour. The length of service on these calls is approximately exponentially distributed with a mean length of 5 min.

(a) What is the probability that a subscriber will have to wait for his long distance call during the peak hours of the day?

(b) If the subscribers will wait and are serviced in turn, what is the expected waiting time?

52. A petrol pump station has 4 pumps. The service times follow the exponential distribution with a mean of 6 min. and cars arrive for service in a Poisson process at the rate of 30 cars per hour.

(a) What is the probability that an arrival would have to wait in line?

(b) Find the average waiting time in the queue, average time spent in the system and the average number of cars in the system.

(c). For what percentage of time would a pump be idle on an average?

53. A one-person barber shop has 6 chairs to accommodate people waiting for a hair-cut. Assume that customers who arrive when all the 6 chairs are full leave without entering the barber shop. Customers arrive at the average rate of 3 per hour and spend an average of 15 min. in the barber's chair.

(a) What is the probability that a customer can get directly into the barber's chair upon arrival?

(b) What is the expected number of customers waiting for a hair-cut?

(b) How much time can a customer expect to spend in the barber shop?

(d) What fraction of potential customers is turned away?

54. Assume that the goods trains are coming in a yard at the rate of 30 trains per day and suppose that the inter-arrival times follow an exponential distribution. The service time for each train is assumed

to be exponential with an average of 36 min. If the yard can admit 9 trains at a time, calculate the probability that the yard is empty and the average queue length.

55. A car park contains 5 cars. The arrival of cars is Poisson at a mean rate of 10 per hour. The length of time each car spends in the car park has negative exponential distribution with a mean of 2 min. How many cars are in the car park on an average and what is the probability of a newly arriving customer finding the car park full and leaving to park his car elsewhere.

56. A stenographer is attached to 5 officers for whom she performs stenographic work. She gets calls from the officers at the rate of 4 per hour and takes on the average 10 min. to attend to each call. If arrival rate is Poisson and service time is exponential, find (a) the average waiting time for an arriving call, (b) the average number of waiting calls and (c) the average time an arriving call spends in the system.

57. A 2-person barber shop has 5 chairs to accommodate waiting customers. Potential customers, who arrive when all the 5 chairs are full, leave without entering the barber shop. Customers arrive at the average rate of 3.7634 per hour and spend an average of 15 min. in the barber's chair. Compute P_0, P_1, P_7, $E(N_q)$ and $E(W)$.

58. A barber shop has 2 barbers and 3 chairs for waiting customers. Assume that customers arrive in Poisson fashion at a rate of 5 per hour and that each barber services customers according to an exponential distribution with a mean of 15 min. Further, if a customer arrives and there are no empty chairs in the shop, he will leave. Find the steady-state probabilities. What is the probability that the shop is empty? What is the expected number of customers in the shop?

59. An automobile inspection station has 3 inspection stalls. Assume that cars wait in such a way that when a stall becomes vacant, the car at the head of the line pulls up to it. The station can accommodate at most 4 cars waiting (7 in the station) at one time. The arrival pattern is Poisson with a mean of 1 car every minute during the peak hours. The service time is exponential with a mean of 6 min. Find the average waiting time and the average number of cars per hour that cannot enter the house because of full capacity.

60. A mechanic repairs 4 machines. The mean time between service requirements in 5 h for each machine and forms an exponential distribution. The mean repair time is 1 h and also follows the same distribution pattern. Machine downtime costs Rs. 25 per hour and the mechanic costs Rs. 55 per day.

    (a) Find the expected number of machines under service and in waiting.

    (b) Determine the expected downtime cost per 8-h day?

    (c) Wouldn't it be economical to engage 2 mechanics, each repairing only 2 machines?

61. Show that, for the case when the service is constant, the P–K formula reduces to $L_s = \rho + \dfrac{\rho^2}{2(1-\rho)}$, where $\mu = \dfrac{1}{E(T)}$ and $\rho = \dfrac{\lambda}{\mu}$ *or* $\lambda E(T)$

62. Given that the service time is an Erlang distribution with parameters m and μ, so that $E(T) = \dfrac{m}{\mu}$ and $V(T) = \dfrac{m}{\mu^2}$, show that the P–K formula reduces to $L_s = m\rho + \dfrac{m(1+m)\rho^2}{2(1-m\rho)}$

63. In a car manufacturing plant, a loading crane takes exactly 10 minutes to load a car into a wagon and again come back to position to load another car. If the arrival of cars is a Poisson stream at an average of 1 every 20 minutes, calculate the average waiting time of a car.
64. Solve the problem in example 23 if T is a constant equal to 10 minutes.
65. Repairing a certain type of faulty machine consists of 5 basic steps that must be performed sequentially. The time taken to perform each of the 5 steps is exponentially distributed with a mean of 5 minutes and is independent of the other steps. If the machines become faulty in a Poisson fashion at an average rate of 2 per hour and if there is only one repairman, what is the average idle time for each faulty machine?

ANSWERS

Exercise 15

Part A

11. 5 min.
12. 0.369
19. 50%
20. $\frac{1}{9}$
21. $\frac{8}{9}, \frac{26}{9}$
22. 2
23. $\frac{4}{9}$
24. $\frac{5}{9}$
26. 10 min.
29. $\frac{81}{121}, \frac{1}{5}$
34. 2.11

Part B

41. (a) 0.43
    (b) 0.3
    (c) 0.097
    (d) 0.3
    (e) 0.16 person per minute
42. (a) 3; 2.25
    (b) 0.25
    (c) 1 h
    (d) 3.2 per hour
    (e) $\frac{3}{4}$ h
    (f) 0.61
    (g) 0.32
43. 3.84 h/service
44. 17.65 min. per patient
45. (a) 30 min.
    (b) 4; 24 min.; 0.0183; 0.2621; 0.8
46. (a) 2.25
    (b) 9 min.
    (c) 0.7135
    (d) 0.8465
47. 2.1739; 0.1739; 21.739 min.; 1.739 min.; $P_0 = 0.1304$, $P_1 = 0.2608$; $P_2 = 0.2602$
48. Single channel is better
49. $\frac{4}{23}$
50. 22.2h; 49 min.
51. (a) 0.48
    (b) 3.2 min.
52. (a) 0.3826
    (b) 3.05 min.; 9.054 min.; 4.53 cars;
    (c) 24.98%

53. (a) 0.2778
    (b) 1.3878
    (c) 43.8 min.
    (d) 3.7%
54. 0.28; 1.55
55. 0.49, 0.0027
56. (a) 12.45 min.
    (b) 0.79 customer
    (c) 22.42 min.
57. 0.36133; 0.33996; 0.00368; 0.2457; 19 min.
58. $P_n = (0.56)(0.625)^n$, for $2 \leq n \leq 5$; $P_1 = 0.35$; $P_0 = 0.28$; 2.956
59. 6.06 cars; 12.3 min.; 30.4 cars
60. (a) 1
    (b) Total cost = Rs. 255
    (c) Total cost with 2 machines = Rs. 270; Use of 2 machines is not economical.
63. 5 min.
64. 0.667 car
65. 100 min.

Chapter 16

SIMULATION

16.0 INTRODUCTION

Almost all real-life problems related to operations research have been represented by mathematical models and the required solutions have been obtained by using mathematical techniques. However, there are some problems which are extremely complex in nature and hence, are not amenable to description by mathematical models and for which scientific solutions cannot be obtained by the mathematical techniques available. Be that as it may, it is found that in a large number of situations, the interest of the decision maker is to get a useful understanding of the truth about the problem, not necessarily its exact solution. Under such circumstances, techniques have been developed which will help the decision maker to determine the effect of all alternative policies without altering the behaviour of the actual problem and to select the best policy. Such techniques are simulation techniques which are used as the last resort when all other methods fail.

16.1 SIMULATION: DEFINITION AND APPLICATION TO QUEUEING PROBLEMS

Simulation is a technique for conducting experiments on a model of a real system which duplicates the characteristics of the real system and which enables the decision maker to collect information about the behaviour of the real system.

In other words, it is a numerical technique for conducting experiments on a digital computer which involves setting up of logical models that describes the behaviour of real systems.

Though simulation techniques can be used to analyse many systems such as inventory problems, we shall mainly discuss its application to queueing problems. We have already seen that a single-server, single queue system can be analysed analytically, if it is of the $(M/M/1) : (\infty/FIFO)$ type or $(M/E_k/1) : (\infty/FIFO)$ type.

However, if the distribution of arrivals is not Poisson and if it changes several times throughout the day, and/or if several queues form in front of several servers who service at different rates, analytical methods are no longer applicable. In fact, realistic queue systems are rarely well-behaved enough for the use of analytic methods. Consequently, simulation is widely used to analyse queueing systems.

Advantages of Simulation

1. Simulation is useful in solving problems relating to a real system when direct experimentation on the system (such as space flight) is impractical or when actual observation of a real system is too expensive and time-consuming.
2. Simulation methods are easier to apply than pure mathematical methods, so non-technical executives can comprehend simulation better.
3. The effect of using new techniques can be observed in the model without actually using them in the real system and hence, spending time and money.
4. The simulation model, once constructed, may be employed as often as required to analyse different situations.

Disadvantages of Simulation

1. Simulation outputs should be carefully interpreted, as they depend on how the model and the underlying assumptions reflect the characteristics of the real system.
2. Simulation is not a precise process; nor is it an optimisation process – it provides only satisfactory approximations involving statistical errors.
3. If the details incorporated in the model are excessive, computer simulations can also be costly.
4. Simulation often provides knowledge of the parts of the system but does not guarantee the knowledge of the system as a whole.

Event Type Simulation

Simulation models constructed by considering changes in the system only when events of the system occur are called *event-oriented models*.

In a queueing system, changes relevant to the analysis can occur only when a customer arrives for service or departs after completion of service. Thus, two events occur in a queueing system, namely, an arrival and a departure. Usually, in event type simulation models, the inter-arrival and service times are considered deterministic and discrete.

16.2 GENERATION OF RANDOM VARIATES (OBSERVATIONS): THE MONTE CARLO TECHNIQUE

A simulation model will generally not be deterministic, but will include some elements of uncertainty. In the queueing models, the inter-arrival and service times are usually probabilistic rather than deterministic. In simulating real systems which involve random characteristics, the main problem is to generate a sequence of numbers which could be considered as typical observations from a given probability distribution. Such generation of a sequence of numbers is called the *Monte Carlo technique* and is usually done in two steps:

1. Generation of random numbers in the interval (0, 1).
2. Generation of random variates from any desired probability distribution.

To generate random numbers as a first step, three methods are normally employed:

1. For hand computations, a table of random numbers can be used. For computer simulations, tables of random numbers in the memory of the computers are used.
2. Some physical procedure such as a spinner or bowl of chips or some sort of electronic device may be used to generate random numbers.
3. Multiplicative congruential method based on modulus arithmetic is used to generate random numbers. The method is explained here.

16.2.1 Congruential Method (Step1 of the Monte Carlo Method)

Definition: Two numbers x and y are said to be (multiplicatively) *congruent modulo m* if $(x - y)$ is an integral multiple of m and is denoted as

$$x \equiv y \text{ (modulo } m) \tag{1}$$

For example, $13 \equiv 3$(modulo 5) and $3 \equiv 13$ (modulo 5)

The formula for generating random numbers is based on (1) and is given by

$$r_n = ar_{n-1} \text{(modulo } m) \tag{2}$$

where a and m are positive integers such that $a < m$ and $n = 1, 2, \ldots$ Formula (2) means that r_n is the remainder (lying between 0 and m) when (ar_{n-1}) is divided by m.

The first random number r_1 is got by taking any large number as r_0 and by using the formula (2). Then, r_2 is got from r_1, r_3 from r_2 and so on. The initially assumed value of r_0 is called the *seed*.

Note: We should choose the values of a and r_0 carefully.

If we assume that $a = 1$, then $r_n = r_0$ for all n.
If we assume that $r_0 = 0$, then $r_n = 0$ for all n.
For example, if we wish to generate b random numbers between 0 and 7 (both excluded), we may take $r_0 = 1$ $(\neq 0)$, $a = 3(\neq 1)$ and of course $m = 7$.

Then, we get

$$\begin{aligned} r_1 &= 3(\text{mod } 7) = 3; \\ r_2 &= 9(\text{mod } 7) = 2; \\ r_3 &= 6(\text{mod } 7) = 6; \\ r_4 &= 18(\text{mod } 7) = 4; \\ r_5 &= 12 \text{ (mod } 7) = 5; \\ r_6 &= 15(\text{mod}7) = 1, \text{ etc.} \end{aligned}$$

Hence, the random numbers are 1, 3, 2, 6, 4, 5; 1, 3, 2, ... After the first 6 numbers, the same sequence repeats itself and the length of the cycle is 6.

Then, the random numbers between 0 and 1 are obtained by dividing these numbers by m (= 7).

Thus, the required random numbers between 0 and 1 are $\frac{1}{7}, \frac{3}{7}, \frac{2}{7}, \frac{6}{7}, \frac{4}{7}$ and $\frac{5}{7}$; viz., 0.1429, 0.4286, 0.2857, 0.8571, 0.5714 and 0.7143

16.2.2 Inverse Transformation Method (Step 2 of the Monte Carlo Method)

When a required sequence of random numbers has been generated, the method used to generate a sequence of random variates from any specified probability distribution is called the inverse transformation method or *the probability transformation method*, the procedure for which involves the following three steps:

Step1: Generate the normalised random numbers $r_1, r_2, \ldots$ that lie in (0, 1).

Step2: Find the cumulative distribution function (cdf) $F(x)$ of the random variable X, whose values are to be generated

Step3: Equate the normalised random number r_k with $F(x_k)$, viz., put $F(x_k) = r_k$. Then, $x_k = F^{-1}(r_k)$; $k = 1, 2, 3, \ldots$ The values of x_k thus obtained will be the required random variates (observations) X, derived from the probability distribution of X.

16.3 PARTICULAR CASES WHEN *x* FOLLOWS SPECIFIED PROBABILITY DISTRIBUTIONS

1. *X follows U (a, b)*

   When X follows the uniform distribution in $a \le x \le b$,

$$f(x) = \frac{1}{b-a}, \quad a \le x \le b$$

$$\therefore \qquad F(x) = \int_a^x \frac{1}{b-a} \cdot dx = \frac{x-a}{b-a}.$$

   Now, we solve the equation $\dfrac{x-a}{b-a} = r$

   Then, $x = a + (b - a)r$. In particular,

   $x_k = a + (b - a)r_k$, where r_k is the typical normalised random number in (0, 1); $k = 1, 2, 3, \ldots$

2. *X follows E(λ).*

   When X follows the exponential distribution with parameter λ,

$$f(x) = \lambda e^{-\lambda x}, \; x \ge 0$$

$$\therefore \qquad F(x) = \int_0^x \lambda e^{-\lambda x} dx$$

$$= \left(e^{-\lambda x}\right)_x^o = 1 - e^{-\lambda x}$$

   Now we solve the equation $1 - e^{-\lambda x} = r$

   viz., $\qquad e^{-\lambda x} = 1 - r$

   or $\qquad x = \dfrac{1}{\lambda}\log\left(\dfrac{1}{1-r}\right)$

   viz., $\qquad x = \dfrac{1}{\lambda}\log\left(\dfrac{1}{r'}\right)$

   or $\qquad x = -\dfrac{1}{\lambda}\log r'$,

   where r' is also a random number in (0, 1). In particular,

$$x_k = -\frac{1}{\lambda}\log(r'_k); \; k = 1, 2, 3, \ldots$$

3. *X follows $E_k(\lambda)$.*

   X follows the k-phased Erlang distribution with mean $\dfrac{k}{\lambda}$, viz., if $X = X_1 + X_2 + \cdots + X_k$, then $X_1, X_2, \cdots X_k$ are independent random variables, each following the same exponential distribution with mean $\dfrac{1}{\lambda}$.

   Hence, a random variate from the Erlang distribution $E_k(\lambda)$ is generated as the sum of k random variates generated from the exponential distribution $E(\lambda)$.

Thus, $$x = \sum_{i=1}^{k} x_i = -\frac{1}{\lambda}\sum_{i=1}^{k} \log r_i$$

or $$x = -\frac{1}{\lambda}\log \prod_{i=1}^{k}(r_i),$$

where $r_1, r_2, \ldots, r_k$ are k random numbers in (0, 1).

4. *X follows N* (μ, σ^2).

If $r_1, r_2, \ldots, r_n$ are random numbers in (0,1), they can also be treated as independent random variates generated from the uniform distribution in (0,1), viz., $r_1, r_2, \ldots r_n$ are independent and identically distributed RVs each with mean $\frac{1}{b+a} = \frac{1}{2}$ and variance $\frac{(b-a)^2}{12} = \frac{1}{12}$.

Then, by Lindeberg–Levy's form of central limit theorem, $\sum_{i=1}^{n} r_i$ follows a normal distribution with mean $\frac{n}{2}$ and variance $\frac{n}{12}$.

$\therefore$ For large values of n, $z = \dfrac{\sum_{i=1}^{n} r_i - \frac{n}{2}}{\sqrt{\frac{n}{12}}}$ follows the standard normal distribution N (0, 1).

Now to generate a random variate X from a normal distribution with mean μ and variance σ^2, we note that $\frac{X-\mu}{\sigma}$ follows N (0, 1) and set

$$z = \frac{x-\mu}{\sigma} = \left(\sum_{i=1}^{n} r_i - \frac{n}{2}\right) \Big/ \sqrt{\frac{n}{12}}$$

$$\therefore \qquad x = \mu + \frac{\sigma}{\sqrt{n/12}}\left\{\sum_{i=1}^{n} r_i - \frac{n}{2}\right\}$$

In practice, we may take $n = 12$

and hence, $$x = \mu + \sigma\left\{\sum_{i=1}^{12} r_i - 6\right\}$$

5. *X follows B(n, p).*

X follows a binomial distribution with parameter n and p whose probability mass function is given by

$$p_x = nC_x p^x (1-p)^{n-x}\ ;\ x = 0, 1, 2, \ldots, n$$

To generate a random variate from the binomial distribution $B(n, p)$, we first generate n random numbers r_i in (0, 1); $i = 1, 2, \ldots, n$. Then, the required random variate x is given by the number of the random variables r_i that are less than or equal to p.

6. *X follows* $P(\lambda)$

X follows a Poisson distribution with parameter (or mean) λ.

Let $x_1, x_2, \ldots, x_n$ be identical exponential RVs with mean $\frac{1}{\lambda}$.

If we define n such that

$$x_1 + x_2 + \cdots x_n \le t \le x_1 + x_2 + \cdots + x_{n+1} \quad (1)$$

then, it is known that the distribution of n will be Poisson with mean λt.

When X follows $E(\lambda)$, we have already proved that $x_i = -\frac{1}{\lambda}\log r_i$, where r_i's are normalised random numbers in (0, 1).
Using this in (1), we get

$$\sum_{i=1}^{n} -\frac{1}{\lambda}\log r_i \le t \le \sum_{i=1}^{n+1} -\frac{1}{\lambda}\log r_i$$

viz.,

$$\sum_{i=1}^{n} \log r_i \ge -\lambda t > \sum_{i=1}^{n+1} \log r_i$$

or

$$\log \prod_{i=1}^{n} (r_i) \ge -\lambda t > \log \prod_{i=1}^{n+1} (r_i)$$

or

$$\prod_{i=1}^{n} (r_i) \ge e^{-\lambda t} > \prod_{i=1}^{n+1} (r_i)$$

7. *X follows a general discrete distribution*
We draw the graph of $F(x)$, the cumulative distributive function of X. The abscissa (x co-ordinate) of the point on the graph whose ordinate (y co-ordinate) is r [viz., the normalised random number in (0, 1)] gives the corresponding random variate x. Alternatively, if X is any discrete R.V., then it can be assumed to take the value j, when the normalised random number r in (0, 1) satisfies

$$F(j-1) < r \le F(j)$$

For example, if the discrete RV X takes the values 0, 1, 2, 3 with probabilities 0.1, 0.3, 0.2 and 0.4 respectively, then the cdf of X is given by

X=x:	0	1	2	3
F(x):	0.1	0.4	0.6	1.0

Then the random variates generated are given by

$$x = 0, \text{ for } 0 < r \le 0.1$$

$$x = 1, \text{ for } 0.1 < r \le 0.4$$

$$x = 2, \text{ for } 0.4 < r \le 0.6$$

and

$$x = 3, \text{ for } 0.6 < r \le 1.0$$

WORKED EXAMPLES 16

Example 1

Suppose that customers arrive at a reservation counter at a constant inter-arrival rate of 1.5 minutes. Service time is assumed to be a constant equal to 2 minutes. Simulate the system, by assuming that the first customer arrives at the counter (and hence, service starts) at $t = 0$ and service goes up to the 20th minute. What is the average waiting time of the customers and that for those customers who must wait in the queue before getting service? Perform event-oriented simulation.

SOLUTION

Type of event (Arrvl / Deptr) (1)	Event occurrence time (2)	Service starting time (3)	Service closing time (4)	Waiting time in the queue (min.) (5)
A_1	0.0	0.0(A_1)	2.0(A_1)	–
A_2	1.5	–	–	–
D_1	2.0	2.0(A_2)	4.0(A_2)	0.5(A_2)
A_3	3.0	–	–	–
D_2	4.0	4.0(A_3)	6.0(A_3)	1.0(A_3)
A_4	4.5	–	–	–
A_5/D_3	6.0	6.0(A_4)	8.0(A_4)	1.5(A_4)
A_6	7.5	–	–	–
D_4	8.0	8.0(A_5)	10.0(A_5)	2.0(A_5)
A_7	9.0	–	–	–
D_5	10.0	10.0(A_6)	12.0(A_6)	2.5(A_6)
A_8	10.5	–	–	–
A_9/D_6	12.0	12.0(A_7)	14.0(A_7)	3.0(A_7)
A_{10}	13.5	–	–	–
A_{11}/D_7	14.0	14.0(A_8)	16.0(A_8)	3.5(A_8)
A_{12}	15.0	–	–	–
D_8	16.0	16.0(A_9)	18.0($_{A9}$)	4.0(A_9)
A_{13}	18.0	18.0(A_{10})	20.0(A_{10})	4.5(A_{10})
				6.0(A_{11})
				5.0(A_{12})
				2.0(A_{13})

$$\text{Average waiting time of a customer} = \frac{35\cdot 5}{13} = 2.7 \text{ minutes}$$

Note: Though the customer A_{11}, A_{12} and A_{13} have not received service, they would have waited in the queue till the end of the 20th time.)

$$\text{Average waiting time of a customer who has to wait before getting service} = \frac{22\cdot 5}{10} = 2.25 \text{ minutes}$$

Example 2

Customers of both sexes arrive at a ration shop at intervals of 2 minutes and form separate queues, one for men and the other for women. The two types of customers are serviced alternately, starting with a woman customer. The service times for women and men are 1.5 minutes and 1.0 minute respectively.

Simulate up to 15 minutes, when service is suspended temporarily and no new customer is allowed into the system afterwards. At the time of suspension of service, how many customers are still waiting in the queues? Find the average waiting time of each type of serviced customer. Assume that the system starts at $t = 0$.

SOLUTION

Type of customers	Actual arrival time (min)	Service starting time (min)	Service closing time (min)	Waiting time (min)
W_1 and M_1	0.0	0.0(W_1)	1.5(W_1)	–
–	–	1.5(M_1)	2.5(M_1)	1.5(M_1)
W_2 and M_2	2.0	2.5(W_2)	4.0(W_2)	0.5(W_2)
–	–	4.0(M_2)	5.0(M_2)	2.0(M_2)
W_3 and M_3	4.0	5.0(W_3)	6.5(W_3)	1.0(W_3)
–	–	6.5(M_3)	7.5(M_3)	2.5(M_3)
W_4 and M_4	6.0	7.5(W_4)	9.0(W_4)	1.5(W_4)
–	–	9.0(M_4)	10.0(M_4)	3.0(M_4)
W_5 and M_5	8.0	10.0(W_5)	11.5(W_5)	2.0(W_5)
–	–	11.5(M_5)	12.5(M_5)	3.5(M_5)
W_6 and M_6	10.0	12.5(W_6)	14.0(W_6)	2.5(W_6)
–	–	14.0(M_6)	15.0(M_6)	4.0(M_6)
W_7 and M_7	12.0	–	–	–
W_8 and M_8	14.0	–	–	–

At the time of suspension of service, 4 customers – 2 women and 2 men are still waiting in the queues.

$$\text{Average waiting time of serviced women customers} = \frac{\text{Total waiting time}}{\text{No. of women customers}}$$

$$= \frac{0.5+1.0+1.5+2.0+2.5}{5}$$

$$= \frac{7\cdot 5}{5} = 1.5\,\text{minutes}$$

$$\text{Average waiting time of serviced men customers} = \frac{1.5+2.0+2.5+3.0+3.5+4.0}{6}$$

$$= \frac{16.5}{6} = 2.75\,\text{minutes}$$

Example 3

Find the random variate x in terms of normalised random number r (viz., $0 < r < 1$), when x follows the probability distributions whose probability density functions are given by

(i) $$f(x) = \begin{cases} 4x^3, \text{ in } 0 \le x \le 1 \\ 0, \text{ otherwise.} \end{cases}$$

(ii) $$f(x) = \begin{cases} \dfrac{2x}{3}, \text{in } 0 \le x \le 1 \\ 1 - \dfrac{x}{3}, \text{in } 1 \le x \le 3. \end{cases}$$

(i) $$f(x) = 4x^3 \text{ in } 0 \le x \le 1$$

$\therefore$ $$F(x) = \int_0^x 4x^3 dx = x^4,$$

for $0 \le x \le 1$

The random variate from the given distribution is given by

$$x^4 = r,$$

viz, $$x = r^{1/4},$$

for $0 < r < 1$

(ii) $$f(x) = \begin{cases} \dfrac{2x}{3}, \text{in } 0 \le x \le 1 \\ 1 - \dfrac{x}{3}, \text{in } 1 \le x \le 3. \end{cases}$$

$\therefore$ $$F(x) = \int_0^x \frac{2x}{3} \text{ dx},$$

if $0 \le x \le 1$

$$= x^2/3,$$

and $0 \le x \le 1$

$$F(x) = \int_0^1 \frac{2x}{3} \text{ dx} + \int_1^x \left(1 - \frac{x}{3}\right) dx,$$

if $1 \le x \le 3$

$$= \frac{1}{3} + \left(x - \frac{x^2}{6}\right)_1^x,$$

if $1 \le x \le 3$

$$= \frac{1}{3} + x - \frac{x^2}{6} - \frac{5}{6},$$

if $1 \le x \le 3$

$$= x - \frac{x^2}{6} - \frac{1}{2},$$

if $1 \le x \le 3$

The random variate from the given distribution is given by

$$\frac{x^2}{3} = r, \text{ if } 0 \le x \le 1$$

and $$x - \frac{x^2}{6} - \frac{1}{2} = r, \text{ if } 1 \le x \le 3$$

viz., $$x = \sqrt{3r},$$

if $$0 \le \sqrt{3r}, \le 1$$

viz., $$\text{if } 0 \le \sqrt{r} \le \frac{1}{\sqrt{3}}$$

viz., $$0 \le r \le \frac{1}{3}.$$

and $x^2 - 6x + (6r+3) = 0$, if $1 \le x \le 3$,

viz., $$1 \le x = \frac{6 \pm \sqrt{36 - 4(6r+3)}}{2} \le 3$$

viz., $$1 \le \frac{6 - \sqrt{36 - 4(6r+3)}}{2} \le 3 \ (\because x \le 3, \text{ the } + \text{ ve sign is omitted})$$

viz., $$1 \le 3 - \sqrt{6 - 6r} \le 3,$$

viz., $$-2 \le \sqrt{6 - 6r}$$

and $$-\sqrt{6 - 6r} \le 0$$

viz., $$2 \ge \sqrt{6 - 6r}$$

and $$\sqrt{6 - 6r} \ge 0$$

viz., $$6 - 6r \le 4$$

and $$6 - 6\text{r} \ge 0$$

viz., $$6r \ge 2$$

and $$6\text{r} \le 6,$$

viz., $$r \ge \frac{1}{3}$$

and $$r \le 1$$

$\therefore$ $$x = 3 - \sqrt{6 - 6r}, \text{if } \frac{1}{3} \le r \le 1$$

$\therefore$ The random variate from the given distribution is given by

$$x = \begin{cases} \sqrt{3r}, \text{ if } 0 \le r \le \dfrac{1}{3} \\ 3 - \sqrt{6-6r}, \text{ if } \dfrac{1}{3} \le r \le 1. \end{cases}$$

Example 4

Simulate a single server (single channel queueing system) until 4 arrivals have been processed through the system. The distributions of inter-arrival times (T) and the inter-service times (S) are given here:

Distribution of T

T	1	2	3
$P(t)$	0.25	0.50	0.25

Distribution of S

S	1	2	3
$P(s)$	0.50	0.25	0.25

Find the cumulative probability distributions of T and S and then generate the inter-arrival times using the random digits 14, 24, 94 and 47 and generate the corresponding service times using the random digits 60, 82, 20 and 41. Find (i) the average waiting time of a customer, (ii) the percentage of idle time of the server and (iii) the average queue length.

SOLUTION

Cumulative prob. distn. of T

T	1	2	3
$p(t)$	0.25	0.50	0.25
$p(t)$	0.25	0.75	1.00
Random No. range	1–25	26–75	76–100

Cumulative prob. distn. of S

S	1	2	3
$p(s)$	0.50	0.25	0.25
$p(s)$	0.50	0.75	1.00
Random No. range	1–50	51–75	76–100

Customer (arrival number	Random number for T	Inter-arrival time(T)	Actual arrival time	Random number for S	Service time (S)	Starting time of service	Closing time of service	Waiting time of server	Waiting time of customer	Queue length
1	14	1	1	60	2	1	3	1	–	–
2	24	1	2	82	3	3	6	–	1	1
3	94	3	5	20	1	6	7	–	1	1
4	47	2	7	41	1	7	8	–	–	–

$$\text{Average waiting time of a customer in the queue} = \frac{2}{4} = \frac{1}{2} \text{ time unit}$$

$$\text{Average W.T. of a customer who has to wait} = \frac{2}{2} = 1 \text{ time unit}$$

$$\text{Average service time} = \frac{7}{4} = 1.75 \text{ time units}$$

$$\text{Average idle time of the server} = \frac{1}{8} \text{ time unit or } 12.5\%$$

$$\text{Average queue length} = \frac{2}{4}.$$

$$= 0.5 \text{ customer}$$

Example 5

Observations of the past data show the following patterns in respect of inter-arrival and service times in a single channel queueing system. Using the random numbers given here, simulate the queue behavior for a period of 60 minutes and estimate the probability of the service being idle and the mean time spent by a customer waiting for the service.

Solution

Distribution of inter-arrival time (T)

Minute	Prob. $p(t)$
2	0.15
4	0.23
6	0.35
8	0.17
10	0.10

Distribution of service time (S)

Minute	Prob. $p(s)$
1	0.10
3	0.22
5	0.35
7	0.23
9	0.10

Random numbers 93, 14, 72, 10, 21, 81, 87, 90, 38, 10 and 39 may be used to generate inter-arrival times and the random numbers 71, 63, 14, 53, 64, 42, 07, 54, 66, 79 and 30 may be used to generate service times.

Generation of inter-arrival times

Minutes	Probability $p(t)$	Cumulative prob. $P(t)$	Random number range
2	0.15	0.15	1–15
4	0.23	0.38	16–38
6	0.35	0.73	39–73
8	0.17	0.90	74–90
10	0.10	1.00	91–100

Generation of service times

Minutes	Probability $p(s)$	Cumulative prob. $P(s)$	Random number range
1	0.10	0.10	1–10
3	0.22	0.32	11–32
5	0.35	0.67	33–67
7	0.23	0.90	68–90
9	0.10	1.00	91–100

Customer (arrival) number	Random number for *T*	Inter-arrival time (*T*)	Actual arrival time	Random number for *S*	Service time (*S*)	Starting time of service	Closing time of service	Waiting time of server (min)	Waiting time of customer (min)	Queue length
0	–	–	9.00(am)	–	–	–	–	–	–	–
1	93	10	9.10	71	7	9.10	9.17	10	–	–
2	14	2	9.12	63	5	9.17	9.22	–	5	1
3	72	6	9.18	14	3	9.22	9.25	–	4	1
4	10	2	9.20	53	5	9.25	9.30	–	5	1
5	21	4	9.24	64	5	9.30	9.35	–	6	1
6	81	8	9.32	42	5	9.35	9.40	–	3	1
7	87	8	9.40	07	1	9.40	9.41	–	–	–
8	90	8	9.48	54	5	9.48	9.53	7	–	–
9	38	5	9.53	66	5	9.53	9.58	–	–	–
10	10	1	9.54	79	7	9.58	10.05	–	4	1
11	45	6	10.00	30	3	10.05	10.08	–	5	1
Total		60m			51m			17m	32m	7

(i) Average queue length $= \frac{7}{11} = 0.64$ customer

(ii) Average WT of a customer in the queue $= \frac{32}{11} \simeq 2.9$ minutes

(iii) Average service time $= \frac{51}{11} = 4.64$ minutes

(iv) Average WT of a customer in the system = (ii) + (iii) = 7.5 minutes

(v) Percentage of idle time of server $= \frac{17 \times 100}{51+17} = \frac{1700}{68}$ or 25%

Example 6

A job is to be processed on two machines M_1 and M_2 continuously without break in that order. The distribution of inter-arrival time of the job at the system (viz., at machine M_1) is as follows:

Time (min)	1	2	3	4
Probability	0.15	0.25	0.25	0.35

The distributions of processing times on the two machines are given here:

Machine(M_1)

Time (min)	1	2	3	4	5
Prob.	0.1	0.2	0.3	0.3	0.1

Machine(M_2)

Time (min)	4	5	6	7
Prob.	0.2	0.3	0.4	0.1

On the basis of 10 simulation runs, determine the average queue lengths in front of M_1 and M_2. Generate the inter-arrival times, service times on M_1 and service times on M_2 by using the random numbers got by the multiplicative congruential method with $a = 2, r_0 = 1, m = 11$; with $a = 2, r_0 = 3$ and $m = 11$ and with $a = 2, r_0 = 5$ and $m = 11$ respectively.

Solution

The random numbers required for generating inter-arrival times are given by $r_n = ar_{n-1}$ (mod m); viz., $r_1 = 2r_0$ (mod 11), with $r_0 = 1$. Hence, $r_2 = 2r_1$ (mod11), $r_3 = 2r_2$ (mod11), etc.

They are 1, 2, 4, 8, 5, 10, 9, 7, 3 and 6.

The random numbers required for getting service times on M_1 are given by $r_n = 2r_{n-1}$ (mod11), with $r_0 = 3$

They are 3, 6, 1, 2, 4, 8, 5, 10, 9, 7.

The random numbers required for getting service times on M_2 are given by $r_n = 2r_{n-1}$ (mod11), with $r_0 = 5$

They are 5, 10, 9, 7, 3, 6, 1, 2, 4, 8.

Generation of inter-arrival times

T	$p(t)$	$P(t)$	R. No. range
1	0.15	0.15	1
2	0.25	0.40	2–4
3	0.25	0.65	5–6
4	0.35	1.00	7–10
–	–	–	–

Generation of service time for M_1

S_1	$p(s_1)$	$P(s_1)$	R. No. range
1	0.1	0.1	1
2	0.2	0.3	2–3
3	0.3	0.6	4–6
4	0.3	0.9	7–9
5	0.1	1.0	10

Generation of service time for M_2

S_2	$p(s_2)$	$P(s_2)$	R. No. range
1	0.2	0.2	1–2
2	0.3	0.5	3–5
3	0.4	0.9	6–9
4	0.1	1.0	10
–	–	–	–

Note: When inter-arrival times are generated, they are 1, 2, 3, 4 according as $0 < r' \leq 0.15$, $0\ 15 < r' \leq 0.40$, $0.4 < r' \leq 0.65$ and $0.65 < r' \leq 1.0$, where r' is the normalized random number, viz., according as $0 < r \leq 1.5$, $1.5 < r \leq 4.0$, $4.0 < r \leq 6.5$ and $6.5 < r \leq 10.0$, where r is the random number generated by the multiplicative congruential method.

Simulation run	Random number for T	Inter arrival time (t)	Actual arrival time at M_1	Service start time at M_1	Random number for S_1	Actual service time S_1 at M_1	Service close time at M_1/ Arrival time at M_2	Service start time at M_2	Random number for S_2	Actual service time S_2 at M_2	Service close time at M_2
1	1	1	1	1	3	2	3	3	5	2	5
2	2	2	3	3	6	3	6	6	10	4	10
3	4	2	5	6	1	1	7	10	9	3	13
4	8	4	9	9	2	2	11	13	7	3	16
5	5	3	12	12	4	3	15	16	3	2	18
6	10	4	16	16	8	4	20	20	6	3	23
7	9	4	20	20	5	3	23	23	1	1	24
8	7	4	24	24	10	5	29	29	2	1	30
9	3	2	26	29	9	4	33	33	4	2	35
10	6	3	29	33	7	4	37	37	8	3	40

Customers 3, 9 & 10 have to wait in the queue in front of M_1.

$\therefore$ Average queue length in front of $M_1 = \frac{3}{10} = 0.3$ customer

Customers 3, 4 & 5 have to wait in the queue in front of M_2

$\therefore$ Average queue length in front of $M_2 = \frac{3}{10} = 0.3$ customer

Total waiting time of customers at $M_1 = 1 + 3 + 4 = 8$ time units

$\therefore$ Average waiting time of customers at $M_1 = \frac{8}{10} = 0.8$ time units

Total waiting time of customers at $M_2 = 3 + 2 + 1 = 6$ time units

$\therefore$ Average waiting time of customers at $M_2 = 0.6$ time units.

Idle time of the server at $M_1 = 1 + 2 + 1 + 1 + 1 = 6$ time units

Idle time of the server at $M_2 = 3 + 1 + 2 + 5 + 3 + 2 = 16$ time units.

Example 7

Generate (i) 7 random observations from the uniform distribution in (2, 6), by using the random numbers 0.45, 0.29, 0.56, 0.96, 0.09, 0.16 and 0.79 (ii) 5 random variates from the exponential distribution with mean 5, by using the random numbers 0.14, 0.30, 0.82, 0.78 and 0.95.(iii) 3 random variates from the 3-phase Erlang distribution with mean 6, by using the random numbers 0.88, 0.31, 0.08, 0.79, 0.21, 0.74, 0.12, 0.75 and 0.31

Solution

The random variates from the uniform distribution are given by $x = a + (b - a)r = 2 + 4r$.

They are 3.80, 3.16, 4.24, 5.84, 2.36, 2.64, 5.16, corresponding to the given normalised random numbers

(i) The random variates from the exponential distribution are given by $x = -\frac{1}{\lambda}\log e^{r}$.

They are $-5\log_e(0.14)$, $-5\log_e(0.30)$, $-5\log_e(0.82)$, $-5\log_e(0.78)$ and $-5\log_e(0.95)$; viz., 9.83, 6.02, 0.99, 1.24 and 0.26.

(ii) The random variates from the Erlang distribution are given by $-\frac{1}{\lambda}\log_e \prod_{i=1}^{3}(r_i)$. Here mean $= \frac{k}{\lambda} = 6$ & so $\lambda = \frac{3}{6}$ or $\frac{1}{2}$

They are $-2\log_e(0.88 \times 0.31 \times 0.08), -2\log_e(0.79 \times 0.21 \times 0.74)$

and $-2\log_e(0.12 \times 0.75 \times 0.31)$

viz., $-2\log_e(0.0218),\ -2\log_e(0.1228), -2\log_e(0.0279)$

viz., 7.65, 4.19 and 9.16

Example 8

Simulate a single server, single channel queueing system until 10 customers have been processed through the system. The inter-arrival time is exponentially distributed with a mean of 10 minutes, while the service time is uniformly distributed between 5 and 10 minutes. To generate exponential random variates, use normalised random numbers obtained by the congruential method with $a = 2$, $r_0 = 5$ and $m = 11$. To generate uniform random variates, use normalised random numbers obtained by the congruential method with $a = 6$, $r_0 = 1$ and $m = 11$. Assume that the first arrival occurs at $t = 0$. What is the average waiting time in the queue per arrival?

SOLUTION

Generation of inter-arrival times

Random no. used (r)	5,	10,	9,	7,	3,	6,	1,	2,	4,	8
Normalised random $r' = \frac{r}{m}$	0.45,	0.91,	0.82,	0.64,	0.27,	0.56,	0.09,	0.18,	0.36,	0.73
Inter-arrival time $x = -10\log_e(1-r')$ (min)	6.0,	24.0,	17.1,	10.2,	3.1,	8.2,	0.9,	2.0,	4.5,	13.1

Generation of service times

Random no. used (r)	1,	6,	3,	7,	9,	10,	5,	8,	4,	2
Normalised random $r' = \frac{r}{m}$	0.09,	0.56,	0.27,	0.64,	0.82,	0.91,	0.45,	0.73,	0.36,	0.18
Service time $x = 5 + (10-5)r'$ (min)	5.5,	7.8,	6.4,	8.2,	9.1,	9.6,	7.3,	8.7,	6.8,	5.9

Arrival (customer) No.	Inter-arrival time (T) min	Actual arrival time (min)	Service starting time	Service time (min)	Service closing time	Idle time of server	Waiting time of customer in the queue	Queue length
0	–	0.0	–	–	–	–	–	–
1	6.0	6.0	6.0	5.5	11.5	6	–	–
2	24.0	30.0	30.0	7.8	37.8	18.5	–	–
3	17.1	47.1	47.1	6.4	53.5	9.3	–	–
4	10.2	57.3	57.3	8.2	65.5	3.8	–	–
5	3.1	60.4	65.5	9.1	74.6	–	5.1	1
6	8.2	68.6	74.6	9.6	84.2	–	6.0	1
7	0.9	69.5	84.2	7.3	91.5	–	14.7	1

(Contd)

Table (*Contd*)

Arrival (customer) No.	Inter-arrival time (T) min	Actual arrival time (min)	Service starting time	Service time (min)	Service closing time	Idle time of server	Waiting time of customer in the queue	Queue length
8	2.0	71.5	91.5	8.7	100.2	–	20.0	1
9	4.5	76.0	100.2	6.8	107.0	–	24.2	1
10	13.1	89.1	107.0	5.9	112.9	–	17.9	1
–	–	–	–	75.3	–	37.6	87.9	6

$$\text{Average queue length} = \frac{6}{10} = 0.6$$

(i) Average WT of a customer in the queue $= \dfrac{87\cdot 9}{10} \simeq 8.8\,\text{min}$.

(ii) Average service time $= \dfrac{75\cdot 3}{10} \simeq 7.5\,\text{min}$.

(iii) Average WT of a customer in the system $= \dfrac{75\cdot 3 + 87\cdot 9}{10} = 16.3\,\text{min}$.

(iv) Percentage of idle time of the service $= \left(\dfrac{37.6}{75.3 + 37.6}\right) \times 100 = 33.3$.

EXERCISE 16

Part A (Short answer questions)

1. Define simulation.
2. Give some advantages of simulation.
3. Mention some disadvantages of simulation.
4. What is event oriented simulation? Explain it with reference to queueing systems.
5. Explain the multiplication congruential method of generating random numbers in (0, 1).
6. Explain the inverse transformation method of generating random variates from a given probability distribution.

Part B

7. Suppose that customers arrive at a reservation counter with an average time of 1.2 time units between one arrival and the next. Service times are assumed to be 2.8 time units. Simulate the system for 12 time units, by assuming that the system starts at $t = 0$. What is the average waiting time per customer? What is the average waiting time per customer who must wait?
8. Solve the previous problem with inter-arrival time = 1.8 minutes and service time = 4 minutes, by simulating up to 16 minutes
9. Solve the problem in worked example 2, assuming that the service starts with a man customer.
10. Find the random variate x in terms of normalised random number r $(0 < r < 1)$, when X follows the distribution with probability density function given by

(i) $f(x)=xe^{-x^2/2}, x\geq 0$

(ii) $f(x)=\begin{cases} x, \text{in } 0\leq x\leq 1 \\ 2-x, \text{ in } 1\leq x\leq 2 \end{cases}$.

11. Find the random variate x in terms of normalised random number r $(0 < r < 1)$, when X follows the distribution with probability density function given by

(i) $f(x)=\frac{x}{8}$ in $0\leq x\leq 4$

(ii) $f(x)=\frac{x}{2}$, in $0\leq x\leq 1;=\frac{1}{2}$,in $1\leq x\leq 2$.

and $=\frac{1}{2}(3-x)$ in $2\leq x\leq 3$

12. Simulate a single channel, single server queueing system until 10 arrivals have been processed through the system. The distributions of inter-arrival times (T) and service times (S) are given here:

Distribution of T

T	1	2	3
$p(t)$	0.25	0.50	0.25

Distribution of S

S	1	2	3
$p(s)$	0.50	0.25	0.25

Generate the inter-arrival time using the random numbers 54, 90, 18, 38, 16, 87, 91, 41, 54 and 11 and generate the corresponding service times using the random numbers 12, 40, 48, 93, 61, 17, 55, 21, 85 and 68. Find (*a*) the average queue length, (*b*) average waiting time of a customer in the queue and in the system, (*c*) average service time and (*d*) the percentage of the idle time of the server.

13. A job is to be processed on two machines M_1 and M_2 in that order.
The distribution of inter-arrival times of the job at the machine M_1 is as follows:

Time(minutes):	1	2	3	4
Probability:	0.4	0.3	0.2	0.1

The distributions of processing times on the two machines are as follows:

Machine M_1

Time(min):	2	4	6	8
Probability:	0.3	0.2	0.3	0.2

Machine M_2

Time(min):	1	3	5	7
Probability:	0.2	0.3	0.2	0.3

On the basis of 10 simulation runs, determine the average queue lengths in front of M_1 and M_2, the average waiting times of customers at M_1 and M_2 and the idle time of servers at M_1 and M_2. Generate the inter-arrival times, service times on M_1 and service times on M_2 by using the random numbers got by the multiplicative congruential method with (i) $a = 2, r_0 = 5$ and $m = 11$, (ii) $a = 2, r_0 = 7$ and $m = 11$ and (iii) $a = 2, r_0 = 9$ and $m = 11$ respectively

14. Simulate a single server, single channel queueing system until 10 customers have been processed through the system. The inter-arrival time is exponentially distributed with a mean of 10 minutes,

while the service time for each customer is a constant, equal to 7.5 minutes. To generate exponential random variates, use normalized random numbers obtained by the congruential method with $a = 2$, $r_0 = 5$ and $m = 11$.

Assuming that the first arrival occurs at $t = 0$, find (*a*) the average queue length, (*b*) the average waiting time of a customer in the queue and in the system, (*c*) the average service time and (*d*) the percentage of idle time of the server.

15. Simulate a single server queueing system until 10 customers have been processed through the system. The inter-arrival times are uniformly distributed in (6, 8) while the service time is distributed according to *a* 4-phase Erlang distribution with mean 24. Use the random numbers 0.45, 0.29, 0.56, 0.96, 0.09, 0.16, 0.79, 0.14, 0.30 and 0.82 to generate the uniform random variates. Use 10 sets of 4 random numbers each
(0.78, 0.95, 0.88, 0.31); (0.08, 0.79, 0.21, 0.74); (0.12, 0.75, 0.31, 0.10);
(0.32, 0.46, 0.61, 0.99); (0.94, 0.64, 0.81, 0.53); (0.18, 0.25, 0.82, 0.98);
(0.61, 0.12, 0.72, 0.62); (0.34, 0.91, 0.95, 0.77); (0.71, 0.23, 0.36, 0.35);
(0.94, 0.63, 0.44, 0.11) to generate the Erlang random variates. Find the average waiting time of a customer in the queue and in the system and the percentage of idle time of the server.

ANSWERS

Exercise 16
Part B

7. 3.4 time units, 3.8 time units
8. 4.4 time units, 4.9 time units
9. 2 men and 2 women customers. For men, the average waiting time = 1.5 min and for women, the average waiting time = 2.25, min.
10. (i) $x = \sqrt{-2\log(1-r)}$, for $0 \le r \le 1$

(ii) $$x = \begin{cases} \sqrt{2r}, \text{ for } 0 < r \le \frac{1}{2} \\ 2 - \sqrt{2(1-r)}, \text{ for } \frac{1}{2} \le r < 1 \end{cases}$$

11. (i) $x = 4\sqrt{r}$, for $0 < r < 1$

(ii) $$x = \begin{cases} 2\sqrt{r}, \text{ for } 0 < r \le \frac{1}{4} \\ 2r + \frac{1}{2}, \text{ for } \frac{1}{4} \le r \le \frac{3}{4} \\ 3 - 2\sqrt{1-r}, \text{ for } \frac{3}{4} \le r < 1 \end{cases}$$

12. (*a*) 0.3 customer; (*b*) 0.5 min. and 2.2 min.; (*c*) 1.7 min.; (*d*) 29.17%
13. 0.9, 0.5 customers; 7.8 and 2.7 time units; 2 and 15 time units.
14. (*a*) 0.6 customer, (*b*) 6.62 min.; 14.12 min., (*c*) 7.5 min., (*d*) 31.7%
15. 55.4 min., 74.3 min.; 35.2%

Bibliography

Gupta PK and Man Mohan, *Problems in Operations Research*, Methods and Solution, Sultan, 1980.

Gupta RK, *Operations Research*, Krishna Prakashan Media (P) Ltd, 2010.

Gupta SC and Kapoor VK, *Fundamentals of Mathematical Statistics*, Sultan Chand and Sons, 1971.

Hadley G, *Linear Algebra*, Addison-Wesley Publishing Company, 1961.

Kanti S, Gupta PK and Man Mohan, *Operations Research*, Sultan Chand and Sons, 1978.

Sharma JN and Goyal JK, *Mathematical Statistics*, Krishna Prakashan Media (P) Ltd, 1963.

Veerarajan T, *Higher Engineering Mathematics*, Shanlax Publications, 2015.

Veerarajan T, *Probability, Statistics and Random Processes*, Tata McGraw-Hill Education Private Limited, 2002.

Wagner HM, *Principles of Operations Research*, Prentice-Hall, 1967.

Index